Physics of Space Plasmas

An Introduction

George K. Parks

University of Washington
Seattle, Washington

LONDON AND NEW YORK

First published 1991 by Westview Press

Published 2019 by Routledge
52 Vanderbilt Avenue, New York, NY 10017
2 Park Square, Milton Park, Abingdon, Oxon OX14 4RN

Routledge is an imprint of the Taylor & Francis Group, an informa business

Library of Congress Cataloging-in-Publication Data
Parks, George K.
Physics of space plasmas/George K. Parks.
p. cm.
"The Advanced book program."
Includes bibliographical references.
1. Space plasmas. I. Title.
QC809.P5P37 1991 523.01'876—dc20 91/370

ISBN 13: 978-0-367-28292-9 (hbk)
ISBN 13: 978-0-367-29838-8 (pbk)

Physics of Space Plasmas

Preface

This textbook was developed to provide seniors and first-year graduate students in physical sciences with a general knowledge of electrodynamic phenomena in space. Since the launch of the first unmanned satellite in 1957, experiments have been performed to study the behavior of electromagnetic fields and charged particles. There is now a considerable amount of data on hand, and many articles, including excellent review articles, have been written for the specialists. However, for students, new researchers, and non-specialists, a need still exists for a book that integrates these observations in a coherent way. This book is an attempt to meet that need by using the theory of classical electrodynamics to unify space observations. The contents of this book are based on classroom notes developed for an introductory space physics course that the author has taught for many years at the University of Washington. Students taking the course normally have had an undergraduate course in electricity and magnetism but they come with very little knowledge about space.

Much of matter in interplanetary and interstellar space and throughout the Universe is very tenuous. The density is only a few particles per cubic centimeter, and matter is most likely to be in an ionized state, in the form of plasma, which consists of electrons and positive ions. Binary collisions, which occur in the more familiar laboratory plasmas, are rare in many regions of space and space plasmas are referred to as "collisionless." Even though the same physical laws determine the behavior of particles in space and in Earth laboratories, certain phenomena are apparently unique to the space environment. This book describes the physics of large-scale electrodynamic forces in space, differing from standard plasma physics books, which emphasize laboratory plasmas.

That space supports many interesting electrodynamic phenomena has been borne out by space experiments. An important discovery is that space is not a vacuum as once thought. Observations have shown that our solar system is pervaded by the flowing solar wind whose temperature is high enough that most of the constituents are ionized. The ionized particles are sources of electric and magnetic fields, and their interactions with planets and comets create magnetospheres, Van Allen radiation belts, auroras, interplanetary shocks, and intense radio emissions. These phenomena in our solar system are manifestations of basic electrodynamic forces in action and are fundamental to the study of how cosmic electrodynamic forces work in other parts of the Universe.

The discipline of space electrodynamics is only about thirty years old. Like any new field, it has undergone search and discovery phases, during which space phenomena were studied at the conceptual level and the physics was qualitative. However, the field has now matured and has entered a more rigorous phase. We are striving to understand and quantify the physics by using the fundamental laws of electricity and magnetism. An understanding of how charged particles move around, currents are generated, and how both interact with electromagnetic fields will be developed in the framework of non-relativistic Maxwell's equations for electromagnetic fields and the Lorentz equation for charged particles. Attention has been given to both fundamental and practical aspects of the subject. Rationalized mks units will be used throughout this book.

This book contains eleven chapters. Chapter 1 gives a brief introduction to large-scale electrodynamic phenomena observed in space, sets the stage, and defines the scope of the book. Formal equations are given in Chapter 2, and the assumptions that underlie our study of electrodynamics in space are discussed. Electromagnetic structures play an important role in space and in Chapter 3 we begin discussing simple properties of stationary electromagnetic fields in the absence of local currents. The dipole magnetic structure will be studied in detail since the dipole geometry represents a standard of reference for many magnetized planets and stars. We then study the motions of single particles in Chapter 4 and develop the first order Lorentz theory of particle motion (guiding center theory), which helps us understand how radiation belts are formed. The next chapter, Chapter 5, examines the physics of a collection of particles treated as a fluid and the formal magnetohydrodynamic (MHD) theory is formulated. Fluid theory enables us to study the origin of solar wind, the interplanetary magnetic field, and plasma convective phenomena in magnetospheres (Chapter 6). The interaction of conducting fluids and magnetic fields produces currents and this topic is introduced in Chapter 7. A practical problem we will study is how currents are formed at boundaries by the solar wind interacting with planetary magnetic fields and ring currents formed during magnetic storms. MHD currents and plasma discontinuities are intimately related and these topics, addressed in Chapter 8, permit us to examine the physics of boundaries. Next, we study wave generation in MHD fluids (Chapter 9). Waves are time-dependent phenomena and this chapter is different from the first eight chapters, which treat only time-stationary phenomena. The MHD fluid medium supports many classes of waves, many more than ordinary fluids. Chapter 10 treats the physics of shocks and the last chapter (Chapter 11) introduces the topic of instabilities—why some phenomena are unstable in space.

Many chapters include a short introductory discussion of experimental results, a formulation of the theory needed to explain the observations, and conclude with a statement about the current status of the subject. Where appropriate, an evaluation is made of how successful the theory is in explaining the observations and how one may further develop the theory.

In selecting what materials to include, choices were made on the basis that this book is meant to be an introductory textbook, aimed at newcomers, seniors, first-year graduate students, scientists in other disciplines, and engineers who are interested in learning about space. Complex mathematics are thus minimized, and the emphasis is on understanding the physical concepts; however, the theory is complete enough to unify and integrate the subject matter.

The problems at the end of the chapters are extensions of the material covered in the chapters. Some of these problems are elementary and the intention here is to provide an idea of the scope of the physical parameters that are involved in space. Other problems test understanding of the theory and concepts developed in the chapters. The references cited at the end of each chapter are annotated to help the reader who might be interested in further reading.

Acknowledgements

It is virtually impossible to be aware of all the papers in this field. An attempt was made to credit the original papers and we apologize in advance for any omissions or errors in citations. The author has benefited from numerous discussions with Dr. Michael McCarthy, who carefully read the text and made many valuable suggestions. It is also a pleasure to acknowledge useful suggestions made by Dr. Yannis Dandouras. Past students who have taken the introductory course, notably Messrs. Ron Elsen, Nels Larson, Dayle Massey, and Ms. Ruth Skoug, asked important questions that clarified certain sections. The author thanks Ms. Kay Dewar for her assistance in the preparation of the figures and Mr. Ed Clinton for his able editing of the text. Ms. Lisa Peterson deserves special thanks for her patience in typing and retyping many different versions and preparing the camera-ready copy of this manuscript.

CONTENTS

1

Electrodynamics in Space

1.1 Introduction

The year 1957 marked a new era for the physics of electromagnetism in space. With the launch of the first unmanned satellite, Sputnik, by the USSR on October 4, 1957, space became accessible to intelligent beings. The United States followed four months later with the launch of Explorer I on January 31, 1958. Before the spacecraft era, electrodynamic phenomena in space were studied primarily by remote means. Spacecraft made possible *in situ* physical measurements in distant regions previously not accessible. This led to important discoveries that clarified and revolutionized our thinking about space.

The sixties were exploratory years and most satellites were instrumented to study Earth, although a few studied other planets and the Sun. The missions in the seventies were more sophisticated and included a number that reached the distant planets in the late eighties. For instance, the Voyager 2 spacecraft, launched on August 20, 1977, first passed by Jupiter in July 1979 and subsequently encountered Saturn in August 1981, and Uranus in January 1986. Neptune encounter occurred on August 24, 1989.

For the first time we have voyaged beyond Earth and have begun studying the neighboring planets and the solar system.

Much of known matter in the Universe exists as plasmas. A plasma is defined as a collection of charged particles possessing certain electrodynamic properties. (To be discussed further in Chapter 2.) Although our measurements have thus far been confined to our solar system, knowledge gained here has dramatically changed our perception of space in general. This is because the electrodynamics in our solar system exemplify similar dynamics that occur in other planets and solar systems and knowledge gained from our solar system plasmas can be applied directly to other plasmas in the Universe. For example, our studies of the Sun, the planets, and their magnetic fields have been providing important data on the physics of stellar and planetary dynamics and dynamo theories. Old notions, models, and theories about space are being continually reworked with new information provided by the satellite experiments.

The motivation to study electrodynamic phenomena in space comes from our desire to learn how natural plasmas work. This chapter will provide a brief overview of global electrodynamic phenomena that occur from the Sun to the outer edge of our solar system. In so doing, we will introduce the language that has been developed in this field, as well as some of the outstanding research problems.

1.2 Solar (Stellar) Wind

The Sun is an ordinary star of average size and temperature with an absolute magnitude of 4.8. Its close proximity to Earth permits us to study in detail its structure and dynamics in a way that is not possible with any other stars. The Sun is about 4.6 billion years old and its life expectancy is estimated at another 5 billion years. This makes our Sun a typical dwarf star just past its middle age. The Sun is about 90% hydrogen, 10% helium, and 0.1% other minor constituents, such as carbon, nitrogen, and oxygen (CNO). These constituents are ionized because nuclear reactions and electrodynamic interactions create temperatures that exceed the binding energies of the atoms. The temperature of the Sun's central core is estimated to be more than $10,000,000\,^{\circ}$K.

The Sun's atmosphere is divided into the photosphere, the chromosphere, and the corona. The photosphere is a layer about 500 km thick. The chromosphere extends to a height of about 2500 km above the photosphere and the corona extends beyond the chromosphere into interplanetary space. The temperature of the photosphere is about $6000\,^{\circ}$K; after an initial decrease just above the photosphere, the temperature reaches about

1,000,000 °K at the base of the corona. The temperature subsequently decreases as the corona expands into interplanetary space.

The solar corona expands into space because the Sun's atmosphere is not in static equilibrium. This expanding corona, called solar wind, was detected by space-borne instruments early in the space era. This important discovery dispelled the long-held view that space was a vacuum. The solar corona expands into the most distant regions of space and our solar system is entirely pervaded by the solar wind.

The existence of the solar wind was predicted from observations of the long cometary tails (Figure 1.1). Astronomers who study comets have noted that cometary tails always point away from the Sun. For a long time, an accepted explanation for these cometary tails was that they were produced by the solar electromagnetic radiation. However, Ludwig Biermann in 1950 showed that the electromagnetic energy was not adequate to explain certain features of cometary tails. Instead, his studies predicted that the Sun emitted particulate energy in addition to the electromagnetic energy.

1.3 Interplanetary Magnetic Field

The Sun is a magnetic star. Magnetic fields on the Sun were first inferred in sunspots by G.E. Hale in 1908. Sunspots have lateral dimensions of about 10^4 km and they support magnetic fields with intensities of a fraction of a tesla to several teslas. The Sun also has a general magnetic field whose average intensity is about 10^{-4} teslas.

Magnetic fields diffuse very slowly in good electrical conductors; therefore, the solar wind will carry out a large fraction of the solar coronal magnetic fields into interplanetary space. The interplanetary magnetic field (IMF) observed beyond the most distant planet of the solar system originates in the Sun's magnetic field and has been transported out with the solar wind. That this indeed occurs has been demonstrated by direct measurements of the interplanetary magnetic field in the solar wind medium.

1.4 Interplanetary Electric Field

The solar wind is a good electrical conductor, and as with ordinary conductors, excess charges do not accumulate locally inside solar wind plasmas. Any excess charges that appear in the solar wind will be neutralized by the flow of the free charges. In the rest frame of the solar wind, the charge density must therefore vanish and so does the electric field. However, this does not mean that space is devoid of electric fields. The solar wind motion across interplanetary magnetic fields will induce a motional electromotive

FIGURE 1.1 A photograph of a comet showing two distinct tails resulting from solar electromagnetic radiation and particle pressures. (Courtesy of the European Space Agency)

force and an electric field will be observed in the reference frame at rest with respect to the solar wind. This motional electric field in space is referred to as a convective electric field, if the plasma motion does not change the magnetic energy.

1.5 Solar (Stellar) Activity

The activity of the Sun varies with a period of about 22 years. Solar flares are an example of solar activity that occurs during solar active years. Solar flares produce high-energy particles by mechanisms as yet not completely understood. The high-energy solar particles then bombard planetary atmospheres and can cause, for example, radio blackouts in polar regions of Earth.

Another example of solar activity is the ejection of coronal mass. Figure 1.2 shows the Sun's transient coronal activity observed in a sequence of pictures taken by a satellite-borne coronograph/polarimeter. The bright loop represents solar material confined by the coronal magnetic field being ejected into space. The outward speed of this particular coronal mass ejection event was estimated at about 270 km/second.

The solar wind is also more disturbed during solar active years. For example, the solar wind velocities and temperatures are higher and the solar wind is accompanied by large fluctuations of the interplanetary magnetic field. A disturbed solar wind will induce intense auroral and magnetic storm activities in planetary magnetospheres.

1.6 Collisionless Shocks

As the solar corona expands into space, the expansion velocity increases and it becomes supersonic. Hence, much like the shock that develops when a supersonic jet breaks the sound barrier in Earth's atmosphere, a "bow" shock forms in front of planets (the planet is supersonic in the solar wind frame). The Mach number of the solar wind is typically 5 to 10.

The density of the solar wind near the vicinity of Earth is very low (a few 10^6 particles per cubic meter) and therefore the collisional mean free path in the solar wind is very long, about 1.5×10^8 km, which is equivalent to the distance from Sun to Earth. This distance is also known as an Astronomical Unit (AU). For all practical purposes, then, the solar wind plasma can be considered collisionless. Shocks in space are therefore produced by a collisionless process. Whereas viscous forces and heat conduction play important roles in the dynamics of ordinary shocks, it is not presently known how momentum is transferred and heat conducted in collisionless shocks.

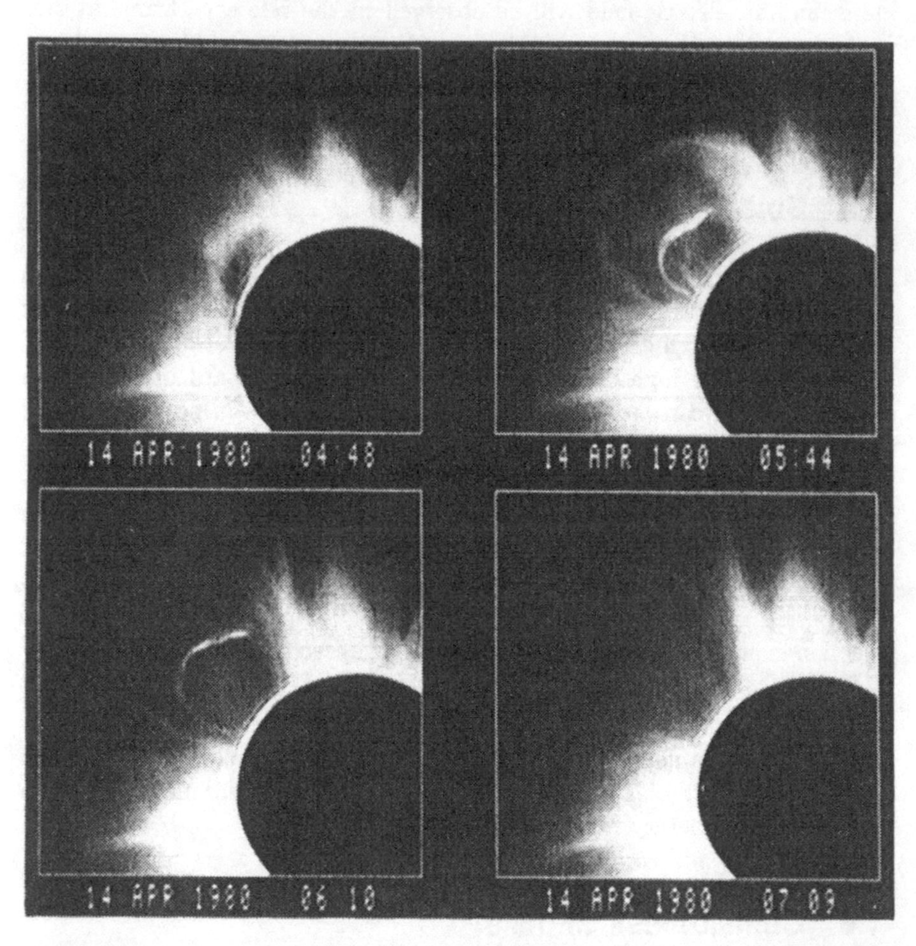

FIGURE 1.2 A coronal mass ejection event in progress as observed by a coronagraph/polarimeter on the Solar Maximum Mission spacecraft. The bright loop-like shape represents the magnetic field. Coronal material is tied to this magnetic field and the outward movement shows that the material is being ejected into space. (Courtesy of National Aeronautics and Space Administration)

Shocks also arise from solar flares and stellar explosions such as supernovas. These larger shocks are important in theories of births and deaths of new and old stars, solar systems, and acceleration of cosmic rays to energies unattainable by accelerators in Earth laboratories. It is important to study the bow shock of Earth because we can ask questions and test our theories about the fundamental questions of space electrodynamics.

1.7 Magnetospheres

The planets, their moons, and the comets are immersed in the magnetized solar wind. With magnetized bodies such as Mercury, Jupiter, Earth, Saturn, Uranus, and Neptune, the interaction induces large-scale currents that can almost confine the planetary magnetic field. A major magnetic structure called a magnetosphere is formed around each of these bodies. With unmagnetized bodies such as Mars, Venus, and the comets, the solar wind interacts with the ionized particles of their atmospheres and induces currents whose magnetic fields then divert the solar wind around them. A magnetic "cavity" formed around an unmagnetized body is called an induced magnetosphere.

Unmagnetized planets and moons without an ionized atmosphere act much like a dielectric obstacle placed in the conducting solar wind. For instance, our Moon is non-magnetic and non-conducting. When the solar wind particles impact the surface of the Moon, they are neutralized and absorbed. A cavity and a wake are formed behind the Moon. However, no bow shock or magnetosheath layer of thermalized shocks exists around the Moon.

The basic elements of a steady state solar wind-magnetosphere system of Earth's environment, deduced from many years of observations, are shown in Figure 1.3. The outer boundary, called the magnetopause, separates the domain of the planetary magnetic field from the solar wind. The inner boundary, located at the base of the ionosphere, separates the conducting from the neutral atmosphere. The size of the magnetosphere is determined by the balance between the planet's magnetic energy and the solar wind flow energy. Magnetospheres are very asymmetric. They are compressed on the side facing the solar wind and elongated in the other direction, forming a magnetic tail. The sunward magnetopause on the equator for Earth is located typically at about 10 earth radii ($R_E \approx 6375$ km). Jupiter's magnetic moment is very large and therefore Jupiter's magnetosphere is very large. Jupiter's magnetopause is thought to extend beyond 50 Jovian radii ($R_J \approx 7.14 \times 10^4$ km).

The magnetic moments of Earth, Jupiter, and Saturn are nearly centered, with a tilt relative to the rotation axis of $< 12°$. The magnetic moments of Uranus and Neptune are displaced from the center of the planets, $0.3R_U$ for Uranus ($R_U = 25,600$ km) and $0.55R_N$ for Neptune ($R_N = 24,765$ km), and they also have a large tilt angle (60° for Uranus and 47° for Neptune). The magnetospheric configurations produced by these offset tilted dipoles are dynamic and change as the planet rotates (configurations will resemble those produced by oblique magnetic rotators).

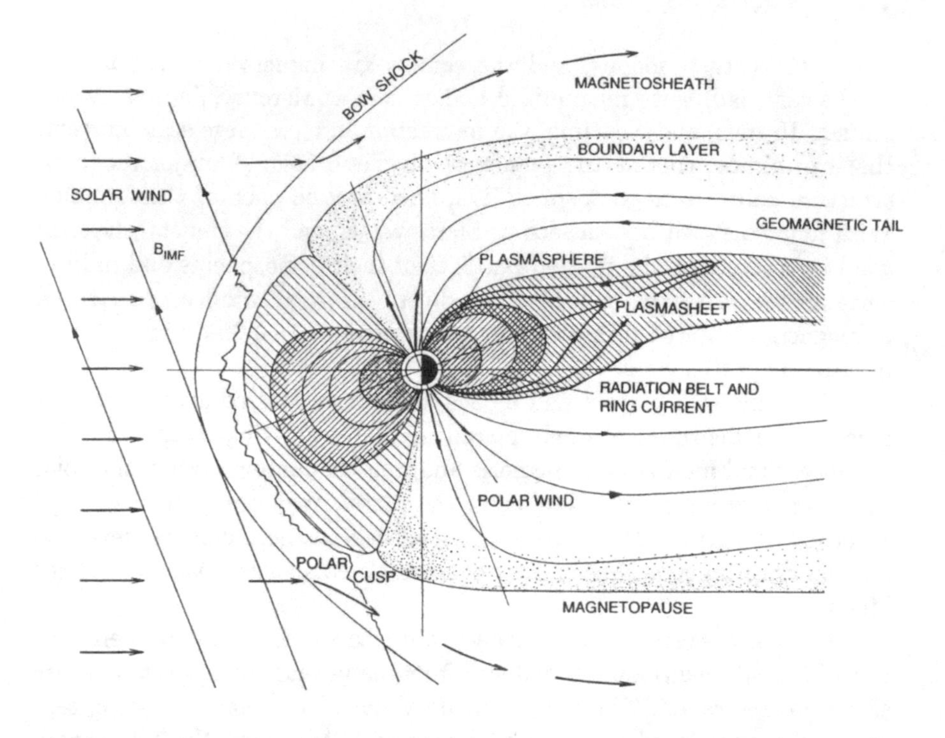

FIGURE 1.3 A schematic diagram of Earth's magnetosphere in the noon-midnight plane. The basic particle and magnetic field features are representative of other planetary magnetospheres although the details can be different.

1.7.1 Magnetic Tail

Like the comets in our solar system, magnetospheres have long tails pointing away from the Sun. Magnetic tails are produced by the interaction of the magnetized solar wind with the planetary magnetic fields. Although this interaction is collisionless, it transfers a part of the solar wind momentum and energy to the planetary magnetic field and "stretches" it in the direction of the solar wind flow. A large fraction of this stored energy in the tail is dissipated into the atmosphere during an aurora. The length of the tail depends on the strength of the magnetic moment of the planet. For Earth, the tail has been observed to extend beyond $200R_E$. Jupiter's magnetic tail is believed to be very long, possibly extending all the way to Saturn's orbit and at times engulfing Saturn's magnetosphere.

1.7.2 Closed and Open Magnetospheres

Researchers often discuss whether a magnetosphere is open or closed. The magnetopause of a closed magnetosphere isolates the planetary magnetic field from interplanetary space. An open magnetosphere has magnetic lines of force one end of which is anchored on the planet and the other on the Sun. This contrast is shown in Figure 1.4. These lines of force are produced by "merging" the planetary and the interplanetary magnetic fields, leading the magnetosphere to an open configuration.

Whether a magnetosphere is open or closed is determined by the flow characteristics of the solar wind at the planetary magnetopause. These two types of magnetospheres use different processes to transport the mass, momentum, and energy across the boundaries. This, in turn, has consequences upon the internal dynamics of the magnetosphere, such as the auroral phenomena, and upon the origin of particles in the magnetosphere.

1.7.3 Van Allen Radiation Belts

Magnetospheres are populated with charged particles and the regions occupied by the energetic particles are called Van Allen radiation belts. Magnetospheric particles come from two primary sources: the solar wind and the planetary ionosphere. However, the particles have energies from less than an electron volt (eV) to millions of eV and they differ from the original ionospheric and solar wind populations. The most energetic particles are cosmic rays and the decay products of neutrons that are produced by the cosmic rays interacting with the atmosphere. Some of the neutrons travel outward into space across magnetic fields and decay into protons and electrons (and anti-neutrinos) and these particles are captured by the planetary magnetic field. The energetic component of the inner magnetosphere of Earth, the inner Van Allen radiation belt, can be understood by this process, which is known as the Cosmic Ray Albedo Neutron Decay (CRAND) contribution.

The origin of the majority of particles in the intermediate- and the low-energy range (tens to thousands of eV) that populate the outer magnetosphere is not well known. There is evidence that these are solar wind and ionospheric particles that become trapped and accelerated by the auroral processes. These processes involve complicated electrodynamic forces not yet completely understood.

1.7.4 Ring Current

The trapped particles in planetary magnetospheres move under the action of the Lorentz force. In addition to the usual cyclotron motion, the particles

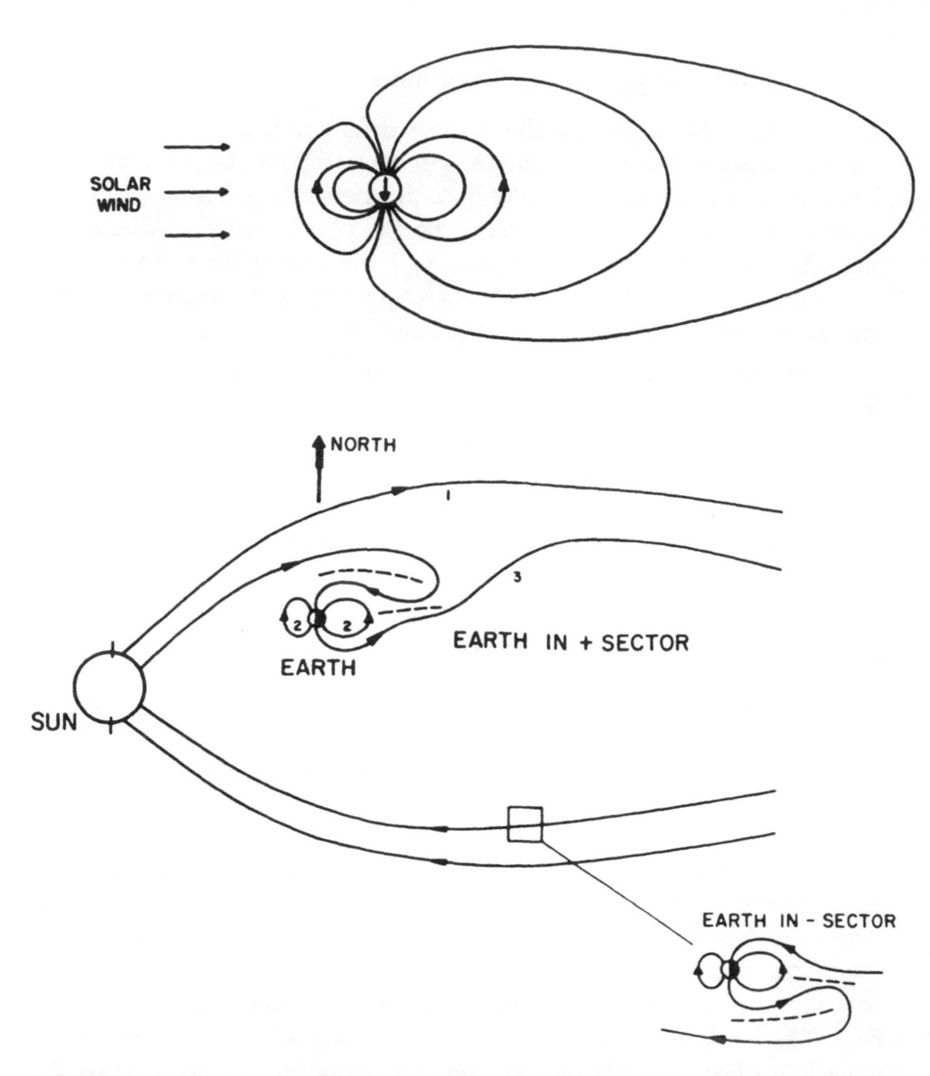

FIGURE 1.4 The upper diagram shows a closed magnetosphere which is a cavity formed by the dipole field modified by the solar wind. The low-latitude field is still dipole-like while the high-latitude field has been modified. The lower diagram shows an open magnetosphere. Here there are three types of magnetic lines of force: (1) an interplanetary magnetic field of solar origin, (2) the dipole of planetary origin, and (3) a "merged" planetary and interplanetary line of force. Dashed lines represent neutral sheets in which the magnetic field vanishes. (From Anderson and Lin, 1969)

travel in the north-south direction. The inhomogeneous planetary magnetic field also forces the trapped particles to drift in the azimuthal direction. This drift motion is charge-dependent and electrons and ions drifting in the opposite direction create a large-scale current, called a ring current. A considerable amount of energy is stored in the ring current during magnetic storms. For instance, the intensity of Earth's ring current exceeds several million amperes during a moderate size magnetic storm. Most of this energy is dissipated into the polar ionosphere during auroral events.

1.7.5 Ionosphere

The solar ultraviolet rays that impinge on a planet ionize a portion of the neutral atmospheric constituent. At altitudes where collisions are infrequent and therefore the ionization recombination is slow, a permanent population of ionized atmosphere called the ionosphere is formed. For Earth, the ionosphere begins around 60 kilometers and extends high up into the magnetosphere. The "boundary" of the equatorial ionosphere, located around 3 to $4R_E$ from the surface of Earth, is called the plasmapause. The region inside the plasmapause is called the plasmasphere and it includes the ionosphere.

Ionospheres are sources as well as sinks of particles. As mentioned earlier, the ionosphere is one of two primary sources of particles for the magnetosphere. However, the ionosphere is also a sink because the particles of the solar wind, the ring current, and the tail are dissipated into the ionosphere during auroras. Understanding the dynamics of polar ionospheres is important in the study of auroras, magnetic storms, and the processes of solar wind energy transfer into the magnetosphere.

1.7.6 Aurora

Auroras represent one of the most dynamic products of a solar wind-magnetized planetary system. Auroral lights are produced when energetic particles are precipitated into the planetary ionospheres. The different auroral colors come from the atmospheric emissions characteristic of the excited constituents (which depend on the planetary composition). A portion of the precipitated energy is also converted to bremsstrahlung X-ray radiation. Hence, like solar flares and X-ray emitting objects, an aurora is an X-ray source. Auroras have been observed on Earth, Jupiter, Saturn, Mercury, and Uranus and our current understanding is that auroras of one form or another should occur with all magnetospheres.

Figure 1.5 shows an example of the global aurora encircling the northern polar region of Earth as imaged by an ultraviolet camera flown on a polar-orbiting spacecraft. Auroras are very dynamic and are a global phenomenon. They are intimately tied to solar activity and the current theory

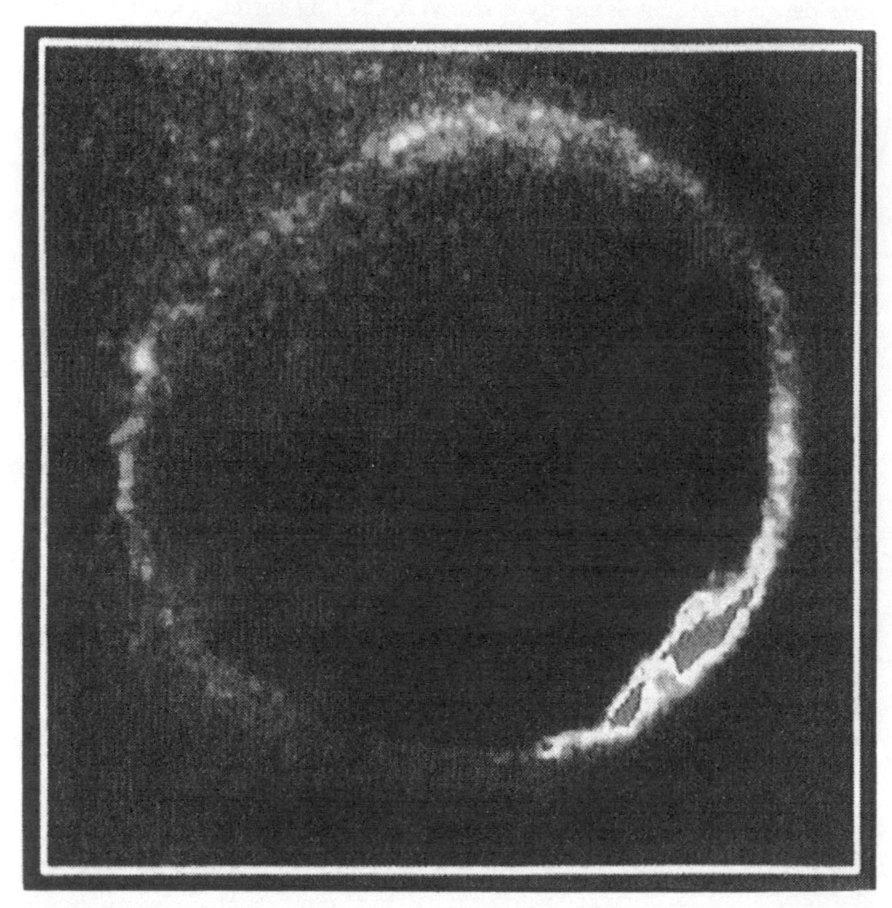

FIGURE 1.5 A global image of the aurora borealis taken by a UV camera flown on the Swedish Viking spacecraft, devoted to the study of the physical processes on auroral lines of force. Auroral intensities are color-coded, with red being the brightest. The local noon-midnight line goes through (roughly) the upper left-hand corner to the bottom right-hand corner. Note that the aurora circles the entire Earth in the polar region. The intense discrete structures centered near local midnight are associated with the breakup of the auroral arc structure. (See back cover for color version of photo. Courtesy of C. Anger and J. Murphree)

of auroras is that they represent the dissipated energy of the dynamic transfer of solar wind energy into the magnetosphere. A typical aurora on Earth dissipates about 10^{14} joules (10^{21} ergs) of energy into the atmosphere and this occurs every few hours during a moderate level of solar activity.

1.7.7 Corotational Electric Fields

We mentioned earlier how the solar wind convective electric field is produced. Let us now discuss another example of a convective electric field, called a corotation electric field, which is induced in the vicinity of a rotating conducting planet with a magnetic field. As in the case of the solar wind electric field, a corotational electric field arises because an electric field will exist in a nearby volume of plasma in planetary magnetospheres unless it moves around with the planet.

Our solar system supports magnetospheres whose plasmas may be partially corotating (Earth), entirely corotating (Jupiter), or not corotating at all (Mercury). The plasma inside the plasmasphere in Earth's magnetosphere is corotating with the planet but the plasma outside the plasmasphere is not corotating because the dominant electric field there is due to the solar wind convective electric field. Thus, only a part of Earth's magnetosphere is corotating. In Jupiter's magnetosphere, observations indicate that all of its plasma may be corotating, from the base of the ionosphere all the way out to the magnetopause boundary, which is some 50 Jovian radii from the center of the planet. This contrasts with Mercury's magnetosphere, which does not have an ionosphere. The plasma in Mercury's magnetosphere may not be corotating at all.

1.7.8 Radio Emissions

The plasmas in the magnetosphere and the ionosphere are dynamic and unstable. As a consequence, electromagnetic and electrostatic waves of many different types are observed. Figure 1.6 shows an example of electrostatic waves observed in the vicinity of Earth. This spectrogram was obtained by a satellite instrument as it was inbound on the sun-lit hemisphere. As is evident, space is extremely noisy. Waves are observed over a broad frequency interval. Different regions support different types of waves. Outside the magnetopause, waves covering a broad frequency range below 10^4 hertz (Hz) are observed. Inside the magnetosphere, wave emissions can be broadbanded or discrete and emissions can include several harmonics (see the interval between 1900 and 2100 Universal Times).

In addition, intense electromagnetic emissions are sporadically observed from the Sun, Jupiter, Saturn, and Earth. Major solar emissions associated with the flare activities frequently include microwave activities. The sporadic emissions from Jupiter and Earth are associated with auroral dynamics and cover the decametric and kilometric wavelengths. These emissions represent the most intense sources of radiation at these wavelengths detected on Earth.

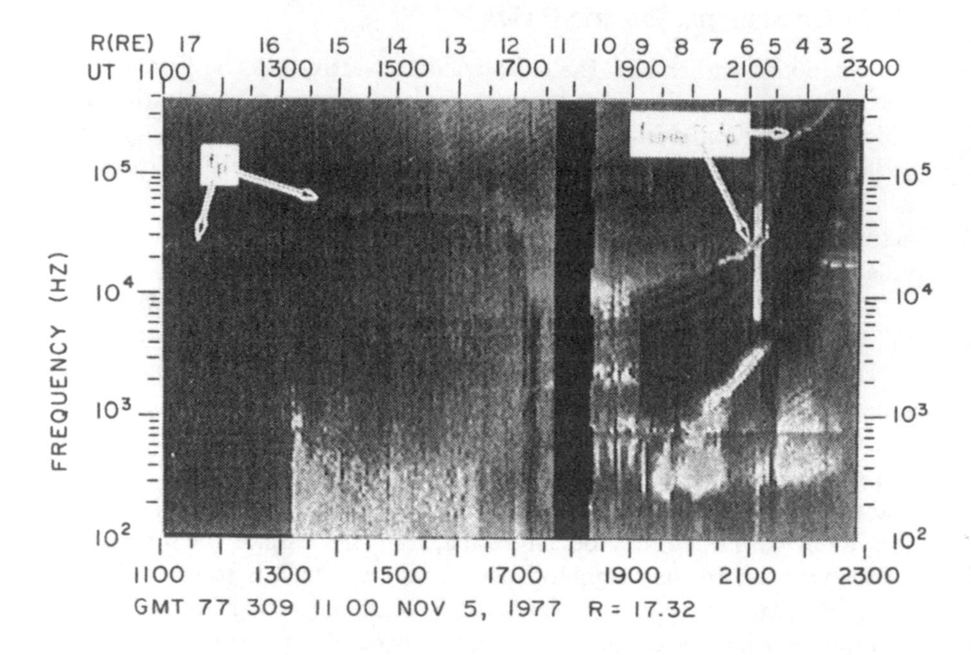

FIGURE 1.6 Electrostatic waves measured by a spacecraft inbound from interplanetary space into Earth's magnetosphere. The shock was encountered around 1300 UT, the magnetopause around 1645 UT, and the plasmapause around 2115 UT. The bands observed between 1915 UT and 2100 UT are cyclotron harmonic waves. The frequency increase of the narrow emission inside the plasmasphere is due to increasing electron density closer to Earth. Interplanetary space and the magnetosphere are generally very noisy. Note data gap between 1745 and 1830 UT. (See back cover for color version of photo. From Gurnett *et al.*, 1979)

Collective plasma processes are involved in producing these waves. Some of the waves have large amplitudes and cannot be described by a linear process. Non-linear turbulent processes may be involved. Detailed theories of these waves are not well developed and much effort is being made to better understand their origin.

1.8 Heliosphere and Heliopause

When the solar wind reaches the outer limits of the solar system, it should encounter the interstellar wind, assuming that other stars behave similarly to our Sun. The boundary that separates the solar system from the interstellar region is called the heliopause, and the region inside it is the heliosphere (Figure 1.7). Depending on the model used, the heliopause is predicted to be anywhere from about 40 AU to about 100 AU from the

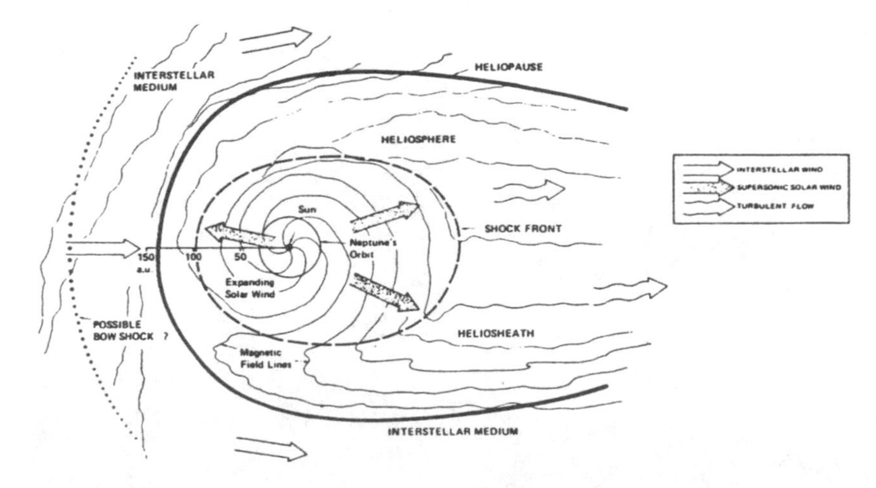

FIGURE 1.7 A schematic diagram of the heliosphere. Many of the features need to be verified by future observations. (Courtesy of NASA Headquarters, Space Physics Division)

Sun. Pioneer 10, which is farthest from Earth, is travelling toward the heliospheric tail. Pioneer 11 and Voyagers 1 and 2 are travelling in the direction of the heliopause and it is predicted that one of these spacecraft will encounter the heliopause before the turn of this century.

1.9 Comparative Magnetospheres

Magnetic structures observed in our solar system and in other celestial configurations are remarkably similar in that each includes a dense central nucleus and an elongated tail (Figure 1.8). The vastly differing magnetospheric dimensions are produced by the differences in the intrinsic strengths of the associated magnetic fields and plasmas. Mercury's magnetosphere is tiny compared to Earth's, while Jupiter's is enormous, an order of magnitude larger than our Sun. The size of a typical pulsar magnetosphere is of the same order of magnitude as Earth's, but its magnetic field is a trillion times stronger. Pulsar plasma is locked in its magnetic field until it is spun up to nearly the speed of light. Note that as the radio galaxy NGC 1265 (shown here with a map of radio emission for comparison) plows through the intergalactic gas and plasma, the ram pressure creates a magnetospheric tail that stretches millions of light years into space (the distance of a million trillion kilometers—10^{18} km—is equivalent to a hundred thousand light years).

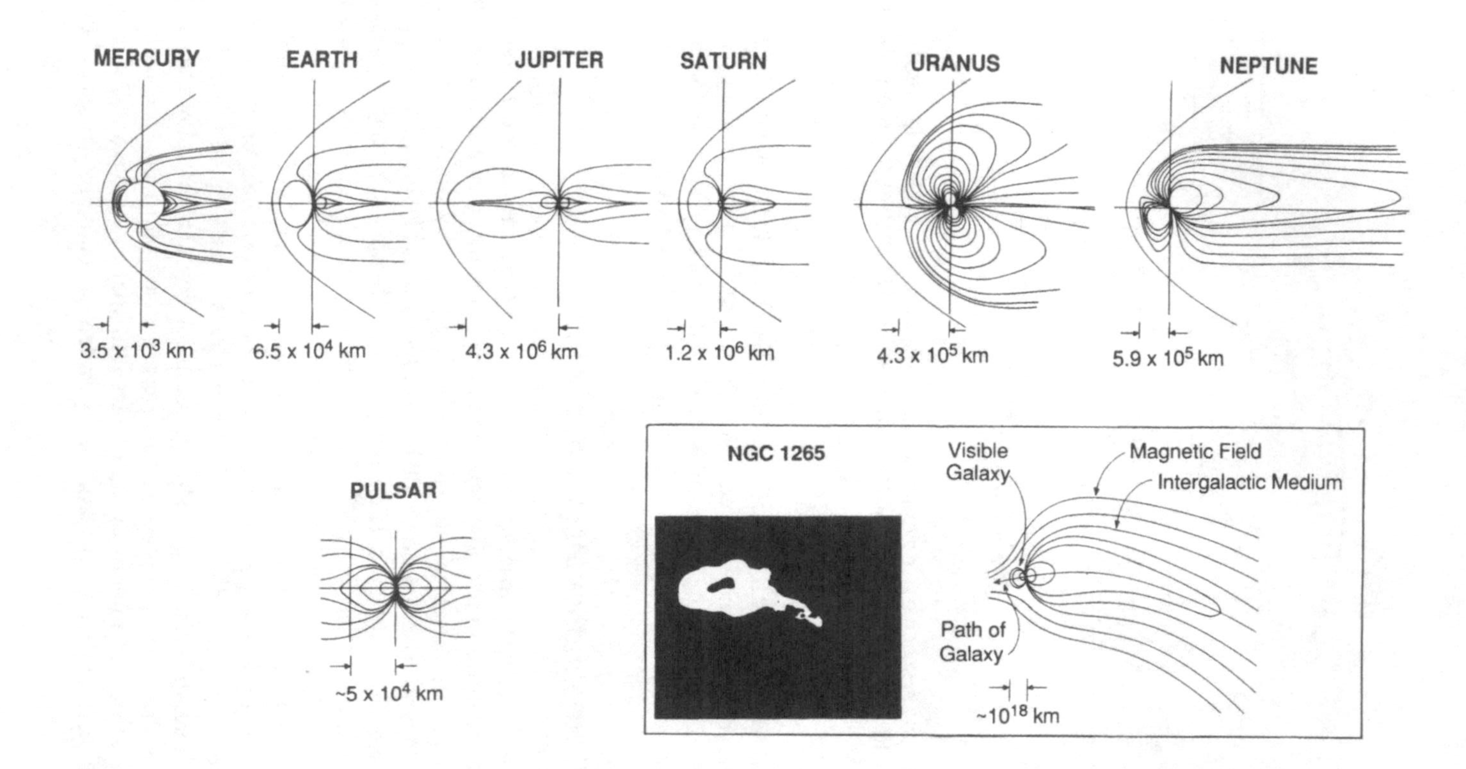

FIGURE 1.8 Fundamental similarities characterize magnetospheric configurations of the planets in the solar system and some celestial objects in the Universe. (Based on earlier drawings from NASA Goddard Space Flight Center)

Large-scale electrodynamic phenomena in our solar system are consequences of basic forces in action, and they are representative of phenomena that occur in other stellar systems. For example, plasma interactions with electromagnetic fields generate magnetic fields in planets, stars, and galaxies and they accelerate particles in planetary magnetospheres, magnetized stars, and quasars. Electromagnetic fields and charged particles obey respectively, the fundamental Maxwell equations and the Lorentz equation of motion. Our objective is to study these equations in a systematic way in order to learn how electromagnetic fields and charged particles behave, with an emphasis on understanding the various electrodynamic phenomena observed in space. Space measurements made in the vicinity of Earth are sufficiently complete to be evaluated by quantitative theories; hence these observations will be referred to often.

Bibliography

Akasofu, S.-I. and L.J. Lanzerotti, The Earth's Magnetosphere, *Physics Today*, **28**, 12, 1975. This article is written for non-specialists, and provides general information about the Earth's magnetosphere.

Alfvén, H., The Theory of Magnetic Storms and Auroras, *Nature*, **167**, 984, 1951. Alfvén has proposed theories of magnetic storms and auroras since 1939. These theories were proposed before the advent of the space age.

Alfvén, H., On the Importance of Electric Fields in the Magnetosphere and Interplanetary Space, *Space Sci. Rev.*, **7**, 140, 1967. This journal publishes review articles submitted by experts in the various disciplines of space research. Interested readers are recommended to scan through the volumes beginning around 1960.

Alfvén, H., Plasma Physics, Space Research, and the Origin of the Solar System, *Science*, **172**, 991, 1971. Nobel lecture delivered on 11 December, 1970. This lecture contains some of Alfvén's ideas on how natural plasmas work.

Alfvén, H., The Plasma Universe, *Physics Today*, **22**, September, 1986. Another article in which the importance of plasma in the Universe is emphasized. This article makes reference to recent results. Other articles referred to in this article should also be read.

Anderson, K.A. and R.P. Lin, Observations of Interplanetary Field Lines in the Magnetotail, *J. Geophys. Res.*, **74**, 3953, 1969. This article shows a method for tracing geomagnetic tail field lines, using solar energetic electrons and the Moon as an absorber.

Biermann, L., Observed Dynamical Processes in Interplanetary Space, in *Plasma Dynamics*, F.H. Clauser, ed., Addison-Wesley Publishing Co., Reading, MA, 1960. An article that summarizes observational results of comets observed in space.

Chapman, S., Historical Introduction to Aurora and Magnetic Storms, *Annales de Geophysique*, **24**, 497, 1968. A good, readable article for beginners.

Cowling, T.G., On Alfvén's Theory of Magnetic Storms and of the Aurora, *Terrest. Magnetism and Atmos. Elec.*, **47**, 209, 1942. This paper criticizes Alfvén's theory. Debates often arise when points of view and interpretations differ among researchers. This paper is an example of such a debate.

Dungey, J.W., Interplanetary Magnetic Field and the Auroral Zones, *Phys. Rev. Letters*, **6**, 47, 1961. One of the earliest papers that suggested the concept of the open magnetosphere.

Gurnett, D.A., R.R. Anderson, B.T. Tsurutani, E.J. Smith, G. Paschmann, G. Haerendel, S.J. Bame and C.T. Russell, Plasma Wave Turbulence at the Magnetopause: Observations from ISEE 1 and 2, *J. Geophys. Res.*, **84**, 7043, 1979. A comprehensive survey of plasma waves observed in the neighborhood of Earth's magnetosphere.

Heikkila, W.J., Aurora, *EOS*, **54**, 764, 1973. This article provides general information about auroras.

Lanzerotti, L., Geospace—The Earth's Plasma Environment, *Astronautics and Aeronautics*, April, 1981. This article can be read by non-specialists. It discusses future NASA programs that will address some of the key questions about solar wind energy transfer into the Earth's magnetosphere.

Ness, N.F., Interaction of the Solar Wind with the Moon, in *Solar Terrestrial Physics*, 1970, E.R. Dyer, ed., D. Reidel Publishing Co., Dordrecht, Holland, 1972. This article summarizes our understanding of how the solar wind interacts with the Moon.

Parker, E.N., *Interplanetary Dynamical Processes*, Interscience Publishers, New York, NY, 1963. This monograph brings together the theory of the solar wind as formulated by Parker.

Roederer, J., The Earth's Magnetosphere, *Science*, **183**, 37, 1974. This article, written for non-specialists, contains a considerable amount of detailed information on magnetospheric particles and fields.

First-hand information on spacecraft encounters with the various planets are contained in a series of "popular" articles published in *Science*. See *Science*, **183**, 301, 1974, and *Science*, **188**, 445, 1975 for Pioneer 10 and 11 encounters with Jupiter; *Science*, **185**, 141, 1974 for the Mariner encounter with Mercury; *Science*, **204**, 945, 1979 and *Science*, **206**, 925, 1979 for Voyagers 1 and 2 encounters with Jupiter; *Science*, **212**, 159, 1981 and *Science*, **215**, 499, 1982 for Voyagers 1 and 2 encounters with Saturn; *Science*, **233**, 39-109, 1986 for the Voyager 2 encounter with Uranus; and *Science*, **246**, 1417-1501, 1989 for Voyager 2 encounter with Neptune.

Plasmas and Fluids, National Research Council, National Academy Press, Washington, D.C., 1986. This book contains a survey of plasma and fluid physics (including space and astrophysical plasmas), its present activities, and recent major achievements. It also identifies future research areas for study. This 318-page book is one of the *Physics Through the 1990s* series of the U.S. National Academy of Sciences. A good book for readers interested in obtaining a general knowledge of the fields of plasmas and fluids.

Plasma Science, special issues on space and cosmic plasmas, published by IEEE, Inc. The nuclear and plasma science society of IEEE has recently published two special issues on space and cosmic plasmas. The articles in IEEE Transactions, *Plasma Science*, **PS-14**, December, 1986 and **PS-17**, April, 1989, discuss the relevance of plasma physics in astrophysical phenomena. The articles are generally for advanced students. A few of the introductory articles can be read by beginning students.

2

Equations and Definitions

2.1 Introduction

As shown in the previous chapter, space is endowed with a rich variety of electrodynamic phenomena. The problem to be studied in space, briefly stated, is how ionized particles interact among themselves and with electromagnetic fields. The formal equations that are required for this study are now presented, with a short discussion of how some of the electrodynamic quantities are measured in space.

2.2 Maxwell Equations

James Clerk Maxwell's unified theory of electromagnetic fields, published in 1864, is fundamental to this subject. The sources of electric and magnetic fields are distributions of charges and currents, which can be either discrete or continuous. With a charge density ρ and a vector current density $\mathbf{J}$, the electrodynamic Maxwell equations in the mks (meter/kilogram/second) system are

$$\nabla \times \mathbf{E} = -\frac{\partial \mathbf{B}}{\partial t} \tag{2.1}$$

$$\nabla \times \mathbf{H} = \mathbf{J} + \frac{\partial \mathbf{D}}{\partial t} \tag{2.2}$$

$$\nabla \cdot \mathbf{B} = 0 \tag{2.3}$$

$$\nabla \cdot \mathbf{D} = \rho \tag{2.4}$$

By traditional usage, **E** and **H** are electric and magnetic field intensities, **D** is the electric displacement and **B** the magnetic induction vector. **E**, **B**, **D** and **H** are vector functions defined in space and time (x, y, z, t) that obey the usual continuity and derivative rules of ordinary calculus. The relationships of these field vectors in vacuum are

$$\mathbf{B} = \mu_0 \mathbf{H} \tag{2.5}$$

and

$$\mathbf{D} = \epsilon_0 \mathbf{E} \tag{2.6}$$

Here, ϵ_0 and μ_0 are constants equal to 8.85×10^{-12} farads/meter and $4\pi \times 10^{-7}$ henrys/meter, respectively. The charge and current densities at any point in space and time are defined by

$$\rho = \frac{\sum_{\Delta V} q_k}{\Delta V} \tag{2.7}$$

and

$$\mathbf{J} = \frac{\sum_{\Delta V} \mathbf{v}_k q_k}{\Delta V} \tag{2.8}$$

The summation is carried over a suitably chosen small volume element ΔV. q_k and $\mathbf{v}_k$ are the charge and velocity vectors, respectively, of the kth particle.

2.3 Lorentz Equation of Motion

The electromagnetic fields originate in charges ρ and currents **J**, and to describe the motion of these charges in space, the set of Maxwell equations must be supplemented with the Lorentz equation of motion. For a non-relativistic particle, $(v \ll c)$, the position $\mathbf{r}_k$ and the velocity $\mathbf{v}_k$ are deduced from the Lorentz equation of motion. The force on the kth particle is

$$m_k \mathbf{a}_k = q_k(\mathbf{E} + \mathbf{v}_k \times \mathbf{B}) \tag{2.9}$$

where $\mathbf{a}_k$ is the acceleration. The Lorentz equation of motion and the electromagnetic equations are coupled through the charge and the current densities ρ and $\mathbf{J}$ and these equations must be solved self-consistently.

These nine equations are sufficient to describe a system of charged particles and electromagnetic fields in space. A complete description of particle motions requires that the position and velocity of all the particles be known. Since the electromagnetic force is long-range, the dynamics of one particle will be influenced by the other particles. Hence, knowledge is required on how the particles interact among themselves as well as with the electromagnetic fields. This study requires as many equations as there are particles. Even in a medium as tenuous as space, the actual number of equations involved will be extremely large. Clearly, it is not practical to solve all of the coupled equations analytically.

2.4 Statistical Concepts

Rather than require an exact knowledge of a system with many particles as discussed above, the behavior of such a particle system can be studied statistically. Plasma physics statistically deals with equilibrium or nonequilibrium properties of a collection of charged particles, called a plasma, by means of the distribution function. Let us begin this discussion with a definition of what is meant by a plasma.

2.4.1 Plasma

A plasma is composed of a collection of discrete ionized particles, but not every collection of charged particles qualifies as a plasma. In its simplest form, a plasma consists of electrons and one species of ions, for example, protons (p^+). A more complex plasma system includes neutral atoms and molecules. Plasma in Earth's ionosphere includes complex molecules such as N_2^+, O^+, and NO^+. The density of the particles in a plasma is required to be sufficiently low so that the short-range binary collisions are negligible. Instead, long-range electromagnetic forces are effective.

A plasma possesses properties that arise from both the individual interactions of the charged particles and those that come from the "collective" behavior of particles. This behavior arises from many particles interacting simultaneously through the long-range Coulomb potential. The collective interactions are best described by the concepts of statistical physics. A plasma system is required to contain a sufficiently large number of particles.

2.4.2 Coulomb Potential and Debye Length

The electrostatic potential at a distance r from a particle with charge q is $\psi = q/4\pi\epsilon_0 r$. This simple relation is modified in the presence of a plasma. Consider the potential in the neighborhood of a positively charged particle in an equilibrium plasma. Here, the positive charge attracts electrons toward it while ions are repelled and its electrostatic field therefore becomes shielded from the plasma. Similarly, an electron attracts ions and repels electrons, again shielding itself from the plasma. (Unless stated otherwise, ions will be assumed positive throughout this text). This is an example of collective behavior of plasma and the net effect is that the electrostatic field becomes confined to a short range. A short-range electrostatic potential can be written as

$$\psi = \frac{q}{4\pi\epsilon_0 r} e^{-r/\lambda_D} \tag{2.10}$$

where r is the distance from the charge and λ_D, the e-folding distance, is called the Debye length, defined by

$$\lambda_D^2 = \frac{kT_e\epsilon_0}{n_0 q_e^2} \tag{2.11}$$

k is the Boltzmann constant ($k = 1.38 \times 10^{-23}$ Joules/°K), T_e is the electron temperature in degrees Kelvin, ϵ_0 is the dielectric constant of free space, n_0 is the equilibrium density of the plasma with a temperature T_e, and q_e is the charge of electrons ($q_e = 1.6 \times 10^{-19}$ coulombs). Equation (2.10) shows us that for $r \ll \lambda_D$, the potential reduces to the simple Coulomb potential. For $r > \lambda_D$, ψ decreases exponentially and the potential around a point charge is effectively screened out. In plasma, $\lambda_D \ll L$, where L is the characteristic dimension of a plasma system. Equation (2.10) is referred to as the Debye screening potential and it arises from the kinetic properties of plasma.

2.4.3 Neutrality of a Plasma

When an electron with an average energy kT_e moves away from a positive ion, leaving behind an electron depleted region, the potential it sets up is kT_e/q_e. λ_D corresponds to the average distance an electron departs from the equilibrium position. Equation (2.10) shows that this potential decreases very rapidly for $r > \lambda_D$ and leaves the bulk of plasma free of large electric potentials. Thus, in regions outside λ_D, the plasma is "unaffected" and the plasma remains neutral where the ion number density n_i is on the average equal to electron density n_e.

2.4.4 Plasma Parameter

The number of particles in a Debye sphere is

$$N_D = \frac{4\pi n_0 \lambda_D^3}{3} \tag{2.12}$$

The plasma parameter is defined as

$$g = \frac{1}{N_D} \tag{2.13}$$

The description of plasma is significant only if $g \ll 1$, for then we have a large enough number of particles to apply the concepts of statistics and Debye shielding is meaningful. The condition $g \ll 1$ is also known as the plasma approximation.

2.4.5 Collisionless Plasma

When $g \ll 1$, it also defines the condition that the binary collisions will be rare. To see this, use (2.11), (2.12), and (2.13) and obtain

$$g \sim \frac{n_0^{1/2}}{T^{3/2}} \tag{2.14}$$

Since the collision frequency decreases with decreasing density n and with increasing temperature T, smaller g corresponds to less collisions and in the limit $g \to 0$, the plasma becomes collisionless. In many regions of space, the density is low and the temperature is high, and the plasma approaches the collisionless limit.

2.4.6 Plasma as Ideal Gas

Frequently, a plasma is treated as consisting of ideal particles. Ideal gas particles do not interact among themselves. However, the particles can have charge densities, which are a source of electric fields, and they can move around, giving rise to currents. g is also a measure of the ratio of interparticle potential energy to the mean kinetic energy of the plasma particles. Hence, when $g \ll 1$, the interparticle forces can be neglected and the plasma behaves almost like an ideal gas. In later chapters, when plasma equations are formulated under the fluid approximation, the ideal gas behavior of plasma permits us to invoke the simple ideal gas equation of state.

2.4.7 Plasmas in Nature

Nature's plasmas cover a broad range of temperatures and densities. Table 2.1 shows typical parameters that are characteristic of some of the plasmas in our solar system.

TABLE 2.1
Parameters of Some of Nature's Plasmas

	T(°K)	n (m^{-3})
Stellar interior	10^7	10^{33}
Solar corona	10^6	10^{12}–10^{13}
Photosphere	6×10^3	10^{20}
Solar wind/Interplanetary space	10^5	10^6–10^7
Earth's core	6×10^3	2×10^{29}
Earth's ionosphere	10^3–10^4	10^8–10^{12}
Earth's magnetosphere	10^7–10^9	10^5–10^7

2.5 Statistical Equations

The motion of a particle of mass m is defined by its position $\mathbf{r}$ and its velocity $\mathbf{v}$. Each particle can therefore be represented by a point $(\mathbf{r}, \mathbf{v})$ in space called "phase" space. (Note that this is not the same phase space defined by variables $\mathbf{r}$ and $\mathbf{p}$.) This space is a six-dimensional space with coordinates (x, y, z, v_x, v_y, v_z). The probability density of points in this $(\mathbf{r}, \mathbf{v})$ space at time t is proportional to the distribution function $f(\mathbf{r}, \mathbf{v}, t)$. $f(\mathbf{r}, \mathbf{v}, t)\, d\mathbf{r}\, d\mathbf{v}$ represents the expected number of particles at time t in $(\mathbf{r}, \mathbf{v})$ space with coordinates $\mathbf{r}$ and $\mathbf{r} + d\mathbf{r}$ and velocity $\mathbf{v}$ and $\mathbf{v} + d\mathbf{v}$.

2.5.1 Boltzmann-Vlasov Equation

The distribution function $f(\mathbf{r}, \mathbf{v}, t)$ is a function of seven independent variables. The total time derivative of f is

$$\frac{df}{dt} = \frac{\partial f}{\partial t} + \frac{\partial f}{\partial x}\frac{dx}{dt} + \frac{\partial f}{\partial y}\frac{dy}{dt} + \frac{\partial f}{\partial z}\frac{dz}{dt} + \frac{\partial f}{\partial v_x}\frac{dv_x}{dt} + \frac{\partial f}{\partial v_y}\frac{dv_y}{dt} + \frac{\partial f}{\partial v_z}\frac{dv_z}{dt} \tag{2.15}$$

Recognizing that dx/dt, dy/dt, and dz/dt are velocity components, v_x, v_y, and v_z and that dv_x/dt, dv_y/dt, and dv_z/dt are acceleration components a_x, a_y, and a_z, the above equation can be rewritten as

$$\frac{df}{dt} = \frac{\partial f}{\partial t} + \mathbf{v} \cdot \frac{\partial f}{\partial \mathbf{r}} + \mathbf{a} \cdot \frac{\partial f}{\partial \mathbf{v}} \tag{2.16}$$

To fully appreciate the meaning of this equation, consider a case where f is a function of only time and coordinates, $f = f(\mathbf{r}, t)$. Then, df/dt is defined by the first two terms on the right and the total time derivative of f is given by

$$\frac{df}{dt} = \frac{\partial f}{\partial t} + \mathbf{v} \cdot \frac{\partial f}{\partial \mathbf{r}} \tag{2.17}$$

In fluid dynamics, the derivative $d/dt = \partial/\partial t + (\mathbf{v} \cdot \partial/\partial \mathbf{r})$ is called the convective derivative (see Chapter 5 for further discussion). df/dt can be viewed as the time derivative of f taken in a "fluid" frame of reference moving with a velocity of $\mathbf{v}$ relative to a rest frame. $\partial f/\partial t$ represents the rate of change of f at a fixed point in space and $(\mathbf{v} \cdot \nabla)f$ represents the change of f measured by an observer moving in the fluid frame into a region where f is inhomogeneous.

Return now to a space defined by the coordinates $(\mathbf{r}, \mathbf{v})$. The convective derivative in this six-dimensional space must take into account the variation of f relative to the coordinate $\mathbf{v}$ in addition to $\mathbf{r}$. The third term of (2.16), $(\mathbf{a} \cdot \partial f/\partial \mathbf{v})$, represents this additional contribution.

To describe the physical behavior of particles in the $(\mathbf{r}, \mathbf{v})$ space, consider the motion of density of points, assuming that the velocity and acceleration of each particle are finite in the six-dimensional phase space. In the absence of collisions, these points move in continuous curves and $f(\mathbf{r}, \mathbf{v}, t)$ obeys the continuity equation (also called the Liouville equation)

$$\frac{\partial f}{\partial t} + \nabla_{\mathbf{r},\mathbf{v}} \cdot [(\dot{\mathbf{r}}, \dot{\mathbf{v}})f] = 0 \tag{2.18}$$

where $\nabla_{\mathbf{r},\mathbf{v}}$ is a six-dimensional operator and $(\dot{\mathbf{r}},\ \dot{\mathbf{v}})$ is a velocity vector in the $(\mathbf{r},\ \mathbf{v})$ space (the dots represent the time derivative). In Cartesian coordinates, (2.18) becomes

$$\frac{\partial f}{\partial t} + \frac{\partial}{\partial x_i}(f\dot{x}_i) + \frac{\partial}{\partial v_i}(f\dot{v}_i) = 0 \tag{2.19}$$

Now expand (2.19) and obtain

$$\frac{\partial f}{\partial t} + \left(\frac{\partial \dot{x}_i}{\partial x_i} + \frac{\partial \dot{v}_i}{\partial v_i}\right) f + \frac{\partial f}{\partial x_i}\dot{x}_i + \frac{\partial f}{\partial v_i}\dot{v}_i = 0 \tag{2.20}$$

Since $\dot{x}_i$ and x_i are independent variables, $\partial \dot{x}_i/\partial x_i = 0$. Also, $\partial \dot{v}_i/\partial v_i = 0$ because, from the Lorentz force equation,

$$\begin{aligned} \frac{\partial \dot{v}_i}{\partial v_i} &= \frac{1}{m}\frac{\partial F}{\partial v_i} \\ &= \frac{q}{m}\frac{\partial}{\partial v_i}(\mathbf{v} \times \mathbf{B})_i \\ &= 0 \end{aligned} \tag{2.21}$$

Hence, (2.19) reduces to

$$\frac{\partial f}{\partial t} + \frac{\partial f}{\partial x_i}\dot{x}_i + \frac{\partial f}{\partial v_i}\dot{v}_i = 0 \tag{2.22}$$

Equation (2.22) is identical to (2.16) if $df/dt = 0$. Equation (2.22) was derived assuming collisions are absent. This is the famous Boltzmann equation derived by L. Boltzmann before the turn of the century. In collisionless plasmas, $df/dt = 0$ and (2.22) is an equation of continuity for f in the $(\mathbf{r},\ \mathbf{v})$ space. The density of points in motion remains unchanged when plasma is collisionless. Note that in arriving at (2.22), we found that

$$\frac{\partial \dot{x}_i}{\partial x_i} + \frac{\partial \dot{v}_i}{\partial v_i} = 0 \tag{2.23}$$

This equation shows that the divergence of the six-dimensional velocity vector vanishes in the $(\mathbf{r}, \mathbf{v})$ space. In hydrodynamics, this behavior is described as incompressible, and one can say that particles in $(\mathbf{r}, \mathbf{v})$ space behave as if they are incompressible.

If the particles are only subject to electromagnetic forces, (2.22) written in vector form becomes

$$\frac{\partial f}{\partial t} + \mathbf{v} \cdot \frac{\partial f}{\partial \mathbf{r}} + \frac{q}{m}(\mathbf{E} + \mathbf{v} \times \mathbf{B}) \cdot \frac{\partial f}{\partial \mathbf{v}} = 0 \tag{2.24}$$

This equation was first studied by A.A. Vlasov in 1945 and is therefore also called the Vlasov equation. Equation (2.24) is a differential equation with seven variables (x, y, z, v_z, v_y, v_z, t) and (2.24) and the electrodynamic Maxwell equations must be solved self-consistently.

2.5.2 Macroscopic Plasma Parameters

Observable properties of a system of particles are obtained through quantities averaged over a distribution function. In general, a plasma system will consist of many particle species. Define f^α ($\mathbf{r}$, $\mathbf{v}$, t) as a one-particle distribution function for particle species α. The average of a physical quantity $M(\mathbf{r}, \mathbf{v}, t)$ defined in terms of the distribution function is

$$< M^\alpha(\mathbf{r},t) > = \frac{\int M f^\alpha(\mathbf{r},\mathbf{v};t)d^3v}{\int f^\alpha(\mathbf{r},\mathbf{v};t)d^3v} \tag{2.25}$$

where the integration limits are $-\infty$ and $+\infty$ (unless otherwise stated, the limits of integration are always from $-\infty$ to $+\infty$). Given the distribution function, macroscopic physical parameters are obtained by "integrating out" the velocity space. Equation (2.25) is basic and the definition comes from probability theory.

Density The number density of particles for species α is obtained for real space by integrating the phase space density $f(\mathbf{r}, \mathbf{v}, t)$ over the velocity,

$$n^\alpha(\mathbf{r},t) = \int f^\alpha(\mathbf{r},v,t)d^3v \tag{2.26}$$

The integral includes all of the particles since the limit extends from $-\infty$ to $+\infty$. Note that the density always appears as a normalizing factor in the computation of averages.

Average Velocity The average velocity of particle species α at r at time t is defined by

$$\mathbf{u}^\alpha = < \mathbf{v}^\alpha(\mathbf{r},t) > = \frac{1}{n^\alpha(\mathbf{r},t)} \int \mathbf{v} f^\alpha(\mathbf{r},\mathbf{v},t)d^3v \tag{2.27}$$

Equation (2.27) represents the "bulk flow" of particle species α. This flow velocity is different from the velocities associated with the particles' thermal energy.

Kinetic Energy The average kinetic energy density of particle species α at r at time t is

$$\epsilon^\alpha(\mathbf{r},t) = \left\langle \frac{1}{2}mv^2 \right\rangle^\alpha = \frac{1}{n^\alpha(\mathbf{r},t)} \int \frac{1}{2} m^\alpha v^2 f^\alpha d^3v \tag{2.28}$$

The average kinetic energy can be equated to the thermal energy if the distribution function is a Maxwellian (the plasma is then in equilibrium). In this case, the average kinetic energy per particle per degree of freedom is $kT/2$, where k is the Boltzmann constant and T is the temperature.

Charge Density Charges from all of the different species contribute to the charge density. The charge density at $\mathbf{r}$ at time t is

$$\rho(\mathbf{r}, t) = \sum_{\alpha} \int q^{\alpha} f^{\alpha} d^3 v = \sum_{\alpha} q^{\alpha} n^{\alpha}(\mathbf{r}, t) \tag{2.29}$$

where the summation is carried over all particle species. Because a plasma tends to charge neutrality, free charges normally do not accumulate and the average charge density vanishes in plasmas.

Current Density The current density at $\mathbf{r}$ at time t is

$$\mathbf{J}(\mathbf{r}, t) = \sum_{\alpha} q^{\alpha} \int \mathbf{v} f^{\alpha} \, d^3 v = \sum_{\alpha} q^{\alpha} n^{\alpha} < \mathbf{v}^{\alpha} > \tag{2.30}$$

where again the summation is carried over all the particle species. Equations (2.29) and (2.30) are the source terms for the electromagnetic fields in the Maxwell equations.

Many other macroscopic physical parameters can be derived by performing the averages as prescribed. Here just a few examples of these parameters were computed. The averaged parameters, referred to as moments of the distribution function, provide information on the macroscopic behavior of a system of particles. These parameters become variables in the large-scale fluid description of space.

2.5.3 Macroscopic Plasma Equations

The partial differential equation which describes the behavior of f is very difficult to solve exactly except for a few simple cases. However, it turns out that approximate methods are available for solving (2.24). One method is to take moments of (2.24), which involves multiplying the equation by the various powers of the velocity vector $\mathbf{v}$ and integrating over the velocity space. This procedure "destroys" information but we gain general knowledge about the behavior of the particles that is quite useful.

The zeroth order moment is obtained by multiplying (2.24) with $v^0 = 1$ and integrating over the velocity. This yields as the first term

$$\begin{aligned} \int \frac{\partial f}{\partial t} d^3 v &= \frac{\partial}{\partial t} \int f d^3 v \\ &= \frac{\partial n}{\partial t} \end{aligned} \tag{2.31}$$

where n is the density defined by (2.26). The second term is

$$\begin{aligned}\int \mathbf{v}\cdot\frac{\partial f}{\partial \mathbf{r}}d^3v &= \int \mathbf{v}\cdot\nabla f\, d^3v \\ &= \nabla\cdot\int \mathbf{v} f d^3v - \int f\nabla\cdot\mathbf{v}\, d^3v \\ &= \nabla\cdot(n\mathbf{u}) \end{aligned} \tag{2.32}$$

where the average velocity $\mathbf{u}$ is defined through (2.27). The second line uses the vector identity $\nabla\cdot f\mathbf{v} = (\mathbf{v}\cdot\nabla)f + f\nabla\cdot\mathbf{v}$ and, since the fluid is incompressible, the second term vanishes. The third term is

$$\begin{aligned}\int (\mathbf{E}+\mathbf{v}\times\mathbf{B})\cdot\frac{\partial f}{\partial \mathbf{v}}d^3v &= \int (\mathbf{E}+\mathbf{v}\times\mathbf{B})\cdot\nabla_v f d^3v \\ &= \int \nabla_v\cdot f(\mathbf{E}+\mathbf{v}\times\mathbf{B})d^3v - \int f\nabla_v\cdot(\mathbf{E}+\mathbf{v}\times\mathbf{B})d^3v \\ &= \int f(\mathbf{E}+\mathbf{v}\times\mathbf{B})d^2v - \int f\nabla_v\cdot(\mathbf{E}+\mathbf{v}\times\mathbf{B})d^3v \\ &= 0 \end{aligned} \tag{2.33}$$

In going from the first to the second line, we used the same vector identity as in (2.32). The first term of the third line is a surface integral obtained from integrating the divergence term through use of the divergence theorem. To carry out the surface integral, we need to define the behavior of the distribution function f at large velocities. Since there are no particles with infinite velocities (recall the integration limits are $-\infty$ to $+\infty$), $f \to 0$ as $v \to \infty$. Thus, the surface integral vanishes. The second integral also vanishes because $\mathbf{E}$ and $\mathbf{B}$ do not depend on v and, for an electromagnetic force, (2.21) shows that the integrand must vanish.

The non-vanishing terms (2.31) and (2.32) yield

$$\frac{\partial n}{\partial t} + \nabla\cdot(n\mathbf{u}) = 0 \tag{2.34}$$

which is an equation of continuity (in real space). The zeroth order moment of the Vlasov-Boltzmann equation yields an equation which states that the number of particles is conserved. The time rate of change of the number density per unit volume is balanced by the loss of particles (particles leaving the volume). Here, $n\mathbf{u}$ is defined as the number flux.

The first-order moment is obtained by multiplying (2.24) with v and integrating over the velocity. This will yield three moment equations corresponding to the three velocity coordinates. For any component v_i, the first term is

$$\int v_i \frac{\partial f}{\partial t} d^3v = \frac{\partial}{\partial t} \int f v_i d^3v$$
$$= \frac{\partial}{\partial t}(nu_i) \tag{2.35}$$

where use was again made of (2.27). Note that $\partial v_i/\partial t = 0$ because v_i and t are independent variables. The second term yields

$$\int v_i v_j \frac{\partial f}{\partial x_i} d^3v = \frac{\partial}{\partial x_i} \int v_i v_j f d^3v$$
$$= \frac{\partial}{\partial x_i}(n < v_i v_j >) \tag{2.36}$$

where $<>$ means the quantity has been averaged over the velocity space. Note that $\partial v_i/\partial x_i = \partial v_j/\partial x_i = 0$. The third term yields

$$\int v_i(\mathbf{E} + \mathbf{v} \times \mathbf{B})_j \cdot \nabla_v f d^3v = \int v_i F_j \frac{\partial f}{\partial v_j} d^3v$$
$$= \int F_j \frac{\partial}{\partial v_j}(v_i f) d^3v - \int F_j f \frac{\partial v_i}{\partial v_j} d^3v$$
$$= -\int F_j f \delta_{ij} d^3v$$
$$= -n < F_i > \tag{2.37}$$

where $F_j = (q/m)(\mathbf{E} + \mathbf{v} \times \mathbf{B})_j$ and δ_{ij} is the Kroenecker delta and is equal to 1 if $i = j$ and is equal to 0 otherwise. The first term on the second line involves the surface integral and it vanishes for the same reason stated in deriving (2.33). The final expression for the first moment is

$$\frac{\partial}{\partial t}(nmu_i) + \frac{\partial}{\partial x_i}(nm < v_i v_j >) = qn < \mathbf{E} + \mathbf{v} \times \mathbf{B} >_i \tag{2.38}$$

This equation states that momentum is conserved. The first term describes the time rate of change of momentum per unit volume and the second term the loss of momentum due to particles leaving the volume. The right-hand term thus represents the total change of the momentum density, which is equal to the electromagnetic force density.

The zeroth order moment yielded the particle conservation equation, the first order the momentum conservation, and one could predict that the second order moment would yield the energy conservation equation. If we multiply (2.24) with v^2 and proceed as before, the first two terms become

$$\frac{\partial}{\partial t} \int v^2 f d^3v = \frac{\partial}{\partial t}(n < v^2 >) \tag{2.39}$$

and

$$\frac{\partial}{\partial x_i}\int v_i v^2 f d^3v = \frac{\partial}{\partial x_i}(n < v_i v^2 >) \tag{2.40}$$

The third term, after integration by parts, becomes

$$\int F_i v^2 \frac{\partial f}{\partial v_i} d^3v = -\int F_i f \frac{\partial v^2}{\partial v_i} d^3v \tag{2.41}$$

Now note that

$$\frac{\partial}{\partial v_i}(v_k v_k) = 2v_k \delta_{ik} \tag{2.42}$$

Hence, the right-hand side of (2.41) can be rewritten as

$$\begin{aligned}\int F_i f \frac{\partial v^2}{\partial v_i} &= -2\int F_i f v_k \delta_{ik} d^3v \\ &= -2n < F_i v_i >\end{aligned} \tag{2.43}$$

The energy conservation equation becomes

$$\frac{\partial}{\partial t}\left(\frac{mnu^2}{2}\right) + \frac{\partial}{\partial x_i}\left(n\left\langle\frac{mv_i v^2}{2}\right\rangle\right) = n < F_i v_i > \tag{2.44}$$

The time rate of change of the kinetic energy density (first term) plus the loss of kinetic energy due to particles leaving the volume (second term) equals the total work done by the electromagnetic field (right-hand term). Note that the subindices in these equations represent the individual (i, j, k) components and we use the summation convention of summing over repeated indices.

These conservations are useful but they contain more unknowns than the number of equations. For example, the unknowns of (2.34) are n and $\mathbf{u}$ and we have only three equations. Similarly, for the momentum equation (2.38), the unknowns include $< v_i v_j >$ in addition to n and $\mathbf{u}$. The unknowns always exceed the number of equations and this number is not reduced from higher moments because each order introduces new unknowns. Therefore these conservation equations do not form a closed set and certain assumptions must be made before they can be solved.

These equations are macroscopic in the sense that they involve macroscopic parameters (density n, flow velocity $\mathbf{u}$, energy $1/2mv^2$). The macroscopic parameters are physically meaningful since they can be directly measured or can be obtained from the distribution function. The macroscopic equations form the basis for studies of plasma phenomena in the fluid approach. The discussion and use of these equations will be postponed until Chapter 5 where we begin our studies of space from the point of view of fluids.

2.6 Lorentz Transformation of Electromagnetic Fields

The application of electrodynamic equations to space requires that we have access to electromagnetic field quantities. Vector magnetic field measurements have been made on board spacecraft since the beginning of the space era. For instance, various components of the low-frequency magnetic field $\mathbf{B}$ ($\mathbf{r}$, t) are directly measured by flux-gate magnetometers (the magnetic field in mks units is in teslas). Electric field measurements in space are more recent. An electric field antenna basically consists of an insulated boom that is several hundred meters long with conducting spheres on each end. Information on the low-frequency electric field $\mathbf{E}$ ($\mathbf{r}$, t) is obtained by measuring the potential difference between the spheres and dividing this voltage by the length of the antenna (the electric field in mks units is in volts/meter).

Electromagnetic field measurements are made in frames of reference moving relative to the plasma frame. To correctly interpret such field data requires that we understand the effects of performing measurements in frames of reference in relative motion. We rely on the hypothesis that, given two inertial reference systems that are considered equivalent, the principle of relativity asserts that the laws of electromagnetism are identical in these systems. H.A. Lorentz in 1904 derived the theory of transformation for space and time that, extended to electromagnetic fields, yields the relationship of electromagnetic fields in the different reference systems. We now present a short discussion of how this theory works. (The Lorentz transformation theory of electromagnetic fields is treated in many advanced textbooks of electrodynamics; see Jackson, 1975, cited at the end of this chapter.) The applications of the Lorentz theory will show how currents produce magnetic fields, how changing magnetic fields induce electromotive forces, and how moving charges are deflected perpendicular to the magnetic field direction.

An electromagnetic field consists of both electric ($\mathbf{E}$) and magnetic ($\mathbf{B}$) fields. What one sees, however, depends on the state of the observer. For example, an electric field measured in one frame of reference may appear as both electric and magnetic fields in another reference system. This peculiarity is evident even in non-relativistic cases. Let S be an inertial frame of reference and S' another inertial frame which is moving relative to S with a constant velocity $\mathbf{V}$. Let $\mathbf{E}$ and $\mathbf{B}$ denote measurements made in S and $\mathbf{E}'$ and $\mathbf{B}'$ the same fields measured in S'. Now, orient the two axes so that $\mathbf{V}$ is along x and x'.

Lorentz transformation relates the quantities measured in S and S'. The components of the electric $\mathbf{E}'$ and $\mathbf{B}'$ measured in S' in terms of $\mathbf{E}$

and $\mathbf{B}$ of the same fields measured in S are

$$\begin{aligned} E'_x &= E_x \\ E'_y &= \gamma(E_y - VB_z) \\ E'_z &= \gamma(E_z + VB_y) \end{aligned} \tag{2.45}$$

and

$$\begin{aligned} B'_x &= B_x \\ B'_y &= \gamma\left(B_y + \frac{V}{c^2}E_z\right) \\ B'_z &= \gamma\left(B_z - \frac{V}{c^2}E_y\right) \end{aligned} \tag{2.46}$$

Here, V is the speed of S' relative to S along the x-direction, $\gamma = (1 - V^2/c^2)^{-1/2}$, and c is the speed of light (the speed of light is constant in both frames of reference). The inverse transformations are

$$\begin{aligned} E_x &= E'_x \\ E_y &= \gamma(E'_y + VB'_z) \\ E_z &= \gamma(E'_z - VB'_y) \end{aligned} \tag{2.47}$$

and

$$\begin{aligned} B_x &= B'_x \\ B_y &= \gamma\left(B'_y - \frac{V}{c^2}E'_z\right) \\ B_z &= \gamma\left(B'_z + \frac{V}{c^2}E'_y\right) \end{aligned} \tag{2.48}$$

Equations (2.45) and (2.46) can be written in a compact vectorial form,

$$\begin{aligned} \mathbf{E}' &= \mathbf{E} + \gamma\mathbf{V} \times \mathbf{B} + \frac{\gamma - 1}{V^2}\mathbf{V} \times (\mathbf{E} \times \mathbf{V}) \\ \mathbf{B}' &= \mathbf{B} + \gamma\mathbf{E} \times \frac{\mathbf{V}}{c^2} + \frac{\gamma - 1}{V^2}\mathbf{V} \times (\mathbf{B} \times \mathbf{V}) \end{aligned} \tag{2.49}$$

These equations show that

$$B^2 - \frac{E^2}{c^2} = B'^2 - \frac{E'^2}{c^2} \tag{2.50}$$

and

$$\mathbf{E} \cdot \mathbf{B} = \mathbf{E}' \cdot \mathbf{B}' \tag{2.51}$$

are Lorentz invariants. These invariants indicate that an electromagnetic field that is purely magnetic in S may be in part magnetic and in part electric in S'. Also, if the electric field is more intense in S than in S', the magnetic field is also more intense in S than in S'. Finally, if $\mathbf{E}$ and $\mathbf{B}$ are orthogonal in S, they are orthogonal in S'.

A question of interest is whether electric or magnetic parts of an electromagnetic field can be eliminated by some suitable choice of transformation variables. Suppose $\mathbf{E}$ and $\mathbf{B}$ are measured in S and we wish to find in S' an electromagnetic field that is entirely magnetic. Consider for simplicity a non-relativistic case, $V/c \ll 1$, and $\gamma \approx 1$. Then, from the first equation of (2.49), we immediately obtain

$$0 \approx \mathbf{E} + \mathbf{V} \times \mathbf{B} \tag{2.52}$$

and taking the cross product with $\mathbf{B}$, one obtains

$$\mathbf{V} \approx \frac{\mathbf{E} \times \mathbf{B}}{B^2} \tag{2.53}$$

since $\mathbf{V} \cdot \mathbf{B} = 0$. Note that $\mathbf{E}$, $\mathbf{B}$, and $\mathbf{V}$ are all perpendicular to each other. Thus, $\mathbf{E}' = 0$ if S' moves relative to S with a velocity given by (2.53). A requirement posed by the Lorentz invariant (2.50) is that, in this case, the magnitude of the electric field must be less than the magnitude of the magnetic field in S.

Suppose our requirement now is for the field in S' to be purely electric, $\mathbf{B}' = 0$. One then easily finds that

$$\mathbf{V} \approx c^2 \frac{\mathbf{E} \times \mathbf{B}}{E^2} \tag{2.54}$$

The requirement of the Lorentz invariant in this case is that the intensity of the magnetic field must be less than that of the electric field in S, $|\mathbf{B}| < |\mathbf{E}|/c$. Note that in both of the above examples, $\mathbf{V}$ is perpendicular to $\mathbf{E}$ and $\mathbf{B}$.

A practical application of these results is that if S' is the solar wind frame moving with $\mathbf{V}$ relative to a stationary frame (a spacecraft reference frame, for example), $\mathbf{E}' \approx 0$ in the solar wind frame if $\mathbf{V}$ is given by (2.53). Another application would suppose that we are initially given the information that $\mathbf{E}$ and $\mathbf{B}$ are perpendicular to each other in some frame of reference S. What can we say about the fields in another frame of reference? The above results imply that we can always find a reference system S' in which the field is either purely electric or magnetic if S' moves with a velocity given by (2.53) or (2.54). Finding such a reference system also requires that $|\mathbf{E}|/c < |\mathbf{B}|$ or $|\mathbf{B}| < |\mathbf{E}|/c$ depending on whether the field is purely magnetic or electric in S'.

We now apply the Lorentz transformation theory to demonstrate that currents give rise to magnetic fields and that changing magnetic fluxes induce electromotive force. Consider a charge distribution in S given by

$E_x = 0$, $E_y = \sigma_s/4\pi\epsilon_0$, $E_z = 0$, where σ_s is the surface charge density and let $B_x = 0$, $B_y = 0$, $B_z = 0$. Our interest is to ask what electromagnetic fields are observed in an S' frame that is moving with a velocity V relative to S in the x-direction. The answer is obtained from the transformation equations (2.45) and (2.46). In S', $E'_x = 0$, $E'_y = \gamma\sigma_s/4\pi\epsilon_0$, $E'_z = 0$, $B'_x = 0$, $B'_y = 0$, and $B'z = -\gamma\beta\sigma_s/4\pi\epsilon_0$. The observer in S' thus measures an increase in the intensity of the electric field which is attributed to the Lorentz contraction. The observer in S' also sees a magnetic field which is not in S. The presence of the magnetic field is attributed to currents produced by the motion of the charges relative to the inertial system S'.

Consider now another situation in which a uniform electromagnetic field in the S frame is given by $E_x = 0$, $E_y = 0$, $E_z = 0$, $B_x = 0$, $B_y = B_y$, and $B_z = 0$. In S' frame, the Lorentz transformation yields $E'_x = 0$, $E'_y = 0$, $E'_z = \gamma VB_y$, $B'_x = 0$, $B'_y = \gamma B_y$, and $B'_z = 0$. Two important results illustrated by this example are (1) the magnetic field in S' has been amplified by γ, and (2) there is an electric field in S' which is in the z-direction.

The magnetic field in the limit $\gamma \approx 1$ is the same in the two frames of reference. However, in this limit, there is still an electric field in S' even though there is no electric field in S. This electric field in the S' frame can perform work. For example, if there is a piece of conductor aligned in the z-direction, the electromotive force will drive the electrons into motion. An observer in the S frame is aware of this electromotive force because the work performed on the electrons by the electromotive force can be observed. This observer attributes the electromotive force to the motion of the conductor relative to S through a region where there is a magnetic field.

A final comment concerns the magnetic field in S' that in general is amplified by a factor γ. Since the source of magnetic fields is current, this result implies that the source of current in S' is larger than the one in S. The Lorentz transformation from S to S' amplifies the current density by a factor γ, $\mathbf{J}' = \rho'\mathbf{V} = \gamma\rho\mathbf{V} = \gamma\mathbf{J}$.

2.7 Charged Particles

There are two ways to measure the current density. Equation (2.3) indicates that, in principle, $\mathbf{J}$ can be obtained if one measures the curl of the magnetic field and the time variation of the electric field. Another way is to perform direct measurements of particle fluxes and identify the charges of the particles (2.8). Measurements of currents in plasmas require that the various particles be detected from the lowest to the highest energy possible. Thus, a number of different particle detection techniques are required to perform this task. For the very low-energy particles ($\sim$ 0 eV to 500 eV), Langmuir probes and retarding potential analyzers are used. Electrostatic analyzers in combination with windowless channel multipliers are used to detect particles with energies $\sim$ 0.005 keV to 30 keV. For particles with energies $>$ 20 keV, solid-state semi-conductor detectors are often used.

Langmuir probes directly measure the total current that flows from the plasma to the probe. The probes are designed to detect either electrons or ions in a specified range of energies. Electrostatic analyzers measure energy per charge of the particles (electrons or ions) from which the differential flux and the distribution function of the measured particles can be computed. Solid-state detectors detect differential particle fluxes of electrons or ions, which can be converted to currents if the charged state of the particles is also measured.

2.7.1 Particle Flux

Let N represent the total number of particles of a given type (for example, electrons of a given energy) in a volume V. The number density of these particles is then $n = N/V$. Let $\mathbf{U}$ be the velocity of these particles. Then the omni-directional flux J_0 of these particles is defined at any point in space as the number of these particles which arrive uniformly from all directions and which traverse a test sphere of a unit cross-sectional area. $J_0 = n\mathbf{U} \cdot \hat{\mathbf{a}}$ where $\hat{\mathbf{a}}$ is a unit vector normal to the test sphere of unit area. The magnitude J_0 is equal to the number of particles crossing the unit area per second (number/m^2 - sec).

2.7.2 Differential Particle Flux

The concept of flux can be made more specific by denoting, for example, the direction from which the particles are arriving. In the case of omni-directional flux defined above, we can introduce $j_0 = J_0/4\pi$, which defines that portion of the omni-directional flux which arrives at the test sphere per unit solid angle from a particular direction. $J_0/4\pi\, d\Omega$ then equals the number of particles arriving at the test sphere within the solid angle $d\Omega$, and the number of particles incident on a plane of cross-sectional area dA oriented at an angle θ with respect to the direction of incidence is $(J_0/4\pi) d\Omega\, dA \cos\theta$. Note that the integration of this expression over all angles recovers the omni-directional flux J_0. However, in space the distribution of particles is neither uniform in energy nor in direction. It is therefore necessary to introduce a more general definition of particle flux.

The particle detectors in space directly measure particle fluxes of a given species over a designated energy interval, arriving from a certain direction of space. We now define the differential directional flux which retains this additional information. Figure 2.1 shows an area element dA,

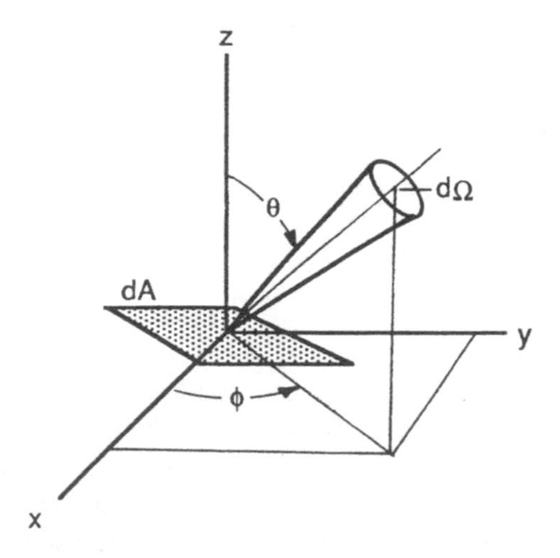

FIGURE 2.1 Definition of differential flux.

whose normal $\mathbf{n}$ is defined by the spherical coordinates θ and ϕ. Let $d\Omega$ be the solid angle centered about this normal direction. Consider now only those particles with energy E and $E + dE$ whose direction of incidence is within the solid angle cone $d\Omega$. The number of particles passing through an area element dA at location $\mathbf{r}$ made over the time interval t and $t + dt$ is defined as $dN(\mathbf{r}, t, E, \theta, \phi)$ and the differential flux $\mathbf{j}$ is defined by

$$dN = \mathbf{j} \cdot \mathbf{n} dA \, dE \, d\Omega \, dt \tag{2.55}$$

$\mathbf{j}$ is a function of $(\mathbf{r}, t, E, \theta, \phi)$ and it is measured in units of $\#/(m^2 - \text{ster} - keV - sec)$.

2.7.3 Differential Flux j and the Distribution Function f

There is a relationship between the differential flux j and the distribution function f. Let the particles in $d\Omega$ with energy dE occupy a momentum space volume $p^2 \, dp \, d\Omega$ where p is the magnitude of the momentum corresponding to the energy E. That is, in the non-relativistic limit $p^2 = 2mE$. Now consider the number of particles that during dt traverse an area dA perpendicular to the direction of motion. These particles occupy at time t a volume in a real space given by $U \, dt \, dA$ where U is the speed of the particles. The distribution function f multiplied by the volume in momentum space and the real space gives the total number of particles and this must be equal to (2.55). Hence, conservation of particles requires that

$$j \, dA \, d\Omega \, dE \, dt = f \, p^2 \, dp \, d\Omega \, U \, dt \, dA \tag{2.56}$$

Since $p = mU$, $dp = mdU$, and $pdp = mdE$, the above equation reduces to

$$j = p^2 f \tag{2.57}$$

This relation shows that the measurement of the differential flux is equivalent to a measurement of f.

2.8 How to Study Plasma Phenomena in Space

To solve the coupled Maxwell-Lorentz equations, many investigations today use super-computers. This approach has the potential to accommodate the large number of equations needed to solve the problems. However, the actual number of equations used in these computer simulation experiments is much less than the required number for real situations. Nevertheless, the results one obtains in these studies are important, albeit less than complete.

A first step toward describing and understanding the motion of a system of charged particles is to study analytically how the *individual* particles move around in given electromagnetic fields. Even then, the Lorentz equation is not easily solved except for a few simple cases. In complex electromagnetic field geometries, the Lorentz equations can be solved by perturbation methods. The first-order perturbation theory is also known as the guiding center theory. This approach, first introduced by Hannes Alfvén in the 1940s as an approximate method, provides an average picture of particle motion in complex field geometries. This study is instructive and can organize particle motions in radiation belts (Chapter 4).

A better approach is to use the concept of the distribution function and the tools of statistical mechanics to study the motion of a large number of particles. This approach would involve solving the Boltzmann transport equation, which, as mentioned above, is extremely difficult to solve (and beyond the scope of this book). Instead, we will develop the macroscopic hydromagnetic theory to illustrate the collective effects of many particles interacting as an ensemble (Chapter 5). It is possible to integrate and unify under this simpler theory certain classes of large-scale structures and dynamics of space plasma phenomena.

Bibliography

Aggson, T.L., Probe Measurements of Electric Fields in Space, in *Atmospheric Emissions*, B.M. McCormac and A. Omholt, eds., Van Nostrand Reinhold, New York, NY, 1969. One of the first articles demonstrating that electric fields can be measured in space.

Alfvén, H. and C.G. Fälthammar, *Cosmical Electrodynamics, Fundamental Principles*, 2nd ed., Oxford University Press, Oxford, England, 1963. This book is a new edition of the original book by H. Alfvén published in 1950 under the same title.

Chapman, S. and T.G. Cowling, *The Mathematical Theory of Non-Uniform Gases*, 2nd ed., Cambridge University Press, Cambridge, England, 1952. This classic text gives basic and detailed properties of the Boltzmann equation. It is for intermediate and advanced students.

Hundhausen, A.J., J.R. Asbridge, S.J. Bame, H.E. Gilbert and I.B. Strong, Vela 3 Satellite Observations of Solar Wind Ions: A Preliminary Report, *J. Geophys. Res.*, **72**, 87, 1967. This article reports observations of details of solar wind ion composition.

Ichimaru, S., *Basic Principles of Plasma Physics, A Statistical Approach*, W.A. Benjamin Publishing Co., Reading, MA, 1973. An excellent book on plasma physics for advanced students.

Jackson, J.D., *Classical Electrodynamics*, 2nd ed., John Wiley and Sons, Inc., New York, NY, 1975. A good reference book for electrodynamic theory, including the theory of Lorentz transformation of electromagnetic fields.

Krall, N.A. and A.W. Trivelpiece, *Principles of Plasma Physics*, McGraw Hill Book Co., New York, NY, 1973. An excellent plasma physics text for intermediate and advanced students.

Landau, L.D. and E.M. Liftschitz, *Statistical Physics*, Pergamon Press, Ltd., London, England, 1958. A good reference for topics in statistical mechanics.

Mozer, F.S., R.B. Torbert, U.V. Fahleson, C.G. Fälthammar, A. Gonfalone and A. Pedersen, Measurements of Quasi-Static and Low-Frequency Electric Fields with Spherical Double Probes on the ISEE-1 Spacecraft, *IEEE Transactions on Geoscience Electronics*, **GE-16**, No. 3, 1978. This article describes an instrument that was flown on the International Sun Earth Explorer (ISEE) spacecraft to measure electric fields.

Ness, N.F., Magnetometers for Space Research, *Space Sci. Rev.*, **11**, 459, 1970. This article contains designs of magnetometers flown in space.

Ögelman, H. and J.R. Wayland, eds., *Introduction to Experimental Techniques of High Energy Astrophysics*, NASA Publication, Washington, D.C., 1970. This book mainly discusses techniques for detecting energetic particles.

Thompson, W.B., *An Introduction to Plasma Physics*, Addison-Wesley Publishing Co., Inc., Reading, MA, 1962. One of the early publications on plasma physics. For introductory to advanced students.

Questions

1. Compute the Debye length and the number of particles in the Debye sphere for
   a. The solar wind. The solar wind electron energy is typically 10 eV and the average density is $5 \times 10^6 (\text{m})^{-3}$.
   b. Earth's ionosphere at 100 km altitude. The electron density is 10^{12}m^{-3} and the average energy of the particles is 1 eV.
   c. The plasma sheet. The average energy is 1 keV and the density is 10^6m^{-3}.
   d. The corona of the Sun. The average energy of the particles here is 100 eV and the density is 10^{14}m^{-3}.
2. Verify that Equation (2.49) yields (2.45) and (2.46).

3. A Maxwellian velocity distribution function is given by

$$f(\mathbf{v}) = n\left(\frac{m}{2\pi kT}\right)^{3/2} \exp\left(-\frac{m\mathbf{v}\cdot\mathbf{v}}{2kT}\right)$$

Compute the average velocity of particles described by this distribution function.

4. Compute the average kinetic energy for particles described by the Maxwellian distribution function.

5. Write down the expression of a Maxwellian distribution function that is flowing with a velocity $\mathbf{u}$. Sketch and discuss this distribution function.

3

Electromagnetic Fields in Space

3.1 Introduction

Electric and magnetic fields are associated with planets, stars, pulsars, and exotic objects such as neutron stars. Observations of electromagnetic fields in space normally rely on indirect means, but in our solar system the fields have been directly measured by instruments carried on spacecraft. The behavior of electromagnetic fields is governed by Maxwell's equations. However, to understand how these fields are produced and what mechanisms transport these fields into other regions of space, knowledge is required of the behavior of charged particles (Chapter 4), the collective properties of charged particles (Chapter 5), and the concept of currents in plasma media (Chapter 7). We must therefore defer to these later chapters a comprehensive discussion of electromagnetic fields in space plasmas.

Our primary interest is to study time-stationary electromagnetic fields in the region external to the planetary and stellar bodies. This simplifies the problem considerably since it permits us to drop the source terms in Maxwell's equations. We treat space as a "vacuum," empty of particles, but electromagnetic fields occupy this space. Physically this means that the

TABLE 3.1
Magnetic Parameters of the Sun and the Planets

	Distance (AU)	*Radius (10^3 km)*	*Dipole Moment (Amp-m^2)*	*Tilt (°)*	*Polarity*	*Spin*	R_m/R_p
Sun	0	700	1.8×10^{14}	var.	NR	27d	—
Mercury	0.4	2.49	4.5×10^{19}	≈ 10	N	58.6d	1.6
Venus	0.7	6.10	$0.8–6 \times 10^{19}$	< 90	R	243d	1.1
Earth	1.0	6.37	8.05×10^{22}	11.5	N	24h	11
Mars	1.5	3.38	2×10^{19}	< 20	R	24.5h	1.4
Jupiter	5.2	71.4	1.5×10^{27}	10	R	10h	50
Saturn	9.5	60.4	2.2×10^{26}	1	N	10h39m	40
Uranus	19.2	23.8	3.8×10^{24}	~ 60	N	17.3h	18
Neptune	30.0	22.2	2×10^{24}	~ 47	R	16h	26.5

AU = Astronomical Unit $\approx 1.5 \times 10^8$ km; d = day; h = hour; N means normal from north to south; R means reversed from south to north; Sun's field reverses every 11 years; tilt is between rotational and magnetic axis; R_m/R_p is the ratio between the subsolar magnetopause and the planetary radius. Note that the unit of the dipole moment is also given as tesla-m^3. In cgs system, the unit of the magnetic dipole moment is gauss-cm^3. (From Siscoe, 1979, and Ness et al., 1989)

electric and magnetic fields produced locally by the particles and currents in space are small and negligible compared to the fields of the planets and stars.

A particular magnetic configuration, the magnetic dipole field, is studied because the dipole provides a standard of reference for many magnetized planets and celestial bodies (see Table 3.1). The dipole is also an example of axisymmetric fields, which are analytically tractable. To illustrate how a dipole is modified by the presence of other magnetic fields in space, we will examine the topological features of planetary dipoles immersed in the interplanetary magnetic field (IMF). This study will reveal some basic properties of large-scale magnetic structures, which can be compared and evaluated against observations of real magnetic structures made in the vicinity of Earth. The magnetic field discussion concludes with a short introduction of non-dipolar field geometries that are also encountered in space.

Electric fields are more difficult to measure in space than magnetic fields. One difficulty is that the intensities of electric fields in space can be fairly small, only a few 10^{-3} volts/meter. Such fields must be measured on a spacecraft whose motion through interplanetary media induces electric fields in the spacecraft frame of reference that are comparable to the intensity of the natural field.

Another important piece of information about measuring electric fields is that space is a good electrical conductor because it contains charged particles. In good conductors, free charges do not normally accumulate locally. Hence even though space contains many charged particles, electric fields in the rest frame do not originate from free charges. Instead, large-scale electric fields in space are predominantly associated with the electromotive forces (EMF) of charged particles, induced by the motions of charged particles across magnetic fields and by the collective interactions of charged particles with magnetic fields. The discussion of EMF in space requires that we understand the behavior of electromagnetic fields measured in various frames of reference. Since measurements are normally made on vehicles that are rapidly moving through a medium which itself may be moving, the theory of Lorentz transformation discussed in Chapter 2 is applicable.

This chapter begins with a brief discussion of the magnetic fields of Sun and Earth. We will stick to our goal of studying simplified problems. Thus, while the concept of motional electric fields is introduced, the physics of plasma interaction with electromagnetic fields is not discussed.

3.1.1 Solar and Interplanetary Magnetic Fields

Sunspots support magnetic fields with intensities ranging from about 0.1 T to several T. Sunspots generally occur in pairs aligned on constant latitudes within a region approximately $\pm 35°$ from the solar equator. Sunspots are believed to be generated by a magnetic buoyancy effect, which can be explained by magnetohydrodynamic theory. The number of sunspots observed on the Sun varies from day to day and increases with solar activity. The disruption of sunspots usually coincides with flare eruptions during which time solar particles are accelerated to high energies.

Beginning around 1952, H.D. and H.W. Babcock observed the weaker solar magnetic fields by means of the Zeeman effect and have now established that the Sun has a general dipole field of about 10^{-4} teslas distributed over the visible photosphere. The dipole tilt with respect to the rotation axis is variable and is a function of the solar cycle. When first observed in 1955, the dipole direction was from south to north (opposite to the direction of Earth's dipole observed today), but the Sun's dipole field reversed direction in 1958. Apparently, this reversal occurs every 11 years with the general evolution of the sunspot cycle.

A configuration representative of large- and small-scale magnetic fields of a quiet Sun is shown in Figure 3.1. On the solar scale, the lines of force are radial over much of the Sun's surface (because of the flow of the solar wind).

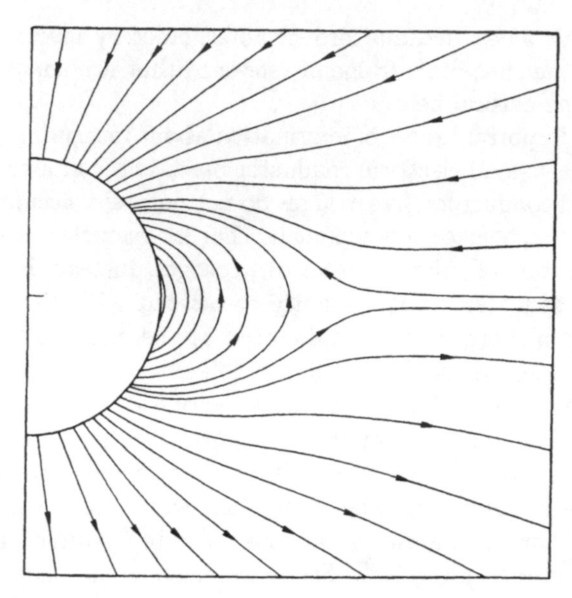

FIGURE 3.1 A schematic diagram to illustrate that the Sun's magnetic field includes dipole, radial, and field reversal (neutral sheet) geometries.

On the local scale, there are dipole-like lines of force and field reversal regions with neutral sheets. On an even smaller scale, in active sunspots, for example, magnetic field geometry becomes very complicated and includes twisted, tangled, and sheared magnetic lines of force. The IMF originates from magnetic fields of the Sun (star) that convect out with the solar (stellar) wind. The IMF has been observed beyond the most distant planet of the solar system. A theory of how solar and stellar fields are convected in plasmas will be discussed in Chapter 6 after the magnetohydrodynamic theory has been formulated. The IMF drastically modifies the shape of planetary magnetic fields and it is this aspect that we will be concerned with in this chapter.

3.1.2 Magnetic Field of Planet Earth

The magnetic field of Earth has been studied more than that of any other planet in our solar system. Magnetic studies of Earth date back to the year 1600 when William Gilbert established that Earth possessed an intrinsic magnetic moment. With the establishment of the first magnetic observatory in 1838, continuous magnetic recordings were made, which permitted systematic studies of the geomagnetic field. Today, there are more than 200 observatories distributed over the land masses of Earth.

Geomagnetic data obtained during the last century show that Earth's main field can be roughly characterized by a dipole whose pole is displaced about 11.5° from the geographic pole and intersects the surface at 78.5°N 291°E (geographic). The source of the main field is internal to the solid Earth. These observations further established that there is a slow westward drift (0.2 to 0.3° per year) of the pole and that the equatorial magnetic field strength is decreasing (at the present time) at the rate of about 20 to 30 nanoteslas per year (1 tesla = 1 weber/square meter = 10^4 gauss). This represents about 0.1% of the total field strength on the equator at the surface of Earth, which is about 31,000 nanoteslas. Paleomagnetic studies, moreover, reveal that the geomagnetic field has reversed directions several times during the course of Earth's history. The current dipole direction is from north to south. The last reversal occurred around one million years ago.

Observations of the Magnetopause Measurements of the magnetic field in space began in 1958. Space measurements provide information about the magnetic field that cannot be obtained on the ground. The first important result from space showed that Earth's magnetic field in the distant regions behaves very differently from that predicted for a dipole field. The deviation from the dipole is due to the presence of new current sources external to the solid Earth. These external currents are large-scale and quite strong. Figure 3.2 shows an example from one of the first magnetic measurements made in space. The data come from the sunward hemisphere near local noon. In this graph, B represents the scalar magnetic field intensity, which is plotted as a function of the radial distance measured in earth-radii ($R_E \approx 6375$ km) from the center of Earth. The solid line represents the r^{-3} dependence of the dipole field. Note that Earth's measured magnetic field intensity closely follows the dipole field behavior until about $5R_E$, after which the observed field intensity is stronger. An abrupt change in both intensity and direction observed around $8R_E$ is due to the crossing of the magnetopause boundary by the spacecraft near the subsolar point.

Observations of the Geomagnetic Tail The data that established the existence of the geomagnetic tail are shown in Figure 3.3. The magnetic field intensity B, the polar latitude angle θ, and the azimuth angle ϕ are plotted as a function of time and spacecraft distance measured from the center of Earth. The spacecraft was inbound toward Earth, and the measurements cover the night side region from about $31R_E$ to $8R_E$ distances. Earth's magnetic field from $31R_E$ to about $16R_E$ is parallel to the Earth-Sun direction and pointing away from Earth ($\theta = 0°, \phi = 180°$), forming a tail. At approximately $16R_E$, ϕ abruptly changes direction and is very close to 0° and the magnetic field now is directed toward Earth. The abrupt change

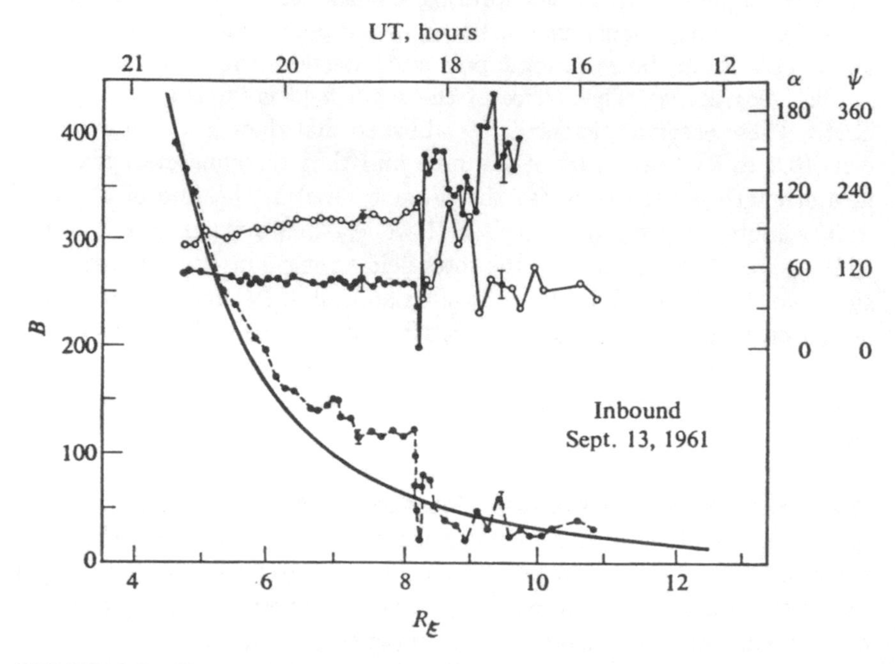

FIGURE 3.2 The magnetic field data were measured by a magnetometer flown on the Explorer XII satellite. The data shown come from a satellite pass in the sunward direction on September 13, 1961. | **B** | is the magnetic field intensity in nanoteslas. α is the angle between the vector **B** and the spin axis of the spacecraft. Ψ is the dihedral angle between the plane that contains the **B** vector and the spin axis and the plane that contains the spin axis and the satellite sun-line. (From Cahill and Amazeen, 1963)

of ϕ from 180° to 0° resulted as the spacecraft crossed the magnetic equator across which the magnetic field pointed in the opposite direction.

The schematic diagram of Earth's magnetosphere shown in Figure 1.3 was derived from many repeated magnetic measurements of the type shown in Figures 3.2 and 3.3. Figure 1.3 is representative of Earth's magnetosphere on any given day.

3.2 Representations of Electromagnetic Fields

The electromagnetic fields obey Maxwell's equations. The purpose of this section is to develop the basic tools that are needed to study the large-scale electromagnetic structures in space. Since our interest here is mainly to learn about static structures, we will ignore the time-dependent properties of electromagnetic fields.

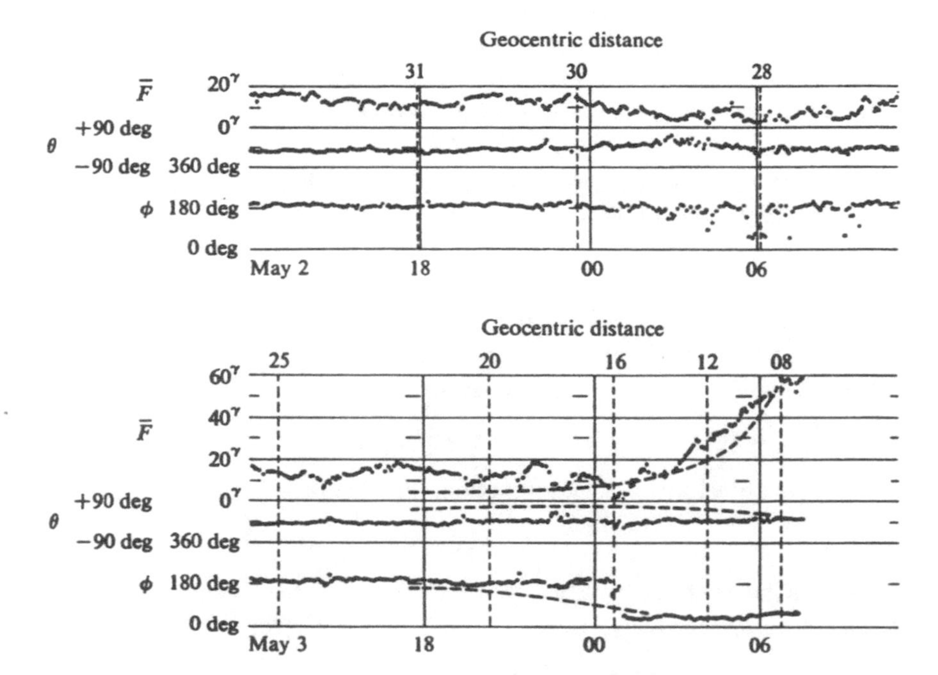

FIGURE 3.3 Magnetic field measurements on the night side of Earth. The direction of the magnetic field is very closely aligned along the Sun-Earth line from about $16R_E$ to about $31R_E$, suggesting a tail geometry. The abrupt change of direction observed at about $16R_E$ is due to the crossing of the magnetic equator. The intensity of the magnetic field, denoted here by $\mathbf{F}$, is measured in nanoteslas (the symbol γ is defined as 10^{-5} gauss = 10^{-9} teslas). (From Ness *et al.*, 1963)

3.2.1 Magnetic Field

The two Maxwell equations that govern the behavior of the magnetic field are

$$\nabla \cdot \mathbf{B} = 0 \tag{3.1}$$

$$\nabla \times \mathbf{H} = \mathbf{J} + \frac{\partial \mathbf{D}}{\partial t} \tag{3.2}$$

These equations have already been introduced in Chapter 2. However, we can let the displacement current $\partial \mathbf{D}/\partial t = 0$ for static problems. Then, given the current density $\mathbf{J}$ everywhere in space, the solution of (3.2) is given by

$$\mathbf{B}(\mathbf{r}) = \frac{\mu_0}{4\pi} \int \frac{\mathbf{J}(\mathbf{r}') \times (\mathbf{r} - \mathbf{r}')}{| \mathbf{r} - \mathbf{r}' |^3} d^3r' \tag{3.3}$$

which is the well-known Biot-Savart's law. (The relation $\mathbf{B} = \mu_0 \mathbf{H}$ has been used.) Here, $\mathbf{r}'$ refers to the position where the current density $\mathbf{J}$ is located and the magnetic field is evaluated at $\mathbf{r}$. Computations of the magnetic field $\mathbf{B}$ require knowledge of the source term, $\mathbf{J}$.

Vector Potential The divergence-free property of $\mathbf{B}$ permits one to write $\mathbf{B}$ in terms of the curl of a vector,

$$\mathbf{B} = \nabla \times \mathbf{A} \tag{3.4}$$

where $\mathbf{A}$ is defined as the vector potential. Equation (3.4) is not the most general form. Since the curl of a gradient of any arbitrary scalar also vanishes, one can write (3.4) more generally as

$$\mathbf{B} = \nabla \times \mathbf{A} + \nabla \times (\nabla \chi) \tag{3.5}$$

where χ is an arbitrary scalar function. The transformation

$$\mathbf{A} \rightarrow \mathbf{A} + \nabla \chi \tag{3.6}$$

is called a gauge transformation. Since the divergence of a curl of any vector vanishes, (3.5) satisfies (3.1). Use of (3.5) in (3.2) yields

$$\begin{aligned} \nabla \times \mathbf{B} &= \nabla \times (\nabla \times (\mathbf{A} + \nabla \chi)) \\ &= \nabla \times (\nabla \times \mathbf{A}) + \nabla \times (\nabla \chi) \\ &= \nabla \times (\nabla \times \mathbf{A}) \\ &= \nabla (\nabla \cdot \mathbf{A}) - \nabla^2 \mathbf{A} \\ &= -\nabla^2 \mathbf{A} \end{aligned} \tag{3.7}$$

where the choice of a coulomb gauge $\nabla \cdot \mathbf{A} = 0$ has been used. Hence, the vector potential $\mathbf{A}$ obeys the Poisson equation

$$\nabla^2 \mathbf{A} = -\mu_0 \mathbf{J} \tag{3.8}$$

whose solution can be written formally as

$$\mathbf{A}(\mathbf{r}) = \frac{\mu_0}{4\pi} \int \frac{\mathbf{J}(\mathbf{r}') d^3 r'}{\mid \mathbf{r} - \mathbf{r}' \mid} \tag{3.9}$$

We can solve for $\mathbf{A}$ and use (3.4) to obtain $\mathbf{B}$.

Scalar Potential If the current density vanishes in some region of space, (3.2) becomes $\nabla \times \mathbf{B} = 0$. In this case, we can introduce a scalar potential Ψ such that

$$\mathbf{B} = -\nabla \Psi \tag{3.10}$$

Substitution of (3.10) into (3.1) shows

$$\begin{aligned}\nabla \cdot \mathbf{B} &= -\nabla \cdot (\nabla \Psi) \\ &= -\nabla^2 \Psi \\ &= 0\end{aligned} \tag{3.11}$$

Ψ obeys Laplace's equation and can be solved by using the appropriate boundary conditions. The surfaces on which Ψ is constant are called equipotential surfaces. The equipotential surfaces are orthogonal to magnetic field surfaces.

Euler Potential An alternative way to write $\mathbf{A}$ in terms of two scalar functions α and β is

$$\mathbf{A} = \alpha \nabla \beta + \nabla \chi \tag{3.12}$$

which then permits $\mathbf{B}$ to be written as

$$\begin{aligned}\mathbf{B} &= \nabla \times \mathbf{A} \\ &= \nabla \times (\alpha \nabla \beta + \nabla \chi) \\ &= \nabla \alpha \times \nabla \beta + \alpha \nabla \times (\nabla \beta) + \nabla \times (\nabla \chi) \\ &= \nabla \alpha \times \nabla \beta\end{aligned} \tag{3.13}$$

where the last step follows from the fact that the curl of a gradient vanishes. Note that $\mathbf{A} \cdot \mathbf{B} = 0$. The parameters α, β, and χ are sometimes referred to as Euler potentials or Clebsch variables.

The $(\alpha\beta)$ representation of $\mathbf{B}$ is useful as it provides another way to obtain information about the magnetic field. According to (3.13),

$$\begin{aligned}\mathbf{B} \cdot \nabla \alpha &= 0 \\ \mathbf{B} \cdot \nabla \beta &= 0\end{aligned} \tag{3.14}$$

$\nabla \alpha$ and $\nabla \beta$ are thus perpendicular to $\mathbf{B}$ and α and β are constant along $\mathbf{B}$. $\alpha =$ constant and $\beta =$ constant surfaces are orthogonal to their gradients and tangent to $\mathbf{B}$. Hence, $\mathbf{B}$ can be conveniently defined in terms of a pair of $(\alpha\beta)$ values.

We can use (3.5) with the help of (3.9) and (3.3), or use (3.13) to obtain information about $\mathbf{B}$. Depending on the nature of the problem to be examined, one form may have an advantage over the other. For instance, in space it is not always possible to know how the current density is distributed. On the other hand, the $(\alpha\beta)$ variables, while in general not easily obtained, are available for some axisymmetric magnetic geometries. Another advantage of the $(\alpha\beta)$ representation will become more evident in a later chapter, when we discuss the concept of field lines and "motions" of such field lines in the context of magnetohydrodynamic theory.

Field Lines and Flux Tubes The concept of lines of force or field lines is used extensively in the description of magnetic fields. A line of force of the magnetic field $\mathbf{B}$ or a magnetic field line is defined as a curve which is tangent everywhere to the field intensity $\mathbf{B}$. If $d\mathbf{l}$ is an arc length, the lines of force are defined by the differential equation

$$d\mathbf{l} \times \mathbf{B} = 0 \tag{3.15}$$

The solution of this equation will in general yield two parameters that label a line of force. Equation (3.15) is valid in any medium, whether or not the medium contains a current.

If C is a closed curve spanned by an open surface S with a unit normal $\mathbf{n}$, then the flux of the magnetic field $\mathbf{B}$ across S is defined by

$$\Phi = \int \mathbf{B} \cdot \mathbf{n}\, dS = \int_C \mathbf{A} \cdot d\mathbf{l} \tag{3.16}$$

where Stoke's law has been used and $d\mathbf{l}$ is an element of arc along C. A flux tube is defined as an assemblage of magnetic field lines passing through a closed curve. The divergence theorem requires that the flux of the magnetic field through two ends of a tube be equal. Φ is therefore constant along a flux tube.

In the discussion of magnetic fields, it is sometimes stated that a line of force is uniquely defined and, from $\nabla \cdot \mathbf{B} = 0$, that $\mathbf{B}$-lines are never-ending, and either close on themselves or extend to infinity. These statements are not always correct and one must exercise caution. A line of force is not uniquely defined. It may have finite length and may not be closed, especially if a singular point exists in which the magnetic intensity vanishes and the direction is not defined. There are no general criteria that determine whether a line of force forms a closed curve. These concepts will be dealt with further in our discussion of magnetic field structures in interplanetary-planetary systems and later when magnetic fields in magnetohydrodynamic media are discussed.

3.2.2 Electric Field

As mentioned in the Introduction, the concept of electric fields in space requires understanding the physics of the collective interaction of charged particles with magnetic fields. This theory is developed in Chapter 5. Here, we limit the discussion to some common notions of electric fields.

The two Maxwell equations that describe the behavior of electric fields are

$$\nabla \cdot \mathbf{E} = \frac{\rho}{\epsilon_0} \tag{3.17}$$

and

$$\nabla \times \mathbf{E} = -\frac{\partial \mathbf{B}}{\partial t} \tag{3.18}$$

where ρ is the charge density, and ϵ_0 is permittivity of free space. In writing (3.17) the constitutive relation $\mathbf{D} = \epsilon_0 \mathbf{E}$ has been used.

Scalar Potential Since $\mathbf{B} = \nabla \times \mathbf{A}$ where $\mathbf{A}$ is the vector potential, one can write (3.18) as

$$\nabla \times \left(\mathbf{E} + \frac{\partial \mathbf{A}}{\partial t} \right) = 0 \tag{3.19}$$

The vector $\mathbf{E} + \partial \mathbf{A}/\partial t$ has a zero curl and therefore it can be written as a gradient of a scalar,

$$\mathbf{E} + \frac{\partial \mathbf{A}}{\partial t} = -\nabla \Phi \tag{3.20}$$

where Φ is defined as the electromagnetic scalar potential function. It is interesting to note that while $\mathbf{B}$ is derived from the vector potential only, $\mathbf{E}$ requires both a vector and a scalar potential.

Electrostatic Field: Coulomb Potential For the static case, we set $\partial \mathbf{A}/\partial t = 0$ and insert (3.20) into (3.17), yielding

$$\nabla^2 \Phi = -\frac{\rho}{\epsilon_0} \tag{3.21}$$

which is the familiar Poisson equation. Thus, given the charge distribution ρ, a particular solution of (3.21) is

$$\Phi(\mathbf{r}) = \frac{1}{4\pi\epsilon_0} \int \frac{\rho(\mathbf{r}')d^3r'}{|\,\mathbf{r} - \mathbf{r}'\,|} \tag{3.22}$$

where $\mathbf{r}'$ is the location of the charge density and the potential is evaluated at $\mathbf{r}$. Equation (3.22) is the familiar Coulomb potential for the charge density ρ and the gradient of Φ yields the electrostatic field.

Although space contains many charged particles, free charges are not maintained because, as mentioned earlier, space is a good conductor (see Sections 1.4 and 2.6). Hence, (3.22) is not as useful when considering electric fields in space.

Faraday's Law of Induction Michael Faraday established in 1831 that an electromotive force (EMF) is induced when magnetic fluxes change in time.

Defining the electromotive force as $\int \mathbf{E} \cdot d\mathbf{l}$ and using the definition of magnetic flux (3.16), Faraday's law of induction is written as

$$\oint_C \mathbf{E} \cdot d\mathbf{l} = -\frac{d}{dt} \int \mathbf{B} \cdot \mathbf{n}\, dS \tag{3.23}$$

Here, the line integral is for an arbitrarily shaped contour (circuit) enclosing the surface S, which can be moving in space, and $d\mathbf{l}$ is the arc element. The negative sign comes from Lenz's law, which states that induction is in such a direction as to oppose the change of magnetic flux. Equation (3.18) represents the differential form of (3.23).

The EMF is proportional to the total time derivative d/dt of the magnetic flux. This means that the EMF is induced by changing either the magnetic flux in time or the shape of the contour, or both. When the circuit is in motion, motional contribution to the EMF becomes important. To fully appreciate the significance of motional EMF requires understanding how charged particles behave as conducting fluids and how such fluid motions interacting with magnetic fields induce electric fields. Equation (3.23) can be written as

$$\oint_C (\mathbf{E} + \mathbf{v} \times \mathbf{B}) \cdot d\mathbf{l} = -\frac{\partial}{\partial t} \int \mathbf{B} \cdot \mathbf{n}\, dS \tag{3.24}$$

where the motional term is clearly identified. Note that the right-hand side is now a partial derivative $\partial/\partial t$. This representation has separated the temporal and the spatial contributions in the term d/dt. This motional EMF is the primary source of large-scale electric fields in space. As shown later, the left-hand side of (3.24) represents the total electric field measured in a coordinate frame that is moving with a velocity $\mathbf{v}$ relative to the stationary (laboratory) frame (in the non-relativistic limit, $\mathbf{v} \ll c$, where c is the speed of light).

3.3 Magnetic Fields in Space

Consider a planetary dipole field in a vacuum immersed in a constant IMF with no solar wind. The required equations to be studied are (3.1) and (3.2). Our model is steady-state and, moreover, does not support a current $\mathbf{J}$. Hence (3.2) reduces to $\nabla \times \mathbf{B} = 0$. Since this model does not permit generation of local currents, the effects arising from local currents cannot be studied. $\mathbf{B}$ is then represented by (3.10).

3.3.1 The Dipole Magnetic Field

Let **M** be the dipole moment of a magnetized planet. Consider a spherical coordinate system in which the z-axis is parallel but opposite (as for Earth) to the direction of **M** (Figure 3.4). Outside the solid planet, the magnetic field is described by $\nabla \times \mathbf{B} = 0$. The magnetic field **B** is obtained from a scalar magnetic potential. For the dipole field, the potential is represented by (for $r > 0$).

$$\begin{aligned}\Psi &= -\frac{\mu_0}{4\pi}\mathbf{M}\cdot\nabla\frac{1}{r} \\ &= -\frac{\mu_0 M}{4\pi}\frac{\cos\theta}{r^2} \\ &= -\frac{\mu_0 M}{4\pi}\frac{\sin\lambda}{r^2}\end{aligned} \qquad (3.25)$$

Here we use the relation $\mathbf{M} = -M\hat{\mathbf{z}}$, where $\hat{\mathbf{z}}$ is a unit vector along the z-direction. Also note that $\theta = \pi/2 - \lambda$ (θ is colatitude and λ is the latitude). The gradient operation of (3.25) gives the following expression for

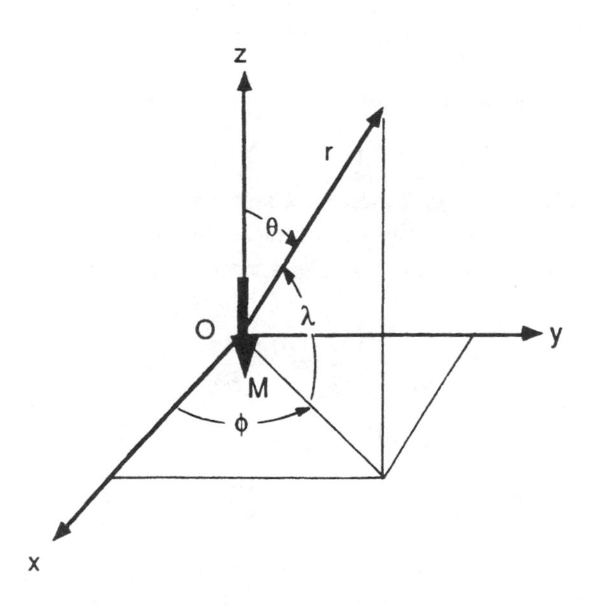

FIGURE 3.4 Geometry of Earth's dipole in spherical coordinate system. This is a dipole-centered coordinate system in which the dipole is located at the origin and is parallel to the $-z$-direction.

the components of **B**.

$$\begin{aligned} B_r &= -\frac{\partial \Psi}{\partial r} = -\frac{\mu_0 M}{2\pi}\frac{\sin\lambda}{r^3} \\ B_\lambda &= -\frac{1}{r}\frac{\partial \Psi}{\partial \lambda} = \frac{\mu_0 M}{4\pi}\frac{\cos\lambda}{r^3} \\ B_\phi &= -\frac{1}{r}\frac{1}{\cos\lambda}\frac{\partial \Psi}{\partial \phi} = 0 \end{aligned} \tag{3.26}$$

$B_\phi = 0$ because a dipole field is symmetric about ϕ. Since the intensity of the magnetic field is given by $|\mathbf{B}| = \sqrt{B_r^2 + B_\lambda^2 + B_\phi^2}$, the magnitude of **B** at distance r and at a latitude λ is

$$B(r, \lambda) = \frac{\mu_0 M}{4\pi r^3}(1 + 3\sin^2\lambda)^{1/2} \tag{3.27}$$

Note that the magnetic field intensity decreases as r^{-3}. The field intensity is smallest at the equator and increases toward the pole.

Dipole Field Equation As mentioned earlier, **B** is tangent to $d\mathbf{l}$ along which the magnetic field is defined. The equations of the lines of force are determined by the solutions of the differential equation (3.15). In Cartesian coordinates, this reduces to

$$\frac{dx}{B_x} = \frac{dy}{B_y} = \frac{dz}{B_z} \tag{3.28}$$

Equation (3.28) has two integrals,

$$f(x, y, z) = c_1 \quad \text{and} \quad g(x, y, z) = c_2 \tag{3.29}$$

$f(x, y, z)$ and $g(x, y, z)$ each represent a surface that contains the magnetic field B. The intersection of the two surfaces defines a specific line of force.

The differential equation for a magnetic field obtained in spherical coordinates is

$$\frac{dr}{B_r} = \frac{r\,d\theta}{B_\theta} = \frac{r\sin\phi\,d\phi}{B_\phi} \tag{3.30}$$

For a dipole (Figure 3.5), (3.30) with the help of (3.26) reduces to

$$d\phi = 0 \quad \text{and} \quad \frac{1}{r}\frac{dr}{d\lambda} = \frac{B_r}{B_\lambda} = -\frac{2\sin\lambda}{\cos\lambda} \tag{3.31}$$

The second equation in (3.31) can be written as

$$\frac{dr}{r} = \frac{2\,d(\cos\lambda)}{\cos\lambda} \tag{3.32}$$

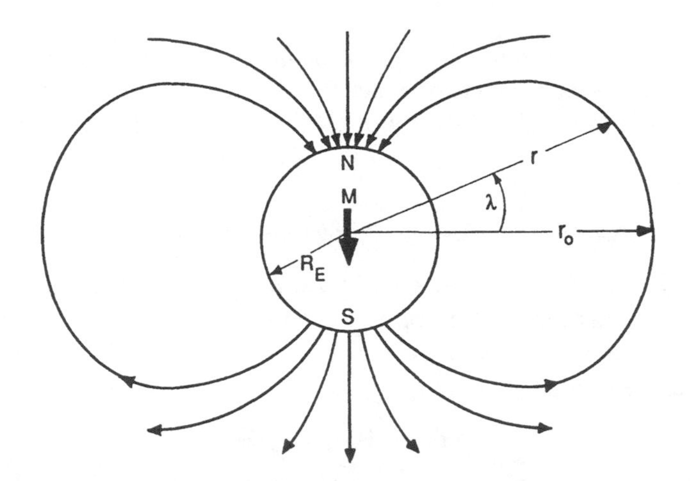

FIGURE 3.5 Schematic drawing of the locus of a dipole line of force.

Integration of (3.31) then yields the dipole field equation

$$\phi = \phi_0 \quad \text{and} \quad r = r_0 \cos^2 \lambda \tag{3.33}$$

where ϕ_0 and r_0 represent the constants of integration which are the magnetic longitude and the radial distance of the line of force at $\lambda = 0$, respectively. The dipole is symmetric about the z-axis. Axisymmetric fields such as the dipole are invariant under rotation about the axis of symmetry.

3.3.2 Interplanetary Magnetic Field

A magnetometer carried on a spacecraft continues to detect weak but significant magnetic fields outside planetary magnetospheres. The intensity of the IMF near Earth's orbit varies from a few nanoteslas at quiet times to more than 10 to 20 nanoteslas accompanying solar disturbances. The IMF is predominantly in the ecliptic plane. However, there is also a weak but significant component in the direction perpendicular to the ecliptic plane. In general,

$$\mathbf{B}_{IMF} = B_x\hat{\mathbf{x}} + B_y\hat{\mathbf{y}} + B_z\hat{\mathbf{z}} \tag{3.34}$$

where B_x and B_y are in the ecliptic plane and B_z is perpendicular to it. $\hat{\mathbf{x}}$, $\hat{\mathbf{y}}$, and $\hat{\mathbf{z}}$ are unit vectors. $\hat{\mathbf{x}}$ points toward the Sun, $\hat{\mathbf{z}}$ is perpendicular to the ecliptic plane, and $\hat{\mathbf{y}} = \hat{\mathbf{z}} \times \hat{\mathbf{x}}$. This coordinate system is known as the geocentric solar ecliptic (GSE) system.

The presence of the IMF directly affects the shape and size and the general topology of the planetary magnetic field. We now examine the topological changes that occur in the planetary dipole fields when they are superposed with different IMF geometries. The source of the IMF is the Sun, and in interplanetary space it is represented by a potential field. (A magnetohydrodynamic theory of IMF will be found in Chapter 6.)

3.3.3 Superposition of the Dipole Field and the IMF

Assume as before a spherical coordinate frame of reference in which the dipole is located at the origin. Since both the dipole and IMF are potential fields, the total field due to these can be obtained by adding the two fields. The superposition of the dipole and a constant IMF gives

$$\mathbf{B}_T = \mathbf{B}_d + \mathbf{B}_{IMF} \tag{3.35}$$

where the subscripts T and d stand for total and dipolar. The boundary conditions require that for large distances ($r \to \infty$), $\mathbf{B}_T \to \mathbf{B}_{IMF}$, and for small distances ($r \to 0$), $\mathbf{B}_T \to \mathbf{B}_d$.

Northward IMF Let us illustrate the effect of the interplanetary field by assuming that $\mathbf{B}_{IMF}$ has a positive constant magnitude B_0 along z, $\mathbf{B}_{IMF} = B_0\hat{\mathbf{z}}$. This IMF geometry is also referred to as northward IMF. The scalar magnetic potential for this IMF in the reference frame of a planet is $\Psi_{IMF} = -B_0 z = -B_0 r \sin\lambda$. The ($r$, λ) components of the magnetic field obtained from $-\nabla\Psi$ are

$$\begin{aligned} B_{IMF}(r) &= -\frac{\partial \Psi}{\partial r} = B_0 \sin\lambda \\ B_{IMF}(\lambda) &= -\frac{1}{r}\frac{\partial \Psi}{\partial \lambda} = B_0 \cos\lambda \end{aligned} \tag{3.36}$$

These IMF components are now superposed on Earth's dipole field (Figure 3.6). The (r, λ) components of the total field are

$$\begin{aligned} B_T(r) &= -\frac{\mu_0 M \sin\lambda}{2\pi r^3} + B_0 \sin\lambda \\ B_T(\lambda) &= \frac{\mu_0 M \cos\lambda}{4\pi r^3} + B_0 \cos\lambda \end{aligned} \tag{3.37}$$

where (3.36) has been added to (3.26).

The total intensity of the superposed magnetic field at r and λ is given by

$$B_T^2(r,\lambda) = \frac{\mu_0^2}{16\pi^2}\frac{M^2}{r^6}(1 + 3\sin^2\lambda) + B_0^2 + \frac{\mu_0 M B_0}{2\pi r^3}(1 - 3\sin^2\lambda) \tag{3.38}$$

This equation reduces to that of a dipole (3.38) if $B_0 = 0$. Note that (3.37) satisfies the prescribed boundary condition. At large distances from a planet ($r \to \infty$), $B_T \to B_0$ and at small distances ($r \to 0$), $B_T \to B_d$, since the first term dominates over the other two terms close to a planet.

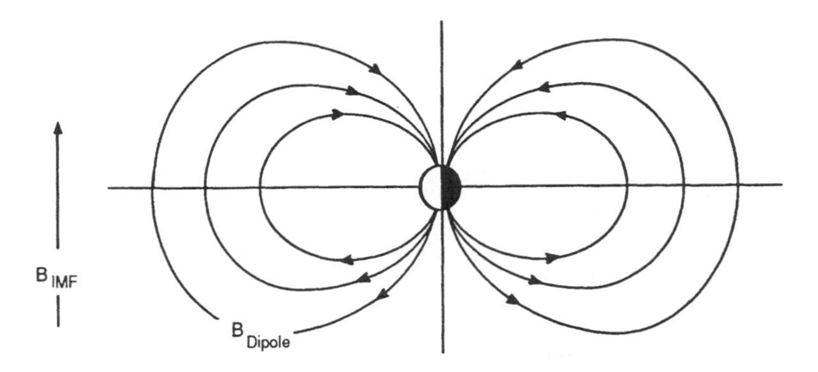

FIGURE 3.6 Superposition of a dipole field and northward interplanetary magnetic field.

The equation of the magnetic field that represents the superposed field geometry can be obtained as before. Use of (3.37) in (3.30) will yield the differential equation, $d\phi = 0$, and

$$\frac{1 + \dfrac{4\pi B_0 r^3}{\mu_0 M}}{1 - \dfrac{2\pi B_0 r^3}{\mu_0 M}} \frac{dr}{r} = -\frac{2\sin\lambda}{\cos\lambda} d\lambda \tag{3.39}$$

Integration of $d\phi = 0$ yields $\phi = \phi_0$ and integration of (3.39) yields

$$\frac{r^3}{\left(1 - \dfrac{4\pi B_0}{\mu_0 M} r^3\right)^3} = \frac{r_0^3}{\left(1 - \dfrac{4\pi B_0}{\mu_0 M} r_0^3\right)^3} \cos^6\lambda \tag{3.40}$$

or, taking the cube root, we obtain

$$\frac{r}{1 - \dfrac{4\pi B_0}{2\mu_0 M} r^3} = \frac{r_0}{1 - \dfrac{4\pi B_0}{2\mu_0 M} r_0^3} \cos^2\lambda \tag{3.41}$$

Here, as before, r_0 is the equatorial crossing distance at $\lambda = 0$. Again note that if $B_0 = 0$ the dipole field equation will be recovered.

Magnetic Neutral Points A magnetic neutral point is defined as a point at which the magnetic field intensity vanishes. Let $B_T(r) = 0$ in (3.37). This equation is satisfied if

$$r^* = \left(\frac{\mu_0 M}{2\pi B_0}\right)^{1/3} \tag{3.42}$$

Setting $B_T(\lambda) = 0$ in (3.37), this equation is satisfied only if $\lambda = \pm\pi/2$. Thus, we find that in the case of $\mathbf{B}_{IMF} = B_0\hat{\mathbf{z}}$, there are two neutral points located at $\lambda = \pm\pi/2$ at a distance of $r^* = (\mu_0 M/2\pi B_0)^{1/3}$ away from the center of the planet.

Equation (3.42) defines a sphere with a radius $r^* = (\mu_0 M/2\pi B_0)^{1/3}$. The total magnetic field intensity at r^* is defined by substituting (3.42) into (3.38). We then obtain

$$B_T(r^*, \lambda) = \frac{3}{2} B_0 (1 - \sin^2 \lambda)^{1/2} \tag{3.43}$$

Figure 3.7 shows a plot of $B_T(r, \lambda)$ from (3.38) where we have included in this figure a surface described by $B_T(r^*, \lambda)$ given by (3.43). The two neutral points are on $B_T(r^*, \lambda)$ where $\lambda = \pm\pi/2$. This figure shows that in regions $r < r^*$, the magnetic field is dipole-like and the field is anchored in the solid Earth. For $r > r^*$, the magnetic field geometry is IMF-like, although in the vicinity of $r \approx r^*$ the magnetic field is curved. As mentioned earlier, the IMF is solar in origin and the field is anchored at the Sun.

Southward IMF In the case of superposition of the dipole field with a purely southward pointing IMF, $\mathbf{B}_{IMF} = -B_0\hat{\mathbf{z}}$, one finds that the (r, λ) components of the total field are represented by

$$B_T(r) = -\frac{\mu_0 M \sin \lambda}{2\pi r^3}\left(1 + \frac{2\pi B_0 r^3}{\mu_0 M}\right)$$

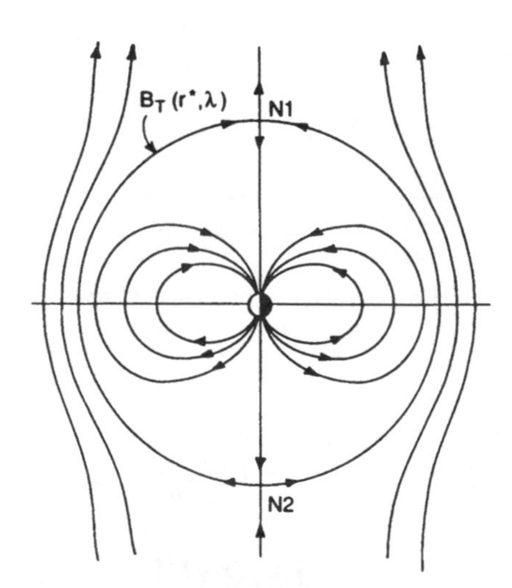

FIGURE 3.7 The topology resulting from the superposition of $\mathbf{B}_{IMF} = B_0\hat{\mathbf{z}}$ with a dipole field. N_1 and N_2 are neutral points.

and

$$B_T(\lambda) = \frac{\mu_0 M \cos\lambda}{4\pi r^3}\left(1 - \frac{4\pi B_0 r^3}{\mu_0 M}\right) \quad (3.44)$$

The expression for the total intensity obtained from (3.44) is

$$B_T^2(r,\lambda) = \frac{\mu_0^2}{16\pi^2}\frac{M^2}{r^6}(1+3\sin^2\lambda) + B_0^2 - \frac{\mu_0 M B_0}{2\pi r^3}(1-3\sin^2\lambda) \quad (3.45)$$

The equation of the total magnetic field for this case is

$$\frac{r}{1+\frac{4\pi B_0}{2\mu_0 M}r^3} = \frac{r_0}{1+\frac{4\pi B_0}{2\mu_0 M}r_0^3}\cos^2\lambda \quad (3.46)$$

Magnetic Neutral Line The total field vanishes at $r^{*3} = \mu_0 M/4\pi B_0$ according to (3.44). This occurs at $\lambda = 0$, which means that we have a *circle* at a distance $r^* = (\mu_0 M/4\pi B_0)^{1/3}$ on the equator where $B_T = 0$. The distance to the neutral line here is smaller by a factor of $2^{1/3}$ than to the neutral point for $\mathbf{B}_{IMF} = B_0\hat{\mathbf{z}}$ (see (3.42)).

The magnetic field topology that results in this case is quite different from the previous case of the northward IMF. Figure 3.8 shows a cross-sectional view of the magnetic field. There are three categories of magnetic lines of force here. Type 1 is the IMF and is solar in origin. Type 2 is the dipole field, and the ends are anchored in the planet. The dipole field is inside the neutral line. The type 3 magnetic field has one end anchored

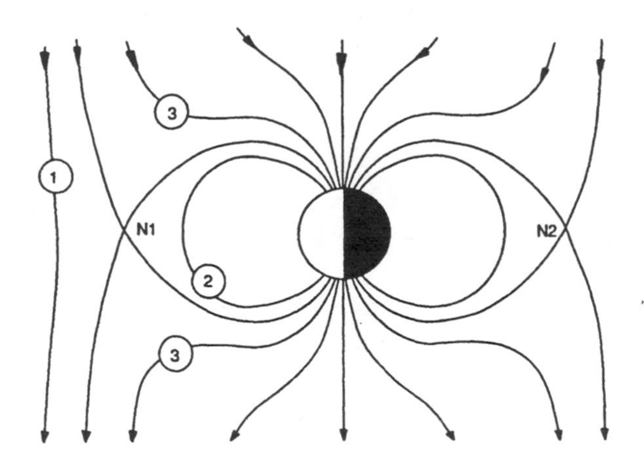

FIGURE 3.8 Topology of the magnetic field when a dipole and IMF southward field are superposed. The neutral line goes through N_1 and N_2. The field line labeled 1 is the IMF, 2 is the dipole, and 3 has one end anchored on the planet and the other on the Sun.

in the planet and the other end in interplanetary space. The type 3 field is a new feature which cannot exist if B_{IMF} is wholly northward. The existence of a type 3 magnetic field requires the presence of southward IMF. It indicates that the high-latitude region of the planet is now open into space.

3.3.4 Comparison With Observations

The magnetic properties derived above are basic and describe the regions of space near a magnetized planet where the effects of local currents can be ignored. Let us evaluate what we have just learned by comparing the derived features with features that are actually observed in space. A real magnetosphere necessarily includes (a) a boundary that separates the dipolar field regions from the interplanetary space, (b) magnetic cusps that locate the last planetary dipole line of force on the sunlit hemisphere and the first line of force that is in the anti-solar direction, and (c) a magnetic tail (Figure 1.3).

Magnetopause Boundary In both of the examples studied, magnetic boundaries were found. These boundaries "confined" the planetary dipolar field as in the real magnetospheres, but we must realize that they are physically quite different. The real magnetosphere is confined by a large-scale current that does not exist in our model. In a real situation, the location of a magnetospheric boundary is determined by a balance between the energy densities in the solar wind and the IMF, and the planetary dipole field and the particles it contains. Ignoring the particles, our model has placed our magnetic boundary further away than where, for example, Earth's magnetopause is observed.

Cusps and Polar Cap The closest feature in our model that corresponds to the cusp is the location where type 3 magnetic lines of force originate. The magnetic field in Earth's cusp maps to the ionosphere at around 78°, and this location seems to be close to what our model predicts. The region poleward of where type 3 magnetic lines of force begin is called the "polar cap," and the existence of such a region has been established by polar orbiting satellites. Polar cap lines of force are believed to map out into interplanetary space. Evidence for this comes from observations of direct entry of solar protons following a major proton-producing solar flare. These protons cause polar blackouts for radio wave propagation and are called polar cap absorption events. Our simple model predicts that the polar cap reduces to a point when $\mathbf{B}_{IMF}$ is completely northward. This prediction is consistent with observations that show that the size of the polar cap of Earth becomes smaller as $\mathbf{B}_{IMF}$ switches from southward to northward direction. The polar cap is smallest when the solar wind is quiet for a long time and accompanied by a prolonged period of northward $\mathbf{B}_{IMF}$.

Closed and Open Magnetospheres The existence of magnetic neutral points and lines will become important when particles are included. The existence of neutral points gives rise to the concept of open and closed magnetospheres. The closed magnetospheric model is a classical view that stipulates that the magnetic boundary isolates the planetary environment from the interplanetary space. The magnetosphere is closed off from the direct entry of the solar wind particles, and solar wind particles can only enter the magnetosphere proper by diffusing in. An open magnetospheric model takes an opposite view. It invokes a magnetic merging process (to be further discussed below), which opens the magnetosphere at neutral points and lines. Hence, the solar wind can directly enter the magnetosphere.

Magnetic Tail Our model does not produce a tail. The existence of a magnetic tail requires that we produce a sheet-like current on the night side of the magnetosphere. This current sheet on Earth is represented by a plasma sheet (see Figure 1.3). One of the fundamental questions in solar wind-planetary magnetic field interaction concerns the processes and dynamics of magnetic tail formation.

3.4 Inhomogeneous Magnetic Fields

The dipole field is inhomogeneous in space. We now introduce a topic that will be important when considering motions of particles in inhomogeneous magnetic fields. Suppose a homogeneous magnetic field $\mathbf{B}_0$ that is in the z-direction is slightly perturbed. Before the perturbation, each field line of $\mathbf{B}_0$ is of the form

$$\begin{aligned} x &= x_0 \\ y &= y_0 \end{aligned} \tag{3.47}$$

The perturbation will change the shape of $\mathbf{B}_0$. If the perturbation is "small," the shape of the magnetic field at $\mathbf{r}$ can be estimated by Taylor expansion of the magnetic field about the origin ($x_0 = 0, y_0 = 0$),

$$\mathbf{B}(\mathbf{r}) = \mathbf{B}_0 + (\mathbf{r} \cdot \nabla_0)\mathbf{B} + \cdots \tag{3.48}$$

where higher-order terms have been ignored. By small, it is meant that

$$|\,(\mathbf{r} \cdot \nabla)\mathbf{B}\,| \ll |\,\mathbf{B}_0\,| \tag{3.49}$$

The perturbed field is no longer homogeneous as it involves $\nabla\mathbf{B}$ which is a tensor

$$\nabla\mathbf{B} = \begin{pmatrix} \partial B_x/\partial x & \partial B_x/\partial y & \partial B_x/\partial z \\ \partial B_y/\partial x & \partial B_y/\partial y & \partial B_y/\partial z \\ \partial B_z/\partial x & \partial B_z/\partial y & \partial B_z/\partial z \end{pmatrix} \tag{3.50}$$

with nine components. However, the condition $\nabla\cdot\mathbf{B} = 0$ means that the sum of the diagonal terms must vanish and the number of unknowns is reduced to eight independent components. Let us examine each of the components to obtain a clearer meaning and a physical picture of the perturbed field.

3.4.1 Converging and Diverging Magnetic Fields

Examine first the diagonal terms, $\partial B_x/\partial x$, $\partial B_y/\partial y$, and $\partial B_z/\partial z$. Since the field is mainly in the z-direction, we can write

$$\begin{aligned} B_x &= x\left(\frac{\partial B_x}{\partial x}\right)_0 \\ B_y &= y\left(\frac{\partial B_y}{\partial y}\right)_0 \\ B_z &= B_0 + z\left(\frac{\partial B_z}{\partial z}\right)_0 \end{aligned} \tag{3.51}$$

B_x and B_y arise from the perturbation and we consider them to be first-order terms as contrasted from the original field B_0 which is considered zeroth order. Hence, we can ignore $z(\partial B_z/\partial z)_0$ compared to B_0. The equations of the perturbed lines of force are

$$\begin{aligned} \frac{dx}{dz} &= \frac{B_x}{B_z} \\ \frac{dy}{dz} &= \frac{B_y}{B_z} \end{aligned} \tag{3.52}$$

Combining (3.51) and (3.52) yields

$$\begin{aligned} \frac{dx_1}{dz} &= \frac{1}{B_0}\left(\frac{\partial B_x}{\partial x}\right)_0 x_0 \\ \frac{dy_1}{dz} &= \frac{1}{B_0}\left(\frac{\partial B_y}{\partial y}\right)_0 y_0 \end{aligned} \tag{3.53}$$

where the subindices "1" means first-order terms and "0" associated with the differential means that it is evaluated at (x_0, y_0). Integration of these

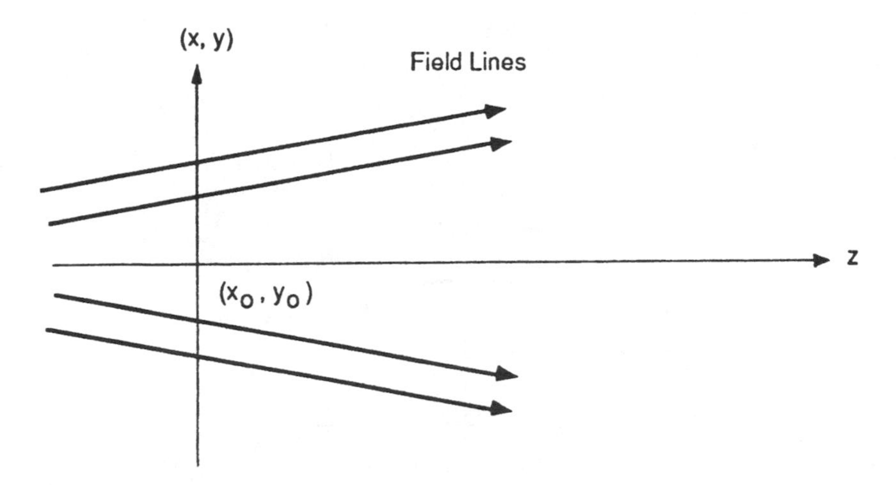

FIGURE 3.9 Converging and diverging magnetic field lines about (x_0, y_0). The principal field is in the z-direction.

equations yields

$$\begin{aligned} x_1 &= \frac{1}{B_0}\left(\frac{\partial B_x}{\partial x}\right)_0 x_0 z + \text{constant} \\ y_1 &= \frac{1}{B_0}\left(\frac{\partial B_y}{\partial x}\right)_0 y_0 z + \text{constant} \end{aligned} \tag{3.54}$$

These field lines depend on z and they diverge or converge from (x_0, y_0) as shown in Figure 3.9. The effect of the diagonal terms in $\nabla\mathbf{B}$ is to produce a converging or diverging bundle of field lines.

3.4.2 Curved Magnetic Fields

The effects of $\partial B_x/\partial z$ and $\partial B_y/\partial z$ are now examined. The zeroth order equation of the lines of force is again a straight line. The integration of the first-order terms

$$\begin{aligned} \frac{dx_1}{dz} &= \frac{1}{B_0}\left(\frac{\partial B_x}{\partial z}\right)_0 z \\ \frac{dy_1}{dz} &= \frac{1}{B_0}\left(\frac{\partial B_y}{\partial z}\right)_0 z \end{aligned} \tag{3.55}$$

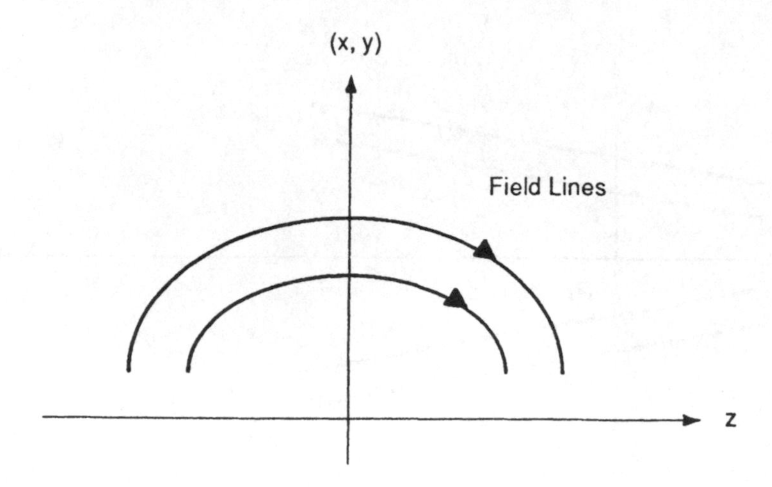

FIGURE 3.10 Curved magnetic field lines are due to perturbations (unspecified) of the principal field which is in the z-direction.

yields the solutions

$$x_1 = \frac{1}{B_0}\left(\frac{\partial B_x}{\partial z}\right)_0 \frac{z^2}{2} + \text{constant}$$
$$y_1 = \frac{1}{B_0}\left(\frac{\partial B_y}{\partial z}\right)_0 \frac{z^2}{2} + \text{constant} \tag{3.56}$$

where it is seen that the shape depends on z^2. These magnetic lines are curved as shown in Figure 3.10.

3.4.3 Gradient of the Magnetic Field

The terms now being examined are $\partial B_z/\partial x$ and $\partial B_z/\partial y$. In this case, the first-order corrections give

$$\frac{dx_1}{dz} = \frac{dy_1}{dz} = 0 \tag{3.57}$$

and the lines of force remain as straight lines. The shape of the original field lines remain as straight lines but these field lines have a gradient in the perpendicular direction (Figure 3.11).

3.4.4 Shear and Twist of Magnetic Fields

The two remaining terms in the $\nabla\mathbf{B}$ tensor are $\partial B_x/\partial y$ and $\partial B_y/\partial x$. These terms give rise to twisting and shearing of the magnetic lines. Twisted and

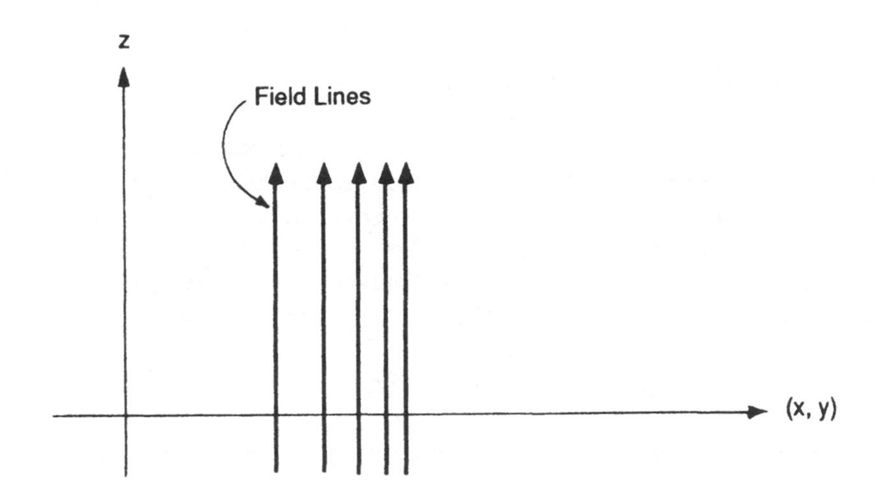

FIGURE 3.11 Gradient of magnetic field lines.

sheared field lines will introduce a component of the magnetic field **B** in the transverse direction. Twisted magnetic fields are produced, for example, by a plasma in a magnetic field that rotates. Sheared magnetic fields are produced, for example, at a boundary across which there is a plasma flow discontinuity (Figure 3.12).

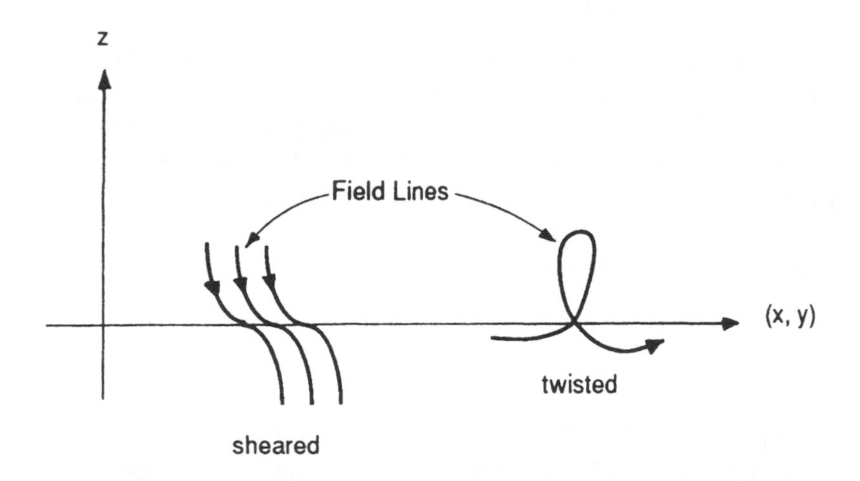

FIGURE 3.12 Sheared and twisted magnetic field lines.

3.5 Rotational Magnetic Fields

The study of magnetic fields in a vacuum is quite revealing, but this simplified approach cannot completely explain realistic magnetic structures in space. A better approach must include local currents produced by particles of the solar wind and magnetospheres. The topic of currents will be treated in Chapter 7 after particles have been studied. It is, however, instructive to introduce this subject in a limited way by studying the properties of magnetic fields whose curls do not vanish. These magnetic fields are associated with local currents but we will study only the magnetic fields these currents produce and not the currents.

Currents and magnetic fields are related by one of the Maxwell equations given by

$$\nabla \times \mathbf{H} = \mathbf{J} + \frac{\partial \mathbf{D}}{\partial t} \tag{3.58}$$

In the limit that the conduction current $\mathbf{J}$ is much larger than the displacement current $\partial \mathbf{D}/\partial t$ (see Chapter 5), the second term can be ignored and (3.58) is reduced to

$$\nabla \times \mathbf{H} = \mathbf{J} \tag{3.59}$$

The magnetic field equation here is different from the ones we have thus far studied since $\nabla \times \mathbf{H} \neq 0$. $\mathbf{J}$ takes into account all of the currents, including ones produced locally.

3.5.1 Stretched Dipole: Parabolic Geometry

If a planetary magnetosphere supports a ring current produced by the motion of trapped particles, or if the dipole is stretched by the solar wind as in the case of Earth's dipole at the flanks, the dipole geometry will become more elongated. The exact shape of the distorted dipole magnetic field depends on the details of the current distribution that modifies the dipole. We consider here a parabolic magnetic field geometry that has been considered for studies of magnetic tail dynamics for regions not too distant from the planet.

Consider a two-dimensional magnetic field represented by

$$\mathbf{B} = -z\hat{\mathbf{x}} + \hat{\mathbf{z}} \tag{3.60}$$

where $\hat{\mathbf{x}}$ and $\hat{\mathbf{z}}$ are unit vectors in a Cartesian coordinate system. It can be easily verified from (3.15) that the equations of the field lines are

$$x = -\frac{1}{2}z^2 + \text{constant} \tag{3.61}$$

which are a family of parabolas (Figure 3.13). This magnetic field is supported by a current in the y-direction. A parabolic geometry adequately describes the magnetic field in the near-Earth tail regions, where field lines are closed.

3.5.2 Magnetic Tails

The parabolic geometry discussed above is not suitable for describing the distant tail. The magnetic field on the night side of a planet very far from the planet is nearly parallel to the solar wind flow direction. The normal component (in the z-direction) becomes very weak or vanishes and the tail supports a magnetic neutral sheet embedded in the plasma sheet. The magnetic neutral sheet separates the oppositely directed magnetic fields in the northern and southern hemispheres. The magnetic field intensity away from the neutral sheet slowly increases and becomes nearly constant across the plasma sheet boundary.

There are several ways to model a magnetic field that includes the neutral line or sheet. The simplest model considers a one-dimensional magnetic

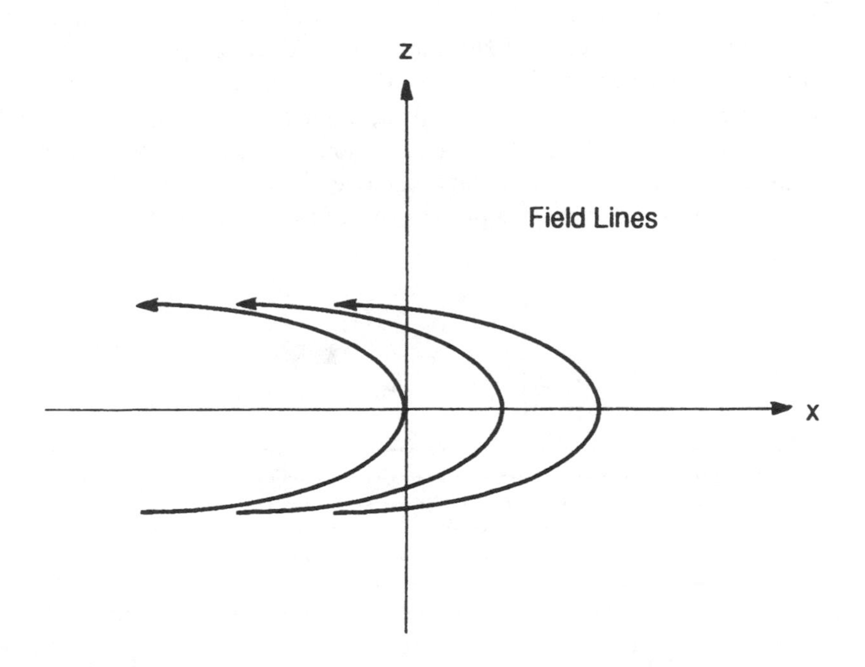

FIGURE 3.13 A parabolic magnetic field can be used to study the dynamics of particles in the near-Earth plasma sheet.

field given by

$$\begin{aligned}
B_x &= B_0 && z \geq L \\
B_x &= \frac{B_0 z}{L} && L \geq z \geq -L \\
B_x &= -B_0 && z \leq -L
\end{aligned} \tag{3.62}$$

where $z = 0$ is the location of the neutral line, B_0 is the value of the magnetic field in the lobe (a lobe is a region adjacent to the plasma sheet where the plasma density is very much reduced), and L is the half-thickness of the plasma sheet in the north-south direction (we use the geocentric coordinate system in which x is positive pointing toward the Sun, y toward dusk, and z toward north). Equation (3.62) models the magnetic neutral line in the plasma sheet to be straight in the x-direction and the field increases linearly with x in the $|z|$-direction and reaches a value B_0 at the plasma sheet boundary. The field is constant in the lobe (Figure 3.14).

Another frequently used model considers the magnetic field of the form

$$\mathbf{B} = B_0 \tanh\left(\frac{z}{L}\hat{\mathbf{x}}\right) \tag{3.63}$$

which is also one-dimensional but as in (3.62) permits variation of the intensity in the z-direction. This field configuration is also known as a Harris field (named after E.G. Harris, who first used this form to study the equilibrium behavior of the magnetic field in the plasma sheet). The Harris field increases slightly slower than (3.62) as one moves away from the neutral line ($z = 0$). At the plasma sheet boundary ($z/L = 1$), the field magnitude is $0.76B_0$. Since the function $\tanh(z/L)$ varies continuously across the plasma sheet boundary, the boundary here is not as sharp as in the previous model from the perspective of the magnetic field.

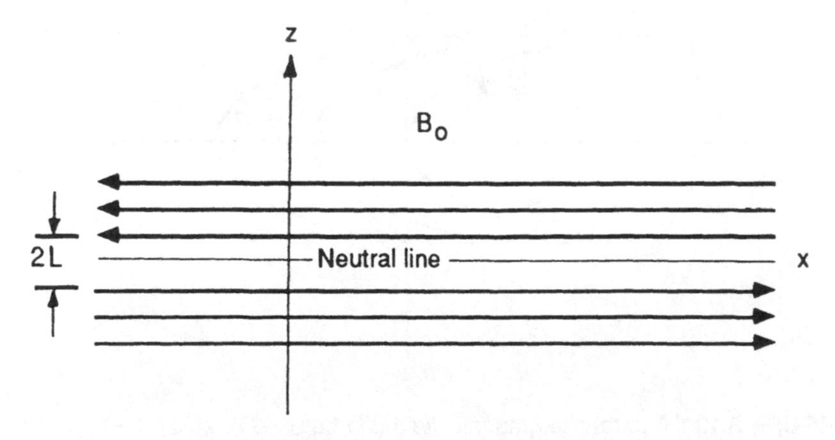

FIGURE 3.14 A distant magnetic tail modeled after the Harris field.

Observations indicate that while the tail field is dominated by the component in the x-direction, a small but non-negligible magnetic field component exists in the direction normal to the neutral line (z-direction). Equation (3.62) can be modified by the addition of a small magnetic field component in the z-direction. Then

$$\mathbf{B} = B_0 \left(\frac{z}{L} \hat{\mathbf{x}} + \delta \hat{\mathbf{z}} \right) \tag{3.64}$$

describes the behavior of the magnetic field in the plasma sheet. We leave the lobe field unchanged. Here, δ is assumed to be small, $\lesssim 0.1$, as required by observations. Equation (3.64) is more realistic and although δ is small, it significantly affects the particle trajectories in the neighborhood of the neutral line (see Chapter 4).

Two questions arise about a magnetic-tail field geometry. One is the current that is required to maintain the tail geometry. Although it is not exactly known how the current is produced and maintained, the current must be embedded in the neutral line (sheet) and the particles responsible for the current must come from the plasma sheet. These particles must move in such a way that they produce a net drift in the $+y$-direction to produce the desired magnetic-tail geometry (Figure 3.14). The other question is about the stability of such tail currents. This problem, although complicated, is very important for many geophysical, solar, and astrophysical applications. It is closely related to the concept of magnetic field merging.

3.5.3 Magnetic Neutral Points, Lines, and Sheets

Various magnetic field configurations support regions in which the magnetic field vanishes. These magnetic null regions can be points, lines, or sheets, and they occur when magnetic field lines intersect. The point where field lines intersect is a singular point in the equation of the lines of force that define the magnetic field. At a singular point, a magnetic field can branch into two or more lines and the direction of a line of force is no longer unique. When the magnetic field intensity vanishes at a singular point, it is called a magnetic neutral point. We showed earlier that neutral points and lines are formed, for example, when an IMF and Earth's dipole are superposed. The existence of such neutral points, lines, and sheets is important in magnetic field "merging" or "reconnection" theories, which describe physically how two magnetic field lines intersect.

A schematic picture of a reconnection process in a two-dimensional plane is shown in Figure 3.15. The field lines labeled 1 and 2 are original field lines. Now suppose the two field lines closest to each other are permitted to merge. The merged field lines are labeled 3 and 4. Note that

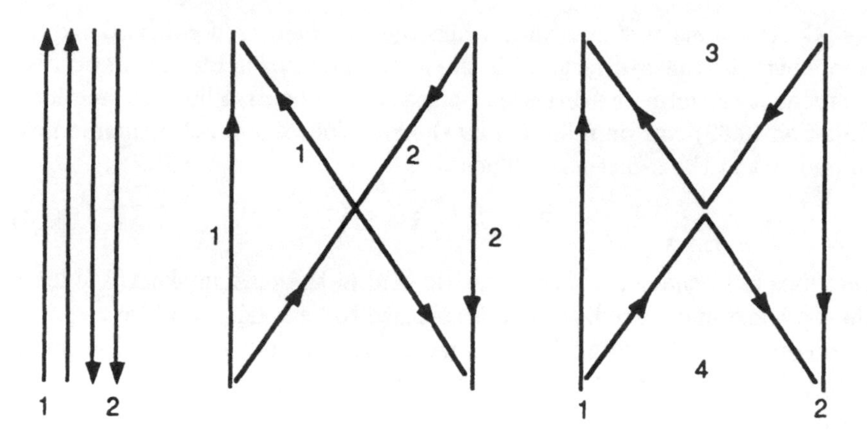

FIGURE 3.15 Geometry of magnetic field merging. The oppositely directed magnetic field lines (left) are allowed to merge (center) which leads to global change in the original geometry (right). Note the X-type neutral point.

merging changes the global geometry of the original magnetic configuration and closed field lines can become open field lines.

Examine now the magnetic field geometry, including an x-type neutral point, in the immediate vicinity where two field lines are merging. A magnetic field of the form

$$\mathbf{B} = x\hat{\mathbf{x}} - y\hat{\mathbf{y}} \tag{3.65}$$

is adequate to describe the geometry to include such a neutral point. The equation of the field lines is a hyperbola

$$xy = \text{constant} \tag{3.66}$$

in the xz-plane (Figure 3.16). The neutral point is located at the origin.

Consider now a more general case of a three-dimensional magnetic field given by

$$\mathbf{B} = \alpha x\hat{\mathbf{x}} + \beta y\hat{\mathbf{y}} - (\alpha + \beta)z\hat{\mathbf{z}} \tag{3.67}$$

where α and β are positive constants. Consider first the case that one of the coefficients vanishes, for example $(\alpha + \beta) = 0$. Equation (3.67) then reduces to

$$\mathbf{B} = \alpha(x\hat{\mathbf{x}} - y\hat{\mathbf{y}}) \tag{3.68}$$

The equations of the line of force are defined by

$$\begin{aligned} xy &= \text{constant} \\ z &= \text{constant} \end{aligned} \tag{3.69}$$

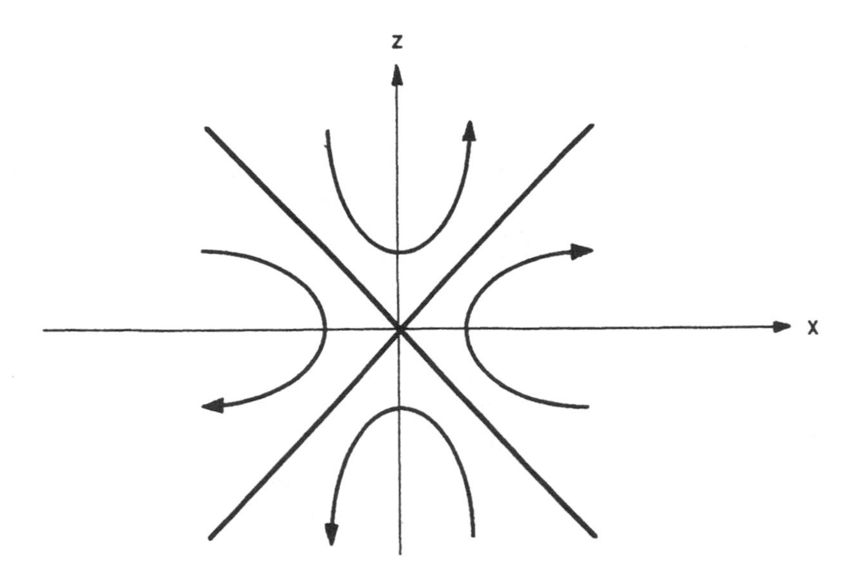

FIGURE 3.16 A hyperbolic geometry is often used in magnetic field merging studies.

which are equations of hyperbolas in the xy-plane and the z-axis is a singular line at which the lines of force intersect at right angles.

Consider now a less restrictive case where all of the coefficients do not vanish, but we let $\alpha = \beta$ in (3.67). The equations of the lines of force for this case are described by

$$\frac{x}{y} = \text{constant}$$

$$(x^2 + y^2)z = \text{constant} \tag{3.70}$$

which are equations of hyperboloids. In the general case $\alpha \neq \beta$, the lines of force remain hyperboloids and the magnetic field pattern is obtained by squeezing the field configuration of (3.70). The amount of squeezing depends on the ratio of α to β.

3.6 Electric Fields in Space

As stated in the introduction to this chapter, electric fields in space come predominantly from the motional electromotive force. This problem can only be understood completely after we have studied how electric and magnetic fields behave in plasma. Here, we briefly discuss how the flow of magnetized solar wind and the rotation of magnetized planetary or stellar objects through a plasma medium induce electric fields.

3.6.1 EMF Induced by the Flowing Solar Wind

The expansion of a magnetized solar wind across IMFs induces a motional electric field in the frame at rest with respect to the Sun. The electric field in the frame moving with the solar wind will be negligibly small. Lorentz transformation theory then shows that the induced electric field for a non-relativistic solar wind is given by

$$\mathbf{E} = -\mathbf{U}_{sw} \times \mathbf{B}_{IMF} \tag{3.71}$$

where $\mathbf{U}_{sw}$ and $\mathbf{B}_{IMF}$ are the solar wind velocity and the IMF, respectively. Equation (3.71) is referred to as the solar wind convective electric field. Magnetospheres immersed in a flowing solar wind interact with this convective electric field by mechanisms not completely understood, resulting in energization of plasmas inside the magnetospheres. Convective electric fields are often invoked as a source of energy for auroras.

3.6.2 EMF Induced by Planetary and Stellar Rotation

The motion of ionized atmospheres of a magnetized planet or star will induce an electric field and this motional electric field can be measured in the rest frame. If $\boldsymbol{\omega}$ is the rotation vector,

$$\mathbf{E} = -(\boldsymbol{\omega} \times \mathbf{r}) \times \mathbf{B} \tag{3.72}$$

where $\mathbf{r}$ is the position where the field is measured. This electric field is referred to as the corotational electric field. Equation (3.72) assumes that the ionized atmosphere is corotating with the planet. This equation can be written out explicitly for planetary and stellar magnetic fields that are a centered dipole. Let

$$\begin{aligned} \boldsymbol{\omega} &= \omega\hat{\mathbf{z}} \\ \mathbf{B} &= \frac{\mu_0 M}{4\pi r^3}(-2\sin\lambda\hat{\mathbf{r}} + \cos\lambda\hat{\boldsymbol{\lambda}}) \end{aligned} \tag{3.73}$$

where $\hat{\mathbf{r}}$, $\hat{\boldsymbol{\lambda}}$, and $\hat{\mathbf{z}}$ are unit vectors. Now define B_0 as the intensity of the magnetic field measured at the planetary or stellar surface and where $r = R_0$ at $\lambda = 0$. Then, $B_0 = \mu_0 M/4\pi R_0^3$ and

$$\begin{aligned} \mathbf{E} &= -\omega r \sin\theta(\hat{\mathbf{z}} \times \hat{\mathbf{r}}) \times \mathbf{B} \\ &= -\frac{B_0 \omega R_0^3 \cos\lambda}{r^2}(2\sin\lambda\hat{\boldsymbol{\lambda}} + \cos\lambda\hat{\mathbf{r}}) \end{aligned} \tag{3.74}$$

On the equatorial plane, the electric field is directed radially inward.

Magnetospheres can have different degrees of corotation. Consider a magnetosphere that is entirely corotating, that is the plasma from the ionosphere to the magnetopause is rotating with the planetary magnetic field.

Jupiter's magnetosphere falls in this class. Jupiter's corotational electric field is very strong and its magnetosphere may be corotating out to the magnetopause. In such magnetospheres, the effects of the corotational electric field must be taken into account in the dynamics of particles in the entire magnetosphere.

Consider now a magnetosphere that is partially corotating. Earth's magnetosphere falls in this class. For Earth, only the plasma inside the plasmapause is corotating with the planet. Here, the dynamics of particles inside the plasmasphere are affected by the corotational electric field, but the particles outside the plasmasphere are affected mainly by the solar wind electric field.

The last example is a magnetosphere that is not corotating at all. Observations indicate that Mercury has no ionosphere and its magnetosphere may not be corotating. In this case, the dynamics of particles in the magnetosphere are affected totally by the solar wind electric field.

3.7 Concluding Remarks

The planetary magnetic fields were characterized as a centered dipole to facilitate computations. The centered dipole approximation is reasonable for many planets like Earth, whose higher multipoles of the core field are much smaller than the dipole component. However, for the planets Uranus and Neptune, the magnetic moments are offset and they have large tilt angles relative to the rotation angle. As shown in Table 3.1, the tilt angle for Uranus is about 60° and for Neptune is around 47°. A schematic diagram illustrating Neptune's magnetic field is shown in Figure 3.17. Magnetic topologies on these planets change as the planet rotates and their description requires quadrupole and possibly even higher moments.

The Lorentz motions of particles are affected by the inhomogeneous properties of a magnetic field. The idealized dipole field serves as a basic reference for the study of magnetospheric particles. Inhomogeneous magnetic fields give rise to particle trapping and create radiation belts. Even though Earth's magnetic field departs more and more from the dipole with increasing distance from the solid planet, the dynamics of the particles can be adequately described by using the dipole approximation to geocentric distances of about six Earth radii.

Most of our knowledge of Earth's magnetic field comes from data obtained by magnetometers distributed over the surface of Earth. However, measurements from the surface alone are not adequate because external currents (external to the core) are important. Currents run laterally and vertically in the ionosphere, and this external contribution must be correctly accounted for. Researchers are now active in creating realistic mod-

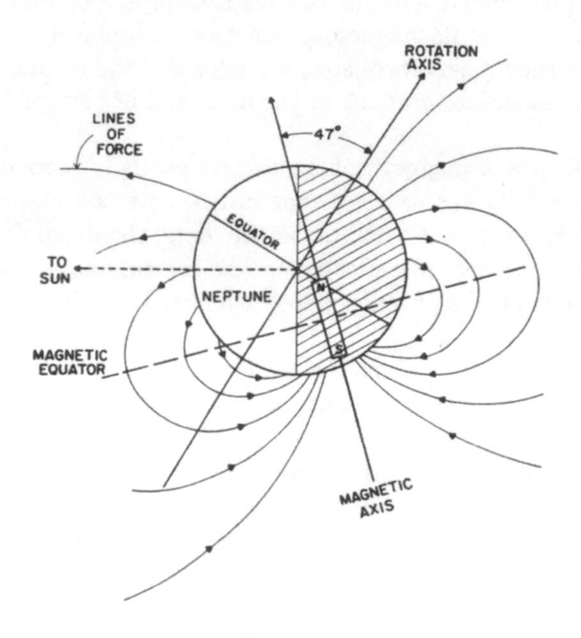

FIGURE 3.17 A schematic diagram to illustrate the geometrical properties of Neptune's magnetic moment. (From Ness *et al.*, 1989)

els that can accurately predict the shape of the magnetosphere (including the tail) for various solar wind conditions. One of the main challenges is to construct a model that will correctly predict where the ionospheric field lines map to in the distant magnetosphere.

In the presence of an IMF, the shape of the planetary dipole in the outer regions is very much modified. When two fields intersect, a field line can split or branch points can form which are singularities in the equations that define the lines of force. These features of magnetic fields are more common than not in space, especially in the neighborhood of magnetized bodies. Merging of magnetic field lines is a picturesque way to describe what might happen with magnetic fields in space and laboratory plasmas. While this is useful, the reader is cautioned that Maxwell's equations of electrodynamics say nothing about merging.

The concept of magnetic field merging is widely used to describe how magnetic energy is converted to particle energy. For example, merging is invoked as the origin of high-energy particles produced in solar flares and it has been suggested that merging in magnetic tails across the neutral sheet produces energetic auroral particles. As mentioned previously, merging between an IMF and an oppositely directed planetary magnetic field on the sunward magnetopause boundary leads to an open magnetospheric configuration that permits the solar wind energy to be transported into the magnetosphere. This open geometry contrasts with a closed one in which the planetary magnetic field is completely confined and no solar wind directly enters the magnetosphere.

The concept of magnetic field lines in plasmas is a convenience, restricted to only ideal fluids with infinite conductivity (Chapter 5). Hence, the concept of merging must be used cautiously when describing real situations. If one considers an intersection of two field lines in an ideal fluid, a fluid with infinite conductivity, the field may not behave regularly and the field need not vanish. Although formal theories are available to study the behavior of magnetic surfaces in the neighborhood of nonregular singular points, this is beyond the scope of the present chapter.

The topic of electric fields in space was briefly discussed and the concepts of convective and corotating electric fields were introduced. Note again that these concepts are not defined by the Maxwell equations. The electric field and its relation to plasma motions are correctly given by the Lorentz transformation. The description of these fields in terms of convection and corotation is, however, a convenient way to picture how electric fields are generated in space. A more complete discussion of electric fields in plasmas cannot be given until we have studied how a group of particles can be treated as a fluid and how such conducting fluids interact with magnetic fields. These are the central topics to be formulated and studied in Chapter 5.

Bibliography

Cahill, L.J., Jr. and P.G. Amazeen, The Boundary of the Geomagnetic Field, *J. Geophys. Res.*, **68**, 1835, 1963. This article discusses the first measurements of the magnetic field away from the solid Earth and the observations of the magnetopause boundary.

Chapman, S. and J. Bartels, *Geomagnetism*, Clarendon Press, Oxford, England, 1940. A monumental two-volume work that characterized and organized Earth's magnetic field. A classic and still a major reference book.

Jackson, J.D., *Classical Electrodynamics*, 2nd ed., John Wiley and Sons, Inc., New York, NY, 1975. Another excellent book on the fundamentals of electromagnetic fields.

Knecht, D.J. and B.M. Shuman, The Geomagnetic Field, in *Handbook of Geophysics and Space Environments*, ADA167000, A.S. Jursa, ed., National Technical Information Service, Air Force Geophysics Laboratory, Springfield, VA, 1985. Contains a good summary of geomagnetic fields, including a discussion of disturbed geomagnetic fields.

Landau, L.D. and E.M. Liftshitz, *Electrodynamics of Continuous Media*, Pergamon Press, Ltd., New York, NY, 1960. Basic concepts of magnetic fields are presented at the graduate level.

McDonald, K.L., Topology of Steady Current Magnetic Fields, *Am. J. Phys.*, **22**, 586, 1954. This article discusses in detail (with many examples) why magnetic field lines need not be closed.

Mead, G.D. and D.B. Beard, Shape of the Geomagnetic Field Solar Wind Boundary, *J. Geophys. Res.*, **69**, 1169, 1964. This is one of the early articles that discusses how one needs to include the surface currents at the magnetopause boundary to explain the geomagnetic field distortions observed in space.

Merrill, R.T. and M.W. McElhinny, *The Earth's Magnetic Field*, Academic Press, London, England, 1983. This is a good book in which to obtain the history, origin and planetary perspective of Earth's magnetic field.

Morozov, A.I. and L.S. Solov'ev, The Structure of Magnetic Fields, in *Reviews of Plasma Physics, Volume 2*, M.A. Leontovich, ed., Consultants Bureau, a division of Plenum Publishing Corp., New York, NY, 1966. A very thorough review of magnetic structures and theories for studying them. Although the emphasis is on fusion, the methods are applicable to general studies. For advanced students.

Ness, N.F., C.S. Searce and J.B. Seek, Initial Results of the IMP-1 Magnetic Field Experiment, *J. Geophys. Res.*, **69**, 3531, 1964. First observations of the magnetic field on the dark hemispheres that reported the existence of the geomagnetic tail.

Ness, N.F., M.H. Acuña, L.F. Burlaga, J.E.P. Connerney, R.P. Lepping and F.M. Neubauer, Magnetic Fields at Neptune, *Science*, **246**, 1473, 1989. First observations of the magnetic field at Neptune. Magnetic field data have been interpreted in terms of an offset tilted dipole.

Parker, E.N., *Cosmical Magnetic Fields, Their Origin and Their Activity*, Oxford University Press, New York, NY, 1979. Parker has made many original contributions to the study of magnetic fields. This monograph brings together Parker's ideas about this topic. The physics of large-scale magnetic fields in highly conducting fluids are examined in great detail. The material is advanced and the contents will be better understood after the reader has been introduced to the fluid theory.

Siscoe, G.L., Towards a Comparative Theory of Magnetospheres, in *Solar System Plasma Physics*, E.N. Parker, C.F. Kennel and L.J. Lanzerotti, eds., North Holland Publishing Co., Amsterdam, the Netherlands, 1979. The information in Table 3.1 comes from this article.

Stern, D.P., Euler Potentials, *Am. J. of Phys.*, **38**, 494, 1970. This article presents a detailed discussion of Euler potentials.

Williams, D.J. and G.D. Mead, Nightside Magnetospheric Configuration as Obtained from Trapped Electrons at 1100 Kilometers, *J. Geophys. Res.*, **70**, 3017, 1965. A discussion is presented on the existence of a current sheet to account for the geomagnetic tail structure.

Observations and models of Jupiter's magnetic field are contained in a Conference Proceedings entitled, *Jupiter*, ed. by T. Gehrels, University of Arizona Press, Tucson, Arizona, 1976.

For information on properties of the magnetic field of Mercury, see Ness, N.F., K.W. Behannon, R.P. Lepping, Y.C. Whang, and K.H. Schatten, Magnetic Field Observations near Mercury: Preliminary Results from Mariner 10, *Science*, **185**, 151, 1974.

For information on modeling of planetary magnetic fields, especially Earth's magnetic field, see *Quantitative Modeling of Magnetospheric Processes*, W.P. Olson, ed., American Geophysical Union, Washington, D.C., 1979.

Questions

1. Show that the dipole field equations satisfy $\nabla \cdot \mathbf{B} = 0$ and $\nabla \times \mathbf{B} = 0$.
2. On a dipole field, show that the angle α between r and B is related to λ.

$$\text{ctn}\lambda = 2\tan\alpha$$

$(\pi/2 - \alpha)$ is defined as inclination of the magnetic field.

3. Show that the arc length of a dipole line is given by:

$$\frac{ds}{d\lambda} = r_0 \cos\lambda(1 + 3\sin^2\lambda)^{1/2}$$

where ds is the differential arc length.

4. Show that for a dipole, the magnetic field along the arc changes according to

$$\frac{dB}{ds} = \frac{3\mu_0 M}{4\pi r_0^4} \frac{\sin\lambda}{\cos^8\lambda} \frac{(3 + 5\sin^2\lambda)}{(1 + 3\sin^2\lambda)}$$

5. Find the total volume the dipole field occupies from the surface of Earth to the boundary located at $10R_E$.
6. Earth's surface field measured on the equator is about 0.31×10^{-4} teslas. Show that the magnetic moment of Earth is 7.8×10^{22} amp-m^2.
7. The total magnetic energy contained in a volume V is obtained from

$$\int_V \frac{B^2}{2\mu_0} dV$$

Determine the amount of energy of Earth's dipole field external to the solid Earth.

8. Estimate the distance r of Equation (3.42) for Earth if $B_{IMF} = 10^{-8}$ teslas. Compare this number to the observed geomagnetic boundary distance on the sunward side, which is $\sim 10R_E$.
9. Show the missing steps in going from Equation (3.39) to Equation (3.41).
10. Show that for B_{IMF} southward ($B_{IMF} = -B_0\hat{\mathbf{z}}$) the total magnetic intensity is given by Equation (3.45) and that the equation of the magnetic field lines is given by Equation (3.46).
11. Estimate the latitude on Earth's surface from which the geomagnetic field is open into space for $\mathbf{B}_{IMF} = -B_0\hat{\mathbf{z}}$.
12. Determine the amount of magnetic energy for a magnetosphere that includes the dipole and IMF. Do this problem for both northward and southward IMF. Is the energy difference between these two configurations sufficient to produce aurora?
13. Suppose the dipole makes an angle α with respect to Earth's rotational axis. Show that

$$\begin{aligned}
B_r &= -\frac{\mu_0 M}{2\pi r^3}(\cos\alpha\sin\lambda + \sin\alpha\cos\lambda) \\
B_\lambda &= \frac{\mu_0 M}{4\pi r^3}(\cos\alpha\cos\lambda - \sin\alpha\sin\lambda) \\
B_\phi &= 0
\end{aligned}$$

14. Verify that the vector potential for a dipole field can be written as

$$\mathbf{A} = -B_0 \left(\frac{R_E}{r} \right)^3 r \sin\theta \hat{\boldsymbol{\phi}}$$

where B_0 is the equatorial magnitude of the field ($\theta = \pi/2$), $\hat{\boldsymbol{\phi}}$ is the unit vector in the azimuthal direction, and the other symbols have their usual meaning.

15. Derive a set of Euler parameters $\alpha = \alpha(r, x)$ and $\beta = \beta(r, x)$ in spherical coordinates for a northward IMF, $\mathbf{B} = B_0\hat{\mathbf{z}}$. Do the same problem for a dipole field.

16. Derive the Euler parameters of a southward IMF superposed on the dipole.

17. Derive the coordinates of the neutral points for a magnetic topology formed by

$$\mathbf{B} = B_0\hat{\mathbf{x}} + B_0\hat{\mathbf{y}} + B_0\hat{\mathbf{z}}$$

superposed on the dipole field.

18. Show
    a. That the Harris-type field supports a current density.
    b. Derive an expression of a vector potential for this magnetic field.

19. What are the current densities for the magnetic field given by Equations (3.62) and (3.65)?

20. Consider a magnetic field given by $\mathbf{B} = y\mathbf{x} + \alpha x\hat{\mathbf{y}}$ where $\alpha > 1$. Derive
    a. The equations of the field lines, and
    b. The current density.

21. Show that $\mathbf{E}$ given by Equation (3.74) and $\mathbf{B}$ given by Equation (3.73) satisfy the relation $\mathbf{E} \cdot \mathbf{B} = 0$. Discuss the meaning of this relation.

22. Show that $\mathbf{E}$ given by Equation (3.74) is a potential field. Derive an expression for the potential function.

4

Particles in Space

4.1 Introduction

The preceding chapter considered elementary cases of electromagnetic fields in the absence of any charged particles. However, it is known that the vast regions of space are populated by tenuous but significant amounts of charged particles. For example, close to a planet, there are ionospheric particles. Beyond the ionosphere, there are trapped particles of the Van Allen radiation belts, and the magnetosphere is enveloped in the expanding solar wind.

Particles in space can travel long distances before they encounter another particle. A single particle thus gyrates around the magnetic field many times before a collision occurs. Space is, in this sense, a collisionless medium. If the particles do not collide with each other it would seem reasonable to study the behavior of a single particle, neglecting the presence of the other particles. However, as mentioned in the earlier chapter, the existence of the long-range electromagnetic forces precludes this possibility. Characterization of electrodynamic phenomena in space requires accurate information on the charge and current distributions and this in turn requires knowledge of the instantaneous position and velocity of all the particles. The procedure for solving the coupled Maxwell-Lorentz equations was described in Chapter 2. However, because there are as many equations as there are particles, this approach is not suited for analytical studies.

A somewhat less ambitious approach is to study the particles in the average sense. This requires knowledge of the distribution function of the particles, performing the averages as prescribed in Chapter 2, and subsequently studying the macroscopic equations (topic of Chapter 5). The macroscopic equations describe the motion of an ensemble of particles that contains many individual particles. A natural question then concerns the motion of the individual particles. What motions are executed by single particles in electromagnetic fields?

The purpose of this chapter is to develop an understanding of particle motions from the perspective of a single particle. The emphasis is on understanding the origin and behavior of particles in magnetospheres. We begin with a short discussion of particles observed in Earth's radiation belts. Then the Lorentz equation of motion is studied with the goal of obtaining information on particle motions in dipole magnetospheres. Unfortunately, analytical solutions to the Lorentz equation are available only for simple magnetic field geometries, such as the homogeneous field. The Lorentz equation with inhomogeneous magnetic fields is in general non-linear and exact solutions are not always available.

The study of charged particles in a dipole magnetic field dates back to 1895 when Kristian Birkeland, a Norwegian physicist, was performing experiments with cathode rays. Birkeland noted that electrons were guided toward the magnetic pole. He thus surmised that perhaps auroras were produced in a similar way. Quantitative study of the Lorentz equation with a dipole field began around 1904 with C. Störmer, another Norwegian. Störmer was fascinated with Birkeland's experiment and he sought to mathematically explain the particle trajectories observed in Birkeland's terella experiment. Störmer numerically solved the Lorentz equation with a dipole field. However, it has become apparent that these Störmer solutions are better suited for understanding the behavior of high-energy cosmic ray particles and they are not as useful for understanding the lower-energy particles in magnetospheres.

For the lower-energy magnetospheric particles, a more useful method is to study the Lorentz equation, using the guiding center (GC) approximation, a method originally used by H. Alfvén. This method approximates the particle orbits in inhomogeneous magnetic fields to first-order by a perturbation technique yielding information on particle orbits in the average sense. Hence, the guiding center theory is also known as first-order orbit theory. The guiding center theory provides a valid picture of the behavior of the particles when the distances the particles travel are much larger than the gyroradii of the individual particles. This is the case for most magnetospheric particles.

As in other branches of physics, we will find the concept of invariants useful and extremely helpful in describing adiabatic particle motions in complex electromagnetic fields. Fortunately, the guiding center theory provides three invariants of motion that can be used to define trapped and precipitated particles in radiation belts. We will apply these invariants to particles in Earth's radiation belt, and show how the guiding center theory organizes their motions in a systematic way.

This chapter concludes with a short discussion of particle motions in magnetic fields that include neutral points, lines, or sheets. Since particles approaching the neutral point may not behave adiabatically, we return to the original Lorentz equation to examine these problems. As mentioned in Chapter 3, magnetic neutral regions exist in between two oppositely directed magnetic fields and they may have important applications in astrophysical, geophysical, and solar electrodynamic phenomena. In Earth's magnetosphere, magnetic neutral regions may be found at the subsolar magnetopause boundary and in the geomagnetic tail.

4.2 Discovery of Earth's Radiation Belt

Although cosmic rays and high-energy solar flare particles were known before the space age, the existence of radiation belts was not known until James Van Allen and his colleagues discovered Earth's radiation belt in 1958. It is now known that radiation belts also exist in the magnetospheres of Jupiter, Mercury, Saturn, Neptune, and Uranus. Figure 4.1 shows the first radiation belt particle data obtained from Earth by the Explorer I spacecraft. The particle intensity is plotted as a function of the altitude from the surface of Earth. The three sets of curves represent the intensity profiles obtained at different longitudes. The bottom curve of Figure 4.1 shows the same data plotted as a function of the magnetic field intensity where measurements were made. The fact that the particle intensity can be organized by the magnetic field has been interpreted to mean that the particles are charged and controlled by Earth's geomagnetic field.

4.2.1 Inner and Outer Radiation Belts

Explorer IV was the first spacecraft that provided information about the Van Allen radiation belts over an extended region in space. Figure 4.2 summarizes the observations. The early experiments were designed to detect high-energy particles, > 30 MeV protons and > 1.6 MeV electrons (curves a and b). Observations of these high-energy particles showed two

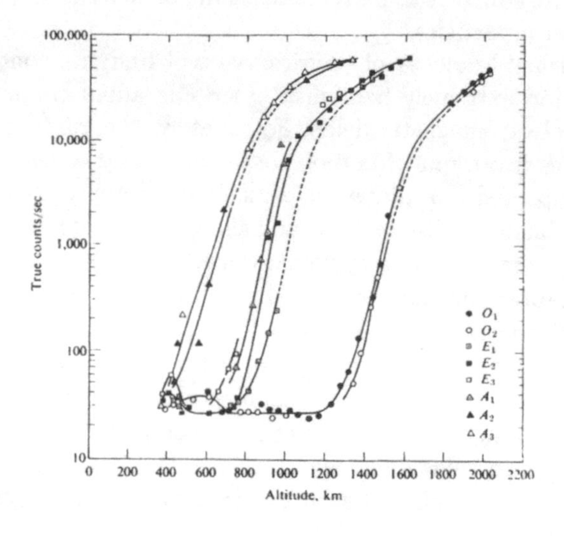

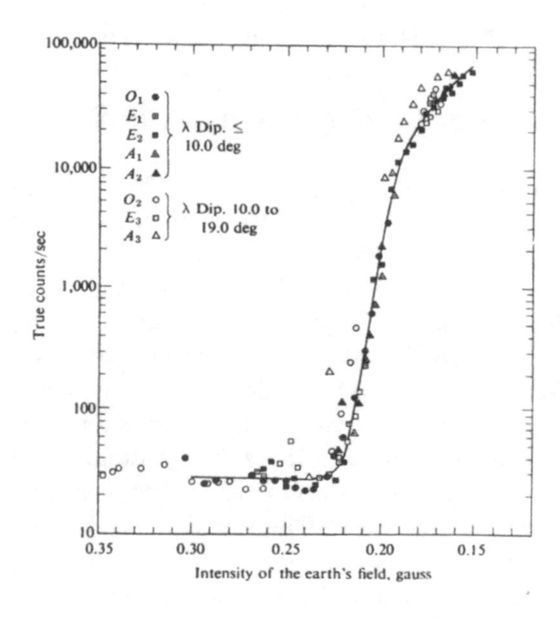

FIGURE 4.1 First particle data obtained in space by the Explorer I spacecraft. The top figure shows that the particle intensity increases rapidly with the altitude. The three sets of curves are obtained from three different geographic longitudes (0_1 and 0_2 over Asia; E_1, E_2 and E_3 over Africa; and A_1, A_2 and A_3 over South America). The bottom figure is the same data now plotted as a function of the magnetic field intensity. This plot organizes the data indicating the importance of the magnetic field. (From Yoshida *et al.*, 1960)

zones of maximum, one located between the ionosphere and about $2R_E$ and the other from about $3R_E$ to regions past $4R_E$. This double-peak distribution led to the concept of two radiation zones, the inner and the outer radiation belts. Subsequent measurements of lower-energy particles (> 40 keV), however, show that the radiation is "continuous" from ionospheric heights to the magnetopause (curve d).

4.2.2 Sources of the Inner and Outer Radiation Belts

Cosmic rays impinging on the atmosphere interact with atmospheric constituents and produce many secondary particles. These secondaries are a major source of the high-energy particles of the inner radiation belt. The height at which maximum secondaries are produced is roughly 70 millibars (this height corresponds to the mean interaction length of cosmic rays in Earth's atmosphere). The secondaries include neutrons that are uncharged, and therefore not affected by Earth's magnetic field. As mentioned earlier, the neutrons in transit decay into protons and electrons (and anti-neutrinos), and these particles are the primary source of the energetic component of the inner radiation belt.

The region beyond the inner radiation belt is known as the outer radiation belt. The outer radiation belt is bounded on the sunward side by a particle-trapping boundary that coincides roughly with the magnetopause position, which is nominally at $\sim 10R_E$. The trapping boundary on the dark hemisphere is closer to Earth by about 1–$2R_E$. The outer radiation belt particles have energies of a few eV to hundreds of keV. Both the solar wind and the ionosphere contribute to this population. The low-energy component most likely comes from the ionosphere. The higher-energy components are solar wind particles that have been accelerated and remnants of the plasma sheet particles that have been injected into the outer radiation belt during auroral disturbances (more on this later).

4.2.3 Plasma Sheet

The plasma sheet of Earth mapped to ionospheric heights occupies a region of the geomagnetic latitude from 68° to around 70° (these latitudes correspond to low- and high-latitude boundaries of the plasma sheet). These magnetic lines of force extend into the dark side of the magnetosphere, and the plasma sheet particles populate the geomagnetic tail (Figure 4.3). Both the ionospheric and the solar wind particles contribute to the plasma sheet. A typical plasma sheet electron energy is ≈ 1 keV and the proton energy is about 10 keV. The density of plasma sheet particles is $\approx 10^6$–$10^7 \mathrm{m}^{-3}$. Energetic electrons with energies > 40 keV also populate the geomagnetic

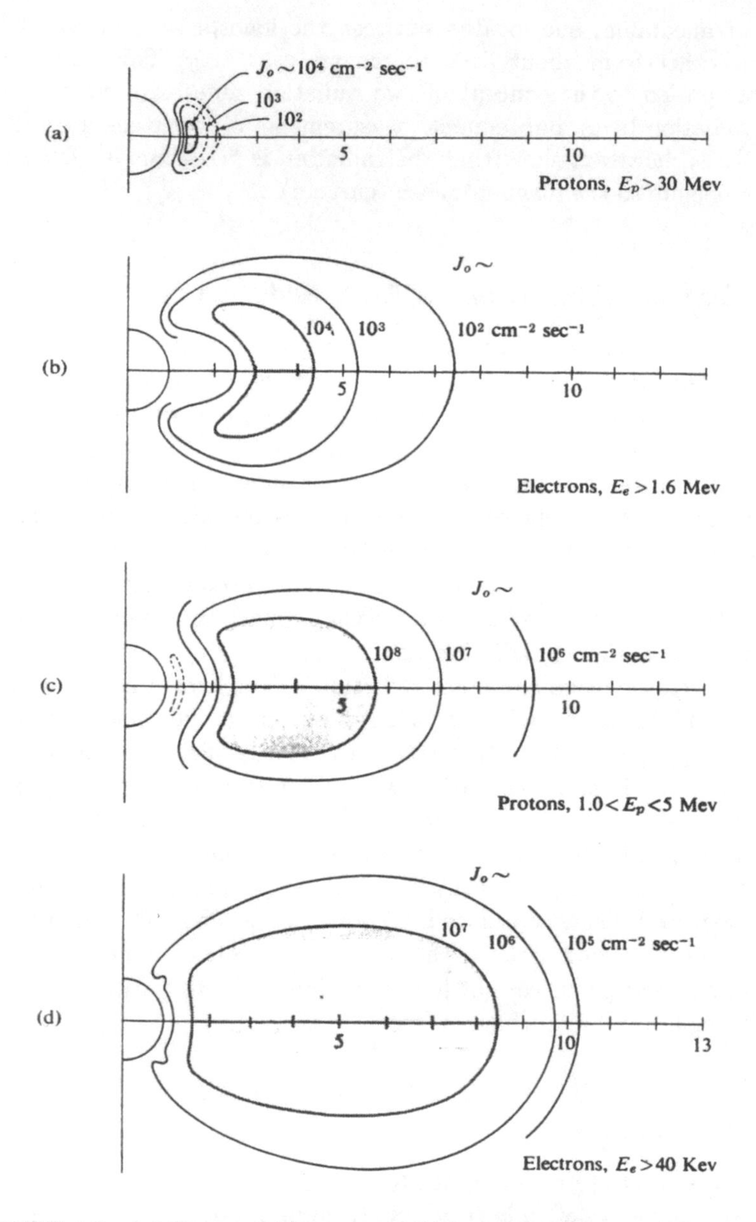

FIGURE 4.2 (a) The approximate spatial distribution of trapped protons of E > 30 MeV. (b) The approximate spatial distribution of trapped electrons of E > 1.6 MeV. (c) The approximate spatial distribution of protons of 0.1 < E < 5 MeV. (d) The approximate spatial distribution of trapped electrons of E > 40 keV. The fluxes at distance $\approx 5R_E$ vary considerably from day to day because of geomagnetic activity. (From Van Allen, 1968)

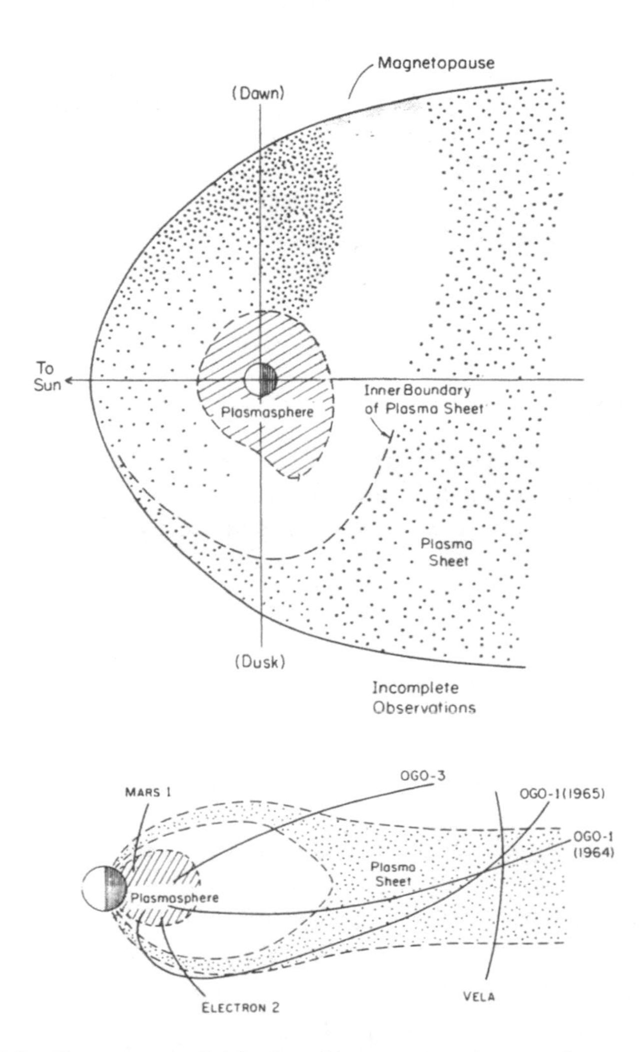

FIGURE 4.3 The average distribution of low-energy electrons in a meridian plane (a) and the equatorial plane (b) on the night side of the magnetosphere. Also shown schematically are the orbits of the principal spacecraft that have explored this region. (From Vasyliunas, 1972)

tail. A large fraction of the plasma sheet particle population is precipitated into the high-latitude ionosphere during geomagnetic activities, creating luminous auroral displays. This precipitation is also accompanied by an injection of plasma sheet particles into the outer radiation belt, increasing the intensity of the trapped radiation.

4.3 Lorentz Equation of Motion

Charged particles in space move in diverse electromagnetic field geometries. To gain an understanding of the behavior of these particles, we begin by examining the Lorentz force equation for particles in uniform electric and magnetic fields. This will provide important information on the basic motions of charged particles, which is needed later in the study of particles in complex field geometries. This exercise assumes that electric and magnetic fields are given (rather than self-consistently calculated).

4.3.1 Motion in a Uniform Magnetic Field

A single charged particle of mass m and velocity $\mathbf{v}$ in a magnetic field $\mathbf{B}$ in the absence of an electric field is examined to define some of the properties of a charged particle. The Lorentz equation of motion is

$$m\frac{d\mathbf{v}}{dt} = q(\mathbf{v} \times \mathbf{B}) \tag{4.1}$$

The force is perpendicular to $\mathbf{v}$ and therefore the magnetic force does no work on the particle. Take the dot product of (4.1) with $\mathbf{v}$ and noting that $\mathbf{v} \cdot (\mathbf{v} \times \mathbf{B}) = 0$, obtain

$$m\mathbf{v} \cdot \frac{d\mathbf{v}}{dt} = \frac{d}{dt}\left(\frac{1}{2}mv^2\right) = 0 \tag{4.2}$$

The kinetic energy of the particle is thus conserved. This result is quite general, and is valid for any arbitrary magnetic field.

A convenient way to study (4.1) is to separate the total velocity $\mathbf{v}$ into two principal directions, one parallel to and the other perpendicular to $\mathbf{B}$, $\mathbf{v} = \mathbf{v}_{\parallel} + \mathbf{v}_{\perp}$. (4.1) then becomes

$$\frac{d\mathbf{v}_{\parallel}}{dt} + \frac{d\mathbf{v}_{\perp}}{dt} = \frac{q}{m}(\mathbf{v}_{\perp} \times \mathbf{B}) \tag{4.3}$$

since $\mathbf{v}_{\parallel} \times \mathbf{B} = 0$. (4.3) splits into two equations, one describing the parallel and the other the perpendicular motion of the particles,

$$\begin{aligned} \frac{d\mathbf{v}_{\parallel}}{dt} &= 0 \\ \frac{d\mathbf{v}_{\perp}}{dt} &= \frac{q}{m}(\mathbf{v}_{\perp} \times \mathbf{B}) \end{aligned} \tag{4.4}$$

Gyrofrequency, Gyroradius, and Guiding Center Assume a planar static magnetic field given by $\mathbf{B} = (0, 0, B)$. For this simple geometry, the components of the second equation of (4.4) are

$$
\begin{aligned}
m\frac{dv_x}{dt} &= qBv_y \\
m\frac{dv_y}{dt} &= -qBv_x
\end{aligned}
\tag{4.5}
$$

To solve these coupled equations let $u = v_x + iv_y$ where $i^2 = -1$. Then (4.5) can be written as a single equation

$$
\frac{du}{dt} + i\omega_c u = 0 \tag{4.6}
$$

where $\omega_c = qB/m$ is defined as the gyrofrequency (note that ω_c is always positive and is also called the cyclotron and Larmor frequency). The solution of (4.6) is

$$
u = v_\perp e^{-i\omega_c t} \tag{4.7}
$$

$v_\perp$ is the magnitude of the velocity perpendicular to **B**, which is obtained from the relation

$$
v^2 = v_\parallel^2 + v_\perp^2 \tag{4.8}
$$

The gyroradius (also called the cyclotron and Larmor radius) of the particle is defined as

$$
r_c = \frac{v_\perp}{\omega_c} = \frac{mv_\perp}{qB} \tag{4.9}
$$

r_c is always positive and it represents the length of a constant vector rotating with a frequency ω_c in the xy plane.

The coordinates of the tip of this vector are obtained by solving (4.5), whose components are

$$
\begin{aligned}
\frac{dx}{dt} &= v_\perp \cos\omega_c t \\
\frac{dy}{dt} &= -v_\perp \sin\omega_c t
\end{aligned}
\tag{4.10}
$$

In writing (4.10), use was made of the relation $e^{-i\omega_c t} = \cos\omega_c t - i\sin\omega_c t$. Integration of (4.10) yields

$$
\begin{aligned}
x - x_0 &= \frac{v_\perp}{\omega_c}\sin(\omega_c t + \phi) \\
y - y_0 &= \frac{v_\perp}{\omega_c}\cos(\omega_c t + \phi)
\end{aligned}
\tag{4.11}
$$

These equations describe a particle moving in a circle in the xy-plane with $v_\perp/\omega_c$ as its radius. x_0, y_0, z_0, and ϕ are determined from the initial conditions. The complete trajectory of the particle must fold in the motion along the z-direction. Since $v_\parallel$ is constant according to (4.4), the z-motion is

$$z - z_0 = v_\parallel t \tag{4.12}$$

The trajectory of a particle in a uniform magnetic field along the z-axis is a helix with its axis parallel to **B**. If the motion of the particle is viewed by an observer moving along **B** with a velocity $v_\parallel$, the motion is a circle with its center at (x_0, y_0). The point $(x_0, y_0, z_0 + v_\parallel t)$ describes the locus of the center of the circle and is called a guiding center (Figure 4.4).

Pitch-angle The pitch-angle of a particle is defined as the angle between the velocity vector of the particle **v** and the magnetic field **B** and is usually denoted as α. The pitch-angle α is defined by

$$\alpha = \tan^{-1} \frac{v_\perp}{v_\parallel} \tag{4.13}$$

Note that $v_\parallel = v \cos\alpha$ and $v_\perp = v \sin\alpha$ where v, $v_\parallel$ and $v_\perp$ are magnitudes of $\mathbf{v}$, $\mathbf{v}_\parallel$ and $\mathbf{v}_\perp$. When $\alpha = \pi/2$, $v_\parallel = 0$ and when $\alpha = 0$, $v_\perp = 0$ (Figure 4.5).

Magnetic Moment In a coordinate system moving with $v_\parallel$, the particle motion is totally in a plane perpendicular to **B**. This perpendicular motion

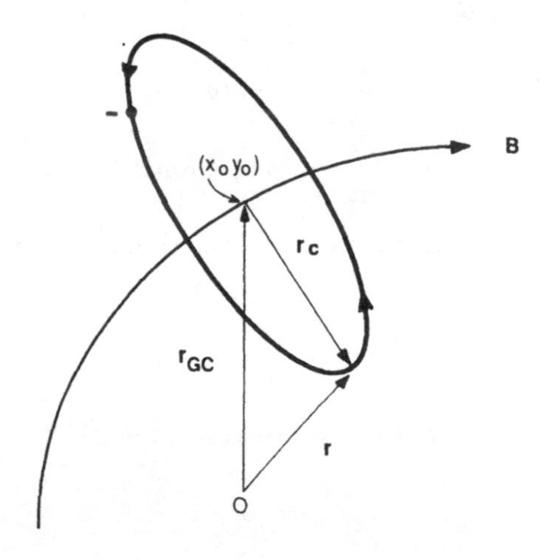

FIGURE 4.4 Definition of orbit parameters for a negative particle gyrating about a magnetic line of force.

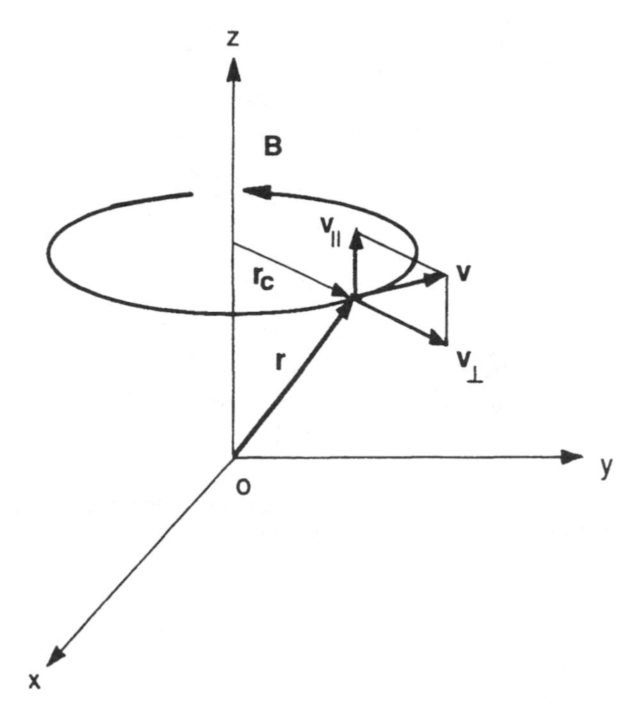

FIGURE 4.5 Definition of $v_{\|}$ and $v_{\perp}$.

gives rise to a circulating current

$$I = \frac{q\omega_c}{2\pi} = \frac{1}{2\pi}\frac{q^2 B}{m} \tag{4.14}$$

The magnetic moment of this "ring current" is

$$\mu = I\pi r_c^2 = \frac{1}{2}\frac{q^2 r_c^2 B}{m} \tag{4.15}$$

The magnetic moment can also be written in terms of the total magnetic flux within the circular path

$$\mu = \frac{1}{2\pi}\frac{q^2}{m}\Phi \tag{4.16}$$

where $\Phi = \pi r_c^2 B$. The gyrating particles generate circulating currents which in turn generate magnetic fields. Figure 4.6 shows that the direction of the magnetic field generated by positive and negative particles is the same. This comes about because positive and negative particles gyrate in opposite directions and currents are defined as positive in the direction of positive particles' velocity (against the negative particles' velocity). Inside r_c, the generated field opposes the applied **B** while outside r_c the field

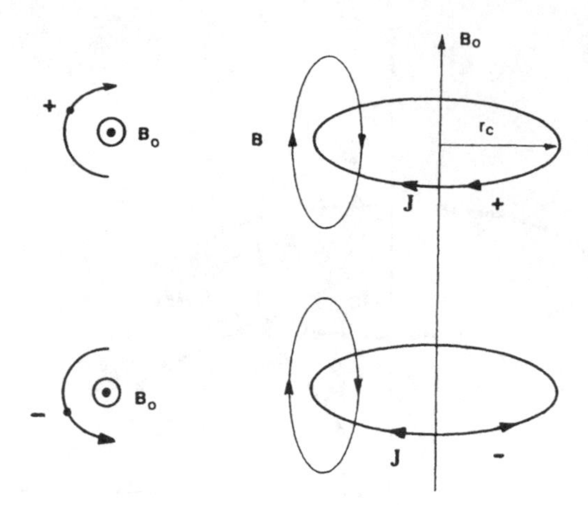

FIGURE 4.6 $+q$ and $-q$ particles gyrating to produce ring currents. **B** represents the magnetic field of the ring current. $\mathbf{B}_0$ is the ambient field.

is in the same direction as **B**. This gives rise to diamagnetic effects (to be discussed further).

Another useful expression for μ is

$$\mu = \frac{1}{2}\frac{mv_\perp^2}{B} \tag{4.17}$$

where the magnetic moment is defined in terms of the kinetic energy of the particles moving in the direction perpendicular to **B**. In deriving (4.17), (4.9) and (4.15) were used. The magnetic moment can be written as a vector quantity

$$\boldsymbol{\mu} = -\frac{\varepsilon_\perp}{B^2}\mathbf{B} \tag{4.18}$$

where $\varepsilon_\perp$ is $mv_\perp^2/2$. The $-$ sign is needed to indicate that the magnetic moment is in the opposite direction of the applied magnetic field. Equation (4.18) can be generalized when there are many particles. Let n be the number of particles per unit volume. Then we define the magnetic moment per unit volume as

$$\mathbf{M} = n\boldsymbol{\mu} = -\frac{n\varepsilon_\perp}{B^2}\mathbf{B} \tag{4.19}$$

The significance of (4.19) will become clearer when diamagnetic properties of plasma are discussed in later chapters. Diamagnetism arises from diamagnetic currents that are given by $\nabla \times \mathbf{M}$ (Chapter 7).

4.3.2 Motion in a Uniform Electric Field

The motion of a charged particle in a uniform electric and magnetic field is now examined. Let $\mathbf{E} = \mathbf{E}_{\parallel} + \mathbf{E}_{\perp}$ where $\parallel$ and $\perp$ again refer to directions parallel and perpendicular to $\mathbf{B}$. Then the equations of motion in the two principal directions are

$$m\frac{d\mathbf{v}_{\parallel}}{dt} = q\mathbf{E}_{\parallel}$$

$$m\frac{d\mathbf{v}_{\perp}}{dt} = q(\mathbf{E}_{\perp} + \mathbf{v}_{\perp} \times \mathbf{B}) \tag{4.20}$$

Parallel Electric Field The solutions of (4.20) in the direction parallel to $\mathbf{B}$ are

$$v_{\parallel} = \left(\frac{qE_{\parallel}}{m}\right) t + v_{0\parallel} \tag{4.21}$$

and

$$z = \frac{(qE_{\parallel}/m)t^2}{2} + v_{0\parallel} t + z_0 \tag{4.22}$$

where $v_{0\parallel}$ and z_0 are initial values. These solutions indicate that the particles are accelerated along the lines of force $\mathbf{B}$. When collisions are ignored, the action of $E_{\parallel}$ will increase $\mathbf{v}_{\parallel}$ to very large values. This leads to the phenomenon of "runaway" particles.

Perpendicular Electric Field We now solve (4.20) for a special case of static electric and magnetic field geometry. Consider for simplicity the orthogonal geometry, $\mathbf{B} = (0, 0, B)$ and $\mathbf{E} = (E_x, 0, 0)$. In this case, the components of the Lorentz equations are

$$\begin{aligned}\frac{dv_x}{dt} &= \omega_c v_y + \frac{\omega_c E_x}{B}\\ \frac{dv_y}{dt} &= -\omega_c v_x\\ \frac{dv_z}{dt} &= 0\end{aligned} \tag{4.23}$$

where ω_c is the cyclotron frequency qB/m. Differentiate the second equation and use the first equation to obtain

$$\frac{d^2 v_y}{dt^2} + \omega_c^2 v_y = -\frac{\omega_c^2 E_x}{B} \tag{4.24}$$

This equation is a second-order differential equation in v_y (note the right-hand side is not zero). The solution of this equation is obtained by transforming it into a new frame of reference,

$$v_y = u - \frac{E_x}{B} \tag{4.25}$$

Substitution of this new variable into the above equation reduces it to

$$\frac{d^2u}{dt^2} + \omega_c^2 u = 0 \tag{4.26}$$

whose well-known solution is

$$u = A\cos\omega_c t + C\sin\omega_c t \tag{4.27}$$

The complete solution of (4.24) is

$$v_y = A\cos\omega_c t + C\sin\omega_c t - \frac{E_x}{B} \tag{4.28}$$

A and C are constants of integration determined from initial conditions. Differentiate this equation and use the second equation of (4.23) to obtain a solution for v_x. Let the initial velocities be $\mathbf{v}(t = 0) = (v_{xo}, v_{yo}, v_{zo})$. Then, the complete solution of (4.23) is

$$\begin{aligned} v_x &= v_{xo}\cos\omega_c t + \left(v_{yo} + \frac{E_x}{B}\right)\sin\omega_c t \\ v_y &= \left(v_{yo} + \frac{E_x}{B}\right)\cos\omega_c t - v_{xo}\sin\omega_c t - \frac{E_x}{B} \\ v_z &= v_{zo} \end{aligned} \tag{4.29}$$

If $E_x = 0$, the above solutions reduce to the solution of a particle in the static magnetic field that we studied earlier, where we found that a particle simply gyrates around the magnetic field. In the presence of a constant static electric field perpendicular to **B**, the gyration solution is modified by the term E_x/B (4.25). A particle now drifts in the y-direction (which is perpendicular to both **E** and **B**) while it is gyrating around **B**. In the frame moving with this drift, one sees only the gyration motion.

We now can study a more general problem when $\mathbf{E} = (E_x, E_y, E_z)$ and $\mathbf{B} = (0, 0, B)$. We proceed in the same manner as above and find that in this case the solutions of motion in the perpendicular direction are obtained by a change of variables from $\mathbf{v}$ to

$$\mathbf{v} = \mathbf{u} + \frac{\mathbf{E} \times \mathbf{B}}{B^2} \tag{4.30}$$

where $\mathbf{u}$ satisfies the equation $d\mathbf{u}/dt = (q/m)\mathbf{u} \times \mathbf{B}$. The general solution now consists of motions in the directions parallel and perpendicular to **B**.

For the parallel motion, we simply recover (4.22). For the perpendicular motion, the particle drifts in the direction perpendicular to both **E** and **B**($\mathbf{E} \times \mathbf{B}/B^2$). (Compare the results obtained here with the results of the Lorentz transformation studied in Chapter 2.)

This $\mathbf{E} \times \mathbf{B}$ drift (pronounced "e cross b drift") is an example of many classes of guiding center drifts that arise when a constant force is applied perpendicular to **B**. For the example studied, the perpendicular force is the electric field and this force gives rise to a drift perpendicular to the applied force and the magnetic field. It is important to understand the physics of these drifts since natural particles are often in an environment where there are constant forces perpendicular to **B**.

4.4 Guiding Center Drift Equation

The above is an example of problems that involve forces acting on particles in the direction perpendicular to the magnetic field. In the special case of a constant electrostatic field, the Lorentz equation yields an exact analytical solution. However, there are other types of forces that do not yield exact analytical solutions. In these problems, we must solve the equations by approximation.

Let a particle in a static magnetic field be subject to an arbitrary constant force **F**. The particle motion is then described by the equation

$$\frac{d\mathbf{v}}{dt} = \frac{q}{m}(\mathbf{v} \times \mathbf{B}) + \frac{\mathbf{F}}{m} \tag{4.31}$$

As before, separate this equation into components parallel and perpendicular to **B**.

$$\frac{d\mathbf{v}_{\parallel}}{dt} = \frac{\mathbf{F}_{\parallel}}{m}$$

$$\frac{d\mathbf{v}_{\perp}}{dt} = \frac{\mathbf{F}_{\perp}}{m} + \frac{q}{m}(\mathbf{v}_{\perp} \times \mathbf{B}) \tag{4.32}$$

According to (4.32) the particle is accelerated along **B** by the constant force $\mathbf{F}_{\parallel}$. The effect of constant $\mathbf{F}_{\perp}$ on the perpendicular motion is illustrated in Figure 4.7. As the particle gyrates around **B**, $\mathbf{F}_{\perp}$ is along $\mathbf{v}_{\perp}$ during half of the particle's orbit, and the particle here will therefore be accelerated. $\mathbf{F}_{\perp}$ decelerates the particle during the other half of the orbit. Since the magnetic field is constant, the gyroradius must change. The gyroradius will increase when the particles are accelerated and will decrease when decelerated. Such changes of gyroradius will create constant particle drifts perpendicular to **B**.

Let us derive an expression for this drift velocity. Define $\mathbf{W}_D$ as the constant drift velocity and let $\mathbf{v}_{\perp} = \mathbf{u} + \mathbf{W}_D$. Then, the equation of the

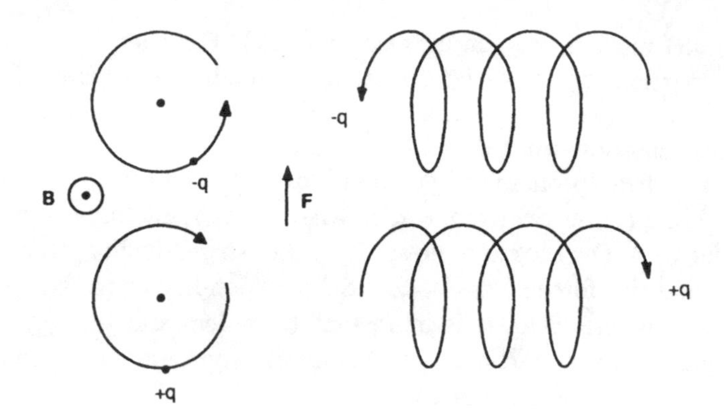

FIGURE 4.7 Particles acted on by a constant $\mathbf{F}_\perp$ result in a drift motion perpendicular to $\mathbf{B}$ and $\mathbf{F}_\perp$.

perpendicular motion (4.32) becomes

$$\frac{d\mathbf{u}}{dt} = \frac{q}{m}\mathbf{W}_D \times \mathbf{B} + \frac{q}{m}\mathbf{u} \times \mathbf{B} + \frac{\mathbf{F}_\perp}{m} \tag{4.33}$$

since $d\mathbf{W}_D/dt = 0$. The idea is now to transform (4.33) to a coordinate frame moving with $\mathbf{W}_D$ so that in this frame the particle motion is described by a purely cyclotron motion

$$\frac{d\mathbf{u}}{dt} = \frac{q}{m}\mathbf{u} \times \mathbf{B} \tag{4.34}$$

We see from (4.33) that this requires $\mathbf{W}_D$ to satisfy the equation

$$q\mathbf{W}_D \times \mathbf{B} + \mathbf{F}_\perp = 0 \tag{4.35}$$

Take the cross product of (4.35) with $\mathbf{B}$ and use the vector identity $(\mathbf{a} \times \mathbf{b}) \times \mathbf{b} = (\mathbf{a} \cdot \mathbf{b})\mathbf{b} - b^2\mathbf{a}$. Since $\mathbf{W}_D$ is perpendicular to $\mathbf{B}$, the solution of the above equation is

$$\mathbf{W}_D = \frac{1}{q}\frac{\mathbf{F} \times \mathbf{B}}{B^2} \tag{4.36}$$

The results just derived can be summarized as follows. The perpendicular motion of a charged particle in a uniform magnetic field $\mathbf{B}$ under the influence of a constant external force $\mathbf{F}$ is described by a superposition of a gyration around a guiding center with the cyclotron frequency and the constant drift motion of this guiding center. The guiding center drift obeys (4.36) and depends on the form of $\mathbf{F}$. Note that the guiding center is a fictitious point and its motion does not obey the laws of mechanics (more on this later). If we now include the parallel motion (4.32), the guiding center is also accelerated in the direction parallel to $\mathbf{B}$. The parallel motion behaves as if the magnetic field is not present.

The concept of the guiding center is useful because it allows us to study complicated motions without actually solving the exact particle motions. One studies instead the guiding center motion. In a frame of reference moving with the guiding center, we know that the particles are near it and gyrating about it. In particle motions involving distances much larger than the gyroradius (as is the case in space), it is more instructive to gain knowledge about the guiding center than the details of the exact motion.

4.4.1 Definition of the Guiding Center

The concept of the guiding center motion for the special case of a homogeneous magnetic field was introduced in Section 4.3. We now expand the picture to a more general case.

A guiding center is a point defined by

$$\mathbf{r}_{GC} = \mathbf{r} - \boldsymbol{\rho} \tag{4.37}$$

where $\mathbf{r}$ is the position vector of the particle and $\boldsymbol{\rho}$ is the radius of curvature which is a vector from the position of the particle to the center of gyration. $\boldsymbol{\rho}$ can be expressed in terms of the momentum of the particle. In the plane perpendicular to $\mathbf{B}$, we have

$$m\omega_c^2 \boldsymbol{\rho} = q\mathbf{v} \times \mathbf{B} \tag{4.38}$$

Since $\omega_c^2 = q^2B^2/m^2$, we can write (4.38) as

$$\boldsymbol{\rho} = \frac{\mathbf{p} \times \mathbf{B}}{qB^2} \tag{4.39}$$

where $\mathbf{p}$ is the momentum of the particle. (4.37) then becomes

$$\mathbf{r}_{GC} = \mathbf{r} - \frac{\mathbf{p} \times \mathbf{B}}{qB^2} \tag{4.40}$$

Note that if the particle is gyrating about a homogeneous magnetic field, $\mathbf{r}_{GC}$ coincides with the center of gyration (Figure 4.4).

To obtain a better physical picture of the guiding center motion, consider a particle in a homogeneous $\mathbf{B}$ that experiences an impact. During a small time interval $\Delta t \ll T_c$, (T_c is the cyclotron period) the particle will experience a large force $\mathbf{f}$. Assume that this force in the direction perpendicular to $\mathbf{B}$ is such that $\mathbf{f} \gg | q\mathbf{v} \times \mathbf{B} |$. Upon impact, the momentum of the particle will change from $\mathbf{p}$ to $\mathbf{p}' = \mathbf{p} + \Delta\mathbf{p}$ where $\Delta\mathbf{p} = \int \mathbf{f}\,dt$ and the limits of this integral are from t to $t + \Delta t$. During the brief impact, $\mathbf{r}$ does not change very much, but $\mathbf{p}$ does. Hence, from (4.40) we note that the guiding center position $\mathbf{r}_{GC}$ must change.

$$\Delta\mathbf{r}_{GC} \approx \frac{\Delta\mathbf{p} \times \mathbf{B}}{qB^2} \tag{4.41}$$

This result can be generalized to a continuous force $\mathbf{f}$. In this case, the guiding center position must change continuously. From (4.37), obtain

$$\frac{d\mathbf{r}_{GC}}{dt} = \frac{d\mathbf{r}}{dt} + \frac{(d\mathbf{p}/dt \times \mathbf{B})}{qB^2} \tag{4.42}$$

Now, $d\mathbf{p}/dt = q\mathbf{v} \times \mathbf{B} + \mathbf{F}$ where $\mathbf{F}$ is a non-magnetic force. Equation (4.42) then becomes

$$\begin{aligned}\frac{d\mathbf{r}_{GC}}{dt} &= \mathbf{v} + \frac{(\mathbf{F} + q\mathbf{v} \times \mathbf{B}) \times \mathbf{B}}{qB^2} \\ &= \mathbf{v} + \frac{[\mathbf{F} \times \mathbf{B} + q(\mathbf{v} \times \mathbf{B}) \times \mathbf{B}]}{qB^2} \\ &= \mathbf{v}_{\parallel} + \frac{\mathbf{F} \times \mathbf{B}}{qB^2}\end{aligned} \tag{4.43}$$

where the last step follows from the expansion of the vector triple product; note that $\mathbf{v} = \mathbf{v}_{\parallel} + \mathbf{v}_{\perp}$. Hence if the force is continuous, the guiding center motion can be viewed as a continuous series of small impacts with each impact given approximately by (4.41). Equation (4.43) is the total velocity of the guiding center with $\mathbf{F}_{\parallel} = 0$.

4.4.2 Examples of Guiding Center Drifts

As examples of guiding center drifts, consider the drifts of a particle due to the gravitational and electric fields. The electric field in space can be static or time-dependent. These two types of electric fields give rise to two completely different drifts.

Gravity Field A particle in a gravity field experiences a gravitational force. The component of the gravity perpendicular to $\mathbf{B}$ gives rise to

$$\mathbf{W}_g = \frac{m}{q}\frac{\mathbf{g} \times \mathbf{B}}{B^2} \tag{4.44}$$

This drift is charge-dependent and positive and negative particles drift in the opposite direction. Hence this guiding center drift gives rise to a current. Note also that $\mathbf{W}_g$ is mass-dependent and for a given gravitational field, heavier masses drift faster. This result is contrary to intuition. This peculiar feature corroborates the statement made earlier that the guiding center motion does not always obey the laws of mechanics.

Static Electric Field A charged particle in a uniform crossed electric and magnetic field is the case under consideration ($\mathbf{E} \cdot \mathbf{B} = 0$). As shown in Section 4.3.2, this problem can be solved exactly by solving the Lorentz equation. According to the guiding center formulation, the external force here is $\mathbf{F} = q\mathbf{E}$. This means the particle motion consists of a superposition of a gyromotion and a drift of the guiding center, whose magnitude and direction are given by

$$\mathbf{W}_E = \frac{\mathbf{E} \times \mathbf{B}}{B^2} \tag{4.45}$$

This drift is charge-, mass-, and energy-independent. Apparently all particles drift identically in such an electric field.

This problem is examined from another point of view to get a better understanding of this drift. The Lorentz force of a particle is

$$\mathbf{F} = q\mathbf{E} + q\mathbf{v} \times \mathbf{B} \tag{4.46}$$

Suppose the particle velocity is given by

$$\mathbf{v} = \frac{\mathbf{E} \times \mathbf{B}}{B^2} \tag{4.47}$$

Equation (4.46) then becomes

$$\begin{aligned} \mathbf{F} &= q\mathbf{E} + q\frac{\mathbf{E} \times \mathbf{B}}{B^2} \times \mathbf{B} \\ &= q\mathbf{E} + q\frac{[(\mathbf{E} \cdot \mathbf{B})\mathbf{B} - B^2\mathbf{E}]}{B^2} = 0 \end{aligned} \tag{4.48}$$

where we assume that $\mathbf{E} \cdot \mathbf{B} = 0$. Equation (4.48) shows that in crossed electric and magnetic fields, a particle experiences no net force if it drifts with a velocity $\mathbf{v} = \mathbf{E} \times \mathbf{B}/B^2$. Here, the electric and magnetic forces are balanced. The concept of $\mathbf{E} \times \mathbf{B}/B^2$ drift is also important when the motion of many particles in conducting fluids is considered. In this case, the electric field vanishes in the fluid frame if a fluid consisting of "cold" particles (zero thermal energy) moves with a velocity $\mathbf{E} \times \mathbf{B}/B^2$. These results are consistent with the theory of Lorentz transformation for $v \ll c$.

Time-Dependent E: Polarization Drift Let $\mathbf{E}$ be perpendicular to $\mathbf{B}(\mathbf{E} \cdot \mathbf{B} = 0)$, uniform in space, but slowly varying in time. The Lorentz equation must now allow for the time variation of the electric field

$$\frac{d\mathbf{v}}{dt} = \frac{q\mathbf{E}(t)}{m} + \frac{q}{m}(\mathbf{v} \times \mathbf{B}) \tag{4.49}$$

If as before we let $\mathbf{v} = \mathbf{u} + \mathbf{W}_E$ where $\mathbf{W}_E = \mathbf{E} \times \mathbf{B}/B^2$, the Lorentz equation of motion will yield

$$m\frac{d\mathbf{u}}{dt} + m\frac{d\mathbf{W}_E}{dt} = q\mathbf{u} \times \mathbf{B} \tag{4.50}$$

since $\mathbf{W}_E$ is time-dependent and $q(\mathbf{E} \times \mathbf{B}) \times \mathbf{B}/B^2 = -q\mathbf{E}$ ($\mathbf{B}$ is static here).

Now let $\mathbf{u} = \mathbf{u}_1 + \mathbf{W}_p$ where $\mathbf{W}_p$ is the polarization drift. Our objective is to obtain an expression for $\mathbf{W}_p$.

Substitution of $\mathbf{u} = \mathbf{u}_1 + \mathbf{W}_p$ into (4.50) yields

$$m\frac{d\mathbf{u}_1}{dt} + m\frac{d\mathbf{W}_p}{dt} + m\frac{d\mathbf{W}_E}{dt} = q\mathbf{u}_1 \times \mathbf{B} + q\mathbf{W}_p \times \mathbf{B} \tag{4.51}$$

Let

$$m\frac{d\mathbf{W}_E}{dt} = q\mathbf{W}_p \times \mathbf{B} \tag{4.52}$$

and if this equation is crossed with $\mathbf{B}$ assuming that $\mathbf{W}_p \cdot \mathbf{B} = 0$, we obtain

$$\mathbf{W}_p = -\frac{m}{qB^2}\frac{d\mathbf{W}_E}{dt} \times \mathbf{B} \tag{4.53}$$

Use of (4.52) reduces (4.51) to

$$m\frac{d\mathbf{u}_1}{dt} + m\frac{d\mathbf{W}_p}{dt} = q\mathbf{u}_1 \times \mathbf{B} \tag{4.54}$$

If $md\mathbf{W}_p/dt$ in (4.54) can be neglected, we obtain the usual cyclotron motion for $\mathbf{u}_1$. Then the total motion of the particle will be the circular motion $\mathbf{u}_1$ superimposed by two drifts

$$\mathbf{v} = \mathbf{u}_1 + \mathbf{W}_E + \mathbf{W}_p \tag{4.55}$$

where $\mathbf{W}_E$ is the same as (4.45) and $\mathbf{W}_p$, an additional drift, called the polarization drift arises when the electric field is time-dependent. Let us now find the conditions under which $m(d\mathbf{W}_p/dt)$ can be ignored.

Compare the relative magnitudes of the second term on the left-hand side of (4.54) against the term on the right-hand side. This yields

$$\begin{aligned}\frac{|md\mathbf{W}_p/dt|}{|q\mathbf{u}_1 \times \mathbf{B}|} &= \frac{m^2}{q^2B^2}\frac{|d^2\mathbf{W}_E/dt^2 \times \mathbf{B}|}{|\mathbf{u}_1 \times \mathbf{B}|} \\ &= \frac{\nu^2}{\omega_c^2}\frac{|\mathbf{W}_E|}{|\mathbf{u}_1|}\end{aligned} \tag{4.56}$$

where ω_c is the cyclotron frequency, d/dt has been replaced by ν which is the characteristic frequency of $\mathbf{E}$, and we used (4.53). Equation (4.56) requires that

$$\left(\frac{|\mathbf{W}_E|}{|\mathbf{u}_1|}\right)\left(\frac{\nu^2}{\omega_c^2}\right) \ll 1 \tag{4.57}$$

if $m(d\mathbf{W}_p/dt)$ is to be ignored. This is usually satisfied by particles in magnetospheres when $\mathbf{W}_E/\mathbf{u}_1$ is small and when the electric field is slowly varying in time. Note that the polarization drift is in the direction of $\mathbf{E}$ and is mass- and charge-dependent. Ions and electrons drift in opposite directions and this gives rise to polarization currents.

4.5 Motion in an Inhomogeneous Magnetic Field

Consider now a particle in a slightly inhomogeneous magnetic field. By slightly, we mean that the variation of the magnetic field is much smaller than the value of the average magnetic field intensity inside the particle's orbit. In this case, the magnetic field at the particle can be expanded in a Taylor's series,

$$\mathbf{B}(\mathbf{r}) = \mathbf{B}_0 + (\mathbf{r} \cdot \nabla_0)\mathbf{B} + \ldots \tag{4.58}$$

where $\mathbf{B}_0$ is the magnetic field at the guiding center and $\mathbf{r}$ is the vector from the guiding center to the instantaneous position of the particle. ∇_0 in (4.58) means that the operation is to be performed relative to 0 which is the guiding center. In (4.58), we ignore higher-order terms and require further that

$$| \mathbf{B}_0 | \gg | (\mathbf{r} \cdot \nabla_0)\mathbf{B} | \tag{4.59}$$

(4.59) requires that $\mathbf{B}$ at $\mathbf{r}$ differ only slightly from $\mathbf{B}_0$ at the guiding center. Hence, the particle orbit is nearly a circle, but it no longer closes on itself because of the small magnetic inhomogeneity (Figure 4.8).

The equation of the particle motion (4.1) in an inhomogeneous field with $\mathbf{B}$ given by (4.58) becomes

$$\frac{d\mathbf{v}}{dt} = \frac{q}{m}(\mathbf{v} \times \mathbf{B}_0 + \mathbf{v} \times (\mathbf{r} \cdot \nabla_0)\mathbf{B}) \tag{4.60}$$

where the first term on the right represents the motion in homogeneous $\mathbf{B}_0$ and the second term, which can be viewed as a correction term, arises from the perturbation of the orbit due to the inhomogeneous $\mathbf{B}$.

Let us designate the unperturbed orbit as zeroth order and the perturbed as first order and we ignore terms higher than the first order. The second term in (4.60) is clearly higher than first order since it involves the product of first-order terms. In order to reduce this term to first order, we

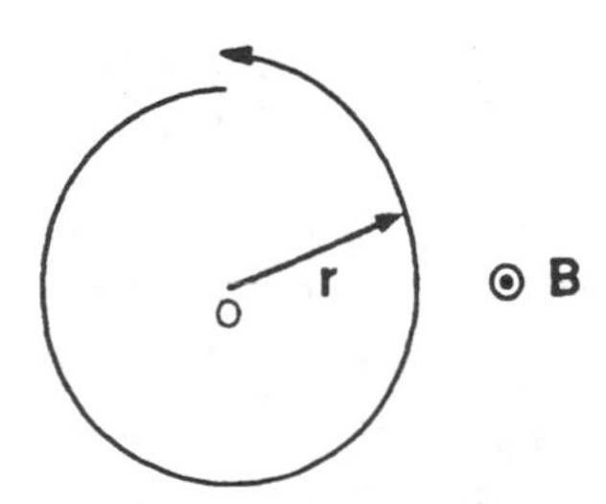

FIGURE 4.8 Particle orbit in a slightly inhomogeneous magnetic field.

replace $\mathbf{v}$ with $\mathbf{v}_0$ (which is a solution with $\mathbf{B}_0$), and $\mathbf{r}$ with $\mathbf{r}_c$. Then (4.60) becomes, to first order,

$$\frac{d\mathbf{v}}{dt} = \frac{q}{m}[\mathbf{v} \times \mathbf{B}_0 + \mathbf{v}_0 \times (\mathbf{r}_c \cdot \nabla_0)\mathbf{B}] \tag{4.61}$$

Equation (4.61) is still complex but it can now be treated by using the guiding center concept. The second term in (4.61) corresponds to the external force in (4.31). However, this force is not constant because $\mathbf{r}_c$ is rotating. Therefore, the force must be averaged over a gyration orbit. Thus,

$$\mathbf{F} = < q\mathbf{v}_0 \times (\mathbf{r}_c \cdot \nabla_0)\mathbf{B} > \tag{4.62}$$

where the symbol $< >$ means average. The perturbed force involves the gradient of the magnetic field evaluated at the guiding center.

We have already shown in Chapter 3 (see (3.50)) that the most general form of the magnetic inhomogeneity can be written as

$$\nabla\mathbf{B} = \begin{pmatrix} \partial B_x/\partial x & \partial B_x/\partial y & \partial B_x/\partial z \\ \partial B_y/\partial x & \partial B_y/\partial y & \partial B_y/\partial z \\ \partial B_z/\partial x & \partial B_z/\partial y & \partial B_z/\partial z \end{pmatrix} \tag{4.63}$$

With the condition that $\nabla \cdot \mathbf{B} = 0$, the sum of the diagonal terms must vanish and there remain eight independent components. We now examine how each of these magnetic inhomogeneities affects the particle motion.

4.5.1 Force Along Converging and Diverging Magnetic Fields

Let us begin with the diagonal terms, $\partial B_x/\partial x$, $\partial B_y/\partial y$ and $\partial B_z/\partial z$. We assume the magnetic field is mainly in the z-direction. Hence we can write the local magnetic field as

$$\begin{aligned} B_x &= \left(\frac{\partial B_x}{\partial x}\right)_0 x \\ B_y &= \left(\frac{\partial B_y}{\partial y}\right)_0 y \\ B_z &= B_0 + \left(\frac{\partial B_z}{\partial z}\right)_0 z \end{aligned} \tag{4.64}$$

These equations show that the originally straight magnetic field B_0 in the z-direction will now diverge (converge). For example, if $\partial B_x/\partial x$ is positive, the magnetic field lines in the xz-plane diverge as shown in Figure 4.9.

The component of the Lorentz force along z is given by

$$F_z = q(v_x B_y - v_y B_x) \tag{4.65}$$

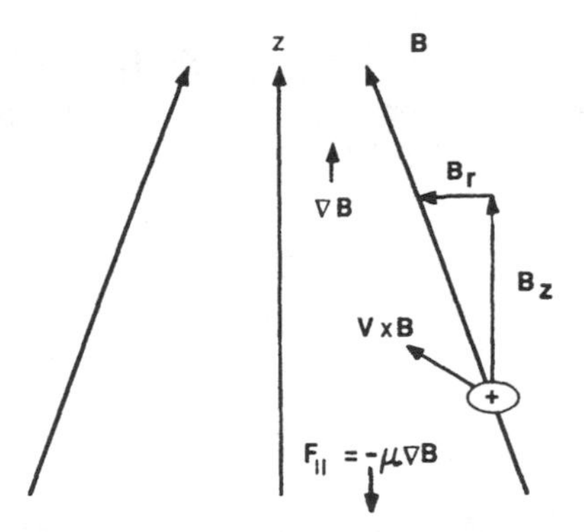

FIGURE 4.9 Force on the guiding center in converging or diverging lines of force. Particle here has a + charge.

From (4.64), this can be written as

$$F_z = q\left[v_x\left(\frac{\partial B_y}{\partial y}\right)_0 y - v_y\left(\frac{\partial B_x}{\partial x}\right)_0 x\right] \tag{4.66}$$

Introduce now polar coordinates $x = r\cos\theta$ and $y = r\sin\theta$. Then

$$\begin{aligned} \frac{dx}{dt} &= \frac{dr}{dt}\cos\theta - r\sin\theta\frac{d\theta}{dt} \\ \frac{dy}{dt} &= \frac{dr}{dt}\sin\theta + r\cos\theta\frac{d\theta}{dt} \end{aligned} \tag{4.67}$$

Substituting these equations into (4.66) yields

$$\begin{aligned} F_z &= q\left[\left(\frac{dr}{dt}\cos\theta - r\sin\theta\frac{d\theta}{dt}\right)\frac{\partial B_y}{\partial y}r\sin\theta\right. \\ &\quad \left. - \left(\frac{dr}{dt}\sin\theta + r\cos\theta\frac{d\theta}{dt}\right)\frac{\partial B_x}{\partial x}r\cos\theta\right] \end{aligned} \tag{4.68}$$

Now average this equation over θ and obtain

$$< F_z > = -\frac{q}{2}r^2\frac{d\theta}{dt}\left[\left(\frac{\partial B_x}{\partial x}\right)_0 + \left(\frac{\partial B_y}{\partial y}\right)_0\right] \tag{4.69}$$

where $< \cos^2\theta > = < \sin^2\theta > = 1/2$ and $< \cos\theta\sin\theta > = 0$. This equation can be rewritten as

$$< F_z > = \frac{q}{2}rv_\theta\left(\frac{\partial B_z}{\partial z}\right)_0 \tag{4.70}$$

since $rd\theta/dt = v_\theta$ and $\nabla\cdot\mathbf{B} = 0$. A more convenient form is to write this equation noting that $v_\theta = \mp v_\perp$. The sense of v_θ is defined by a right-hand rule and it is positive in the counterclockwise direction. Electrons and ions rotate clockwise and counterclockwise, respectively. Therefore, the negative sign is for positive particles, the positive sign is for electrons, and $r \approx |r_c|$. Then for a positive particle, (4.70) becomes

$$\begin{aligned} <F_z> &\approx -\frac{q}{2}v_\perp r_c \frac{\partial B_z}{\partial z} \\ &\approx -\frac{mv_\perp^2}{2B}\left(\frac{\partial B}{\partial z}\right) \\ &\approx -\mu\frac{\partial B}{\partial s} \end{aligned} \tag{4.71}$$

where s is along $\mathbf{B}$. In general, the force on a particle along $\mathbf{B}$ can be written as

$$\mathbf{F}_\parallel = \frac{m d\mathbf{v}_\parallel}{dt} = -\mu\frac{\partial B}{\partial s} = -\mu\nabla_\parallel B \tag{4.72}$$

This parallel force is always acting against the particle travelling along a converging magnetic field geometry. Eventually, a particle may be reflected by this force.

4.5.2 Curvature Drift of the Guiding Center

We showed in Chapter 3 that the terms $\partial B_x/\partial z$ and $\partial B_y/\partial z$ give rise to curved field lines. The magnetic field lines in the xz-plane for positive $\partial B_x/\partial z$ are shown in Figure 4.10. A particle travelling on such curved field lines experiences a centrifugal force and this gives rise to a guiding center drift. If $v_\parallel^2$ represents the square of the average velocity of particles along the direction of $\mathbf{B}$, the centrifugal force is

$$\mathbf{F}_c = mv_\parallel^2\frac{\boldsymbol{\rho}}{\rho^2} \tag{4.73}$$

where $\boldsymbol{\rho}$ is the radius of curvature. $\mathbf{F}_c$ is perpendicular to $\mathbf{B}$ and the guiding center drift due to $\mathbf{F}_c$ is

$$\mathbf{W}_c = \frac{mv_\parallel^2}{qB^2}\frac{\boldsymbol{\rho}\times\mathbf{B}}{\rho^2} \tag{4.74}$$

Consider now an arc element ds along $\mathbf{B}$ and define the unit vectors $(\mathbf{e}_1, \mathbf{e}_2, \mathbf{e}_3)$ of a local orthogonal coordinate system where $\mathbf{e}_1$ is parallel to $\mathbf{B}$, $\mathbf{e}_2$ is normal to $\mathbf{B}$ (along the $-\boldsymbol{\rho}$ direction), and $\mathbf{e}_3 = \mathbf{e}_1 \times \mathbf{e}_2$. The radius of curvature of the field is defined by

$$\frac{\mathbf{e}_2}{\rho} = \frac{\partial \mathbf{e}_1}{\partial s} \tag{4.75}$$

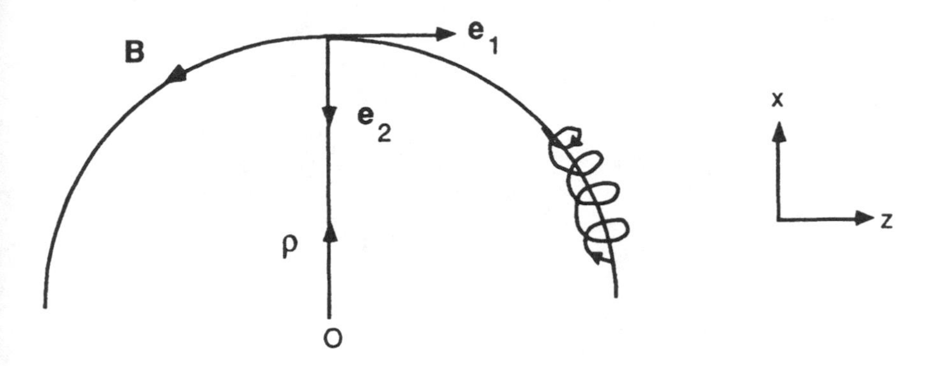

FIGURE 4.10 Curved magnetic field geometry in the xz-plane.

Since $\mathbf{e}_1 = \mathbf{B}/B$ and $\mathbf{e}_2 = -\boldsymbol{\rho}/\rho$, the above equation can also be written as

$$\begin{aligned}
\frac{\boldsymbol{\rho}}{\rho^2} &= -\frac{\partial}{\partial s}\left(\frac{\mathbf{B}}{B}\right) \\
&= -\frac{1}{B}\frac{\partial \mathbf{B}}{\partial s} + \frac{\mathbf{B}}{B^2}\frac{\partial B}{\partial s} \\
&= -\frac{1}{B}\frac{\partial \mathbf{B}}{\partial s} \\
&= -\frac{1}{B^2}(\mathbf{B}\cdot\nabla)\mathbf{B}
\end{aligned} \tag{4.76}$$

where we have assumed that the magnetic intensity is constant along the magnetic field, $\partial B/\partial s \approx \partial B_z/\partial s = 0$, and we have used the relation $\partial/\partial s = (\mathbf{e}_1 \cdot \nabla)$. We now substitute (4.76) into (4.74) and obtain

$$\mathbf{W}_c = \frac{mv_\parallel^2}{qB^4}\mathbf{B}\times(\mathbf{B}\cdot\nabla)\mathbf{B} \tag{4.77}$$

$\mathbf{W}_c$ is both charge- and energy-dependent. When single particles are in a magnetic field in which the local volume currents can be assumed not present ($\mathbf{J} = 0$), the vector relation $(\nabla \times \mathbf{B}) \times \mathbf{B} = (\mathbf{B}\cdot\nabla)\mathbf{B} - \nabla(B^2/2)$ yields $(\mathbf{B}\cdot\nabla)\mathbf{B} = \nabla(B^2/2)$. This permits the curvature drift (4.77) to be rewritten as

$$\begin{aligned}
\mathbf{W}_c &= \frac{mv_\parallel^2}{qB^4}\mathbf{B}\times(\mathbf{B}\cdot\nabla)\mathbf{B} \\
&= \frac{mv_\parallel^2}{qB^4}\mathbf{B}\times\nabla\left(\frac{B^2}{2}\right)
\end{aligned} \tag{4.78}$$

For a given magnetic field curvature, the drift velocity is a function of only the parallel energy; electrons and positive ions of the same energy have

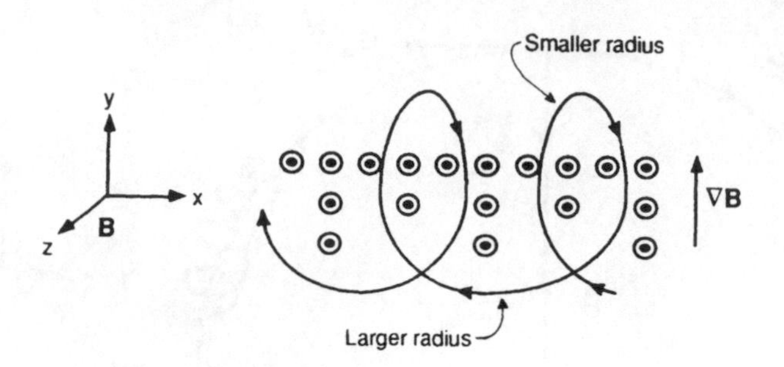

FIGURE 4.11 Effects of ∇B on guiding center motion.

the same drift speed but they drift in the opposite direction. This drift therefore gives rise to a current.

4.5.3 Gradient Drift of the Guiding Center

The terms $\partial B_z/\partial x$ and $\partial B_z/\partial y$ are being examined (recall that we are still assuming $B_z \approx B$). We showed in Chapter 3 that in this case the lines of force remain straight but the field varies in the x- and y-direction. Suppose $\partial B_z/\partial x = 0$ and $\partial B_z/\partial y$ is positive. Then in Figure 4.11, B is stronger in the $+y$-direction and weaker in the $-y$-direction. The cyclotron motion of a particle in this field geometry shows that the radius of the orbit in the $+y$-direction is smaller (radius of curvature is smaller) than in the $-y$-direction. Hence, this motion gives rise to a drift in the x-direction.

We next evaluate the Lorentz force, $\mathbf{F} = q\mathbf{v} \times \mathbf{B}$, compute the average $<\mathbf{F}>$, and determine the guiding center drift resulting from the magnetic field gradient. The symmetry of the orbit indicates that a particle spends as much time moving up as down, hence $< F_x >= 0$. To compute $< F_y >$, use the Taylor expansion of the magnetic field about $(x_0 y_0)$.

$$\begin{aligned} F_y &= -qv_x B_z(y) \\ &\approx -qv_x \left[B_0 + y\left(\frac{\partial B_z}{\partial y}\right) + \ldots \right] \end{aligned} \tag{4.79}$$

This expansion is valid if $r_c \ll L$ where r_c is the cyclotron radius and L is the scale length of the gradient $\partial B_z/\partial y$. Use the zeroth-order solution for v_x and y (from (4.10))

$$\begin{aligned} v_x &= v_\perp \cos\omega_c t \\ y &= \left(\frac{v_\perp}{\omega_c}\right) \cos\omega_c t \end{aligned} \tag{4.80}$$

in (4.79), which then becomes

$$F_y \approx -qv_\perp \cos\omega_c t \left[B_0 + \frac{v_\perp}{\omega_c}\cos\omega_c t \left(\frac{\partial B_z}{\partial y}\right) + \ldots \right] \tag{4.81}$$

To perform the average, note the first term involves $< \cos\omega_c t >$, which averages out to zero over one cyclotron period. The second term involves $<\cos^2\omega_c t>$, which is equal to 1/2 over a cyclotron period. Hence

$$\begin{aligned} <F_y> &= -\frac{qv_\perp^2}{2\omega_c}\frac{\partial B_z}{\partial y} \\ &= -\frac{qv_\perp r_c}{2}\frac{\partial B_z}{\partial y} \end{aligned} \tag{4.82}$$

where $r_c = v_\perp/\omega_c$. Equation (4.82) is used in $\mathbf{W}_D = \mathbf{F}\times\mathbf{B}/qB^2$ to obtain

$$\mathbf{W}_{\nabla B} = -\frac{v_\perp r_c}{2B^2}\frac{\partial B_z}{\partial y}(\hat{\mathbf{y}}\times\mathbf{B}) \tag{4.83}$$

Since $\mathbf{B}\approx B\hat{\mathbf{z}}$, $\hat{\mathbf{y}}\times\mathbf{B} = (\hat{\mathbf{y}}\times\hat{\mathbf{z}})B = \hat{\mathbf{x}}B$. The magnetic field gradient $\partial B_z/\partial y$ thus results in a drift of the guiding center in the x-direction. Equation (4.83) can be written more generally, noting that the choice of the gradient is arbitrary. Hence, $\partial B_z/\partial y$ can be replaced by ∇B. The general expression of the guiding center drift due to the gradient of the magnetic field is written as

$$\begin{aligned} \mathbf{W}_{\nabla B} &= \frac{v_\perp r_c}{2}\frac{\mathbf{B}\times\nabla B}{B^2} \\ &= \frac{mv_\perp^2}{2}\frac{\mathbf{B}\times\nabla(B^2/2)}{qB^4} \end{aligned} \tag{4.84}$$

where use was made of the relations $v_\perp r_c/2B^2 = v_\perp^2/2B^2\omega_c = mv_\perp^2/2qB^3$ and $B\nabla B = \nabla B^2/2$. Equation (4.84) is charge-dependent and positive and negative particles drift in opposite directions, giving rise to a current.

Note that the curvature drift (4.78) has the same form as (4.84) and therefore these equations can be added to yield

$$\mathbf{W}_D = \frac{m}{qB^4}\left(v_\parallel^2 + \frac{v_\perp^2}{2}\right)\mathbf{B}\times\frac{\nabla B^2}{2} \tag{4.85}$$

This equation represents the total guiding center drift of a particle in an inhomogeneous magnetic field that has curvature and gradient.

4.5.4 Shear and Twist of Magnetic Fields

The two remaining terms in the $\nabla\mathbf{B}$ tensor are $\partial B_x/\partial y$ and $\partial B_y/\partial x$, which are associated with twisting and shearing of the magnetic lines of force. Twisted field lines will introduce a component of the magnetic field $\mathbf{B}$

around the particle orbit, which, crossed with $\mathbf{v}_{\|}$, yields a force in the radial direction. This force will change the Larmor radius and distort the circular orbit but will not produce (to first-order) any drifts in the perpendicular direction.

4.5.5 Summary of Guiding Center Drifts

Let us summarize here the zeroth and first-order (guiding center) drifts that arise in particle motions.

Gravitational

$$\mathbf{W}_g = \frac{m}{q}\frac{\mathbf{g} \times \mathbf{B}}{B^2} \tag{4.44}$$

Electric Field

$$\mathbf{W}_E = \frac{\mathbf{E} \times \mathbf{B}}{B^2} \tag{4.45}$$

Polarization

$$\mathbf{W}_p = \frac{m}{qB^2}\frac{\partial \mathbf{E}}{\partial t} \tag{4.57}$$

Curvature

$$\mathbf{W}_c = \frac{mv_{\|}^2}{qB^4}[\mathbf{B} \times (\mathbf{B} \cdot \nabla)\mathbf{B}] \tag{4.78}$$

Gradient

$$\mathbf{W}_{\nabla B} = \frac{mv_{\perp}^2}{2qB^3}(\mathbf{B} \times \nabla B) \tag{4.84}$$

4.5.6 Currents Due to Guiding Center Drifts

The concept of currents is not meaningful in the context of a single particle. However, when there are many particles all taking part in the guiding center motion, currents are generated if the guiding center drift is charge-dependent (the only drift that is not charge-dependent is the electric field drift). The general expression of the current density is

$$\mathbf{J} = n^+ q^+ \mathbf{W}^+ + n^- q^- \mathbf{W}^- \tag{4.86}$$

where $n^{\pm}$ is the number density of positive and negatively charged particles and $\mathbf{W}^{\pm}$ is the appropriate guiding center drift velocities of these particles.

The topic of currents will be discussed further in the context of plasma treated as a magnetized conducting fluid (Chapter 7).

4.6 Adiabatic Invariants

The concept of invariants is extremely powerful in describing the motion of particles in magnetic fields. Consider the motion of a particle described by a pair of variables (p_i, q_i) that are generalized momenta and coordinates. Then, for each coordinate q_i that is periodic, the action integral J_i

$$J_i = \int p_i dq_i \tag{4.87}$$

integrated over a complete period of cycle (oscillation) of q_i with specified initial conditions is an invariant (constant) of motion. (Proof of this statement may be found in Northrop, 1961, cited at the end of the chapter.)

This action integral will remain invariant even if some property of the system is allowed to change. However, it is required that the change be slow (adiabatic change) as compared to relevant periods of the system and the change must not be related to the periods. This statement implies that if we start out in some state of motion and then allow an adiabatic change to occur in some property so that we end up in a different state, the motion in the final state will be such that it will have the same values of the action integrals as in the initial state.

For particles in magnetic fields, an adiabatic invariant is associated with each of the three types of motion discussed above—the gyration motion around **B**, the longitudinal motion along **B**, and the drift motion perpendicular to **B**.

4.6.1 First Adiabatic Invariant

The first invariant is associated with the cyclotron motion of the particle. The existence of this invariant can be shown elegantly by use of canonical momentum of charged particles in magnetic fields. We will not use this approach here. Instead, we derive the first invariant by use of a more familiar equation. Multiply each side of (4.72) by $v_\parallel$. Since $v_\parallel = ds/dt$, this equation becomes

$$\frac{d}{dt}\frac{mv_\parallel^2}{2} = -\mu\frac{\partial B}{\partial s}\frac{ds}{dt} = -\mu\frac{dB}{dt} \tag{4.88}$$

The last step follows because B is not time-dependent and $dB/dt = \partial B/\partial t + (\mathbf{v}\cdot\nabla)B = v\partial B/\partial s$. Conservation of the total energy of the particle requires

$$\frac{d}{dt}\left(\frac{1}{2}mv_{\parallel}^2 + \frac{1}{2}mv_{\perp}^2\right) = \frac{d}{dt}\left(\frac{1}{2}mv_{\parallel}^2 + \mu B\right) = 0 \tag{4.89}$$

where use was made of the definition $\mu B = mv_{\perp}^2/2$. Now combine (4.88) and (4.89) and obtain

$$-\mu\frac{dB}{dt} + \frac{d}{dt}(\mu B) = 0 \tag{4.90}$$

The differentiation of the second term results in

$$B\frac{d\mu}{dt} = 0 \tag{4.91}$$

Since $B \neq 0$, (4.91) states that the magnetic moment μ is independent of time and is a constant in the guiding center motion. The constancy of the magnetic moment implies that the total magnetic flux enclosed by the motion must also remain constant. This constancy of the first invariant of guiding center motion leads to magnetic trapping of particles. μ is conserved as long as the perturbation time scale is much longer than the cyclotron period.

4.6.2 Second Adiabatic Invariant

The longitudinal invariant is associated with the $v_{\parallel}$ motion. Since charged particles travelling in the direction of B behave as if the magnetic field is not there, the canonical momentum is simply $mv_{\parallel}$. The action integral for this motion is usually represented by J (not to be confused with the current density)

$$J = \int mv_{\parallel}\, ds \tag{4.92}$$

where ds is an element of the guiding center path along $\mathbf{B}$ and the integral is evaluated over a complete path, s_1 to s_2. Some definitions of J exclude the mass m of the particle in the integral, which is permitted since m is a constant in non-relativistic formulation.

Before we prove (4.92) for a charged particle, it is instructive to first examine a mechanical problem. Consider the motion of a particle of mass m and velocity $v_{\parallel}$ trapped between two walls separated initially by a distance $2D$ (Figure 4.12). Let the walls approach each other with a velocity U and let $U \ll v_{\parallel}$. The momentum of the particle before colliding with the wall is $p = mv_{\parallel}$. After collision with the wall, the momentum is $p' = m(v_{\parallel} + U)$. Hence, the change of momentum after colliding with both

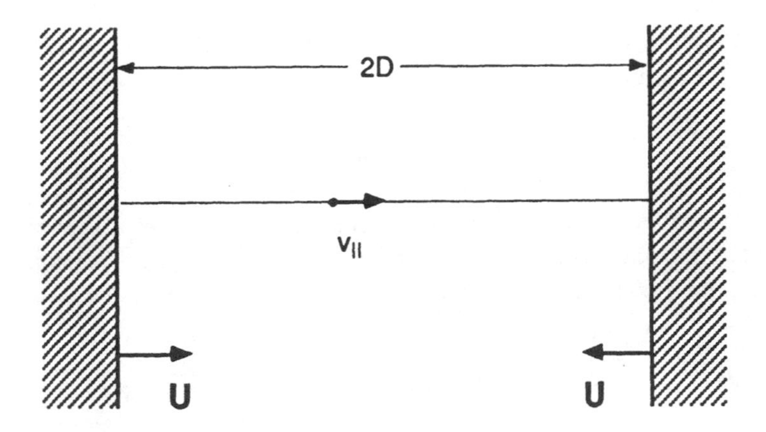

FIGURE 4.12 Particle colliding with moving walls.

walls is

$$\Delta p_{\parallel} = 2mU \tag{4.93}$$

The interval between collisions is $\Delta t = D/v_{\parallel}$, and thus the average rate of momentum change is

$$\frac{\Delta p}{\Delta t} = \frac{2mUv_{\parallel}}{D} \tag{4.94}$$

Now using $2U = -dD/dt$ ($-$ sign because the walls are approaching each other) and $p_{\parallel} = mv_{\parallel}$, (4.94) can be rewritten as

$$\frac{dp_{\parallel}}{dt} = -\frac{p_{\parallel}}{D}\frac{dD}{dt} \tag{4.95}$$

where $dp_{\parallel}/dt \approx \Delta p_{\parallel}/\Delta t$. This equation implies

$$\frac{d}{dt}(p_{\parallel}D) = 0 \tag{4.96}$$

and therefore

$$p_{\parallel}D = \text{constant} \tag{4.97}$$

This example is a special case of a particle in a variable potential trough that is slowly changing in time. The quantity $p_{\parallel}D$ is a constant of motion. Let us now return to the problem of the charged particle.

Suppose s_1 and s_2 are mirror points where the particles are reflected. The mirror points (like the walls in the mechanical problem) are no longer stationary. This means **B** is slowly varying in space and time in the particle reference frame. The question now is whether J is a constant of motion. J is a function of the total energy E (not to be confused with the electric field),

field), the path length s, and time t. From the definition $E = mv_{\|}^2/2 + \mu B$ solving for $v_{\|}$ and substitution of it in (4.92) yields

$$J = \int_{s_1}^{s} \left[\frac{2}{m}(E - \mu B)\right]^{1/2} ds \tag{4.98}$$

where the integration is from one mirror point, s_1, to an arbitrary point along the particle's path. Let us now compute dJ/dt to learn the conditions required for (4.98) to remain constant.

$$\begin{aligned}
\frac{dJ}{dt} &= \left(\frac{\partial J}{\partial t}\right)_{s,E} + \left(\frac{\partial J}{\partial E}\right)_{s,t} \frac{dE}{dt} + \left(\frac{\partial J}{\partial s}\right)_{E,t} \frac{ds}{dt} \\
&= \int_{s_1}^{s} \left[\frac{2}{m}(E - \mu B)\right]^{-1/2} \left[-\frac{\mu}{m}\frac{\partial B}{\partial t}\right] ds \\
&\quad + \left[v_{\|}\dot{v}_{\|} + \frac{\mu}{m}\frac{\partial B}{\partial t} + \frac{\mu v_{\|}}{m}\frac{\partial B}{\partial s}\right] \int_{s_1}^{s} \left[\frac{2}{m}(E - \mu B)\right]^{-1/2} ds \\
&\quad - v_{\|} \int_{s_1}^{s} \left[\frac{2}{m}(E - \mu B)\right]^{-1/2} \frac{\mu}{m}\frac{\partial B}{\partial s} ds \\
&\quad + v_{\|} \left[\frac{2}{m}(E - \mu B)\right]^{1/2}
\end{aligned} \tag{4.99}$$

Suppose now the limits of integration are between two mirror points. Then $s = s_2$, and since at the mirror points $v_{\|} = 0$, all terms involving $v_{\|}$ vanish. Equation (4.99) becomes

$$\begin{aligned}
\frac{dJ}{dt} &= -\int_{s_1}^{s_2} \frac{\mu}{m} \left[\frac{2}{m}(E - \mu B)\right]^{-1/2} \frac{\partial B}{\partial t} ds \\
&\quad + \left(\frac{\partial B}{\partial t}\right) \int_{s_1}^{s_2} \frac{\mu}{m} \left[\frac{2}{m}(E - \mu B)\right]^{-1/2} ds
\end{aligned} \tag{4.100}$$

If $\partial B/\partial t$ is slowly varying and can be considered constant over s_1 to s_2, $\partial B/\partial t$ can be taken outside of the integral. Then, the two terms in (4.100) cancel, giving

$$\frac{dJ}{dt} = 0 \tag{4.101}$$

The second adiabatic invariant applies to motions of particles along the magnetic field direction. This invariant implies that the particle will move in such a way as to preserve the total length of the particle trajectory. This invariant is maintained if the perturbation time scale is longer than the transit time of particles between the two mirror points.

4.6.3 Third Adiabatic Invariant

The first invariant is associated with the cyclotron motion. The second invariant is associated with the longitudinal motion. It would seem natural then to conclude that there must be an invariant associated with the drift motion. Physically, one can argue that this invariant should exist fairly simply. The first invariant implies that the total amount of flux enclosed in a gyration remains constant. We carry this concept to the drift motion. The guiding center drift motion conserves the total magnetic flux within its drift path. The actual path may be very complicated, but a particle will always follow a path to enclose the same magnetic flux. The third invariant is conserved as long as the perturbation time scale is longer than the drift times of the particles. Symbolically, the third invariant is $J = \int mW_\perp d\phi$ and $dJ/dt = 0$. Here ϕ is the azimuthal angle. For a complete drift path, the limits of ϕ are from 0 to 2π. (The proof of the third adiabatic invariant is found in Northrop, 1961, cited at the end of the chapter.)

4.7 Trapped Particles in the Magnetosphere

Consider a particle of mass m and velocity $\mathbf{v}$ on a line of force $\mathbf{B}$ in a magnetosphere. Its guiding center will move around in accordance with the conservation theorems discussed above. Let us now follow the particle. Its energy is $mv^2/2$, which can be separated into $mv_\parallel^2/2 + mv_\perp^2/2$. Now recall that $v_\parallel = v\cos\alpha$ and $v_\perp = v\sin\alpha$. In terms of these variables, the expression of the magnetic moment is

$$\mu = \frac{mv_\perp^2}{2B} = \frac{mv^2\sin^2\alpha}{2B} \tag{4.102}$$

A particle will move from point to point on a line of force conserving this invariant. Suppose this particle is initially on the equator and is moving toward the ionosphere. Since this particle is moving toward a region of stronger $|\mathbf{B}|$, it follows from the invariance of μ that $\varepsilon_\perp = mv_\perp^2/2$ must increase. Since energy is conserved, $\varepsilon_\perp$ must increase at the expense of $\varepsilon_\parallel$. Thus, it may happen that for some value of $\mathbf{B}$ (say $\mathbf{B}_R$), $\varepsilon_\parallel = 0$. At this point, all of the particle's energy is in $\varepsilon_\perp$ and it cannot penetrate any further. It must therefore turn around because there is a force ($\mathbf{F} = -\mu\nabla B$) acting on it to push it back. The pitch-angle of the particle at the turning point is 90°. The particle will bounce back and forth between the two conjugate turning points. This particle is trapped (Figure 4.13).

Consider a particle in an initial state defined by its pitch-angle α and magnetic field B. Let this particle move to another state, the point at which the particle is mirroring. Then the conservation of the magnetic moment (action integral) permits us to write

$$\frac{\sin^2 \alpha}{B} = \frac{\sin^2 \alpha(\alpha = \pi/2)}{B_R}$$

$$\sin^2 \alpha = \frac{B}{B_R} \tag{4.103}$$

This equation shows that the pitch-angle α of a particle at any point B is related to the magnitude of B_R at the reflection point. We can therefore calculate B_R given α and vice versa.

4.7.1 Magnetic Mirror Ratio

Refer to the particles in the magnetic mirror geometry shown in Figure 4.13. Consider the particles for which $B_R > B_M$ and $B_R < B_M$. B_M represents the maximum value of B encountered by a particle. In Figure 4.13, B_M is the value of B at the "throat." For a planetary magnetosphere, B_M is the value of B at the ionospheric heights. Then it should be evident that particles with $B_R > B_M$ will escape the "bottle" while the particles with $B_R < B_M$ are trapped. In a symmetric arrangement such as the one being considered, the midway point will have a minimum magnetic field, B_0. We now define

$$R = \frac{B_M}{B_0} \tag{4.104}$$

R is called the mirror ratio and it gives a measure of the ratio between the strongest and the weakest magnetic field encountered by a particle in a trapping geometry.

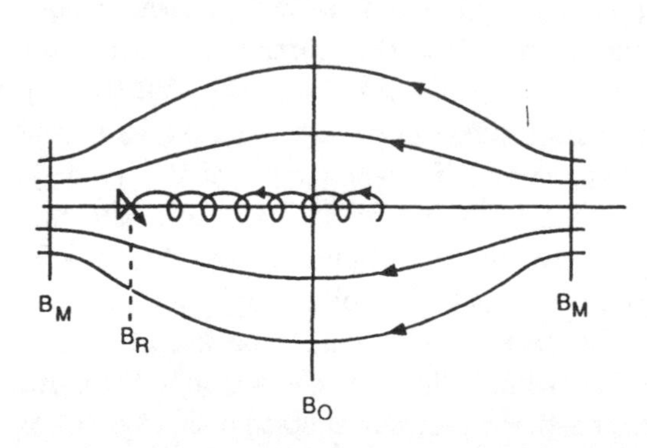

FIGURE 4.13 A magnetic bottle geometry. B_0 is the equatorial field. B_M is the field at the throat. All particles are mirroring in the region where $B < B_M$ are trapped. All particles mirroring where $B > B_M$ will escape the magnetic bottle.

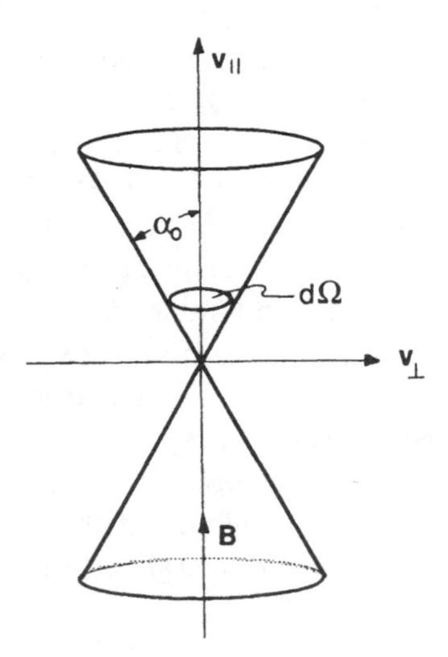

FIGURE 4.14 Definition of loss cone in $v_{\|}v_{\perp}$ space.

4.7.2 Loss Cone

Consider the particles on the equator where $B = B_0$. Let the particles have any arbitrary pitch-angle α. Some of these particles are lost while others are trapped. Let the particles that are lost be those within the solid angle Ω. In the space defined by $v_{\perp}$ and $v_{\|}$, we can draw the following picture (Figure 4.14).

For an isotropic pitch-angle distribution, the probability of particle loss from the magnetic bottle is

$$
\begin{aligned}
P = \frac{\Omega}{2\pi} = \int_o^{\alpha_o} \sin\alpha\, d\alpha &= (1 - \cos\alpha_0) \\
&= 1 - (1 - \sin^2\alpha_0)^{1/2} \\
&= 1 - \left(1 - \frac{B_0}{B_M}\right)^{1/2} \\
&= 1 - \left(1 - \frac{1}{R}\right)^{1/2}
\end{aligned} \tag{4.105}
$$

where (4.103) and (4.104) have been used. This equation states that P is smaller for larger R. Hence the probability for loss of particles is smaller when the magnetic field geometry supports a large mirror ratio. A dipole field has fairly large R and therefore P is small. Note that the loss probability is independent of q, m, or E.

We now define the loss cone for a planetary magnetosphere. In this definition, we must take into account the fact that the planet has an atmosphere. For example, define the equatorial loss cone for Earth as that pitch-angle α_o such that its mirror point is at 100 km from the surface of Earth. This definition is somewhat arbitrary. The reason for the 100-km height is that at this altitude the atmospheric density is large enough that scattering of electrons is highly probable. An electron mirroring at an altitude of 100 km is thus very likely to be "absorbed" by the atmosphere, hence lost from the radiation belt. The equatorial loss cone for a dipole line of force crossing the equator at $6R_E$ is about 3°. All electrons within a 3° equatorial pitch-angle cone are precipitated because they are mirroring below 100 km altitude. Electrons outside of the loss cone mirror at heights above 100 km and they are therefore trapped radiation belt electrons.

4.7.3 Bounce Period in a Dipole Field

The bounce period T_B of trapped particles in a dipole field is now considered. The period of the north-south motion of the guiding center is calculated from

$$T_B = \int \frac{ds}{v_\parallel} \tag{4.106}$$

where ds is the increment of path length along **B** and $v_\parallel$ is the component of the particle's velocity parallel to ds. The integral extends over one complete cycle from one mirror point to the opposite mirror point and back.

In static magnetic fields, the total energy of the particle is conserved. Then

$$\begin{aligned} v_\parallel^2 &= v^2 - v_\perp^2 \\ &= v^2(1 - \sin^2\alpha) \\ &= v^2\left(1 - \frac{B}{B_0}\sin^2\alpha_0\right) \end{aligned} \tag{4.107}$$

assuming the first invariant is conserved, $v^2 = v_0^2$, and the subindex 0 refers to the equatorial plane. Equation (4.107) is relativistically correct.

For a dipole field, $r = r_0\cos^2\lambda$ and ds is

$$ds = r_0\cos\lambda(1 + 3\sin^2\lambda)^{1/2}d\lambda. \tag{4.108}$$

Also, B/B_0 for a dipole is

$$\frac{B}{B_0} = \frac{(1 + 3\sin^2\lambda)^{1/2}}{\cos^6\lambda} \tag{4.109}$$

The integral (4.106) assembled with (4.107) to (4.109) is

$$T_B = \int \frac{r_0 \cos\lambda (1 + 3\sin^2\lambda)^{1/2} d\lambda}{v\left[1 - \sin^2\alpha_0 \dfrac{(1 + 3\sin^2\lambda)^{1/2}}{\cos^6\lambda}\right]^{1/2}} = 4r_0 \frac{I_1}{v} \tag{4.110}$$

where

$$I_1 = \int \frac{\cos\lambda (1 + 3\sin^2\lambda)^{1/2} d\lambda}{\left[1 - \sin^2\alpha_0 \dfrac{(1 + 3\sin^2\lambda)^{1/2}}{\cos^6\lambda}\right]^{1/2}} \tag{4.111}$$

and the integration limits of I_1 are from 0 to λ_{max}, which is the latitude of the mirror point for a particle mirroring in the northern hemisphere. At this point, $v_{\|} = 0$ and λ_m is obtained from the solution of

$$\cos^6\lambda_m - \sin^2\alpha_0 (1 + 3\sin^2\lambda_m)^{1/2} = 0 \tag{4.112}$$

The integral (4.111) is a dimensionless function of λ and for arbitrary α_0 the integral can only be solved numerically. This integral is approximated from the relation (see Hamlin *et al.*, 1961, cited at the end of the chapter)

$$I_1 \approx 1.3 - 0.56 \sin\alpha_0 \tag{4.113}$$

The factor $4r_0/v$ is a function of the total velocity (energy) for a given particle species.

If the distance of closest approach for the particle satisfies the relation

$$\frac{R_{min}}{r_0} < \frac{1}{L} \tag{4.114}$$

then the particle is precipitated. Hence particles are trapped if the distance of the closest approach is larger than this R_{min}. For the particles mirroring at the surface of a planet, λ_{min} is given by

$$\cos^2\lambda_m = \frac{1}{L} \tag{4.115}$$

This equation can be used for particles mirroring just above Earth's surface.

For Earth, $r_0 = LR_E$ where R_E is the radius of Earth and L is a parameter (McIlwain L-parameter, to be discussed below). Figure 4.15 shows curves for Earth's dipole bounce period plotted against the energy for electrons with pitch-angles α_0 corresponding to marginal trapping for L values 1 to 8. For typical magnetospheric electrons ($\sim$ 10–50 keV) on auroral lines of force ($L = 6$) the bounce period is a few seconds. Bounce periods for other kinds of particles can be obtained by scaling these curves. For example, to obtain the bounce period of protons, multiply by $(m_p/m_e)^{1/2}$.

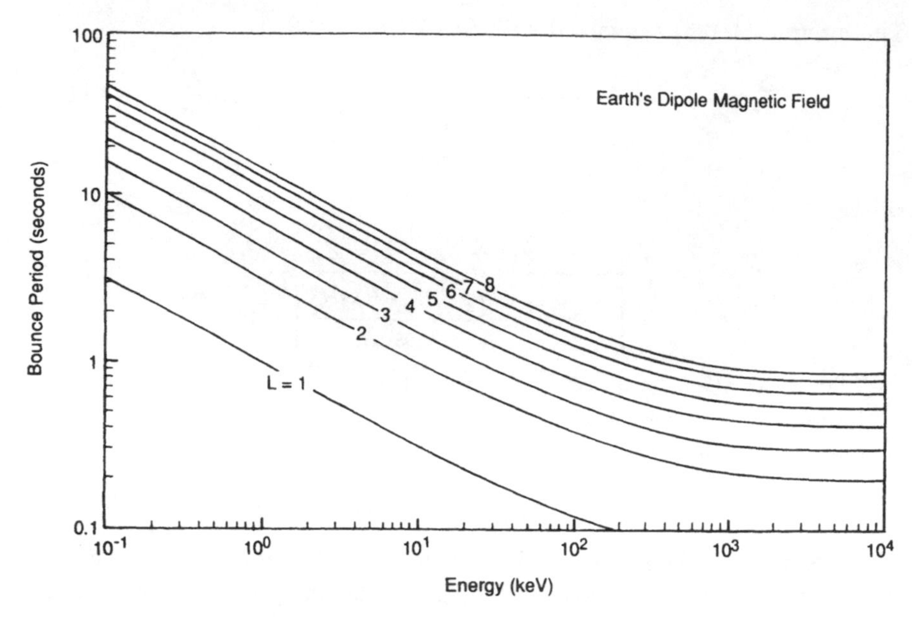

FIGURE 4.15 Bounce period of electrons in a dipole field of Earth for various L-shells.

4.7.4 Drift Period in a Dipole Field

To calculate the guiding center drift period in a dipole field, we note that the drift velocity has two components, the gradient drift and the curvature drift. The two equations (4.77) and (4.84) can be combined and written as

$$\mathbf{W}_D = \frac{1}{\omega_c \rho}\left(v_\parallel^2 + \frac{1}{2}v_\perp^2\right)\frac{\boldsymbol{\rho}\times\mathbf{B}}{\rho B} \tag{4.116}$$

where $\boldsymbol{\rho}$ is the radius of curvature of the field line and ω_c is the cyclotron frequency. For a dipole, ρ in a polar coordinate system is

$$\begin{aligned}\rho &= \frac{[r^2 + (dr/d\theta)^2]^{3/2}}{r^2 + 2(dr/d\theta)^2 - r(d^2r/d\theta^2)} \\ &= r_0 \frac{\cos\lambda(1+3\sin^2\lambda)^{3/2}}{3(1+\sin^2\lambda)}\end{aligned} \tag{4.117}$$

The cyclotron frequency for a dipole is obtained from

$$\omega_c = \omega_0 \frac{(1+3\sin^2\lambda)^{1/2}}{\cos^6\lambda} \tag{4.118}$$

where ω_0 is the cyclotron frequency at the equator, qB_0/m. If we use (4.117), (4.118), and (4.109), the drift velocity (4.116) in a dipole field at latitude λ becomes

$$W_D = \frac{3v^2}{2}\frac{(1+\sin^2\lambda)\cos^5\lambda}{\omega_0 r_0(1+3\sin^2\lambda)^2}\left[2-\sin^2\alpha_0\frac{(1+3\sin^2\lambda)^{1/2}}{\cos^6\lambda}\right] \quad (4.119)$$

Now, the angular drift that occurs in one bounce period is

$$\Delta\phi = \int \frac{W_D}{r\cos\lambda}\frac{ds}{v_\parallel} \quad (4.120)$$

and the angular drift velocity averaged over a bounce is

$$<\omega_D> = \frac{\Delta\phi}{2\pi T_B} \quad (4.121)$$

where $<\omega_D>$ denotes that the angular drift velocity has been averaged over a bounce period. If we use (4.109) and (4.120), (4.121) can be written as

$$\begin{aligned} <\omega_D> &= \frac{3v^2}{2\omega_0 r_0^2}\frac{I_2}{I_1} \\ &= \frac{3\varepsilon}{qB_0R_E^2L^2}\frac{I_2}{I_1} \end{aligned} \quad (4.122)$$

where $\varepsilon = mv^2/2$ and I_2 obtained using (4.119), (4.120), (4.107) and (4.108) is

$$\begin{aligned} I_2 = \int & \frac{\cos^4\lambda(1+\sin^2\lambda)}{\left[1-\sin^2\alpha_0\frac{(1+3\sin^2\lambda)^{1/2}}{\cos^6\lambda}\right]^{1/2}} \\ & \times\left[2-3\sin^2\alpha_0\frac{(1+3\sin^2\lambda)^{1/2}}{\cos^6\lambda}\right]d\lambda \end{aligned} \quad (4.123)$$

The integrals I_1 and I_2 are functions of only λ. An estimate of I_2/I_1 is (see Hamlin *et al.*, 1961, cited at the end of the chapter)

$$\frac{I_2(\alpha_0)}{I_1(\alpha_0)} \approx 0.35 + 0.15\sin\alpha_0 \quad (4.124)$$

The bounce averaged drift period of the guiding center in a dipole field is obtained from

$$<T_D> = <\omega_D>^{-1} \quad (4.125)$$

Equation (4.124) shows that I_2/I_1 varies from 0.35 for $\alpha_0 = 0°$ particles to 0.5 for $\alpha_0 = 90°$ particles. For Earth, we obtain

$$<T_D> \approx \begin{cases} 43.8/L\varepsilon, & \alpha_0 = 90° \\ 62.7/L\varepsilon, & \alpha_0 = 0° \end{cases} \quad (4.126)$$

where T_D is in minutes and ε is in MeV. Figure 4.16 shows the drift periods as a function of the particle energy for these two limits for $L = 1$ to 8. The top panel shows curves for equatorial particles ($\alpha_0 \approx 0°$) and the bottom curves for marginally trapped ($\alpha_0 \approx 90°$) particles. The drift period of typical magnetospheric particles ($\sim$ 50 keV) at $L = 6$ is about an hour. Recall that ions and electrons drift in the opposite directions. However, the drift speed is the same if their energies are equal.

4.8 Earth's Radiation Belt Particles

We now relate the guiding center concepts and the conservation theorems to actual observations of particles in Earth's radiation belt. Let us note that a particle detector onboard a spacecraft does not measure properties of a single particle but rather the flux and the distribution of the particles. Recall from Chapter 2 the definition of the differential flux $j = (\mathbf{r}, E, \theta, \phi, t)$. A natural coordinate in space is to align the direction of the magnetic field with one of the axes. In this system, the angle θ then becomes α, the pitch-angle of the particles, and ϕ is the gyrophase of the particle. Assuming that the gyrophase is uniform, the number of particles with pitch-angles α and $\alpha + d\alpha$ in the energy range E and $E + dE$ crossing an incremental area dA in time dt is

$$\frac{dN}{dA\,dE\,dt} = 2\pi j \sin\alpha\, d\alpha = -2\pi j\, d(\cos\alpha) \tag{4.127}$$

The differential flux of the particles measured by a detector on a spacecraft at position $\mathbf{r}$ is then represented by

$$j = j(\mathbf{r}, \delta, E, t) \tag{4.128}$$

where δ is $\cos\alpha$. We can also write the differential flux as

$$j = j(B, \delta, E, t) \tag{4.129}$$

which refers the flux directly to the intensity of the magnetic field at the location $\mathbf{r}$ where the measurement is made. Writing j as in (4.129) enables one to directly relate the measured particle flux to their pitch-angles.

Trapped and precipitated fluxes are defined as those particles with pitch-angles greater and less than α_{loss} :

$$\begin{aligned} j_T &= j(B, \cos\alpha < \cos\alpha_{loss}, E, t) \\ j_P &= j(B, \cos\alpha > \cos\alpha_{loss}, E, t) \end{aligned} \tag{4.130}$$

where α_{loss} defines the loss cone for a given line of force. Trapped fluxes mirror above the atmosphere, and for Earth this altitude is approximately 100 km from the surface. The precipitated particles mirror below 100 km,

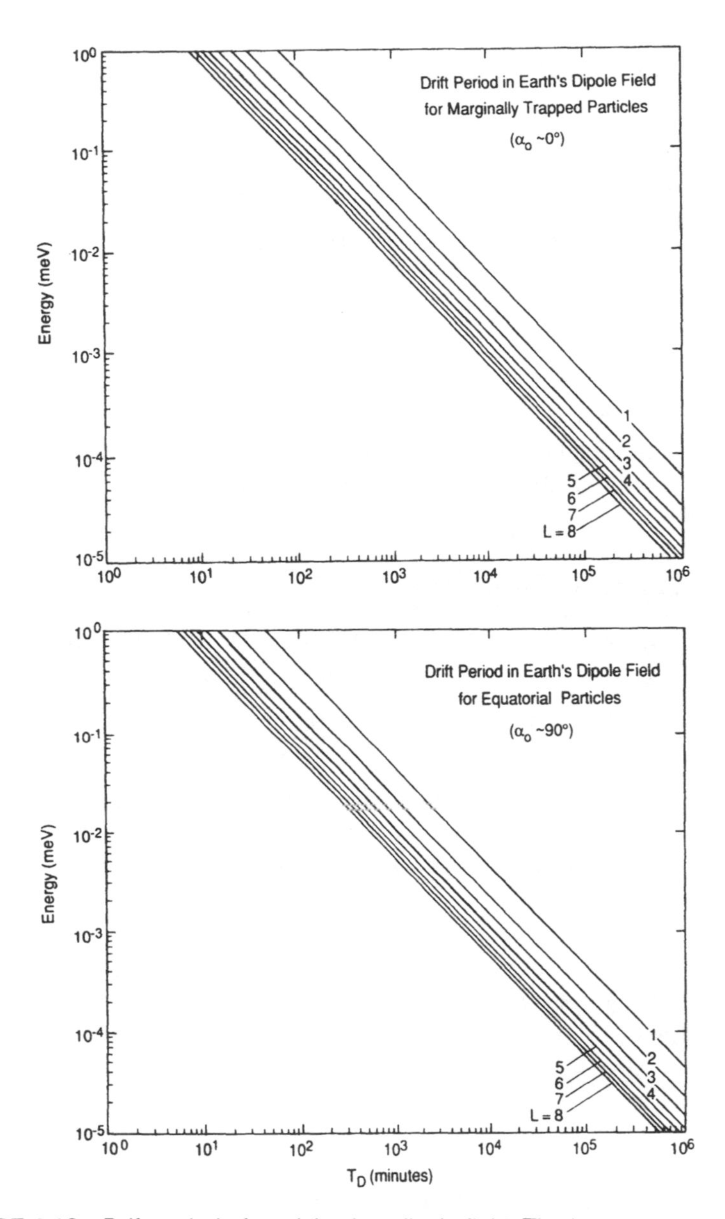

FIGURE 4.16 Drift period of particles in a dipole field. The bottom curves are for $\alpha_0 = 90°$ equatorial mirroring particles. The top curves are for $\alpha_0 \approx 0°$, marginally trapped particles.

and they are lost. These particles interact with atmospheric constituents and produce atmospheric luminosities. Particles frequently precipitate at high geomagnetic latitudes and produce auroral lights.

4.8.1 Energy Spectrum of Inner Radiation Belt Protons

In the early days of the space age, when the radiation environment of Earth was not well understood, experiments were designed to measure omni-directional fluxes. The definition of an omni-directional flux is

$$J(B,E) = 4\pi \int j(B,\delta,E)d(\cos\delta) \tag{4.131}$$

Data obtained from one such particle detector are shown in Figure 4.17. The form of the omni-directional flux distribution of protons measured in the inner radiation belt is very similar to the theoretically calculated proton distribution of CRAND. These results suggest that the cosmic rays are the main source of the high-energy protons in the inner radiation belt (the theoretical CRAND distribution is shown by a solid curve).

4.8.2 Pitch-Angle Distribution

In the absence of collisions, the density of points in the phase space described by the distribution function is conserved. Hence

$$\frac{j(B_1,\delta_1,E_1)}{E_1} = \frac{j(B_2,\delta_2,E_2)}{E_2} \tag{4.132}$$

defines the relationship of the differential fluxes at two positions of a field line. If the fluxes are measured in a static magnetic field, in the absence of parallel electric fields the energy of the particles is conserved and (4.132) reduces to

$$j(B_1,\delta_1) = j(B_2,\delta_2) \tag{4.133}$$

The differential flux is then only a function of the cosine of the pitch-angle and the value of the magnetic field where the measurement is made. For a given line of force, the variables are related by the first invariant

$$\sin^2\alpha_1 = \frac{B_1}{B_2}\sin^2\alpha_2 \tag{4.134}$$

The differential flux at one point B_1 can always be referred to the flux at another point B_2.

Figure 4.18 shows several pitch-angle distributions of fluxes of particles sampled at various altitudes along a line of force above Earth (top panel). These distributions can be transformed to the equator (bottom panel) by use of the flux relationship derived above. This transformation is possible because all of the particles that mirror off the equator must necessarily pass through the equator. Note that all of the data points on the equator lie on a single curve. Note also that the maximum pitch-angle on the equator is about 40° whereas the measured maximum in the ionosphere is 90°. This

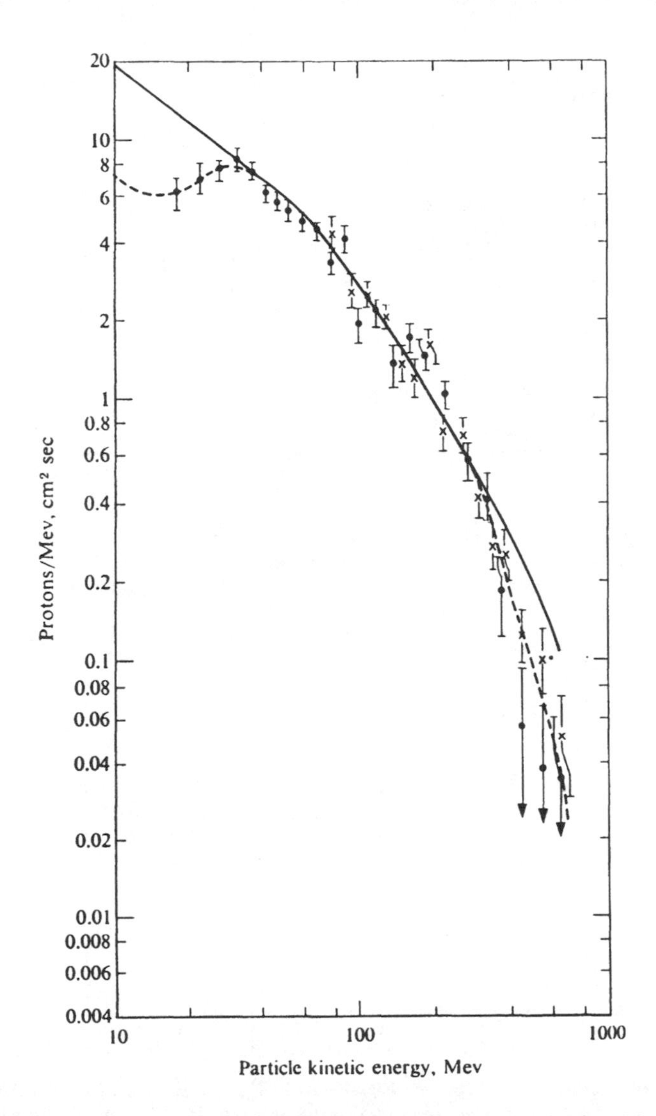

FIGURE 4.17 One of the first energy spectra of protons obtained in the inner radiation belt. The solid curve is derived from the theory of cosmic ray albedo neutron decay. (From Freden and White, 1960)

is because the phase space is largest on the equator and becomes smaller as we move off the equator. For large mirror ratios such as Earth's magnetic field, a 3° pitch-angle on the equator would correspond to about 90° in the ionosphere for a dipole line of force that crosses at $6R_E$.

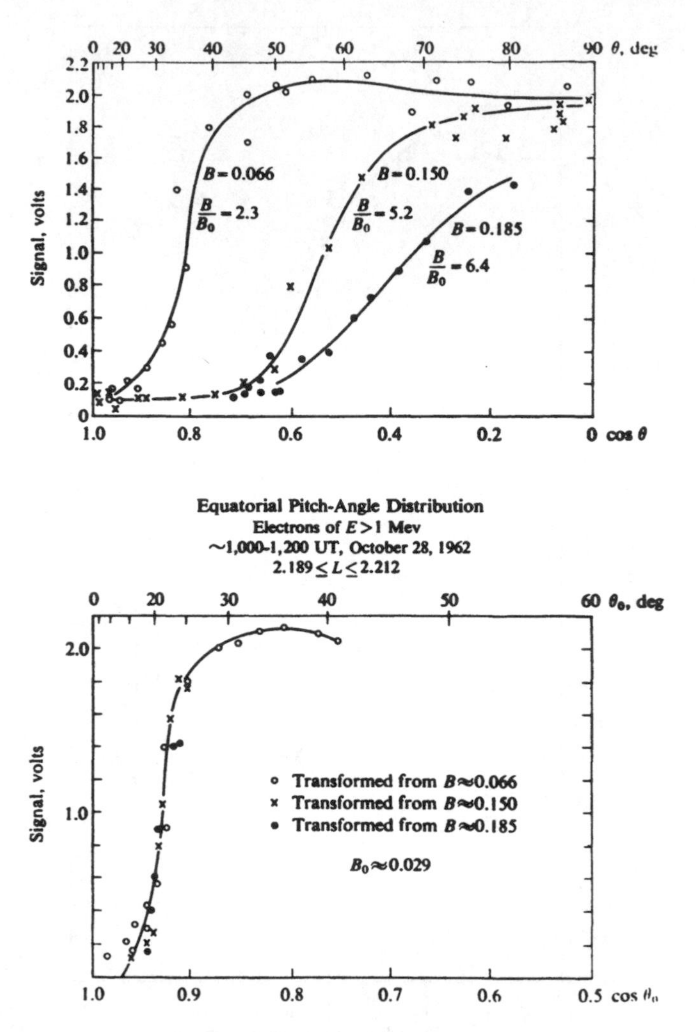

FIGURE 4.18 One of the first demonstrations that used the first adiabatic invariant to organize data in space. The top panel shows data taken at different points along a magnetic line of force. The bottom panel shows the transformation of the data from the top panel onto the equatorial plane. This assumes conservation of the first invariant. (From Katz and Smart, 1963)

4.8.3 Observation of Drifting Particles

A particle detector can be collimated to measure directional fluxes of particles. An example of particles detected in the outer radiation belt is shown in Figure 4.19. The data were obtained by an electrostatic analyzer on board a geostationay satellite during a moderate geomagnetic disturbance.

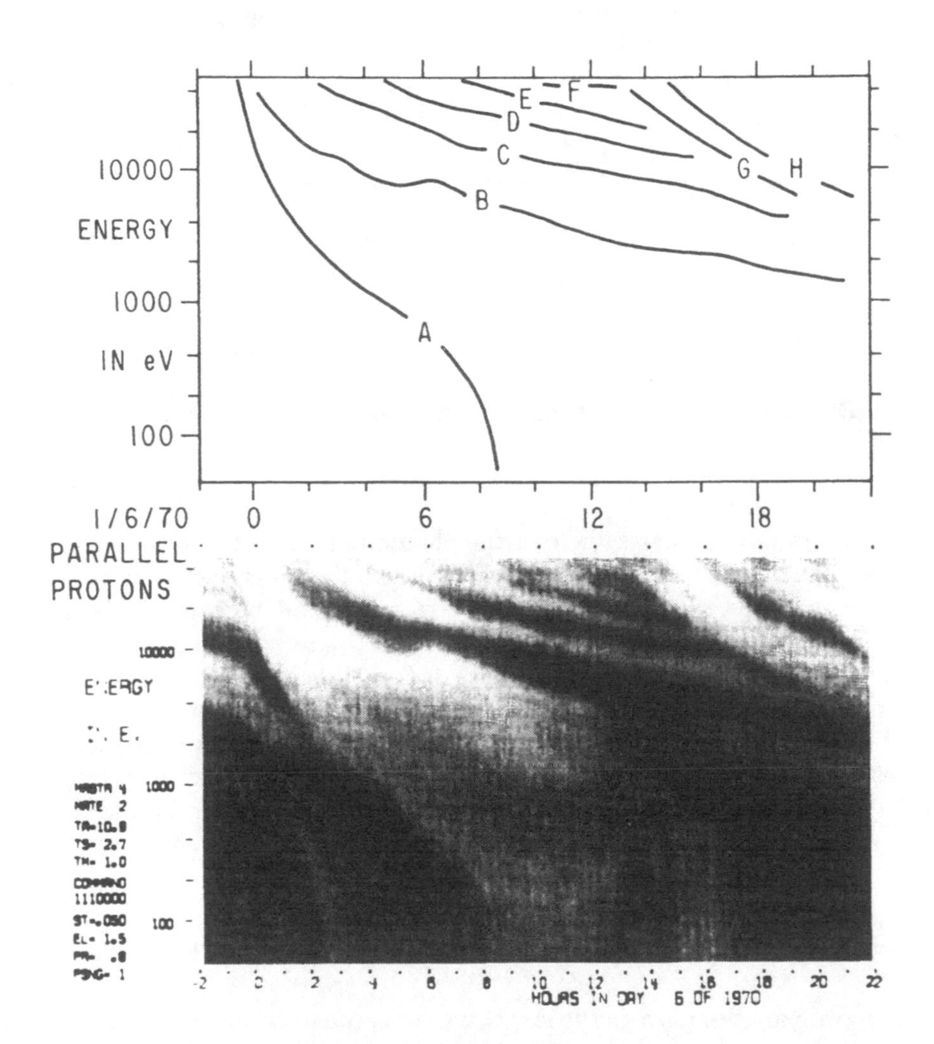

FIGURE 4.19 Low-energy protons obtained by a detector on a geostationary orbit. The data are displayed in the spectrogram format with particle energy as vertical, time of measurement as horizontal, and the grey-shaded intensity forming the third axis (out from the page). Of interest here are curves that show that the energy of the particles is decreasing with time. Energy dispersions arise because the guiding center drifts are energy-dependent. The top panel identifies specific dispersion curves. (From DeForest and McIlwain, 1971)

data shown on the bottom panel are displayed in a three-dimensional spectrogram format. The energy is plotted vertically against the time when the particles are detected, and the intensity forms the third axis (out of the page).

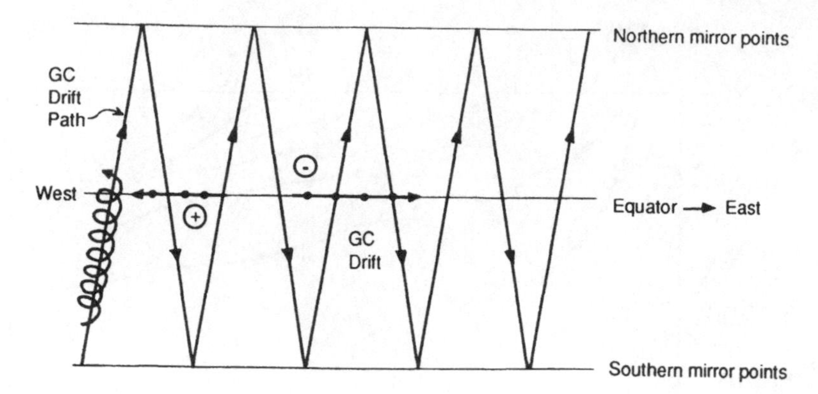

FIGURE 4.20 A mercator projection of guiding center motion in the radiation belt.

A feature of interest here is the observation of energy dispersion. It arises because the particles were injected into the outer radiation belt to the east of the satellite (at the onset of an auroral breakup in the auroral zone around 00 hours) and arrived at the satellite position subsequently by the gradient and curvature drift motions. Since the drift velocity is energy-dependent, higher-energy particles arrived first, followed by the slower lower-energy particles. A fixed detector in space will record the resulting energy dispersion. The energy dispersion curves can be extrapolated and the injection times determined. The top panel identifies some of the dispersion curves that can be used for quantitative analysis.

4.9 Motion of Trapped Particles in Non-Dipole Field

Trapped particles move in the magnetosphere, conserving the three particle invariants. A "mercator projection" of the guiding center drift paths in a planetary magnetic field is schematically shown in Figure 4.20.

The three invariants will be conserved by the particles even in regions where the fields deviate slightly from the dipole configuration. Such deviations will make the guiding center trajectory a little more complicated, but the particles will execute motions with the invariants conserved. This means that if the line of force at one longitude has a different length from a line of force at another longitude, the mirror points of the particles on these two lines will be different, but they will be adjusted according to the second invariant. Distortions of the magnetic field in the azimuth direction result in a non-circular drift path.

If the particle orbits are modified because the real field deviates from the dipole field, how do we identify the lines of force to which the guiding

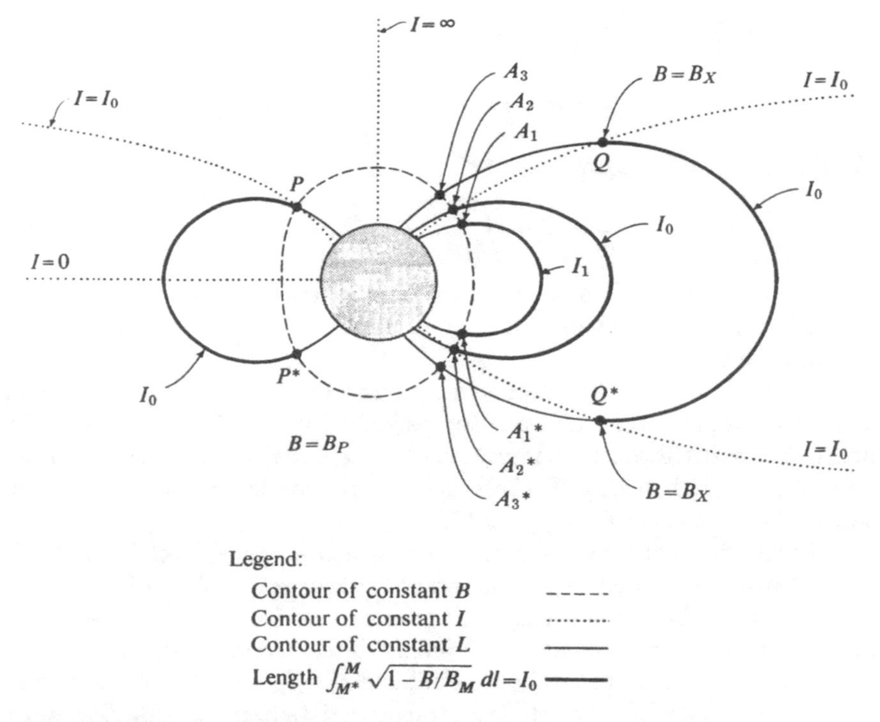

FIGURE 4.21 Guiding center motion traces out mirror points (A_1, A_2, A_3) and I contours. (From O'Brien, 1967)

centers move in time? In Figure 4.21, let a particle be on a line of force labeled I_0 on the left. Its mirror point field is given by $\sin^2 \alpha / B = 1/B_M$. Thus, once the equatorial α is specified, the particle will mirror at $B = B_M$. Now, the particle is also azimuthally drifting. The question is, as it drifts, what factors constrain the motion? The answer is that among all lines of force, the guiding center of the particle will select a line of force such that the second invariant is conserved.

$$\begin{aligned} J &= \int m v_{\|} ds \\ &= \int m v \left(1 - \frac{B}{B_0} \sin^2 \alpha_0\right)^{1/2} ds \end{aligned} \tag{4.135}$$

The total momentum mv of a particle in a static magnetic field is constant and therefore it can be taken out of the integral sign. Let us also integrate this equation between the two mirror points just above the atmosphere. Then (4.135) becomes

$$J = mv \int_{S1}^{S2} \left(1 - \frac{B}{B_M}\right)^{1/2} ds \tag{4.136}$$

We define a new invariant

$$\frac{J}{mv} = I = \int_{S1}^{S2} \left(1 - \frac{B}{B_M}\right)^{1/2} ds \tag{4.137}$$

I is only a function of position, and it is proportional to the length of the line of force. A particle will drift azimuthally onto a line of force conserving this invariant.

In Figure 4.21, in addition to the constant B_M contours, contours of constant I are also shown. When the particle drifts to the right side of the figure, the particle can be on only one line of force. While there are other lines of force with similar I, there is only one line where the particle can satisfy the conditions $B = B_M$ and $I = I_0$.

The guiding center motion of a particle conserves the first (μ) and second (J) invariants in its motion across static magnetic fields. We have also indicated that the guiding center motion can be described by the mirror point magnetic field B_M and $I = J/mv$ if the total momentum of the particle is conserved. Unfortunately, it is difficult to apply the constants of motion to the radiation belt particles. Moreover, data organized by B_M and I are difficult to interpret. For these reasons, Carl McIlwain invented a more physically meaningful parameter, now called the McIlwain L-parameter. The L-parameter is a function of B and I and remains constant on the line of force on which the particle is moving.

Consider the dipole field equation

$$\frac{4\pi B r_0^3}{\mu_0 M} = \frac{(1 + 3\sin^2\lambda)^{1/2}}{\cos^6\lambda} \tag{4.138}$$

and the invariant $I = J/mv$ given by (4.137). Since for the dipole field

$$ds = r_0 \cos\lambda (1 + 3\sin^2\lambda)^{1/2} d\lambda \tag{4.139}$$

and

$$\frac{B}{B_M} = \frac{(1 + 3\sin^2\lambda)^{1/2}\cos^6\lambda_M}{(1 + 3\sin^2\lambda_M)^{1/2}\cos^6\lambda} \tag{4.140}$$

where λ_M represents the mirroring latitude, we can write

$$\frac{I}{r_0} = \int_{S_1}^{S_2} \cos\lambda (1 + 3\sin^2\lambda)^{1/2} \times \left[1 - \frac{\cos^6\lambda_M (1 + 3\sin^2\lambda)^{1/2}}{\cos^6\lambda (1 + 3\sin^2\lambda_M)^{1/2}}\right] d\lambda \tag{4.141}$$

Note that both I/r_0 and the dipole equation (4.138) are only a function of λ. Let $I/r_0 = f_1(\lambda)$ and $4\pi B r_0^3/\mu_0 M = f_2(\lambda)$. Cube I/r_0 and eliminate r_0^3 using the dipole equation (4.138). We then obtain

$$\frac{4\pi B I^3}{\mu_0 M} = f_3(\lambda) \tag{4.142}$$

where $f_3(\lambda) = f_1(\lambda)^3 f_2(\lambda)$. (4.142) can be inverted to obtain

$$\lambda = g\left(\frac{4\pi B I^3}{\mu_0 M}\right) \tag{4.143}$$

where g is a new function that involves B, I, and M. Equation (4.143) is now used in the dipole equation (4.138), which finally becomes

$$\frac{B r_0^3}{M} = \psi\left(\frac{4\pi B I^3}{\mu_0 M}\right) \tag{4.144}$$

or

$$r_0^3 = \xi\left(\frac{4\pi B I^3}{\mu_0 M}\right) \tag{4.145}$$

ξ is a complicated function of B and I but can be computed. The conclusion of this exercise is that in a static dipole magnetic field, there exist functions of B and I that depend only on r_0, hence they are constant along a magnetic line of force.

At the equator of a dipole line of force, $I = 0$ since the limits of integration (mirror points) in (4.137) coincide. Also, the functions $f_1(\lambda)$ and $f_2(\lambda)$ are equal to 1. According to (4.144), then

$$r_0 = L_d = \left(\frac{\mu_0 M}{4\pi B_0}\right)^{1/3} \tag{4.146}$$

where L_d is defined as the McIlwain L-parameter for a dipole line of force and B_0 represents the magnetic field at the equator, which has a minimum value for the dipole. L_d is constant, and for the dipole it simply represents the equatorial crossing distance of a line of force. We can use L_d like r_0 to identify a dipole line of force.

An example of how the L-parameter is used in ordering trapped particles in Earth's radiation belt is shown in Figure 4.22. Constant count-rates (intensity) of > 30 MeV protons measured by Explorer IV are shown in the (B, L) coordinate system. The data are very well organized because the (B, L) coordinates are based on the invariants that govern the motion of trapped particles.

If the magnetic field is different from the dipole, the concept of L is still valid. In this case, B and J must be computed according to the prescriptions given above. In Earth's magnetosphere, L_d is applicable to the line of force out to a distance $\sim 5R_E$. Beyond this distance, the magnetic field deviates from the dipole and computation of L requires knowledge of B along the line of force.

4.10 Particle Motion in Magnetic Neutral Points

We have thus far studied particles whose motions are adequately described by the guiding center (first-order orbit) approximation. For many cases, however, the first-order orbit theory may not be applicable because the magnetic field changes significantly in distances of the order of a gyroradius and the requirement that $\mathbf{B}_0 \gg |(\mathbf{r} \cdot \nabla_0)\mathbf{B}|$ is violated (see (4.59)). In such cases, we must return to the Lorentz equation and try to solve the equation either analytically (if possible) or numerically.

The motion of particles across thin boundaries and magnetic fields that include neutral points is an example of where the guiding center ap-

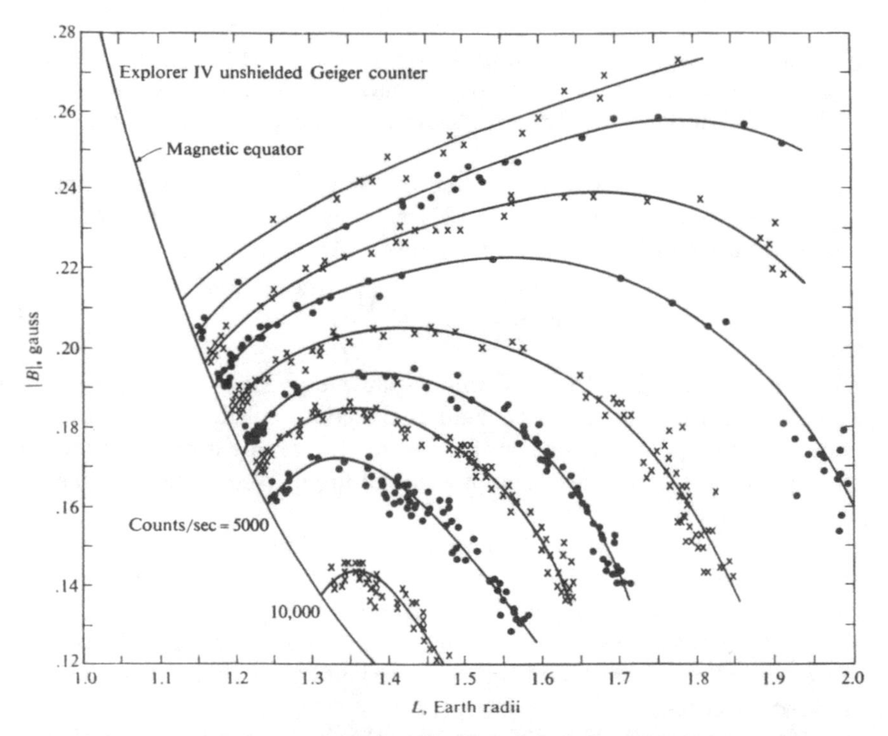

FIGURE 4.22 This plot of particle data is the first that shows that the intensity of radiation belt particles can be organized by the (B, L) coordinates. (From McIlwain, 1961)

proximation may not be valid. Consider a particle that is approaching a magnetic neutral point. The gyroradius of this particle will increase because the magnetic field is becoming weaker and if the particles go through the neutral point where the magnetic moment μ becomes ∞, it clearly does not make sense to talk about the adiabatic invariants.

Let us study the motion of particles in the one-dimensional magnetic tail configuration

$$\begin{aligned} B_x &= B_0 & z \geq L \\ B_x &= \frac{B_0 z}{L} & L \geq z \geq -L \\ B_x &= -B_0 & z \leq -L \end{aligned} \tag{4.147}$$

discussed in Chapter 3. Recall that this model supports a neutral line at $z = 0$. The plasma sheet boundary is located at $\pm L$ and the lobe field intensity is constant at B_0. The component equations of the Lorentz equation of motion using (4.1) are

$$\begin{aligned} \ddot{x} &= 0 \\ \ddot{y} &= \left(\frac{qB_0}{mL}\right) z\dot{z} \\ \ddot{z} &= -\left(\frac{qB_0}{mL}\right) z\dot{y} \end{aligned} \tag{4.148}$$

where the dot means time derivative. Now, multiply the second equation by $\dot{y}$ and the third by $\dot{z}$, add, and obtain

$$\frac{d}{dt}(\dot{y}^2 + \dot{z}^2) = 0 \tag{4.149}$$

This describes the motion of particles in the plane perpendicular to the direction of the magnetic field (yz-plane). Equation (4.149) is just the conservation of energy equation.

Noting that $d(z^2/2)/dt = z\dot{z}$, we can integrate the second equation of (4.148) and obtain

$$\dot{y} = \dot{y}_0 + \left(\frac{qB_0}{2mL}\right)(z^2 - z_0^2) \tag{4.150}$$

Insert this into the third equation of (4.148) and after multiplying it by $2\dot{z}$, integrate exactly and obtain

$$\dot{z} = \dot{z}_0^2 + \left(\frac{qB_0}{mL}\right)\left[\left(\frac{qB_0}{2mL}\right) z_0^2 - \dot{y}_0\right](z^2 - z_0^2) - \left(\frac{qB_0}{2mL}\right)^2 (z^4 - z_0^4) \tag{4.151}$$

In these equations, z_0, $\dot{z}_0$, y_0 and $\dot{y}_0$ are initial values. Equation (4.151) is rather complicated. The equation may be simplified and put in the form

(see Rothwell and Yates, 1984, cited at the end of the chapter)

$$\dot{z}^2 = (1 - k^2 + k^2 z^2)(1 - z^2) \tag{4.152}$$

where $k^2 = qB_0 z_m^2 / 4Ly_0$ and z_m represents the point on the z-axis where the particles turn around.

The general solutions of (4.152) involve elliptic integrals and Jacobi elliptic functions, and there are three different classes of orbits depending on the values of k^2. For example, if $k^2 = 1$, the solutions correspond to orbits originating in the plasma sheet that asymptotically approach the neutral line. The orbits corresponding to $k^2 > 1$ are confined to one side of the neutral line and the orbits for $k^2 < 1$ correspond to trajectories that cross the neutral line. Examples of these orbits for protons in Earth's plasma sheet can be found in Rothwell and Yates, 1984, cited at the end of the chapter.

4.10.1 Neutral Sheet Including a Constant Electric Field

The next improvement in the above model is to include a constant dawn-dusk electric field $\mathbf{E} = E_0 \hat{\mathbf{y}}$. In this case, the Lorentz equation yields

$$\begin{aligned} \ddot{x} &= 0 \\ \ddot{y} &= c_1 z \dot{z} + c_2 \\ \ddot{z} &= -c_1 z \dot{y} \end{aligned} \tag{4.153}$$

where $c_1 = qB_0/mL$ and $c_2 = qE_0/m$. These equations are different from the previous case only in the y-equation where particles are acted on by the constant electric field. The first integrals of the last two equations are obtained in the same way as above, yielding

$$\dot{y} = \dot{y}_0 + \left(\frac{c_1}{2}\right)(z^2 - z_0^2) + c_2 t \tag{4.154}$$

$$\frac{1}{2}(\dot{z}^2 + \dot{y}^2) - c_2 y = \frac{1}{2}(\dot{z}_0^2 + \dot{y}_0^2) - c_2 y_0 \tag{4.155}$$

Equation (4.155) is the conservation of energy equation. As before, the subscript 0 refers to initial values. Substitute (4.154) into the third equation of (4.153) and obtain

$$\ddot{z} = -c_1 z \left[\dot{y}_0 + \left(\frac{c_1}{2}\right)(z^2 - z_0^2) + c_2 t\right] \tag{4.156}$$

Equation (4.156) is nonlinear and the solutions may include chaotic motions. While it is not our purpose to discuss chaotic dynamics here, note that (4.156) can be compared to the general non-linear equation

$$\ddot{z} + 2\gamma\dot{z} + \alpha z + \beta z^3 = f(t) \tag{4.157}$$

This equation (4.156) has no damping, $\gamma = 0$, $\alpha = (c_1 y_0 - c_1^2 z_0^2/2 + c_1 c_2 t)$, $\beta = c_1^2/2$, and $f(t) = 0$. In this case periodic solutions exist where the natural frequency increases with amplitude if $\beta > 0$. (For nonlinear equations involving chaotic dynamics, see Chernikov *et al.*, 1988, and Lichtenberg and Lieberman, 1983, cited at the end of the chapter.)

Physical insight is gained by considering the solution of (4.156) for large time. $c_2 t$ is then positive and will monotonically increase, implying the solution is oscillatory and bounded in $z(t)$. Hence, (4.156) for large time can be approximated by

$$\ddot{z} \approx -c_1 c_2 z t = -\left(\frac{q}{m}\right)^2 \left(\frac{E_0 B_0}{L}\right) z t \tag{4.158}$$

The solution of this differential equation is given by

$$z = \sqrt{t'} Z_{1/3}\left(\frac{2}{3} t'^{3/2}\right) \tag{4.159}$$

where $t' = (B_0 E_0/L)^{1/3}(q/m)^{2/3} t$ and $Z_{1/3}$ is a linear combination of the Bessel function of the first and second kinds, of order one-third. For large time, (4.159) is approximately

$$z \approx -\frac{t^{-1/4}}{(E_0 B_0/L)^{1/12}(q/m)^{1/6}} \left\{ A \cos\left[\frac{2}{3}\left(\frac{q}{m}\right)\left(\frac{B_0 E_0}{L}\right)^{1/2} t^{3/2}\right] + B \sin\left[\frac{2}{3}\left(\frac{q}{m}\right)\left(\frac{B_0 E_0}{L}\right)^{1/2} t^{3/2}\right] \right\} \tag{4.160}$$

where A and B are constants that depend on the initial conditions. (4.160) shows that for large time, the amplitude of oscillation decays as $1/t^{1/4}$.

Taking this z-decay into account, (4.154) can now be integrated to yield

$$y \approx y_0 + \left[\dot{y}_0 - \left(\frac{c_1}{2}\right) z_0^2\right] t + \frac{c_2 t^2}{2} \tag{4.161}$$

Equations (4.161) and (4.155) show that the kinetic energy increases as t^2. Thus, a particle executes a damped oscillation about $z = 0$ while accelerating positive ions in the y-direction and electrons in the $-y$-direction. That this indeed happens can be seen as follows. The electric field in (4.153) accelerates a proton in the y-direction. Then, ignoring the term $c_1 z \dot{z}$, we see that $\dot{y}$ is proportional to t. The third equation of (4.153) then becomes

$$\ddot{z} = -kz \tag{4.162}$$

where $k = c_1 \dot{y}$. This equation describes the motion of an oscillator with a spring constant k. Since k becomes larger with time (because of the constant E-field), the spring becomes stiffer, and the oscillation amplitude decays with time. The oscillation of $z(t)$ is due to the $\mathbf{v} \times \mathbf{B}$ force, which is always directed toward $z = 0$ for z positive or negative because the magnetic field reverses its direction across the neutral sheet. The trajectories of the charged particles in this simple electromagnetic field model remain always in the neutral sheet and their energy increases without bound.

4.10.2 Magnetic Tail with a Small Perpendicular Field

Observations show a small magnetic field component perpendicular to the neutral sheet. We discussed in Chapter 3 that

$$\mathbf{B} = B_0 \left(\frac{z}{L} \hat{\mathbf{x}} + \delta \hat{\mathbf{z}} \right) \tag{4.163}$$

where $\delta \ll 1$ models this magnetic field geometry (note that strictly speaking this magnetic geometry is no longer a neutral sheet). The equations of motion if we use (4.163) become

$$\ddot{x} = c_3 \dot{y} \tag{4.164}$$

$$\ddot{y} = -c_3 \dot{x} + c_1 z \dot{z} + c_2 \tag{4.165}$$

$$\ddot{z} = -c_1 z \dot{y} \tag{4.166}$$

where c_1 and c_2 have already been defined and $c_3 = qB_0\delta/m$. The first integrals of these equations are

$$(\dot{x} - \dot{x}_0) = c_3(y - y_0) \tag{4.167}$$

$$\frac{1}{2}[(\dot{y}^2 - \dot{y}_0^2) + (\dot{z}^2 - \dot{z}_0^2)] = c_2(y - y_0) - c_3 \int \dot{x}\dot{y} dt \tag{4.168}$$

$$(\dot{y} - \dot{y}_0) = \frac{c_1}{2}(z^2 - z_0^2) + c_2 t - c_3(x - x_0) \tag{4.169}$$

Integration of (4.169) yields

$$(y - y_0) = \frac{c_1}{2} \int z^2 dt + \frac{c_2 t^2}{2} - c_3 \int x dt + c_4 t \tag{4.170}$$

where $c_4 = \dot{y}_0 - c_1 z_0^2/2 + c_3 x_0$. Now, substitute (4.170) into (4.167) and obtain

$$\dot{x} = \dot{x}_0 + c_5 t^2 + c_6 t + c_7 \int z^2 dt - c_3^2 \int x dt \tag{4.171}$$

where $c_5 = c_2 c_3/2$, $c_6 = c_3 c_4$, and $c_7 = c_1 c_3/2$. Now, from (4.166) obtain

$$\frac{1}{2}(\dot{x}^2 - \dot{x}_0^2) = c_3 \int \dot{x}\dot{y} dt \tag{4.172}$$

and substituting this into (4.168) yields an energy integral

$$\frac{m}{2}(\dot{x}^2 + \dot{y}^2 + \dot{z}^2) + q\Phi = \frac{m}{2}(\dot{x}_0^2 + \dot{y}_0^2 + \dot{z}_0^2) + q\Phi_0 \qquad (4.173)$$

where $q\Phi$ is the electrostatic potential energy obtained from the relation $\mathbf{E}_0 = -\nabla\Phi$.

Consider a proton incident on the neutral sheet with a small velocity. We now discuss qualitatively some of the behaviors of this particle.

a. The particle will be initially accelerated in the $+y$ direction because of the electric field E_0 (4.165).

b. The particle will gain a velocity $\dot{y}$ proportional to t (4.169). It will also be accelerated in the x-direction proportional to $\dot{y}$ or t (4.164). Hence, $\dot{x}$ is proportional to t^2. This motion is new and not present in the previous model.

c. If $\dot{y} > 0$, motion in the z-direction is oscillatory (4.166). The oscillations will grow if $\dot{y}$ increases in time or damp if $\dot{y}$ decreases with time. As long as $\dot{y} > 0$, the term $c_1 z\dot{z}$ in (4.165) can be considered small.

d. The first term in (4.165) is proportional to t^2 (4.171) so that $\dot{y}$ grows negatively until y goes to zero. Then, $\dot{y}$ decreases, going to zero or even becoming negative. The z-motion will execute a damped motion until $\ddot{y} = 0$ (4.166).

e. After $\ddot{y}$ becomes negative, and until $\dot{y} = 0$, z will execute decreasing oscillatory motion. After $\dot{y} < 0$, z no longer oscillates but increases exponentially (4.166) and the particle is ejected from the neutral sheet. Unlike the previous example, the particles do not remain in the neutral sheet but are ejected.

To see when the particles become ejected from the neutral sheet, note that we can write (4.166) with the help of (4.169) again as

$$\ddot{z} = -kz \qquad (4.174)$$

where $k = c_1^2 z^2/2 + c_1 c_2 t - c_1 c_3 x + c_8$, and $c_8 = c_1 c_3 x_0 - c_1 \dot{y}_0 - c_1 z_0^2/2$. Particle ejection time is obtained by letting $k \to 0$, yielding

$$t = \frac{c_3 x}{c_2} - \frac{c_1 z^2}{2c_2} + \frac{c_8}{c_2} \qquad (4.175)$$

This is the time when k becomes negative and z grows exponentially. The ejection time depends on initial conditions as well as on x and z^2. Now integrate (4.171) and obtain

$$x = x_0 + \dot{x}_0 t + \frac{c_6 t^2}{2} + \frac{c_5 t^3}{3} \qquad (4.176)$$

where we have assumed that the integral over z^2 is small since z is oscillating, and the integral over x is also small because it is multiplied by $c_3^2 = B_0^2\delta^2$. Choose the initial conditions $x = \dot{x}_0 = 0$, $c_4 = c_8 = 0$, which indicates that $\dot{y}_0 = c_1 z_0^2/2$. Note that since $c_6 = c_3 c_4$, $c_6 = 0$.

Combine (4.176) and (4.175) with these initial conditions and obtain the ejection time at the neutral sheet ($z = 0$):

$$t = \frac{\sqrt{6}}{q/m} B_0\delta \tag{4.177}$$

(4.176) shows that at ejection,

$$\begin{aligned} x &= \frac{c_5 t^3}{3} \\ &= \frac{\sqrt{6E_0}}{q/m}(B_0\delta)^2 \end{aligned} \tag{4.178}$$

and

$$\begin{aligned} \dot{x} &= c_5 t^2 \\ &= \frac{3E_0}{B_0\delta} \end{aligned} \tag{4.179}$$

The velocity of the particles is independent of q/m within the approximation made. Thus, protons and electrons are both ejected. Note, however, that electrons are ejected much sooner than protons (4.177).

These results are important because they imply that particles in the neutral sheet are accelerated by the electric field and then the particles interacting with the neutral sheet are ejected toward the planet (x-direction). These ejected plasma sheet particles could be an important source for auroras (see Speiser, 1965, and Sonnerup, 1971 for examples of possible particle trajectories).

4.11 Concluding Remarks

The concepts derived from the guiding center motion of particles developed in this chapter provide a basis for the discussion of motion of stably trapped particles. The drift motions of the trapped particles around the planet contribute to the formation of a large-scale current, called the ring current (the topic of currents is covered in Chapter 7). The "quiet time" ring current is intensified during an auroral event by the injection of freshly accelerated particles from the solar wind and the ionosphere. These particles are injected over a large pitch-angle range, including the loss cone, resulting in particle precipitation into the atmosphere. The particles with larger pitch-angles become trapped, intensifying the ring current. Ionospheric particles are also injected along **B** into the magnetosphere along a "source cone." These ionospheric particles also contribute significantly to the ring current.

The ring current particles of planet Earth have on the average several tens of kiloelectron volts (keV) of energy. The intensity of this current is typically one million amperes during a moderate size aurora and it exceeds several million amperes during a strong magnetic storm. The inner boundary of Earth's ring current is found typically at 4 to $5R_E$ equatorial distance, but it moves closer with increased geomagnetic activity. During the most intense magnetic storms, it is found just $1R_E$ above Earth's surface. The magnetosphere at such times is required to support a very intense convective electric field (see Chapter 5). The ring current is a part of the external current that modifies the planetary magnetic field.

The criteria for the applicability of the guiding center theory are quite strict and we must exercise caution in applying this theory to space. For example, the guiding center theory is applicable in regions of space where the magnetic field is strong and well-defined. In regions where the magnetic field is weak and in the neighborhood of neutral points, the concepts of the guiding center lose their precise meaning and particle description must be obtained by using the exact Lorentz equation of motion. As discussed briefly in this chapter, the Lorentz equation even for the very simple geometries becomes non-linear.

Our final comment concerns the origin of magnetospheric particles. Although it is known that these particles come from the solar wind and the planetary ionospheres, it is not known under what conditions the solar wind particles find their way across the magnetopause boundary and how ionospheric (and solar wind) particles are accelerated to several hundred keV energies. These problems will receive much attention in the 1990s. Sophisticated state-of-the-art particle detectors will be launched on several spacecraft to measure ion composition and charge states of Earth's magnetospheric particles (other detectors will measure electromagnetic fields). These experiments will attempt to differentiate the ions of the solar wind from those of the ionosphere. For example, the ion detectors can resolve the solar wind doubly charged helium (He^{++}) from Earth's singly charged helium (He^{+}). The detectors can also detect oxygen (O^{+}) and nitrogen (N^{+}), which are primarily of Earth's ionospheric origin. By studying the distributions of these ions as a function of time, position, and energy, important information on their origin and dynamics will be obtained.

Bibliography

Alfvén, H. and C.G. Fälthammar, *Cosmical Electrodynamics, Fundamental Principles*, 2nd ed., Oxford University Press, Oxford, England, 1963. This book gives a thorough and clear formulation of the guiding center formalism.

Anderson, K.A., Energetic Particles in the Earth's Magnetic Field, *Annual Review of Nuclear Science*, **16**, 291, 1966. A review article on particle observations in the Earth's magnetosphere during the early years of space research.

Anderson, K.A., H.K. Harris and R.J. Padi, Energetic Electron Fluxes in and beyond the Earth's Outer Magnetosphere, *J. Geophys. Res.*, **70**, 1039, 1965. The first report of energetic electron flux observations to distances $> 30R_E$ in the geomagnetic tail.

Büchner, J. and L. Zelenyi, Regular and Chaotic Charged Particle Motion in Magnetotaillike Field Reversals, 1. Basic Theory of Trapped Motion, *J. Geophys. Res.*, **94**, 11821, 1989. This article develops a systematic theory of trapped nonadiabatic charged particle motion in a two-dimensional magnetotaillike field geometry.

Chen, J. and P. Palmadesso, Chaos and Nonlinear Dynamics of Single-Particle Orbits in a Magnetotaillike Magnetic Field, *J. Geophys. Res.*, **91**, 1499, 1986. This article performs numerical integration of the particle equation of motion in a taillike field geometry.

DeForest, S.S. and C.E. McIlwain, Plasma Clouds in the Magnetosphere, *J. Geophys. Res.*, **76**, 3587, 1971. More recent particle measurements, which show features that are consistent with particle drift phenomena.

Freden, S.C. and R.S. White, Particle Fluxes in the Inner Radiation Belt, *J. Geophys. Res.*, **65**, 1377, 1960. One of the first measurements of particle energy spectra in space.

Hamlin, D.A., R. Karpulus, R.C. Vik and K.M. Watson, Mirror and Azimuthal Drift Frequencies for Geomagnetically Trapped Particles, *J. Geophys. Res.*, **66**, 1, 1961. One of the first articles that published computations of bounce and drift frequencies of particles in a dipole field of Earth.

Harris, K.K., G.W. Sharp and C.R. Chappell, Observations of the Plasmapause from OGO 5, *J. Geophys. Res.*, **75**, 219, 1970. See also Gringauz, K.I., The Structure of the Plasmasphere on the Basis of Direct Measurements, in *Solar Terrestrial Physics, 1970*, E.R. Dyer, ed., D. Reidel Publishing Co., Dordrecht, Holland, 1972. One of the early papers that discussed observations of the plasmaspheric particles.

Hess, W.N., *The Radiation Belt and Magnetosphere*, Blaisdell Publishing Co., Waltham, MA, 1968. This book is for researchers as it contains a vast amount of data obtained by spacecraft experiments during the early era of the space age. The first chapter presents an interesting story about the race for space by the United States and the Soviet Union.

Katz, L. and D. Smart, Measurements on Trapped Particles Injected by Nuclear Detonations, *Space Res.*, **4**, 646, 1963. One of the first papers that showed the first invariant is operable.

McIlwain, C.E., Coordinates for Mapping the Distribution of Magnetically Trapped Particles, *J. Geophys. Res.*, **66**, 3681, 1961. McIlwain was one of the first to recognize that particles in Earth's radiation belt move around, conserving their adiabatic invariants. He then invented a set of parameters that used the concept of the invariants and organized the observed particle distributions in space.

Northrop, T.G., The Guiding Center Approximation to Charged Particle Motion, *Ann. Phys.*, **15**, 209, 1961. Northrop has been working in laboratory fusion problems and this article is one of his first papers dealing with the particle trajectories in electromagnetic fields. Northrop's treatment tends to be more mathematical than physical. He subsequently collected his thoughts on this subject in *The Adiabatic Motion of Charged Particles*, Interscience Publishers, New York, NY, 1963. This monograph gives an advanced treatment of the guiding center formalism, solving the Lorentz equation by the perturbation method.

O'Brien, B.J., Interrelations of Energetic Charged Particles in the Magnetosphere, in *Solar Terrestrial Physics*, J.W. King and W.S. Newman, eds., Academic Press, Inc., London, England, 1967. O'Brien used real data in this early article to show how one applies the first and second invariants.

Roederer, J.G., *Dynamics of Geomagnetically Trapped Radiation*, Springer-Verlag, New York, NY, 1970. This monograph shows how the guiding center theory is applied to particles trapped in Earth's radiation belt. Some subtle effects that arise as a consequence of non-symmetric geomagnetic fields are discussed.

Rothwell, P.L. and G.K. Yates, Global Single Ion Effects Within the Earth's Plasma Sheet, in *Magnetic Reconnection in Space and Laboratory Plasmas*, E.W. Hones, Jr., ed., American Geophysical Union, Washington, D.C., 1984. This article shows numerical results of particles in several neutral sheet models and discusses the importance of the results in the geomagnetic tail of Earth.

Schulz, M. and L.J. Lanzerotti, *Particle Diffusion in the Radiation Belts*, Springer-Verlag, Berlin, Germany, 1974. This monograph consolidates observational and theoretical knowledge of Earth's radiation belt dynamics. The main emphasis is the diffusion process and its role in the radiation belts.

Sonnerup, B.U.Ö., Adiabatic Particle Orbits in a Magnetic Null Sheet, *J. Geophys. Res.*, **76**, 8211, 1971. Examines the same problem as in Speiser's article but emphasizes the exact analytical solutions.

Speiser, T.W., Particle Trajectories in Model Current Sheets, 1. Analytical Solutions, *J. Geophys. Res.*, **70**, 4219, 1965. One of the first articles that examined how particles behave in neutral sheets. Emphasis was given to physical interpretation and application to Earth's geomagnetic tail. Readers interested in details of particles in neutral magnetic field geometries are encouraged to read this article.

Spjeldvik, W.N. and P.L. Rothwell, The Radiation Belts, in *Handbook of Geophysics and Space Environments*, A.S. Jursa, ed., Air Force Geophysics Laboratory, National Technical Information Service, Document Number ADA167000, Springfield, VA, 1985. This article provides a concise summary of the guiding center theory and includes examples of how this theory is applied to observation.

Van Allen, J.A., Particle Description of the Magnetosphere, in *Physics of the Magnetosphere*, R.L. Carovillano, J.F. McClay and H. Radoski, eds., D. Reidel Publishing Co., Dordrecht, Holland, 1968. This is a tutorial article on radiation belt particles and provides a good summary of observations made up to this period. The schematic diagram in Figure 4.2 comes from this article.

Vasyliunas, V.M., Magnetospheric Plasma, in *Solar Terrestrial Physics 1970*, E.R. Dyer and J.G. Roederer, eds., D. Reidel Publishing Co., Dordrecht, Holland, 1972. A survey of low-energy plasma measurements in space.

Yoshida, S., G.H. Ludwig and J.A. Van Allen, Distribution of Trapped Radiation in the Geomagnetic Field, *J. Geophys. Res.*, **65**, 807, 1960. An article that reported the discovery of the radiation belt particles.

There are many excellent articles and books on chaotic dynamics. For an excellent introductory article, readers are referred to: Chaos: How Regular Can it Be? by A.A. Chernikov, R.Z. Sagdeev and G.M. Zaslavsky in *Physics Today*, **27**, November 1988. A thorough treatment of this topic is given by A.J. Lichtenberg and M.A. Lieberman, *Regular and Stochastic Motion*, Springer-Verlag, New York, NY, 1983.

Questions

1. Show that the gravitational force on particles in Earth's ionosphere is negligible compared to the Lorentz force. Assume the particles are 1 eV protons.
2. The rotation of Earth's magnetic field creates an electric field in the rest frame of ionospheric plasma. This corotational electric field is about .015 volts/meter at ionospheric heights. Compute the guiding center drift of the ionospheric particles above the equator due to this electric field.
3. Calculate the guiding center drift of an ionospheric proton on the equator due to Earth's gravitational field. Compare your answer to the one obtained in problem 2.
4. Compute the mass of a magnetized object on which the Lorentz force of a 1 eV proton 100 km away is equal to the gravitational force. Plot the mass as a function of the magnetic field strength, 10^5 teslas (T), 10^3 T, 10^1 T, 10^{-1} T, 10^{-3} T, and 10^{-5} T.
5. Plot the curves showing the behavior of the cyclotron frequency of a proton in a dipole geomagnetic field as a function of the latitude. Do this for the magnetic lines of force whose equatorial crossing distances are 2, 6, and $10R_E$.
6. Plot a curve showing the behavior of the cyclotron radius of 1 keV auroral protons along Earth's dipole line of force whose equatorial crossing is at $6R_E$.
7. Derive an expression for the gradient drift velocity of the particles' guiding centers in a dipole field. Then use this equation to calculate the drift velocity of 90° pitch-angle auroral 1 keV protons on the dipole line of force whose equatorial crossing distance is $6R_E$.
8. Derive an expression for the curvature drift velocity of the particles' guiding centers in a dipole field. Use this equation to calculate the curvature drift of 0° pitch-angle auroral 1 keV protons on the dipole line of force whose equatorial crossing distance is $6R_E$.
9. Calculate the size of the loss cone at the geomagnetic equator for the auroral particles on a dipole line of force whose equatorial crossing distance is $6R_E$. Assume the mirror points are in the ionosphere, 100 km from the surface.
10. Derive an expression showing how the loss cone size varies as one moves off the equator for particles on a dipole line of force. Plot a curve to demonstrate this variation for particles on a dipole line that crosses the equator at $6R_E$.
11. The guiding center drift due to a perpendicular electric field is valid only if the drift velocity is less than the speed of light. Prove this statement.
12. Suppose there is a potential difference between the equator and the ionosphere. Let the potential $\phi = 0$ at the equator and $\phi = \phi$ in the ionosphere where the particles mirror (assume $\phi > 0$ and hence the electrons are accelerated as they travel toward the ionosphere). Use the equation of motion along **B** that includes the potential and

the conservation of the total energy and prove that the magnetic moment of the particle is conserved. Assume that $\nabla\phi = -d\phi/ds = -E_{\|}$ where s is along **B**.

13. From the results of problem 12, derive the expression of the new mirroring latitudes for the electrons of energy ε injected on the equator with pitch-angles α_0.
14. Equation (4.91) shows that the magnetic moment is independent of time if the perturbation time scale is much longer than the cyclotron period. Show that the magnetic moment is constant for fields varying slowly in space. That is, show that

$$\frac{d}{ds}\left(\frac{\varepsilon_{\perp}}{B}\right) = 0$$

where ds is the arc length along **B**.

15. Derive Equation (4.151) by use of Equations (4.150) and (4.148).
16. Consider a pulsar whose mass is 10^{30} kg and the surface magnetic field strength is 10^5 T. The spin period of this pulsar is 10^{-3} seconds. Compute the energy of a proton on this pulsar such that the gravitational force balances the Lorentz force.
17. If the energy of a particle exceeds its rest mass energy (the rest mass of electrons and protons is 0.51 MeV and 980 MeV, respectively), relativistic effects must be taken into account. Derive an expression for the gyroradius and gyrofrequency of a relativistic particle. The relativistic expression of the momentum of a particle is $p = \gamma m_0 v$, where $\gamma = (1-\beta^2)^{1/2}, \beta = v/c$, and m_0 is the rest mass. What is the expression of the magnetic moment for a relativistic particle?
18. Suppose a relativistic particle is in a time-dependent magnetic field so that $\nabla\times\mathbf{E} = -\partial\mathbf{B}/\partial t$. Show that the energy of this particle changes according to

$$\frac{d(\gamma m_0 c^2)}{dt} = \frac{p_{\perp}^2}{2\gamma m_0 c^2}\frac{\partial B}{\partial t}$$

The particle here is accelerated by "betatron acceleration."

19. Show that the relativistic magnetic moment of a particle is conserved.

5

Magnetohydrodynamic Equations and Concepts

5.1 Introduction

The guiding center theory (first-order orbit theory) of single particles that we developed in the preceding chapter helps us understand how particles move around in complex electromagnetic fields. While this theory has given us important insights, it is not adequate when a system contains many particles. The single-particle picture must be augmented with a theory that can provide information on the dynamics of a group of particles.

The purpose of this chapter is to develop a fluid theory to describe the behavior of a group of particles. It is fortunate that under certain restrictions, a collection of charged particles can be treated as a fluid, a magnetohydrodynamic (MHD) fluid. Fluid dynamics deals with macroscopic phenomena where matter is treated generally as a continuum and deformable. Magnetohydrodynamics is a branch of continuum mechanics that deals with the motion of electrically conducting material in the presence of electromagnetic fields. Fluid dynamics treats a group of particles as an entity and assumes the particles are non-interacting. Magnetohydrodynamics also ignores the identity of individual particles and considers only the fluid element. The motion of an ensemble of these particles constitutes a fluid motion.

An important element of MHD theory is that it incorporates the effects that arise from the motion of an electrically conducting fluid across magnetic fields. It is well-known that when a conductor is moved across a magnetic field, an electromotive force appears in the conductor. Currents driven by this electromotive force will then flow in the conductor. The magnetic fields associated with these currents will modify the original magnetic field that created them. As a consequence, the original motion becomes altered. This collective interaction involving motion, currents, and magnetic fields characterizes the general behavior of magnetohydrodynamic fluids.

MHD theory is designed to describe the physics of macroscopic phenomena. The foundations of the macroscopic equations are the moments of the Boltzmann equation. As mentioned in Chapter 2, an inherent difficulty with the moments approach is that there are always more variables than equations. Thus, the problem is indeterminate unless more equations can be obtained. A method used to circumvent this difficulty in practice is to use physical arguments and assume restrictive properties (which can sometimes be very severe) about the nature of the fluid. For example, a charged particle system can sometimes be treated as an electrical conductor. In this case, one has the use of Ohm's law. Another assumption often made is to consider a plasma to be in thermal equilibrium. Equilibrium plasma can be described by a Maxwellian distribution function and, as shown later, this plasma behaves as an ideal gas ($p = nkT$) with an adiabatic equation of state.

Early development of magnetohydrodynamics was closely tied to the discovery of sunspots in 1908. Magnetohydrodynamics provided a forum then and even now for the discussion of how the solar magnetic field is generated and maintained. More recently, *in situ* measurements of space plasmas in our solar system by spacecraft-borne experiments have given direct evidence that many classes of observed large-scale electrodynamic phenomena can be understood if MHD theory is used. This chapter will develop the basic MHD theory that is needed to model, for example, the solar wind, the IMF, and convective motions of plasma inside magnetospheres.

5.2 General Concepts of Fluid Dynamics

The behavior of ordinary fluids and systems containing charged particles differs considerably, but certain concepts and equations that have been developed to describe the ordinary fluids are general and applicable to systems containing charged particles approximated as conducting fluids.

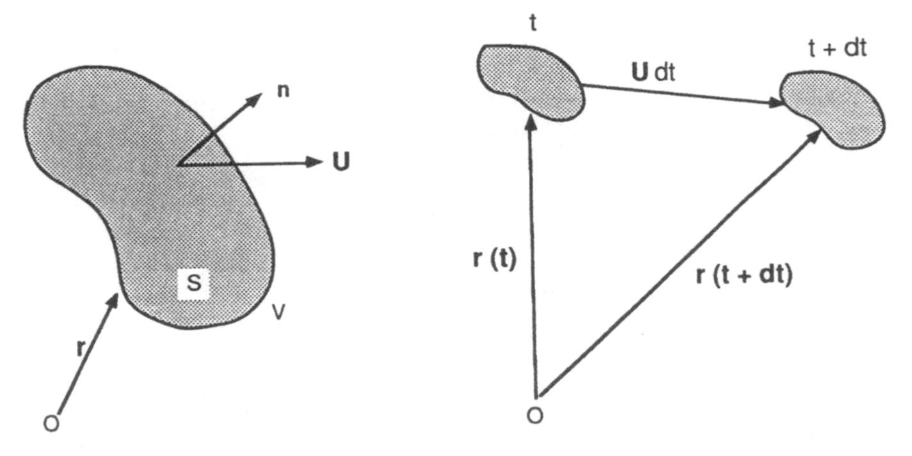

FIGURE 5.1 A schematic diagram to illustrate the parameters of a fluid flow.

5.2.1 Mass Conservation

Like an ordinary fluid medium, an MHD fluid medium is treated as a continuous medium. Fluid dynamics deals primarily with macroscopic phenomena, which means that any small volume in a fluid element contains many particles. Thus, we can define macroscopic parameters such as the density ρ as meaningful in describing a fluid property.

Consider now an arbitrary region of space consisting of an MHD fluid that occupies a volume V bounded by a surface S (Figure 5.1). Let dV be an element of this volume and dS an element of the surface. Let $\mathbf{U}(x, y, z, t)$ be the velocity of a fluid element at a given position in space (x, y, z) and time t. A fluid is displaced in time dt a distance $\mathbf{U}dt$ and the mass of fluid crossing dS per unit time is

$$dm = \rho\mathbf{U} \cdot \mathbf{n}\, dS\, dt \tag{5.1}$$

The total mass of the fluid flowing out of the volume V per unit time is

$$m = \int_S \rho\mathbf{U} \cdot \mathbf{n}\, dS \tag{5.2}$$

where the integral extends over the surface S. This outward flow will result in a decrease of fluid contained in V and the total amount that is diminished per unit time is

$$m = -\int_V \frac{\partial\rho}{\partial t} dV \tag{5.3}$$

where the integral extends over the volume and the negative sign indicates that fluid is being lost. Since there are no sources or sinks for the fluid these two expressions can be equated

$$\int_S \rho\mathbf{U}\cdot\mathbf{n}\,dS = -\int_V \frac{\partial\rho}{\partial t}dV \tag{5.4}$$

Now convert the surface integral term to the volume integral by means of the divergence theorem and obtain

$$\int_V \left(\nabla\cdot\rho\mathbf{U} + \frac{\partial\rho}{\partial t}\right) dV = 0 \tag{5.5}$$

The integrand is continuous and, since this relation holds for any arbitrary volume V, the integrand vanishes,

$$\frac{\partial\rho}{\partial t} + \nabla\cdot\rho\mathbf{U} = 0 \tag{5.6}$$

This equation, known as the equation of continuity, is fundamental in hydrodynamics. The continuity equation states that matter is conserved and it is valid for all fluids, whether the fluid is adiabatic, compressional, isothermal, or turbulent. The vector $\rho\mathbf{U}$ is in the direction of the flow and represents the mass flux density. Matter is conserved in MHD and thus an MHD fluid also obeys the continuity equation.

5.2.2 The Convective Derivative

The density of the fluid $\rho(x, y, z, t)$ depends on time t explicitly and on coordinates (x, y, z) implicitly, since the fluid coordinate changes with time as the particles are displaced. Thus, the total time rate of change of the density is

$$\begin{aligned}\frac{d\rho}{dt} &= \frac{\partial\rho}{\partial t} + \frac{\partial\rho}{\partial x}\frac{dx}{dt} + \frac{\partial\rho}{\partial y}\frac{dy}{dt} + \frac{\partial\rho}{\partial z}\frac{dz}{dt} \\ &= \frac{\partial\rho}{\partial t} + (\mathbf{U}\cdot\nabla)\rho \end{aligned} \tag{5.7}$$

and the second line follows noting that $\mathbf{U} = \mathbf{i}dx/dt + \mathbf{j}dy/dt + \mathbf{k}dz/dt$ and $\nabla = \mathbf{i}\partial/\partial x + \mathbf{j}\partial/\partial y + \mathbf{k}\partial/\partial z$. Here, $(\mathbf{i}, \mathbf{j}, \mathbf{k})$ are unit vectors in the Cartesian coordinate system. In this equation, $d\rho/dt$ is the rate of change of density as one moves with the fluid, and $\partial\rho/\partial t$ is the rate of change of density at a fixed point in space. The term $(\mathbf{U}\cdot\nabla)\rho$ represents the spatial change of ρ in the direction of the flow velocity $\mathbf{U}$.

The total derivative in (5.7) in hydrodynamics is called the convective derivative. This relationship between the rate of change of a variable in moving and fixed frames of reference is a general operational procedure that can be applied to any variable of the fluid, including the variables of MHD fluids.

5.2.3 Streamlines

Flowing fluids can be specified by the velocity of the fluid at every point in the flow region at any given time. The concept of streamlines is useful in describing such flow fields. A streamline is defined as a line in the flow whose tangent is everywhere parallel to the flow velocity **U**. Note that the paths of particles and the streamlines do not necessarily coincide. The tangent to the streamlines is parallel to the velocity of fluid particles at various points in space at any fixed time, while the tangent to the particle paths is parallel to the direction of the velocity of the given fluid particles at all times. However, in steady-state, streamlines are not functions of time and the tangent of a streamline at any point gives the direction of the particle velocity of the fluid at that point.

Let $d\mathbf{l}$ be a differential element along the streamline and **U** be the flow velocity. The definition of streamlines requires that

$$d\mathbf{l} \times \mathbf{U} = 0 \tag{5.8}$$

If U_x, U_y, and U_z are components of the flow vector in a Cartesian coordinate system, the differential equation of a family of streamlines is given by

$$\frac{dx}{U_x} = \frac{dy}{U_y} = \frac{dz}{U_z} \tag{5.9}$$

These differential equations define the velocity fields of the flow. The streamlines in steady-state are fixed and therefore the trajectories of the fluid particles are also fixed. This means that in steady-state the particles on a streamline are constrained to remain on the same streamline, and the particles on different streamlines can never come into contact with each other. There is no "mixing" of the fluid in steady-state. Mixing can occur, however, if flows are time-dependent.

5.2.4 Irrotational Flows

A flow is classified as irrotational if $\nabla \times \mathbf{U} = 0$. For such flows, there is a scalar function Φ such that

$$\mathbf{U} = -\nabla\Phi \tag{5.10}$$

Φ is called a scalar potential of the flow vector **U** (irrotational flows are also referred to as potential flows). The components of $\mathbf{U} = -\nabla\Phi$ in Cartesian coordinates are

$$\frac{\partial \Phi}{\partial x} = U_x$$
$$\frac{\partial \Phi}{\partial y} = U_y$$
$$\frac{\partial \Phi}{\partial z} = U_z \tag{5.11}$$

and the velocity potential Φ is determined from

$$\Phi(x, y, z) = \int (U_x dx + U_y dy + U_z dz) \tag{5.12}$$

The surfaces on which Φ is a constant are called equipotential surfaces. The curves along which the potential is constant are obtained by setting

$$d\Phi(x, y, z) = 0 \tag{5.13}$$

Since $\mathbf{U} = -\nabla\Phi$, the equipotential surfaces are orthogonal to the surfaces of velocity fields.

5.2.5 Incompressible Flows

A flow is classified as incompressible if the divergence of the flow vector $\mathbf{U}$ vanishes, $\nabla \cdot \mathbf{U} = 0$. The flow vector in this case is solenoidal. Incompressible flow fields $\mathbf{U}$ can be described by a vector $\mathbf{w}$ such that

$$\mathbf{U} = \nabla \times \mathbf{w} \tag{5.14}$$

since the divergence of the curl of any vector vanishes. Irrotational and solenoidal vector fields are mathematically important because any vector field $\mathbf{A}$ can be written as $\mathbf{A} = \nabla\Phi + \nabla \times \mathbf{w}$.

The density of incompressible fluid is a constant along a streamline. To see this, expand the divergence term in the equation of continuity (5.6)

$$\frac{\partial \rho}{\partial t} + \nabla \cdot \rho \mathbf{U} = 0$$

$$\frac{\partial \rho}{\partial t} + (\mathbf{U} \cdot \nabla)\rho + \rho \nabla \cdot \mathbf{U} = 0$$

$$\frac{d\rho}{dt} + \rho \nabla \cdot \mathbf{U} = 0 \tag{5.15}$$

where the definition of the convective derivative has been used. It follows then that if $\nabla \cdot \mathbf{U} = 0$, $d\rho/dt = 0$. If a flow is both irrotational and solenoidal, then it can be shown that the flow potential satisfies Laplace's equation:

$$\begin{aligned} 0 &= \nabla \cdot \mathbf{U} \\ &= -\nabla \cdot \nabla\Phi \\ &= -\nabla^2\Phi \end{aligned} \tag{5.16}$$

Standard methods can be used to solve the Laplace equation with specified boundary conditions.

5.2.6 Stream Function and Cauchy-Riemann Relation

In many problems, symmetry in one of the directions permits the reduction of a three-dimensional problem into a two-dimensional one. Consider now a two-dimensional flow field $\mathbf{U} = (U_x, U_y)$. Let C be an arbitrary curve in the xy-plane in which the fluid is flowing and let $d\mathbf{l}$ be a differential line element on C. The amount of fluid crossing C per unit time is the area swept out by the motion

$$\begin{aligned} \mathbf{A} &= \int_C \mathbf{U} \times d\mathbf{l} \\ &= \int_C (U_x dy - U_y dx) \end{aligned} \tag{5.17}$$

If C is a closed curve and the fluid is incompressible, then the net amount of fluid crossing C is zero since as much fluid enters C as leaves it. An incompressible fluid in steady-state is characterized by

$$0 = \int_C (U_x dy - U_y dx) \tag{5.18}$$

where the integration is over any closed curve C. This equation implies that the integrand is an exact differential of the function $\Psi(x, y)$, called the stream function,

$$\Psi = \int (U_x dy - U_y dx) \tag{5.19}$$

where

$$\begin{aligned} \frac{\partial \Psi}{\partial x} &= -U_y \\ \frac{\partial \Psi}{\partial y} &= U_x \end{aligned} \tag{5.20}$$

The Cauchy-Riemann equation relates the stream function to the velocity potential. This relation lends powerful tools for solving two-dimensional problems. If the fluid is incompressible and irrotational, then we find from comparison of (5.20) and (5.11) that

$$\begin{aligned}\frac{\partial \Psi}{\partial x} &= -\frac{\partial \Phi}{\partial y} \\ \frac{\partial \Psi}{\partial y} &= \frac{\partial \Phi}{\partial x}\end{aligned} \tag{5.21}$$

These are the famous Cauchy-Riemann equations. The streamlines $\mathbf{U} = -\nabla\Phi$ are orthogonal to the curves $\Phi(x, y) =$ constant, and tangent to the curves $\Psi(x, y) =$ constant.

5.3 One-Fluid Magnetohydrodynamic Equations

A plasma contains electrons and at least one ion species. Consider such a system of particles moving in an electromagnetic field (unless stated otherwise, the ions are assumed to be protons). This system will require two sets of fluid equations to describe their dynamics, one for the electron fluid motion and the other for the ion motion. Unfortunately, the general solutions of the "two-fluid" equations are not easily obtained because the electron and ion motions are coupled. However, the two-fluid equations can be reduced to a set of "one-fluid" equations, which is much simpler to study. The one-fluid equations, although less complete, still yield important information on the basic behavior of MHD fluids.

5.3.1 Maxwell Equations

Electromagnetic fields are governed by Maxwell equations. Magnetohydrodynamic theory assumes that free charges do not accumulate in the fluid because a system of charged particles is a good electrical conductor. MHD further considers the displacement current to be negligible compared to the conduction current. The "modified" Maxwell equations are

$$\begin{aligned}\nabla \times \mathbf{E} &= -\frac{\partial \mathbf{B}}{\partial t} \\ \nabla \cdot \mathbf{B} &= 0 \\ \nabla \times \mathbf{H} &= \mathbf{J} \\ \nabla \cdot \mathbf{D} &= 0\end{aligned} \tag{5.22}$$

The assumption that the charge density vanishes means the MHD fluid is electrically neutral. Let N be the total number of charged particles contained in a fluid volume V. Define the number density as $n = N/V$. The vanishing of the total charge density ρ_c for a fluid consisting of electrons and one ion species requires that

$$\begin{aligned}\rho_c &= \rho_c^+ + \rho_c^- \\ &= q_i n_i + q_e n_e \\ &= 0\end{aligned} \tag{5.23}$$

where q is the charge and the subindices i and e and the superscripts $+$ and $-$ refer to ions and electrons. For singly charged ions, $q_i = -q_e$, and $n_i = n_e$. The total charge is conserved. Hence the charge density also obeys the conservation equation (5.6).

5.3.2 Equation of Continuity

Each of the electron and ion species conserves mass. According to (5.6), we require a two-component plasma to satisfy

$$\begin{aligned}\frac{\partial \rho_i}{\partial t} + \nabla \cdot \rho_i \mathbf{u}_i &= 0 \\ \frac{\partial \rho_e}{\partial t} + \nabla \cdot \rho_e \mathbf{u}_e &= 0\end{aligned} \tag{5.24}$$

Add these two equations and obtain

$$\frac{\partial}{\partial t}(\rho_i + \rho_e) + \nabla \cdot (\rho_i \mathbf{u}_i + \rho_e \mathbf{u}_e) = 0 \tag{5.25}$$

Now define the mass density of the fluid

$$\begin{aligned}\rho_m &= \rho_m^+ + \rho_m^- \\ &= n_i m_i + n_e m_e\end{aligned} \tag{5.26}$$

and the fluid velocity

$$\mathbf{U} = \frac{n_i m_i \mathbf{u}_i + n_e m_e \mathbf{u}_e}{n_i m_i + n_e m_e} \tag{5.27}$$

Use of (5.26) and (5.27) reduces (5.25) to

$$\frac{\partial \rho_m}{\partial t} + \nabla \cdot \rho_m \mathbf{U} = 0 \tag{5.28}$$

Note that the fluid velocity is like the center of mass velocity, which is the velocity of a fictitious point where all of the fluid mass is centered.

5.3.3 Momentum Equation

We start with the equation of motion for a single particle, given by

$$m\frac{d\mathbf{v}}{dt} = q(\mathbf{E} + \mathbf{v} \times \mathbf{B}) \tag{5.29}$$

Assume now that there are no thermal motions and ignore collisions. Then all particles move together and the fluid equation of motion of the particles is obtained by multiplying (5.29) with n,

$$mn\frac{d\mathbf{u}}{dt} = qn(\mathbf{E} + \mathbf{u} \times \mathbf{B}) \tag{5.30}$$

Here, we have distinguished the velocity of the individual particle $\mathbf{v}$ from the fluid velocity $\mathbf{u}$, which represents an averaged velocity (the particle velocity has been averaged over a distribution function). Note that (5.30) is actually an equation of the force density.

The thermal motion of the particles gives rise to a pressure. Pressure arises from the collective action of the random motion of the particles. If the charged particles are treated as an ideal gas, the gas pressure is related to the temperature T by means of $p = nkT$, where n is the density and k is the Boltzmann constant equal to 1.38×10^{-23}Joules/°K. If the pressure is not uniform, a force arises and this force must be included in the equation of motion. Equation (5.30), including the pressure force, becomes

$$mn\frac{d\mathbf{u}}{dt} = qn(\mathbf{E} + \mathbf{u} \times \mathbf{B}) - \nabla p \tag{5.31}$$

The minus sign in front of ∇p means that the fluid is driven in the opposite direction from the direction of the pressure gradient. Equation (5.31) assumes that the pressure is isotropic, that is, p is a scalar.

A system consisting of electrons and one species of ions requires an equation of motion for each species. These are

$$\begin{aligned} m_i n_i \frac{d\mathbf{u}_i}{dt} &= q_i n_i(\mathbf{E} + \mathbf{u}_i \times \mathbf{B}) - \nabla p_i \\ m_e n_e \frac{d\mathbf{u}_e}{dt} &= q_e n_e(\mathbf{E} + \mathbf{u}_e \times \mathbf{B}) - \nabla p_e \end{aligned} \tag{5.32}$$

Since $n_i = n_e = n$ and $q_i = -q_e = \mid q \mid$, adding the two equations yields

$$n\frac{d}{dt}(m_i\mathbf{u}_i + m_e\mathbf{u}_e) = qn(\mathbf{u}_i - \mathbf{u}_e) \times \mathbf{B} - \nabla p \tag{5.33}$$

where $p = p_i + p_e$ is the total pressure. Define the current density $\mathbf{J}$

$$\mathbf{J} = n_i q_i \mathbf{u}_i + n_e q_e \mathbf{u}_e \tag{5.34}$$

and, using (5.26), (5.27), and (5.34), obtain

$$\rho_m \frac{d}{dt}\mathbf{U} = \mathbf{J} \times \mathbf{B} - \nabla p \tag{5.35}$$

Note that the electric field does not appear explicitly in this one-fluid momentum equation.

Compared to the equation of motion of ordinary gas treated as a fluid, (5.35) contains an additional term, the $\mathbf{J} \times \mathbf{B}$ term. $\mathbf{J} \times \mathbf{B}$ arises from the coupling of the current density to the magnetic field $\mathbf{B}$. It is this $\mathbf{J} \times \mathbf{B}$ force that makes the electromagnetic fluid different from the ordinary gas fluid consisting of only neutral particles. The $\mathbf{J} \times \mathbf{B}$ force is referred to as the electromagnetic stress tensor (more on this later).

It is instructive to pause here a moment and remind ourselves that the equations of continuity and motion are derived by taking the moments of the Boltzmann equation (integrating over all velocities). These two equations apply to all fluids regardless of their nature and, together with the Maxwell equations, they are considered fundamental. These equations contain fifteen independent variables, $\mathbf{E}$, $\mathbf{B}$, $\mathbf{J}$, $\mathbf{U}$, ρ_m, ρ_c, and p but there are only eleven independent equations: two of the Maxwell equations provide six scalar equations (note that only the first and third of the Maxwell equations (5.22) are independent), the mass and charge continuity relations (5.6) provide two equations, and the equation of motion (5.35) provides three more equations.

Solutions to the fluid equations can be determined only if we find four more equations. As mentioned earlier, additional equations are usually obtained by invoking restrictive properties about the nature of the fluid. For example, if one assumes that a fluid behaves like a conductor, Ohm's law can be used and this will provide three more equations. Another assumption often made is to assign a thermodynamic equation of state to a fluid. This provides another equation, giving a total of fifteen independent equations. The "closed" set of MHD equations requires restrictive measures, hence caution must be exercised in their use.

5.3.4 Ohm's Law

An MHD fluid is a good conductor. As with ordinary conductors, currents can flow in an MHD fluid. An empirical law that relates the current density $\mathbf{J}$ to the electric field $\mathbf{E}$ by means of the electrical conductivity (valid in the rest frame of the medium) of the fluid σ is

$$\mathbf{J} = \sigma \mathbf{E} \tag{5.36}$$

This equation is known as Ohm's law. $\mathbf{E}$ is the total electric field here and must include the electric field induced by the motion of the fluid across magnetic fields. Ohm's law then becomes

$$\mathbf{J} = \sigma(\mathbf{E} + \mathbf{U} \times \mathbf{B}) \tag{5.37}$$

where $\mathbf{U}$ is the fluid velocity. Equation (5.37) is referred to as a simple form of Ohm's law. It is an approximation of a generalized Ohm's law (see Chapter 7). Equation (5.37) is applicable in regions of space where the concept of electrical conductivity can be defined and computed. For instance, (5.37) has meaning for plasmas in the Earth's ionosphere and the Sun's photosphere where collisions occur frequently. This equation is not easily applied in the more distant regions where the plasma is tenuous and "collisionless."

Ohm's law permits us to see more clearly why we can neglect the displacement current in MHD theory. The neglect of the displacement current means that MHD deals with low-frequency phenomena. To see this, note that

$$\left|\frac{\partial \mathbf{D}}{\partial t}\right| = \epsilon_0 \left|\frac{\partial \mathbf{E}}{\partial t}\right| \approx \epsilon_0 \frac{|\,\mathbf{E}\,|}{T} \tag{5.38}$$

and since the displacement current is required to be much smaller than the conduction current,

$$\epsilon_0 \frac{|\,\mathbf{E}\,|}{T} \ll \mathbf{J} = |\,\nabla \times \mathbf{H}\,| \approx \frac{H}{L} = \frac{B}{\mu_0 L} \tag{5.39}$$

Now for a very highly conducting fluid, $\mathbf{J}/\sigma \ll 1$ and (5.37) shows that in the limit $J/\sigma \rightarrow 0$, $|\,\mathbf{E}\,| = |\,\mathbf{U} \times \mathbf{B}\,|$. Hence (5.39) is approximated to be

$$\epsilon_0 U \frac{B}{T} \ll \frac{B}{\mu_0 L} \tag{5.40}$$

and, solving for T, we see that

$$T \gg \frac{UL}{c^2} \tag{5.41}$$

where c is the speed of light and $c^2 = 1/\mu_0\epsilon_0$. For most situations, $U/c \ll 1$. Hence (5.41) implies that the displacement current can be ignored if $T \gg L/c$. Physically, this means that a characteristic time variation T of electromagnetic quantities in the fluid must be much longer than the time it takes for light to travel a characteristic distance L.

Ohm's law can also be used to show that free charges do not accumulate in the rest frame of electrical conductors. To demonstrate this, consider the charge conservation equations for ions and electrons given by

$$\begin{aligned} \frac{\partial \rho_c^+}{\partial t} + \nabla \cdot \rho_c^+ \mathbf{u}^+ &= 0 \\ \frac{\partial \rho_c^-}{\partial t} + \nabla \cdot \rho_c^- \mathbf{u}^- &= 0 \end{aligned} \tag{5.42}$$

where $(\rho_c^+, \mathbf{u}^+)$ and $(\rho_c^-, \mathbf{u}^-)$ are ion and electron charge densities and velocities, respectively. As in the case of the mass conservation equation (5.24), the addition of these two equations yield

$$\frac{\partial \rho_c}{\partial t} + \nabla \cdot \mathbf{J} = 0 \tag{5.43}$$

where $\rho_c = \rho_c^+ + \rho_c^-$ is the total charge density, and $\mathbf{J} = \rho_c^+ \mathbf{u}^+ + \rho_c^- \mathbf{u}^-$ is the total current density. Now use Ohm's law, $\mathbf{J} = \sigma \mathbf{E}$ and then, assuming that σ is constant, obtain

$$\begin{aligned} \frac{\partial \rho_c}{\partial t} + \sigma \nabla \cdot \mathbf{E} &= \frac{\partial \rho_c}{\partial t} + \frac{\sigma}{\epsilon_0} \rho_c \\ &= 0 \end{aligned} \tag{5.44}$$

since $\nabla \cdot \mathbf{E} = \rho_c/\epsilon_0$. The solution of this differential equation is $\rho_c = \rho_0 e^{-(\sigma/\epsilon_0)t}$ where ρ_0 is the initial charge at $t = 0$. This equation shows that charge density in a medium of conductivity σ decays away in time. Charge density in plasmas is expected to decay very rapidly since plasma is a good conductor. This is why the MHD formulation (5.22) ignores the charge density.

5.3.5 Equation of State

The description of MHD fluid, like ordinary fluid, requires an equation that relates the variables ρ_m, p, and T. This equation, called an equation of state, is given by a thermodynamic relation, for example, $p = p(\rho_m, t)$. Normal practice is to assume that the fluid is either adiabatic or isothermal,

$$\begin{aligned} \frac{d}{dt}(p \rho_m^{-\gamma}) &= 0 \\ \frac{d}{dt}\left(\frac{p}{\rho_m}\right) &= 0 \end{aligned} \tag{5.45}$$

Here γ is the ratio of specific heats, C_p/C_v, where C_p and C_v are the specific heats at constant pressure and volume, respectively. The value of γ depends on the number of degrees of freedom the particles have in gases. For example, $\gamma = 3$ for field-aligned motions and $\gamma = 2$ for motions perpendicular to the magnetic field direction. Later we will show that $\gamma = 5/3$ for a plasma defined by an isotropic Maxwellian distribution function.

5.3.6 Energy Equation

The MHD fluid has an energy equation, which we now derive. Begin with the momentum equation (5.35), which can be written as

$$\rho_m \frac{d\mathbf{U}}{dt} = -\nabla p + \frac{1}{\mu_0}(\nabla \times \mathbf{B}) \times \mathbf{B} \tag{5.46}$$

Now take the dot product of this equation with $\mathbf{U}$ and obtain

$$\rho_m \mathbf{U} \cdot \frac{d\mathbf{U}}{dt} = -\mathbf{U} \cdot \nabla p + \frac{\mathbf{U}}{\mu_0} \cdot (\nabla \times \mathbf{B}) \times \mathbf{B} \tag{5.47}$$

Let us now examine each of these terms. The term on the left side can be written as

$$\begin{aligned} \rho_m \mathbf{U} \cdot \frac{d\mathbf{U}}{dt} &= \rho_m \mathbf{U} \cdot \left(\frac{\partial}{\partial t} + \mathbf{U} \cdot \nabla \right) \mathbf{U} \\ &= \frac{\rho_m}{2} \frac{\partial U^2}{\partial t} + \frac{\rho_m}{2} \mathbf{U} \cdot \nabla U^2 \\ &= \frac{\partial}{\partial t} \frac{\rho_m U^2}{2} - \frac{U^2}{2} \frac{\partial \rho_m}{\partial t} + \frac{\rho_m}{2} \mathbf{U} \cdot \nabla U^2 \end{aligned} \tag{5.48}$$

Use now the continuity equation $\partial \rho_m / \partial t = -\nabla \cdot (\rho_m \mathbf{U})$ and combine the second and the third terms to obtain

$$\rho_m \mathbf{U} \cdot \frac{d\mathbf{U}}{dt} = \frac{\partial}{\partial t} \frac{\rho_m U^2}{2} + \nabla \cdot \frac{\rho_m U^2}{2} \mathbf{U} \tag{5.49}$$

The first term on the right of (5.47) can be rewritten by considering the MHD fluid as adiabatic. We use the first of (5.45) and obtain

$$\rho_m^{-\gamma} \frac{dp}{dt} - \gamma p \rho_m^{-(\gamma+1)} \frac{d\rho_m}{dt} = 0 \tag{5.50}$$

which reduces to

$$\frac{dp}{dt} - \frac{\gamma p}{\rho_m} \frac{d\rho_m}{dt} = 0 \tag{5.51}$$

But note that $dp/dt = (\partial p/\partial t) + (\mathbf{U} \cdot \nabla) p$ and use once again the continuity equation $(d\rho_m/dt) = -(\rho_m) \nabla \cdot \mathbf{U}$ and obtain

$$(1 - \gamma)(\mathbf{U} \cdot \nabla) p + \frac{\partial p}{\partial t} + \gamma \nabla \cdot (p\mathbf{U}) = 0 \tag{5.52}$$

The second term on the right side of (5.47) can be rewritten as follows:

$$\begin{aligned} \frac{1}{\mu_0} \mathbf{U} \cdot (\nabla \times \mathbf{B}) \times \mathbf{B} &= -\frac{1}{\mu_0} (\mathbf{U} \times \mathbf{B}) \cdot (\nabla \times \mathbf{B}) \\ &= \frac{1}{\mu_0} \mathbf{E} \cdot (\nabla \times \mathbf{B}) \end{aligned} \tag{5.53}$$

Here we assume $\mathbf{E} = -\mathbf{U} \times \mathbf{B}$. According to (5.37), this relation implies $\mathbf{J}/\sigma \ll 1$. For a very large σ, the right-hand term in (5.53) will be small. In the limit $\sigma \to \infty$, $\mathbf{J} \cdot \mathbf{E} = 0$ and MHD formulation ignores any work done by the electric field. Use the vector relation $\nabla \cdot (\mathbf{E} \times \mathbf{B}) = \mathbf{B} \cdot (\nabla \times \mathbf{E}) - \mathbf{E} \cdot (\nabla \times \mathbf{B})$

in (5.53), which then can be written as

$$= -\frac{1}{2\mu_0}\frac{\partial B^2}{\partial t} - \frac{1}{\mu_0}\nabla \cdot (\mathbf{E} \times \mathbf{B}) \tag{5.54}$$

We now use (5.49), (5.52), and (5.54) in (5.47) and obtain the energy conservation relation for an adiabatic MHD fluid as

$$\frac{\partial}{\partial t}\left(\frac{\rho_m U^2}{2} + \frac{p}{\gamma - 1} + \frac{B^2}{2\mu_0}\right)$$

$$+\nabla \cdot \left(\frac{\rho_m U^2}{2}\mathbf{U} + \frac{\gamma}{\gamma - 1}p\mathbf{U} + \frac{\mathbf{E} \times \mathbf{B}}{\mu_0}\right) = 0 \tag{5.55}$$

The terms in the first bracket represent the kinetic energy of the fluid motion, the thermal energy, and the total energy density of the magnetic field. The terms in the second bracket represent the rates at which these various energies are flowing. $\mathbf{E} \times \mathbf{B}$ is the Poynting vector and represents the rate at which the electromagnetic energy is being transported. The reader is reminded that in deriving (5.55) an assumption was made that $\mathbf{E} = -\mathbf{U} \times \mathbf{B}$. The physical condition in which this assumption is valid is restricted and quite special, as we will discuss in a later section.

5.3.7 Summary

We summarize here the set of one-fluid MHD equations when a plasma is assumed to behave as a conductor. These are Maxwell equations:

$$\begin{aligned}
\nabla \times \mathbf{E} &= -\frac{\partial \mathbf{B}}{\partial t} \\
\nabla \cdot \mathbf{B} &= 0 \\
\nabla \times \mathbf{H} &= \mathbf{J} \\
\nabla \cdot \mathbf{D} &= 0
\end{aligned}$$

Continuity equation:

$$\frac{\partial \rho_m}{\partial t} + \nabla \cdot \rho_m \mathbf{U} = 0 \tag{5.28}$$

Equation of motion:

$$\rho_m \frac{d}{dt}\mathbf{U} = \mathbf{J} \times \mathbf{B} - \nabla \mathbf{p} \tag{5.35}$$

Ohm's law:

$$\mathbf{J} = \sigma(\mathbf{E} + \mathbf{U} \times \mathbf{B}) \tag{5.37}$$

and an equation of state:

$$\frac{d}{dt}(p\rho_m^{-\gamma}) = 0$$

$$\frac{d}{dt}\left(\frac{p}{\rho_m}\right) = 0 \tag{5.42}$$

In addition, we have the constitutive relations between the field variables,

$$\mathbf{B} = \mu_0 \mathbf{H} \tag{2.5}$$

$$\mathbf{D} = \epsilon_0 \mathbf{E} \tag{2.6}$$

These equations describe the dynamics of an MHD system. Note that the energy equation (5.55) gives additional information and it is not needed in solving many classes of MHD problems.

In deriving the equation of motion (5.35) from (5.32), we could have included other forces, for instance the force of gravity. Also, the ion species need not be assumed to be singly charged. For ions of charge Z, charge neutrality requires that $q_e n_e = Z q_i n_i$, where Z is the charge number. We also assumed the pressure is isotropic. If we relax this condition, ∇p will be replaced by $\nabla \cdot \mathbf{p}$, where $\mathbf{p}$ is the pressure tensor (see Chapter 7).

The final comment concerns the use of Ohm's law. Ohm's law is an empirical relation obtained in laboratory experiments and is valid if the plasma density is high, so that the concept of conductivity is meaningful. In tenuous plasmas the concept of conductivity is not well-defined, especially where the collision mean free path exceeds the typical MHD scale lengths.

5.4 Magnetic Field and MHD Fluids

The electromagnetic field induced by a moving conductor is described by the first equation of (5.22). Use (5.37) to eliminate $\mathbf{E}$ and obtain

$$\nabla \times \left(\frac{\mathbf{J}}{\sigma} - \mathbf{U} \times \mathbf{B}\right) = -\frac{\partial \mathbf{B}}{\partial t} \tag{5.56}$$

Now use $\nabla \times \mathbf{H} = \mathbf{J}$ and the constitutive relation ($\mathbf{B} = \mu_0 \mathbf{H}$) to obtain

$$\nabla \times \left(\frac{\nabla \times \mathbf{B}}{\mu_0 \sigma}\right) - \nabla \times (\mathbf{U} \times \mathbf{B}) = -\frac{\partial \mathbf{B}}{\partial t} \tag{5.57}$$

Expand the first term on the left side by the vector identity $\nabla \times (\nabla \times \mathbf{B}) = -\nabla^2 \mathbf{B} + \nabla(\nabla \cdot \mathbf{B})$ and since $\nabla \cdot \mathbf{B} = 0$, obtain, after rearranging,

$$\frac{\partial \mathbf{B}}{\partial t} = \frac{1}{\mu_0 \sigma} \nabla^2 \mathbf{B} + \nabla \times (\mathbf{U} \times \mathbf{B}) \tag{5.58}$$

This equation describes how the magnetic field $\mathbf{B}$ varies with time in a medium of conductivity σ that is moving with a velocity $\mathbf{U}$ relative to a fixed observer.

The form of (5.58) is identical to the hydrodynamic equation that describes the behavior of vorticity in incompressible fluids:

$$\frac{\partial \boldsymbol{\omega}}{\partial t} = \nu \nabla^2 \boldsymbol{\omega} + \nabla \times (\mathbf{U} \times \boldsymbol{\omega}) \tag{5.59}$$

$\boldsymbol{\omega}$ is the vorticity ($\boldsymbol{\omega} = \nabla \times \mathbf{U}$) and ν is the kinematic viscosity. The first term of this equation describes the effects of diffusion and the second term represents the convection of the vorticity. By analogy, this interpretation can be carried to the MHD fluid. Thus, in (5.58) $1/\mu_0\sigma$ can be defined as the magnetic viscosity. It is a measure of how fast the magnetic field diffuses out of (or into) the MHD fluid, given the conductivity of the fluid.

Although $\boldsymbol{\omega}$ and $\mathbf{B}$ are described by identical equations, we must be careful not to infer that $\mathbf{B}$ is completely analogous to $\boldsymbol{\omega}$ in hydrodynamics. In hydrodynamics, vorticity and flow are related by $\boldsymbol{\omega} = \nabla \times \mathbf{U}$. No such relation exists for the magnetic field. $\boldsymbol{\omega}$ and $\mathbf{B}$ are mathematically related, however, since both are solenoidal vectors.

5.4.1 Diffusion of the Magnetic Field

If the fluid is at rest, $\mathbf{U}$ is set to zero in (5.58). The time variation of the magnetic field is then described by

$$\frac{\partial \mathbf{B}}{\partial t} = \lambda \nabla^2 \mathbf{B} \tag{5.60}$$

where $\lambda = 1/\mu_0\sigma$. It is instructive to compare this equation with the heat conduction equation

$$\frac{\partial T}{\partial t} = \kappa \nabla^2 T \tag{5.61}$$

where κ is the coefficient of heat conduction, and with the vorticity equation,

$$\frac{\partial \boldsymbol{\omega}}{\partial t} = \nu \nabla^2 \boldsymbol{\omega} \tag{5.62}$$

where ν is kinematic viscosity. All of these partial differential equations have the same form except that $\boldsymbol{\omega}$ and $\mathbf{B}$ are vectors and T is a scalar. These equations, which show how $\mathbf{B}$, T, and $\boldsymbol{\omega}$ change in time relative to spatial changes, describe diffusion of the magnetic field, heat, and vorticity. Examples of applications of the heat and vorticity equations include the famous Rayleigh's problem in ordinary fluid dynamics. Note that the coefficients λ, κ, and ν all have the same dimensions (m^2/sec) and, from the similarity of the equations, that they all play the same role. In MHD fluids, the finite conductivity σ results in Ohmic losses and the currents that are responsible for the magnetic field will decay away. If initially there is a magnetic field trapped in the fluid, it will steadily decay. This implies that magnetic energy is being converted into plasma energy. The total energy (particles plus fields) is conserved (Figure 5.2).

The partial differential equation describing the diffusion process was originally developed by François Fourier in 1822 and is basic to many problems of physics. Assuming that each Cartesian component of the magnetic field diffuses with time from its initial configuration $B_i(\mathbf{r}, 0)$, the general solution of (5.60) can be obtained from the Green's function

$$G(\mathbf{r} - \mathbf{r}', t) = (4\pi\lambda t)^{-3/2} \exp\left[-\frac{(\mathbf{r} - \mathbf{r}')^2}{4\lambda t}\right] \tag{5.63}$$

and the magnetic field for $t > 0$ is obtained from the integral equation

$$B_i(\mathbf{r}, t) = \int\int\int G(\mathbf{r} - \mathbf{r}', t) B_i(\mathbf{r}', 0) d^3\mathbf{r}' \tag{5.64}$$

This equation shows how the magnetic field $B_i(\mathbf{r}', 0)$ at an initial position $\mathbf{r}'$ spreads out in time t with a Gaussian profile with width $(4\lambda t)^{1/2}$.

An estimate of how rapidly or slowly diffusion is occurring can be obtained by letting L be the characteristic spatial scale length of $\mathbf{B}$. Then,

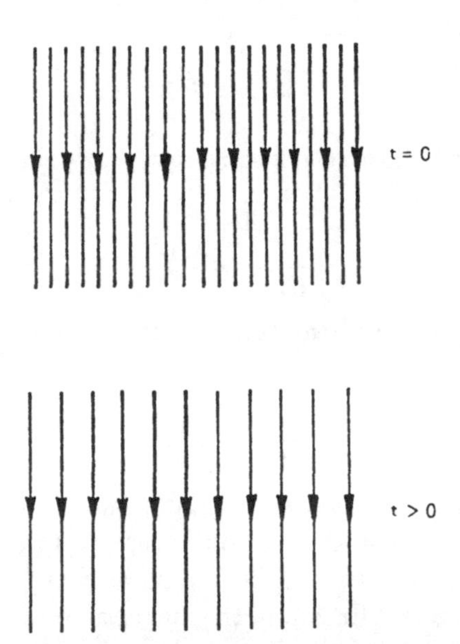

FIGURE 5.2 A schematic diagram to illustrate the concept of magnetic field diffusion in a plasma of finite conductivity.

substitute L^{-2} for ∇^2 and obtain

$$\frac{\partial \mathbf{B}}{\partial t} \approx \pm \frac{1}{\mu_0 \sigma L^2} \mathbf{B} \tag{5.65}$$

where the $\pm$ sign refers to gain or loss of $\mathbf{B}$ with time. The solution of this simplified equation is

$$\mathbf{B} = \mathbf{B}_0 e^{\pm t/t_D} \tag{5.66}$$

where $\mathbf{B}_0$ is the initial value of the magnetic field and

$$t_D = \mu_0 \sigma L^2 \tag{5.67}$$

is the characteristic time for the magnetic field to increase or decay to $1/e$ of its initial value. We mentioned earlier that σ for MHD fluids in space is a large number. When combined with the L^2 dependence, which is also a very large number in space, the diffusion time t_D for magnetic fields can be very long.

5.4.2 Concept of Magnetic Field Annihilation

Instead of the magnetic field configuration shown in Figure 5.2 in which all fields point in the same direction, a configuration often supported by space plasmas includes magnetic fields that are oppositely directed. Consider at time $t = 0$ a magnetic field geometry shown by the top diagram of Figure 5.3. This magnetic geometry differs from the one shown in Figure 5.2 because it includes a magnetic neutral line which is a region where the magnetic field intensity vanishes. Since $\nabla \times \mathbf{B} = \mu_0 \mathbf{J}$, a current exists in this neutral line.

The bottom diagram shows a possible configuration of the magnetic field for $t > 0$. As in the case of Figure 5.2, the weakening of the magnetic field is due to Ohmic dissipative effects and in the present case, the weakening of the magnetic field is due to the dissipation of the current in the neutral line. Since this current was originally created by a plasma system supporting oppositely magnetic fields, another way to describe the weakening of the magnetic field is to say that this field geometry is being destroyed, or oppositely directed fields are being "annihilated." This description invokes the concept of "connection" of oppositely directed field lines (further discussion is given in Chapters 7 and 8).

5.4.3 Magnetic Reynolds Number

A measure of the importance of the diffusion term relative to the transport term is obtained from (5.58) by use of dimensional arguments. Let t represent the characteristic time for magnetic field changes and t_D the magnetic

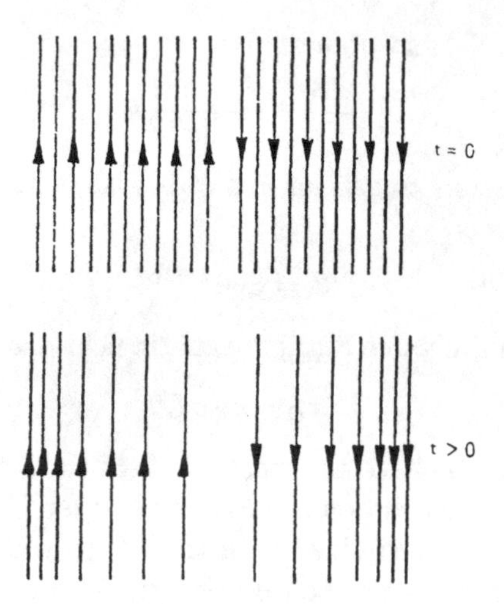

FIGURE 5.3 A plasma system supporting oppositely directed magnetic fields.

diffusion time as defined above. Then (5.58) can be written as

$$\begin{aligned} \frac{1}{t} &= \frac{1}{t_D} + \frac{U}{L} \\ &= \frac{1}{t_D} + \frac{1}{t_T} \end{aligned} \tag{5.68}$$

where t_T is the characteristic transport time defined as L/U. Rewrite this equation as

$$1 = \frac{t}{t_T}\left(1 + \frac{t_T}{t_D}\right) \tag{5.69}$$

In this form, we note immediately that t is approximately equal to t_T if $1 \gg t_T/t_D$. That is, if the transport time is much shorter than the diffusion time for the magnetic field, the diffusion term in (5.58) can be ignored. Hence for this to occur we require that

$$\mu_0 \sigma L U \gg 1 \tag{5.70}$$

or, defining $R_m = \mu_0 \sigma L U$, diffusion can be ignored if

$$R_m \gg 1 \tag{5.71}$$

R_m is defined as the magnetic Reynolds number. For a large class of natural processes involving MHD fluids, R_m is often a very large number. For example, R_m for the solar wind is very large, and the effects of solar magnetic field diffusion during the course of the transit time between the Sun and Earth are ignored. A similar argument applied to Earth's magnetic field will indicate that Earth's magnetic field diffusion into the solar wind in the time-scale of solar wind flow past the Earth can also be ignored. This justifies the statement made earlier that the IMF is primarily of solar origin and the geomagnetic field is confined to the region inside the geomagnetic cavity.

5.4.4 Ideal MHD Fluid ($\sigma = \infty$)

A conducting fluid is considered ideal when its conductivity σ is infinite. Although such a perfectly conducting fluid can never be realized no matter how large σ is, it is useful to study how magnetic fields behave in this limit. If $\sigma = \infty$, then the magnetic field behavior is described by

$$\frac{\partial \mathbf{B}}{\partial t} = \nabla \times (\mathbf{U} \times \mathbf{B}) \tag{5.72}$$

If we compare this formula with one of Maxwell's equations $\partial \mathbf{B}/\partial t = -\nabla \times \mathbf{E}$, we see immediately that the electric field in an infinitely conducting medium is given by

$$\mathbf{E} = -\mathbf{U} \times \mathbf{B} \tag{5.73}$$

Write $\mathbf{E} = \mathbf{E}_{\parallel} + \mathbf{E}_{\perp}$ and $\mathbf{U} = \mathbf{U}_{\parallel} + \mathbf{U}_{\perp}$ where $\parallel$ and $\perp$ refer to directions relative to $\mathbf{B}$. Inserting these into (5.73) then shows that

$$\begin{aligned} \mathbf{E}_{\parallel} &= -\mathbf{U}_{\parallel} \times \mathbf{B} \\ &= 0 \\ \mathbf{E}_{\perp} &= -\mathbf{U}_{\perp} \times \mathbf{B} \end{aligned} \tag{5.74}$$

Hence (5.73) implies that $\mathbf{E}_{\parallel}$, the component of the electric field parallel to $\mathbf{B}$, vanishes. Take the cross product of (5.73) with $\mathbf{B}$, and obtain (omit $\perp$ sign)

$$\begin{aligned} \mathbf{E} \times \mathbf{B} &= -(\mathbf{U} \times \mathbf{B}) \times \mathbf{B} \\ &= -[(\mathbf{U} \cdot \mathbf{B})\mathbf{B} - (\mathbf{B} \cdot \mathbf{B})\mathbf{U}] \\ &= B^2 \mathbf{U} \end{aligned} \tag{5.75}$$

and, since $\mathbf{U} \cdot \mathbf{B} = 0$, obtain

$$\mathbf{U} = \frac{\mathbf{E} \times \mathbf{B}}{\mathbf{B}^2} \tag{5.76}$$

In an ideal conducting fluid with $\sigma = \infty$, the statement that the fluid is in motion is equivalent to saying that there is an electric field in the rest frame given by $\mathbf{E} = -\mathbf{U} \times \mathbf{B}$. Similarly, in such a fluid, the statement that there is an electric field in the rest frame is equivalent to saying that the fluid is moving with a velocity given by $\mathbf{U} = (\mathbf{E} \times \mathbf{B})/B^2$. Thus, the existence of an electric field implies the existence of fluid motion, and vice versa. Formulas (5.73) and (5.76) are completely equivalent.

The relation $\mathbf{E} = -\mathbf{U} \times \mathbf{B}$ can also be obtained from Ohm's law, given by (5.37). Here, since $\sigma = \infty$, $\mathbf{E} + \mathbf{U} \times \mathbf{B}$ must vanish. Otherwise, the medium is required to supply an infinite current density, which is physically not possible. Recall also that the theory of Lorentz transformation in the non-relativistic limit, $|\mathbf{U}| \ll c$, shows that (see Chapter 2) $\mathbf{E}' \simeq \mathbf{E} + \mathbf{U} \times \mathbf{B}$ where $\mathbf{E}$ is the electric field measured in the rest frame, $\mathbf{E}'$ is measured in the frame moving with a velocity $\mathbf{U}$ relative to the rest frame, and $\mathbf{B}$ is the magnetic field (in the limit $|\mathbf{U}| \ll c$, $\mathbf{B}' \approx \mathbf{B}$). The Lorentz theory shows that if the velocity of the moving coordinate frame of reference is given by (5.76), $\mathbf{E}'$ vanishes in that frame ($\mathbf{E}'$ has been transformed away).

$\mathbf{E} = -\mathbf{U} \times \mathbf{B}$ for $\sigma = \infty$ corresponds to the vanishing of $\mathbf{E}'$. An observer moving with the conducting fluid will not measure any electric field. There are no space charges in this moving frame, and the charged particles experience no electrical force in the frame moving with the fluid. However, the motion of the conducting fluid across magnetic fields induces an electric field, and the observer in the rest frame measures an electric field given by $\mathbf{E} = -\mathbf{U} \times \mathbf{B}$ (see also Section 4.3.2).

Some interesting properties are indicated when the MHD flow is in steady-state. Since $\nabla \times \mathbf{E} = 0$, $\mathbf{E}$ can be represented by the electromagnetic scalar potential Φ (3.20), $\mathbf{E} = -\nabla\Phi$ where $\partial \mathbf{A}/\partial t = 0$. For perfectly conducting MHD fluids, $\mathbf{E} = -\mathbf{U} \times \mathbf{B}$, and therefore

$$\nabla\Phi = \mathbf{U} \times \mathbf{B} \tag{5.77}$$

If this equation is scalar-multiplied with $\mathbf{U}$ or $\mathbf{B}$, we obtain

$$\begin{aligned} \mathbf{U} \cdot \nabla\Phi &= 0 \\ \mathbf{B} \cdot \nabla\Phi &= 0 \end{aligned} \tag{5.78}$$

These equations show that $\nabla\Phi$ is perpendicular to $\mathbf{U}$ and $\mathbf{B}$, and Φ is constant along $\mathbf{U}$ and $\mathbf{B}$. The Φ = constant surfaces are equipotential surfaces and, since $\mathbf{U}$ and $\mathbf{B}$ lie on equipotential surfaces, the streamlines and the magnetic field lines are also equipotential lines.

5.4.5 Frozen-in-Field Concept

In ideal fluids, the coupling of the magnetic field and the fluid motion is given by (5.72). As mentioned earlier, this equation is identical in form to the vorticity equation of an ordinary homogeneous inviscid fluid,

$$\frac{\partial \boldsymbol{\omega}}{\partial t} = \nabla \times (\mathbf{U} \times \boldsymbol{\omega}) \tag{5.79}$$

This equation has important applications in ordinary fluid dynamics and leads to the well-known Kelvin-Helmholtz theorems derived before the turn of the 20th century. In essence, the theorems state that

a. Flux of vorticity through any closed contour moving with the fluid is constant.

b. Fluid elements that lie on a vortex line continue to lie on the same vortex line.

Since the forms of the vorticity equation (5.79) and that of the magnetic field equation (5.72) are identical, it seems reasonable to extend the above theorems to MHD and deduce that in a perfectly conducting MHD fluid,

a. The flux of a magnetic field through any closed contour moving with the fluid is constant.

b. Fluid elements that lie on a magnetic field line remain on the same field line.

That these theorems apply to perfectly conducting MHD fluids was first recognized by Hannes Alfvén in 1942. Let us investigate these concepts in some detail, since they are outside the framework of Maxwell equations.

Let C be the contour in the fluid at time t that is moving with a velocity $\mathbf{U}(\mathbf{r}, t)$ as shown in Figure 5.4. The displacement of C to a new position at later time $t + \Delta t$ is indicated by C'. Let S and S' be the surfaces enclosed by C and C'. Now, let $d\mathbf{l}$ be an element of arc on C. This arc element in time Δt moves a distance $\mathbf{U}\Delta t$ and it sweeps out an element of area $d\mathbf{l} \times \mathbf{U}\Delta t$. Consider now the total magnetic flux that enters and leaves these surfaces. Noting that the total magnetic flux at *any* time crossing a closed surface must vanish owing to the divergence theorem, the total flux at time $t + \Delta t$ is

$$-\int_C \mathbf{B}(t+\Delta t)\cdot\mathbf{n}dS + \int_{C'} \mathbf{B}(t+\Delta t)\cdot\mathbf{n}dS'$$

$$+\oint_C \mathbf{B}(t+\Delta t)\cdot d\mathbf{l}\times\mathbf{U}dt = 0 \tag{5.80}$$

where the positive and negative signs come from the definition that the unit normal $\mathbf{n}$ is positive in the outward direction. The first two terms represent the flux through S and S' and the third term the flux through the area swept out by the motion.

Let us now examine the change of the magnetic flux Φ with time as the contour moves with the fluid. The time rate of change of the magnetic

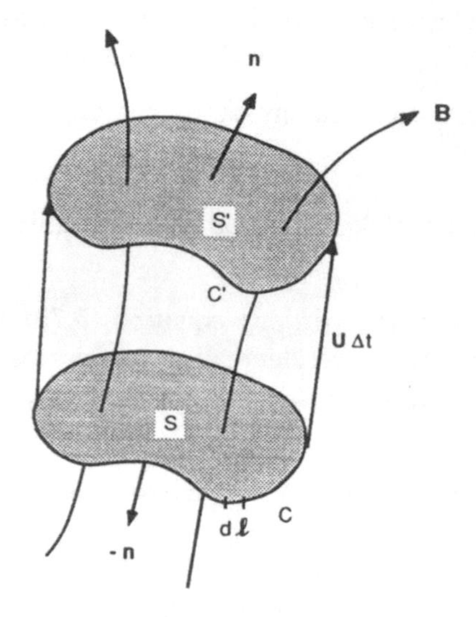

FIGURE 5.4 A schematic diagram showing the motion of an arbitrary contour in a perfect MHD fluid.

flux through the moving contour from t to $t+\Delta t$ is

$$\begin{aligned}
\frac{d\Phi}{dt} &= \lim_{\Delta t \to 0} \frac{\Phi_{C'}(t+\Delta t) - \Phi_C(t)}{\Delta t} \\
&= \lim_{\Delta t \to 0} \frac{\int \mathbf{B}(t+\Delta t)\cdot \mathbf{n}\, dS' - \int \mathbf{B}(t)\cdot \mathbf{n}\, dS}{\Delta t} \\
&= \lim_{\Delta t \to 0} \frac{\int [\mathbf{B}(t+\Delta t) - \mathbf{B}(t)]\cdot \mathbf{n}\, dS}{\Delta t} - \oint_C \mathbf{B}(t+\Delta t)\cdot d\mathbf{l}\times \mathbf{U} \\
&= \int \frac{\partial \mathbf{B}}{\partial t}\cdot \mathbf{n}\, dS - \oint_C (\mathbf{U}\times\mathbf{B})\cdot d\mathbf{l} \\
&= \int \left[\frac{\partial \mathbf{B}}{\partial t} - \nabla\times(\mathbf{U}\times\mathbf{B})\right]\cdot \mathbf{n}\, dS \\
&= -\int \nabla\times(\mathbf{E}+\mathbf{U}\times\mathbf{B})\cdot \mathbf{n}\, dS
\end{aligned} \tag{5.81}$$

In arriving at (5.81), we used (5.80), the definition of time derivative, the vector identity $\mathbf{B}\cdot(d\mathbf{l}\times\mathbf{U}) = d\mathbf{l}\cdot(\mathbf{U}\times\mathbf{B})$, and Stoke's theorem. The integrand in (5.81) vanishes if

$$(\mathbf{E}+\mathbf{U}\times\mathbf{B}) = -\nabla\Psi \tag{5.82}$$

where Ψ is some scalar and $\nabla \times \nabla\Psi = 0$. Equation (5.82) is a necessary and a sufficient condition for

$$\frac{d\Phi}{dt} = 0 \tag{5.83}$$

and it shows that the magnetic flux through any arbitrary contour moving with the fluid is constant in time. This result is valid for any magnetic field that moves with the velocity field $\mathbf{U}(\mathbf{r}, t)$ defined by (5.82).

This constraint of the ideal fluid motion implies that the individual flux tubes must move exactly together with the fluid, since any relative motion between the flux tubes and the fluid will result in the violation of (5.83). To see this, examine the field configuration at any time t and a short time later, $t + dt$, as the fluid is convected. Now consider two neighboring particles on the same magnetic line of force $\mathbf{B}_0$ separated by $d\mathbf{r}_0$ at time t and let these particles move a short time dt later to a new position separated by $d\mathbf{r}$, as shown in Figure 5.5. The velocities of the particles are $\mathbf{U}_0$ and $\mathbf{U}$. Initially, then,

$$d\mathbf{r}_0 \times \mathbf{B}_0 = 0 \tag{5.84}$$

Our objective is to show that the particles remain on the same field line as the field line moves, that is,

$$d\mathbf{r} \times \mathbf{B} = 0 \tag{5.85}$$

Note from Figure 5.5 that

$$d\mathbf{r} = d\mathbf{r}_0 + \mathbf{U}dt - \mathbf{U}_0 dt \tag{5.86}$$

The relationship between $\mathbf{U}$ and $\mathbf{U}_0$ is

$$\mathbf{U} = \mathbf{U}_0 + (d\mathbf{r}_0 \cdot \nabla)\mathbf{U}_0 \tag{5.87}$$

and substituting this into (5.86) shows that the rate of change of the vector $d\mathbf{r}$ is

$$d\mathbf{r} = d\mathbf{r}_0 + (d\mathbf{r}_0 \cdot \nabla)\mathbf{U}_0 dt \tag{5.88}$$

Now the change of the magnetic field in time dt can be written as, to first order,

$$\mathbf{B} = \mathbf{B}_0 + (\mathbf{U} \cdot \nabla)\mathbf{B}_0 dt + \left(\frac{\partial \mathbf{B}_0}{\partial t}\right) dt \tag{5.89}$$

where the second term on the right-hand side comes from the motion across an inhomogeneous magnetic field and the last term from the time variation of the magnetic field at a fixed point in space. Use these two relationships in (5.85) and obtain

$$\begin{aligned}\mathbf{B} \times d\mathbf{r} &= \mathbf{B}_0 \times d\mathbf{r}_0 + \mathbf{B}_0 \times (d\mathbf{r}_0 \cdot \nabla)\mathbf{U}_0 dt \\ &\quad + (\mathbf{U} \cdot \nabla)\mathbf{B}_0 dt \times d\mathbf{r}_0 + (\mathbf{U} \cdot \nabla)\mathbf{B}_0 dt \times (d\mathbf{r}_0 \cdot \nabla)\mathbf{U}_0 dt \\ &\quad + \frac{\partial \mathbf{B}_0}{\partial t} \times d\mathbf{r}_0 dt + \left(\frac{\partial \mathbf{B}_0}{\partial t}\right) dt \times (d\mathbf{r}_0 \cdot \nabla)\mathbf{U}_0 dt \end{aligned} \tag{5.90}$$

$\mathbf{B}_0 \times d\mathbf{r}_0 = 0$ as originally assumed. Let us also ignore terms higher than the first order. Then (5.90) reduces to first order,

$$\mathbf{B} \times d\mathbf{r} = \mathbf{B}_0 \times (d\mathbf{r}_0 \cdot \nabla)\mathbf{U} dt + \left(\frac{\partial}{\partial t} + \mathbf{U} \cdot \nabla\right) \mathbf{B}_0 \times d\mathbf{r}_0 dt \tag{5.91}$$

Now note that $d\mathbf{r}_0 = (\mathbf{B}_0/B_0)dr_0$ and use the vector relation $\nabla \times (\mathbf{U} \times \mathbf{B}_0) = \mathbf{U}(\nabla \cdot \mathbf{B}_0) + (\mathbf{B}_0 \cdot \nabla)\mathbf{U} - \mathbf{B}_0(\nabla \cdot \mathbf{U}) - (\mathbf{U} \cdot \nabla)\mathbf{B}_0$ and, further noting that $\nabla \cdot \mathbf{B}_0 = 0$, rewrite (5.91) as

$$\mathbf{B} \times d\mathbf{r} = \frac{\mathbf{B}_0}{B_0} \times \left[\nabla \times (\mathbf{U} \times \mathbf{B}_0) + \mathbf{B}_0(\nabla \cdot \mathbf{U}) - \frac{\partial \mathbf{B}_0}{\partial t}\right] dr_0 dt \tag{5.92}$$

Since $\partial \mathbf{B}_0/\partial t = -\nabla \times \mathbf{E}_0$, the requirement that (5.92) vanish is that

$$\mathbf{B}_0 \times [\nabla \times (\mathbf{E}_0 + \mathbf{U} \times \mathbf{B}_0)] = 0 \tag{5.93}$$

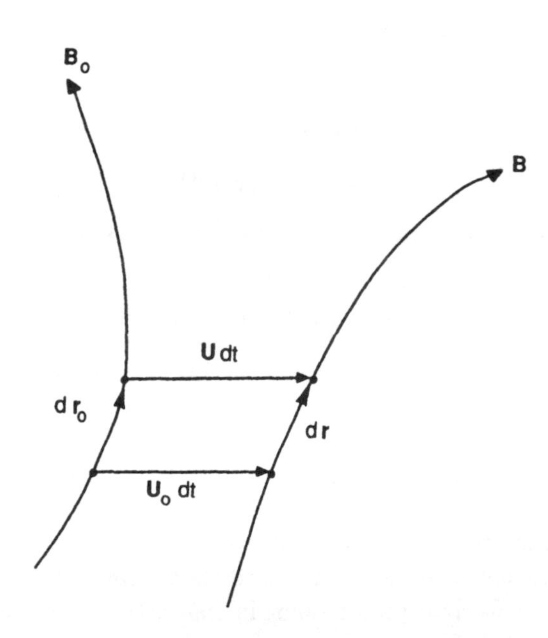

FIGURE 5.5 A schematic diagram showing the motion of a fluid element on a magnetic field line.

This equation vanishes if $\mathbf{E}+\mathbf{U}\times\mathbf{B} = -\nabla\Psi$, which is also a requirement of the flux-preserving condition (see (5.82)). Hence we come to an important conclusion that flux-preserving motion is also "line-preserving" in the sense that a fluid element moving with a velocity field given by (5.82) remains on the same field line (note, however, that the reverse is not true, that is, a line-preservation motion does not necessarily qualify as a flux-preservation motion). Since the line-preservation theorem permits one to say that the particles on a given field line $\mathbf{B}$ will at later time remain on the same field line and, moreover, since the fluid particles and the magnetic field move with the same fluid velocity, one draws a physical picture that the fluid appears to be attached to the magnetic field line, or "frozen" in the magnetic field. Thus, the flux-preservation theorem is also known as the frozen-in-field theorem.

The term magnetic field line, can now be defined because the total magnetic flux is conserved in a perfectly conducting fluid. A magnetic line of force is not a physical entity but represents a concept of abstraction. In general, it is not possible to identify uniquely a field line (see Section 3.2). In a perfect conductor, however, the concept of individual field lines can be meaningful.

In a conducting fluid with infinite conductivity, no electromotive force (EMF) is generated. This can be seen as follows. In (5.81) above, use the $\nabla\times\mathbf{E} = -\partial\mathbf{B}/\partial t$ relation and note that the induced EMF is given by

$$\text{EMF} = \oint_c (\mathbf{E}+\mathbf{U}\times\mathbf{B})\cdot d\mathbf{l} \tag{5.94}$$

However, $\oint \mathbf{E}\cdot d\mathbf{l}$ is just the rate of magnetic flux change through $\mathbf{S}$ and the meaning of $\oint(\mathbf{U}\times\mathbf{B})\cdot d\mathbf{l}$ has already been discussed above. Hence we arrive at

$$\text{EMF} = -\frac{d\Phi}{dt} \tag{5.95}$$

which vanishes for fluids with perfect conductivity. This justifies the statement made earlier in deriving the energy equation for the MHD fluid (5.55) that the MHD formulation, assuming $\sigma = \infty$, ignores any work done by the electric field. There are no Ohmic losses since Joule heating, given by the expression $\mathbf{J}\cdot\mathbf{E}$, vanishes.

If the fluid is not ideal ($\sigma \neq \infty$), we must integrate the full equation (5.58). Then the EMF no longer vanishes and we obtain

$$\frac{d\Phi}{dt} = \frac{1}{\mu_0\sigma}\int \nabla^2\mathbf{B}\cdot d\mathbf{S} \tag{5.96}$$

The right-hand term is sometimes referred to as the "slippage" effect. This slippage arises because the magnetic field is no longer frozen in the fluid.

5.5 Flow of MHD Fluid

The theory of flowing MHD fluids can be developed in the same manner as in ordinary fluids. Let us first study the equation of motion and derive some general theorems about MHD flows. As before, assume that the fluid pressure is isotropic and let us include the possible presence of other forces (such as gravitational). Assume that such forces can be written as $\mathbf{F} = -\rho_m \nabla \Psi$ where Ψ is the potential. The equation of motion then takes the form

$$\frac{\partial \mathbf{U}}{\partial t} + (\mathbf{U} \cdot \nabla)\mathbf{U} + \frac{\nabla p}{\rho_m} + \nabla \Psi = \frac{\mathbf{J} \times \mathbf{B}}{\rho_m} \tag{5.97}$$

This equation can be rewritten, using the vector identity $(\mathbf{U} \cdot \nabla)\mathbf{U} = \nabla U^2/2 - \mathbf{U} \times (\nabla \times \mathbf{U})$, as

$$\frac{\partial \mathbf{U}}{\partial t} + \frac{\nabla U^2}{2} - \mathbf{U} \times (\nabla \times \mathbf{U}) + \frac{\nabla p}{\rho_m} + \nabla \Psi = \frac{\mathbf{J} \times \mathbf{B}}{\rho_m} \tag{5.98}$$

Now take the curl of this equation and, because the curl of a gradient vanishes, (5.98) becomes

$$\frac{\partial \boldsymbol{\omega}}{\partial t} + \nabla \times (\boldsymbol{\omega} \times \mathbf{U}) + \nabla \frac{1}{\rho_m} \times \nabla p = \nabla \times \left(\frac{\mathbf{J} \times \mathbf{B}}{\rho_m} \right) \tag{5.99}$$

where $\boldsymbol{\omega} = \nabla \times \mathbf{U}$ is the vorticity vector of $\mathbf{U}$. If the fluid flow is irrotational, $\nabla \times \mathbf{U} = 0$, $\mathbf{U}$ can be represented by the gradient of a scalar velocity potential function Φ, $\mathbf{U} = -\nabla \Phi$. For irrotational and incompressible flows, (ρ_m = constant), (5.97) can be rewritten as

$$\nabla \left(\frac{\partial \Phi}{\partial t} + \frac{U^2}{2} + \frac{p}{\rho_m} + \Psi \right) = \frac{\mathbf{J} \times \mathbf{B}}{\rho_m} \tag{5.100}$$

For ordinary fluids, the quantity operated on by the gradient vanishes because the right-hand side vanishes. However, for MHD fluids, the terms on the left-hand side are balanced by the $\mathbf{J} \times \mathbf{B}$ term.

5.5.1 Flows Parallel to B

The scalar product of (5.100) with the flow velocity $\mathbf{U}$ gives

$$\mathbf{U} \cdot \nabla \left(\frac{\partial \Phi}{\partial t} + \frac{U^2}{2} + \frac{p}{\rho_m} + \Psi \right) = \mathbf{J} \cdot \left(\frac{\mathbf{B} \times \mathbf{U}}{\rho_m} \right) \tag{5.101}$$

where $(\mathbf{J}\times\mathbf{B})\cdot\mathbf{U} = \mathbf{J}\cdot(\mathbf{B}\times\mathbf{U})$ was used. If $\mathbf{U}\times\mathbf{B} = 0$, that is, if $\mathbf{U}$ is parallel to the direction of $\mathbf{B}$, the right-hand side vanishes and (5.101) yields

$$\mathbf{U}\cdot\nabla\left(\frac{\partial\Phi}{\partial t} + \frac{U^2}{2} + \frac{p}{\rho_m} + \Psi\right) = 0 \tag{5.102}$$

This equation shows $\partial\Phi/\partial t + U^2/2 + p/\rho_m + \Psi$ is constant along $\mathbf{U}$ and $\mathbf{B}$ since $\mathbf{U}$ is parallel to $\mathbf{B}$. In general we can let

$$\frac{\partial\Phi}{\partial t} + \frac{U^2}{2} + \frac{p}{\rho_m} + \Psi = f(t) \tag{5.103}$$

where $f(t)$ is an arbitrary function of time. If the flow is uniform and steady, $\partial\Phi/\partial t = 0$, and $f(t) =$ constant, we obtain Bernoulli's equation

$$\frac{U^2}{2} + \frac{p}{\rho_m} + \Psi = \text{constant} \tag{5.104}$$

In a time-stationary case, the total energy along any given streamline is conserved. The streamline constant may differ from one streamline to another. However, potential flows are assumed uniform and therefore the flow constant is constant throughout the flow.

If the flow and the magnetic field are parallel, one can write

$$\mathbf{B} = \lambda\mathbf{U} \tag{5.105}$$

where λ is an arbitrary scalar function of position and time. Substitution of this equation in $\nabla\cdot\mathbf{B} = 0$ yields

$$\lambda\nabla\cdot\mathbf{U} + (\mathbf{U}\cdot\nabla)\lambda = 0 \tag{5.106}$$

which is a condition for λ. Suppose now the flow is also incompressible. Then, $\nabla\cdot\mathbf{U} = 0$ and (5.106) simplifies to

$$(\mathbf{U}\cdot\nabla)\lambda = 0 \tag{5.107}$$

For incompressible fluids, the requirement for λ is that it be constant along a streamline. Constant λ surfaces are streamline surfaces (surfaces are made up of streamlines). Note that λ is also constant along the field line $\mathbf{B}$ according to (5.105). There are an infinite number of choices for λ. Two possible choices given above are that $\lambda = f(\rho_m)$ and $\lambda = f(U^2/2 + p/\rho_m + \psi)$.

5.5.2 Flows Perpendicular to B

Fluid flows in the direction perpendicular to the magnetic field are coupled to the magnetic field and consequently the electrodynamic effects become important. The momentum equation (5.97) and the continuity equation (5.28) must be supplemented with an equation of state (5.45) and, in addition (for ideal fluids), with

$$\begin{aligned}\frac{\partial \mathbf{B}}{\partial t} &= \nabla \times (\mathbf{U} \times \mathbf{B}) \\ \nabla \cdot \mathbf{B} &= 0 \\ \nabla \times \mathbf{B} &= \mu_0 \mathbf{J}\end{aligned} \tag{5.108}$$

These equations define a class of flow for which $\mathbf{E} + \mathbf{U} \times \mathbf{B} = -\nabla\Psi$ (see (5.82)).

The concept of motion of magnetic field lines or convection of magnetic field lines is now examined in more detail, including their limitations. The term "convection" comes from ordinary fluid dynamics and it describes the motion that occurs in a fluid when the temperature is not uniformly distributed. The temperature gradient causes a convective motion and mixes the fluid so that the temperature in the fluid becomes constant.

In MHD, the term convection describes the bulk motion of the plasma. This large-scale plasma motion is induced because the electric field is being transformed away in the plasma frame of reference. Convective motion in MHD fluids arises as a direct consequence of the fact that free charges are being neutralized. When the fluid has infinite conductivity, the motion conserves the total magnetic flux; and because magnetic fields do not diffuse, the velocity of a fluid element given by (5.76) is also the local velocity of the magnetic field. One then arrives at the notion that magnetic field lines are attached to the material, thus frozen in. The concept of convection of magnetic field lines is only unambiguous for perfectly conducting fluids.

Convection of the magnetic field requires that the EMF must vanish. This occurs if and only if $\nabla \times (\mathbf{E} + \mathbf{U} \times \mathbf{B}) = 0$. Then the necessary and sufficient condition for flux preservation can be written as $\mathbf{E} + \mathbf{U} \times \mathbf{B} = -\nabla\chi$ (see (5.82)) where χ is some scalar. If we take a cross product of $\mathbf{B}$ with this equation, noting that $\mathbf{U} \cdot \mathbf{B} = 0$, we obtain

$$\mathbf{U} = (\mathbf{E} + \nabla\chi) \times \frac{\mathbf{B}}{B^2} \tag{5.109}$$

while a scalar product of $\mathbf{B}$ taken with this equation yields

$$\mathbf{B} \cdot \mathbf{E} + \mathbf{B} \cdot \nabla\chi = 0 \tag{5.110}$$

If the boundary condition is specified for the value of χ at every point of some surface that is not parallel to $\mathbf{B}$, then this equation determines a unique solution for χ at all points. Flux-preserving velocity is then determined from above to within a scalar constant χ on this surface. The flux-preserving velocity is not unique as there are many such velocities.

Formal theory of flux-preserving motion using the Euler potentials will now be given. Our objective is to seek the most general flow field $\mathbf{U}$ that is a solution of (5.108). If we use the relation $\mathbf{B} = \nabla \times \mathbf{A}$ where $\mathbf{A}$ is the vector potential, the first of (5.108) becomes

$$\nabla \times \frac{\partial \mathbf{A}}{\partial t} = \nabla \times [\mathbf{U} \times (\nabla \times \mathbf{A})] \tag{5.111}$$

which upon integration yields

$$\frac{\partial \mathbf{A}}{\partial t} - \mathbf{U} \times \mathbf{B} = \nabla \chi \tag{5.112}$$

where χ is a scalar function (an integration constant), which must satisfy the condition

$$\mathbf{B} \cdot \nabla \chi = \mathbf{B} \cdot \frac{\partial \mathbf{A}}{\partial t} \tag{5.113}$$

since $\mathbf{B} \cdot (\mathbf{U} \times \mathbf{B}) = 0$. Note that the condition $\mathbf{U} \cdot \nabla \chi = \mathbf{U} \cdot \partial \mathbf{A}/\partial t$ is also implied by (5.112). Now choose a gauge such that $\mathbf{A} \cdot \mathbf{B} = 0$. For this choice, we showed in Chapter 3 that the magnetic field can be defined in terms of the Euler parameters α and β such that

$$\mathbf{A} = \alpha \nabla \beta \tag{5.114}$$

and

$$\mathbf{B} = \nabla \alpha \times \nabla \beta \tag{5.115}$$

Recall that a particular set of α and β labels a field line (although this labeling is not unique). Now

$$\begin{aligned} \frac{\partial \mathbf{A}}{\partial t} = \frac{\partial (\alpha \nabla \beta)}{\partial t} &= \frac{\partial \alpha}{\partial t} \nabla \beta + \alpha \frac{\partial \nabla \beta}{\partial t} \\ &= \frac{\partial \alpha}{\partial t} \nabla \beta + \alpha \nabla \frac{\partial \beta}{\partial t} \end{aligned} \tag{5.116}$$

and

$$\begin{aligned} \mathbf{B} \cdot \frac{\partial \mathbf{A}}{\partial t} &= (\nabla \alpha \times \nabla \beta) \cdot \left(\frac{\partial \alpha}{\partial t} \nabla \beta + \alpha \nabla \frac{\partial \beta}{\partial t} \right) \\ &= (\nabla \alpha \times \nabla \beta) \cdot \left(\alpha \nabla \frac{\partial \beta}{\partial t} \right) \\ &= (\nabla \alpha \times \nabla \beta) \cdot \left(\nabla \frac{\alpha \partial \beta}{\partial t} - \frac{\partial \beta}{\partial t} \nabla \alpha \right) \\ &= (\nabla \alpha \times \nabla \beta) \cdot \left(\nabla \frac{\alpha \partial \beta}{\partial t} \right) \\ &= \mathbf{B} \cdot \nabla \frac{\alpha \partial \beta}{\partial t} \end{aligned} \tag{5.117}$$

since $(\nabla \alpha \times \nabla \beta) \cdot \nabla \beta = (\nabla \alpha \times \nabla \beta) \cdot \nabla \alpha = 0$ and use was made of the relation $\alpha \nabla (\partial \beta / \partial t) = \nabla (\alpha \partial \beta / \partial t) - (\partial \beta / \partial t) \nabla \alpha$.

Comparison of (5.117) and (5.113) permits us to define

$$\chi = \alpha\frac{\partial\beta}{\partial t} - \Lambda \tag{5.118}$$

where Λ is an arbitrary scalar function of α and β. With this definition, (5.112) can be re-written with the help of (5.118) as

$$\begin{aligned}
\mathbf{U}\times\mathbf{B} &= \frac{\partial\mathbf{A}}{\partial t} - \nabla\chi \\
&= \frac{\partial\alpha}{\partial t}\nabla\beta + \alpha\frac{\partial}{\partial t}\nabla\beta - \frac{\partial\beta}{\partial t}\nabla\alpha - \alpha\frac{\partial}{\partial t}\nabla\beta + \nabla\Lambda \\
&= \frac{\partial\alpha}{\partial t}\nabla\beta - \frac{\partial\beta}{\partial t}\nabla\alpha + \nabla\Lambda
\end{aligned} \tag{5.119}$$

The solution for $\mathbf{U}$ can be obtained in terms of α, β, and Λ. In order to learn more about the functional relation of Λ upon α and β, form the dot product of (5.119) with $\mathbf{B}\times\nabla\alpha$ and $\mathbf{B}\times\nabla\beta$ and obtain (after some algebra)

$$\frac{\partial\alpha}{\partial t} + (\mathbf{U}\cdot\nabla)\alpha = \frac{d\alpha}{dt} = -\frac{\partial\Lambda}{\partial\beta}$$

and

$$\frac{\partial\beta}{\partial t} + (\mathbf{U}\cdot\nabla)\beta = \frac{d\beta}{dt} = \frac{\partial\Lambda}{\partial\alpha} \tag{5.120}$$

These equations indicate that magnetic field line motion is not unique. There are an infinite number of solutions for $\mathbf{U}$ corresponding to the infinite choices of $\Lambda(\alpha,\beta)$. Note that the flow $\mathbf{U}$ parallel to $\mathbf{B}$ is not defined by this formulation. It can be set equal to zero or any other arbitrary value.

If α and β are convected with the flow, the total derivative $d\alpha/dt$ and $d\beta/dt$ vanish in the frame moving with the fluid velocity. Then

$$\begin{aligned}
\frac{\partial\alpha}{\partial t} + (\mathbf{U}\cdot\nabla)\alpha &= 0 \\
\frac{\partial\beta}{\partial t} + (\mathbf{U}\cdot\nabla)\beta &= 0
\end{aligned} \tag{5.121}$$

A fluid velocity satisfying these equations can be represented by

$$\mathbf{U} = \left[\left(\frac{\partial\beta}{\partial t}\right)\nabla\alpha - \left(\frac{\partial\alpha}{\partial t}\right)\nabla\beta\right]\times\frac{\mathbf{B}}{B^2} \tag{5.122}$$

As mentioned above, this velocity is determined only to a scalar constant Λ, which, for example, can be $\Lambda = \Lambda(\alpha,\beta)$. $\mathbf{B}$ is perpendicular to both $\nabla\alpha$ and $\nabla\beta$, and so $\mathbf{U}$ is perpendicular to $\mathbf{B}$.

5.6 Derivation of the Momentum Equation

A more rigorous derivation of the fluid equation of motion will be given now to augment the discussion of Section 5.3.3. Measurements in space are made in terms of the averaged quantities that obey macroscopic equations. The kth component of the Lorentz equation

$$F_k = \frac{d}{dt}(m\mathbf{v})_k = q(\mathbf{E} + \mathbf{v} \times \mathbf{B})_k \tag{5.123}$$

will be averaged over the distribution function $f(\mathbf{r}, \mathbf{v}, t)$ to describe the motion of the particles in the average sense. When this average is performed according to (2.20), the left-hand side of (5.123) becomes

$$< F_k > = \int \left(\frac{d}{dt} m v_k \right) f(\mathbf{r}, \mathbf{v}, t) d^3 v \tag{5.124}$$

where $< F_k >$ is now force/volume. The total time derivative is a convective derivative (derivative taken along the trajectory) and in indicial notation is

$$\begin{aligned} \frac{d}{dt} &= \frac{\partial}{\partial t} + \mathbf{v} \cdot \nabla \\ &= \frac{\partial}{\partial t} + v_i \frac{\partial}{\partial x_i} \end{aligned} \tag{5.125}$$

Equation (5.124) therefore becomes

$$< F_k > = \int \left[\frac{\partial}{\partial t}(m v_k f) + v_i \frac{\partial}{\partial x_i}(m v_k f) \right] d^3 v \tag{5.126}$$

Now note that

$$\frac{\partial}{\partial x_i}(m v_i v_k f) = v_i \frac{\partial}{\partial x_i}(m v_k f) + (m v_k f) \frac{\partial}{\partial x_i} v_i \tag{5.127}$$

where v_i and x_i are coordinates in $(\mathbf{r}, \mathbf{v})$ space and are considered as independent variables. Hence

$$\frac{\partial v_i}{\partial x_i} = 0 \tag{5.128}$$

and the second term on the right side of (5.127) vanishes. Equation (5.126) then can be written as

$$\begin{aligned} < F_k > &= \frac{\partial}{\partial t} \int (m v_k f) d^3 v + \frac{\partial}{\partial x_i} \int (m v_i v_k f) d^3 v \\ &= \frac{\partial}{\partial t} \Pi_k + \frac{\partial}{\partial x_i} \Pi_{ik} \end{aligned} \tag{5.129}$$

where we have defined

$$\Pi_k = \int m v_k f(\mathbf{r}, \mathbf{v}, t) d^3 v \tag{5.130}$$

as the kth component of the average momentum vector and

$$\Pi_{ik} = \int m v_i v_k f(\mathbf{r}, \mathbf{v}, t) d^3 v \tag{5.131}$$

as the $(ik)^{th}$ component of the momentum transfer tensor. We can also write these equations as $\Pi_k = nm < v_k >$ and $\Pi_{ik} = nm < v_i v_k >$. In these equations, the summation convention of repeated indices is used. The first term of (5.129) represents the local change of momentum and the second term the change of momentum in the reference frame of a volume of particles in motion.

Let us now examine the right-hand term of (5.123). If it is averaged over the distribution function f, we obtain

$$\begin{aligned} \int q(\mathbf{E} + \mathbf{v} \times \mathbf{B}) f d^3 v &= \int q \mathbf{E} f d^3 v + \int q(\mathbf{v} f d^3 v) \times \mathbf{B} \\ &= qn\mathbf{E} + qn <\mathbf{v}> \times \mathbf{B} \end{aligned} \tag{5.132}$$

where it is understood that $\mathbf{E}$ and $\mathbf{B}$ are now averaged fields. A particle system includes at least two species of particles (electrons and protons, for example) and each species will have a separate equation describing its motion. The contributions from each of the species can be taken into account by summing the different species α. Equations (5.129) and (5.132) then become

$$\begin{aligned} \frac{\partial}{\partial t} \sum_\alpha \Pi_k^\alpha + \frac{\partial}{\partial x_i} \sum_\alpha \Pi_{ik}^\alpha &= \sum_\alpha q^\alpha n^\alpha \mathbf{E} + \sum_\alpha n^\alpha q^\alpha <\mathbf{v}>^\alpha \times \mathbf{B} \\ &= \rho_c \mathbf{E} + \mathbf{J} \times \mathbf{B} \end{aligned} \tag{5.133}$$

Here use was made of the definition of the total charge density ρ and the total current density $\mathbf{J}$. With the help of Maxwell's equations we now transform each of the right-hand terms in (5.133). The kth component of $\rho_c \mathbf{E}$ is

$$\begin{aligned} \rho_c \mathbf{E}_k &= \epsilon_0 E_k \frac{\partial}{\partial x_i} E_i \\ &= \epsilon_0 \left[\frac{\partial}{\partial x_i}(E_k E_i) - E_i \frac{\partial}{\partial x_i} E_k \right] \end{aligned} \tag{5.134}$$

Now use the vector relation $\mathbf{E} \times (\nabla \times \mathbf{E}) = \nabla(E^2/2) - (\mathbf{E} \cdot \nabla)\mathbf{E}$ to replace the second term on the right. Since $\nabla \times \mathbf{E} = -\partial \mathbf{B}/\partial t$,

$$\rho_c \mathbf{E}_k = \epsilon_0 \left[\frac{\partial}{\partial x_i}(E_k E_i) - \frac{1}{2}\frac{\partial}{\partial x_k} \mid \mathbf{E}^2 \mid - \left(\mathbf{E} \times \frac{\partial \mathbf{B}}{\partial t} \right)_k \right] \tag{5.135}$$

Now, for the second right-hand term in (5.133), we note that

$$\begin{aligned} \mathbf{J} \times \mathbf{B} &= \frac{1}{\mu_0}(\nabla \times \mathbf{B}) \times \mathbf{B} - \frac{\partial \mathbf{D}}{\partial t} \times \mathbf{B} \\ &= \frac{1}{\mu_0}(\mathbf{B} \cdot \nabla)\mathbf{B} - \nabla \frac{B^2}{2\mu_0} - \frac{\partial \mathbf{D}}{\partial t} \times \mathbf{B} \end{aligned} \tag{5.136}$$

Hence, noting that $(\mathbf{B} \cdot \nabla)\mathbf{B}$ can be written as $B_i \partial B_k/\partial x_i = \partial(B_i B_k)/\partial x_i$ since $\partial B_i/\partial x_i = \nabla \cdot \mathbf{B} = 0$, and that $\nabla(B^2/2\mu_0)$ can be written as $\partial(B^2/2\mu_0)/\partial x_k = \delta_{ik}\partial(B^2/2\mu_0)/\partial x_i$ where δ_{ik} is the Kronecker delta, we obtain

$$(\mathbf{J} \times \mathbf{B})_k = \frac{\partial}{\partial x_i}\left(\frac{B_i B_k}{\mu_0} - \frac{\mid \mathbf{B}^2 \mid}{2\mu_0}\delta_{ik} \right) - \left(\frac{\partial \mathbf{D}}{\partial t} \times \mathbf{B} \right)_k \tag{5.137}$$

The total electromagnetic force density for the kth component is obtained by adding (5.135) and (5.137), which can be written in a compact form,

$$(\rho_c \mathbf{E} + \mathbf{J} \times \mathbf{B})_k = -\frac{\partial T_{ik}}{\partial x_i} - \frac{\partial G_k}{\partial t} \tag{5.138}$$

where

$$T_{ik} = \left(\frac{\epsilon_0 E^2}{2} + \frac{B^2}{2\mu_0} \right)\delta_{ik} - \left(\epsilon_0 E_i E_k + \frac{B_i B_k}{\mu_0} \right) \tag{5.139}$$

is defined as the electromagnetic stress tensor and

$$G_k = \epsilon_0 \mu_0 (\mathbf{E} \times \mathbf{H})_k = \frac{(\mathbf{E} \times \mathbf{H})_k}{c^2} \tag{5.140}$$

is the momentum density of the electromagnetic field (Poynting vector). Use of the above results in (5.133) yields the conservation of the momentum density equation

$$\frac{\partial}{\partial t}\left(\sum_\alpha \Pi_k^\alpha + G_k \right) + \frac{\partial}{\partial x_i}\left(\sum_\alpha \Pi_{ik}^\alpha + T_{ik} \right) = 0 \tag{5.141}$$

If this equation is integrated over a volume V that is bounded by a surface S, the integral equation states that the rate of change of the total momentum (mechanical plus electromagnetic) in this volume is equal to the rate at which the momentum flows out of the surface bounding this volume.

5.6.1 Momentum Equation in Fluid Representation

The thermal velocity, which is associated with the random motion of the particles, is defined as

$$\mathbf{w} = \mathbf{v} - \mathbf{U} \tag{5.142}$$

where $\mathbf{U}$ is the stream velocity

$$\mathbf{U} = \frac{\sum_\alpha m^\alpha n^\alpha <\mathbf{v}>^\alpha}{\sum_\alpha m^\alpha n^\alpha} \tag{5.143}$$

The stream velocity defines the center of mass velocity of the fluid. Define the pressure tensor as

$$p_{ik} = \sum_\alpha n^\alpha m^\alpha <(v_i - U_i)(v_k - U_k)>^\alpha \tag{5.144}$$

Transform (5.144) using (5.142) and (5.143)

$$\begin{aligned} p_{ik} &= \sum_\alpha n^\alpha m^\alpha <(v_i - U_i)(v_k - U_k)>^\alpha \\ &= \sum_\alpha (m^\alpha n^\alpha <v_i v_k>^\alpha - m^\alpha n^\alpha <v_k>^\alpha U_i \\ &\qquad - m^\alpha n^\alpha <v_i>^\alpha U_k + U_i U_k m^\alpha n^\alpha) \\ p_{ik} &= \sum_\alpha \Pi^\alpha_{ik} - U_i U_k \sum_\alpha m^\alpha n^\alpha \end{aligned} \tag{5.145}$$

Equation (5.145) can now be used to rewrite the momentum equation (5.129) in a more representative fluid form,

$$\frac{\partial}{\partial t} \sum_\alpha m^\alpha n^\alpha U_k + \frac{\partial}{\partial x_i}(p_{ik} + U_i U_k \sum_\alpha m^\alpha n^\alpha) = \sum_\alpha n^\alpha q^\alpha <F_k>^\alpha \tag{5.146}$$

where $< F_k > = < (\mathbf{E} + \mathbf{v} \times \mathbf{B})_k >$. Now perform the $\partial/\partial t$ and $\partial/\partial x_i$ operations to obtain

$$\begin{aligned} U_k \frac{\partial}{\partial t} \sum_\alpha m^\alpha n^\alpha + \sum_\alpha m^\alpha n^\alpha \frac{\partial U_k}{\partial t} + \frac{\partial p_{ik}}{\partial x_i} + U_k \frac{\partial}{\partial x_i} \sum_\alpha m^\alpha n^\alpha U_i \\ + \sum_\alpha m^\alpha n^\alpha U_i \frac{\partial U_k}{\partial x_i} \\ = \sum_\alpha n^\alpha q^\alpha <F_k>^\alpha \end{aligned} \tag{5.147}$$

We require conservation of particles

$$\frac{\partial}{\partial t}\sum_\alpha m^\alpha n^\alpha + \frac{\partial}{\partial x_k}\sum_\alpha m^\alpha n^\alpha U_k = 0 \tag{5.148}$$

Hence the first and the fourth terms in (5.147) vanish, with the result

$$\sum_\alpha m^\alpha n^\alpha \left(\frac{\partial U_k}{\partial t} + U_i \frac{\partial U_k}{\partial x_i}\right) + \frac{\partial p_{ik}}{\partial x_i} = \sum_\alpha n^\alpha q^\alpha < F_k >^\alpha \tag{5.149}$$

Since

$$\left(\frac{\partial U_k}{\partial t} + U_i \frac{\partial U_k}{\partial x_i}\right) = \frac{dU_k}{dt} \tag{5.150}$$

we can write (5.149) as

$$\sum_\alpha m^\alpha n^\alpha \frac{dU_k}{dt} + \frac{\partial p_{ik}}{\partial x_i} = \sum_\alpha n^\alpha q^\alpha < F_k >^\alpha \tag{5.151}$$

If p_{ik} is a scalar, (5.151) reduces to the equation of motion (5.35). $\sum m^\alpha n^\alpha$ and $\sum n^\alpha q^\alpha$ are mass and charge densities of the fluid system, ρ_m and ρ_c.

5.6.2 Navier-Stokes Equation

A short discussion of the Navier-Stokes equation, which describes the behavior of viscous fluids, is now presented. Viscosity arises when adjacent fluid elements flowing with different velocities exchange momentum. In ordinary fluid dynamics, the (ik) component of the viscous stress tensor is defined by

$$\sigma_{ik} = \eta\left(\frac{\partial U_i}{\partial x_k} + \frac{\partial U_k}{\partial x_i} - \frac{2}{3}\frac{\partial U_l}{\partial x_l}\delta_{ik}\right) + \zeta\frac{\partial U_l}{\partial x_l}\delta_{ik} \tag{5.152}$$

and the divergence of σ_{ik} is

$$\begin{aligned}\frac{\partial \sigma_{ik}}{\partial x_k} &= \eta\left(\frac{\partial^2 U_i}{\partial x_k \partial x_k} + \frac{\partial}{\partial x_i}\frac{\partial U_k}{\partial x_k} - \frac{2}{3}\frac{\partial}{\partial x_i}\frac{\partial U_l}{\partial x_l}\right) + \zeta\frac{\partial}{\partial x_i}\frac{\partial U_l}{\partial x_l} \\ &= \eta\frac{\partial^2 U_i}{\partial x_k \partial x_k} + \left(\zeta + \frac{1}{3}\eta\right)\frac{\partial}{\partial x_i}\frac{\partial U_l}{\partial x_l}\end{aligned} \tag{5.153}$$

We assume that both η and ζ, the coefficients of viscosity, are constants (note that in general η and ζ are positive numbers that are functions of pressure and temperature). This additional term must be added to our equation of motion if we are considering a viscous conducting fluid. The vectorial form of the equation of motion for a viscous MHD fluid is

$$\rho_m \frac{d\mathbf{U}}{dt} = -\nabla p + \mathbf{J} \times \mathbf{B} + \eta \nabla^2 \mathbf{U} + \left(\zeta + \frac{1}{3}\eta\right)\nabla(\nabla \cdot \mathbf{U}) \tag{5.154}$$

Note that for incompressible fluids $\nabla \cdot \mathbf{U} = 0$ and the last term vanishes. An incompressible viscous fluid is described by η and $\eta/\rho_m = \nu$ is defined as the coefficient of kinematic shear viscosity. ζ is called the second or bulk viscosity.

In space, the effects of viscosity are generally ignored. Although this is justified in many classes of problems, the viscous interaction effects must be included in ionospheres, the lower solar atmosphere, and near boundaries. For example, in the study of solar wind interaction with planetary magnetic fields, the inclusion of viscous forces can produce a magnetic tail and boundary layers. The concept of viscosity is fairly well understood in ordinary fluids, where momentum exchange is achieved by collisions. What is the meaning of viscosity in collisionless fluids? The effects of viscosity on space plasma phenomena have not received much attention and remain one of the many challenges of future research.

5.7 Maxwellian Plasma

Many applications in space physics, geophysics, and astrophysics assume that plasma is in thermal equilibrium. When random collisions occur frequently enough, the local velocity distribution can be approximated by an isotropic Maxwellian,

$$f(\mathbf{r}, \mathbf{v}) = n \left(\frac{m}{2\pi kT} \right)^{3/2} e^{-m(\mathbf{v} - <\mathbf{v}>)^2/2kT} \tag{5.155}$$

where n and T are density and temperature. $<\mathbf{v}>$ is the average velocity, $\mathbf{v} - <\mathbf{v}> = \mathbf{w}$ is the thermal velocity (5.142), and $m(\mathbf{v} - <\mathbf{v}>)^2/2$ is associated with the kinetic temperature.

5.7.1 Pressure Tensor

If the isotropic Maxwellian distribution function is used to compute the pressure tensor given by

$$p_{ij} = \int m v_i v_j f d^3 v \tag{5.156}$$

we obtain

$$p_{ij} = \begin{pmatrix} nkT & 0 & 0 \\ 0 & nkT & 0 \\ 0 & 0 & nkT \end{pmatrix} \tag{5.157}$$

The pressure tensor p_{ij} for a Maxwellian distribution is diagonal and constant, $p_{11} = p_{22} = p_{33} = p_{ii}/3$ and $p_{ij} = 0$ if $i \neq j$. The pressure tensor reduces to a scalar, $p_{ij} = p\delta_{ij}$, where $p = nkT$ and δ_{ij} is the Kronecker delta. This result is not surprising since collisions eliminate anisotropic properties of a plasma. Momentum transfer is then normal to the surface.

Another parameter frequently used in fluid dynamics is temperature. Temperature is an equilibrium concept, and it is meaningful in a plasma defined by the Maxwellian distribution function. Temperature can be defined in terms of the thermal energy $m <(\mathbf{v}- <\mathbf{v}>)^2> /2 = 3/2kT$.

5.7.2 Equations of Continuity and Motion

The behavior of an equilibrium plasma can be further studied by use of the Maxwellian distribution function to compute the moments of the Boltzmann-Vlasov transport equation given by (2.24). The zeroth moment equation yields the continuity equation

$$\frac{\partial n}{\partial t} + \frac{\partial}{\partial x_j}(nu_j) = 0 \tag{5.158}$$

where it is understood that $u_j = <v_j>$ is the velocity averaged over the Maxwellian distribution function.

Compute the first moment by using (2.24), stated here again for convenience,

$$\frac{\partial}{\partial t}(nm <v_i>) + \frac{\partial}{\partial x_j}(nm <v_i v_j>) = \rho_c E_i + (j \times B)_i \tag{5.159}$$

The first term on the left-hand side can be written as

$$\frac{\partial}{\partial t}(nmu_i) = mu_i \frac{\partial n}{\partial t} + mn\frac{\partial u_i}{\partial t} \tag{5.160}$$

Rewriting the second term on the left side of (5.159) as

$$\frac{\partial}{\partial x_j}(nmu_i u_j) = mu_i \frac{\partial}{\partial x_j}(nu_j) + mnu_j \frac{\partial u_i}{\partial x_j} \tag{5.161}$$

and noting that $nm <v_i v_j> = nmu_i u_j + p_{ij}$ (see (5.145)), we can transform the original equation into

$$nm\frac{\partial u_i}{\partial t} + mnu_j \frac{\partial u_i}{\partial x_j} = -\frac{\partial}{\partial x_i}(nkT) + \rho_c E_i + (\mathbf{j} \times \mathbf{B})_i \tag{5.162}$$

recognizing that $\partial n/\partial t + u_i \partial u_i/\partial x_j = 0$ and the relation $p = nkT$ has been used. This is identical to (5.151) with $p_{ij} = p\delta_{ij}$.

5.7.3 Equation of State

The second moment given by (2.39) is

$$\frac{\partial}{\partial t}\left(\frac{nm}{2} <v^2>\right) + \frac{\partial}{\partial x_i}\left(\frac{nm}{2} <v^2 v_i>\right) = \mathbf{J}\cdot\mathbf{E} \tag{5.163}$$

As already mentioned in Chapter 2, the first term of the left-hand side represents the time rate of change of the energy density of the particles and the second, the rate at which energy flows out of the volume. It is convenient at this point to define the heat transfer parameter (heat tensor)

$$q_{ijk} = nm <w_i w_j w_k> \tag{5.164}$$

where $\mathbf{w} = \mathbf{v} - \mathbf{u}$ is the random velocity defined in (5.142). The quantity which appears in the energy conservation equation

$$q_i = \frac{nm}{2} <v^2 v_i> \tag{5.165}$$

represents the rate at which energy flows across a unit area whose normal vector is in the ith direction. For a Maxwellian plasma,

$$\frac{nm}{2} <v^2> = \frac{nm}{2} <\mathbf{v}>^2 + \frac{3}{2} nkT \tag{5.166}$$

and

$$\frac{nm}{2} <v^2 v_i> = \frac{nm}{2} <\mathbf{v}>^2 v_i + \frac{5}{2} nkT <v_i> \tag{5.167}$$

Substitution of these relations in the original energy conservation equation yields

$$\begin{aligned} nm <v_i> \frac{\partial}{\partial t} <v_i> &+ \frac{m}{2} <\mathbf{v}>^2 \frac{\partial n}{\partial t} + \frac{3}{2}\frac{\partial}{\partial t}(nkT) \\ + \ nm <v_i><v_k> \frac{\partial v_i}{\partial x_k} &+ \frac{m}{2} <\mathbf{v}>^2 \frac{\partial}{\partial x_k} n <v_k> \\ + \ \frac{5}{2} <v_i> \frac{\partial}{\partial x_i}(nkT) &+ \frac{5}{2} nkT \frac{\partial}{\partial x_i} <v_i> = J_i E_i \end{aligned} \tag{5.168}$$

Note that the sum of the second and the fifth terms vanishes

$$\frac{m <\mathbf{v}>^2}{2}\frac{\partial n}{\partial t} + \frac{m <\mathbf{v}>^2}{2}\frac{\partial}{\partial x_k}(n v_k) = 0 \tag{5.169}$$

because of the continuity relation. Also, the sum of the first, fourth, and a part of the sixth terms of the left-hand side equals $J_i E_i$

$$\frac{nm}{2} <v_i> \frac{\partial}{\partial t} <v_i> + \frac{nm}{2} <v_i><v_k> \frac{\partial v_i}{\partial x_i}$$
$$+ \frac{1}{2} <v_i> \frac{\partial}{\partial x_i}(nkT) = J_i E_i \tag{5.170}$$

since momentum is conserved. Hence the remaining terms of the energy conservation equation (5.168) are

$$\frac{3}{2}\frac{\partial}{\partial t}(nkT) + \frac{3}{2} <v_i> \frac{\partial}{\partial x_i}(nkT) + \frac{5}{2}(nkT)\frac{\partial}{\partial x_i} <v_i> = 0 \tag{5.171}$$

Further simplification can be made by noting that

$$n\frac{\partial}{\partial x_i} <v_i> = -\frac{\partial n}{\partial t} - <v_i> \frac{\partial n}{\partial x_i}$$
$$= -\frac{dn}{dt} \tag{5.172}$$

Here, d/dt represents the convective derivative taken moving along with the average velocity $<v_i>$. Thus, the conservation of energy equation can be written as

$$\frac{3}{2}\frac{d}{dt}(nkT) - \frac{5}{2}kT\frac{dn}{dt} = 0 \tag{5.173}$$

Since $p = nkT$, this equation becomes

$$3\frac{dp}{p} = 5\frac{dn}{n} \tag{5.174}$$

which upon integration yields

$$p = p_0 \left(\frac{n}{n_0}\right)^{5/3} \tag{5.175}$$

This is an adiabatic law for an ideal gas with the ratio of specific heats

$$\gamma = C_p/C_v = 5/3 \tag{5.176}$$

A Maxwellian plasma behaves as an ideal adiabatic gas. The heat tensor for a Maxwellian gas $q_{ijk} = 0$ for all (ijk) because viscosity has been neglected.

5.7.4 One-Fluid Equations

One-fluid MHD equations are frequently used to study plasma phenomena and below we summarize the set of one-fluid equations for an isotropic Maxwellian plasma. The equations are

$$\frac{\partial n}{\partial t} + \nabla \cdot n\mathbf{U} = 0 \tag{5.177}$$

$$nm\left[\frac{\partial \mathbf{U}}{\partial t} + (\mathbf{U}\cdot\nabla)\mathbf{U}\right] = -\nabla p + \mathbf{J}\times\mathbf{B} \tag{5.178}$$

$$p = p_0\left(\frac{n}{n_0}\right)^{5/3} \tag{5.179}$$

$$\nabla \times \mathbf{E} = -\frac{\partial \mathbf{B}}{\partial t} \tag{5.180}$$

$$\nabla \times \mathbf{B} = \mu_0 \mathbf{J} \tag{5.181}$$

$$\mathbf{J} = nq\mathbf{U} \tag{5.182}$$

There are fourteen unknowns (n, p, $\mathbf{U}$, $\mathbf{E}$, $\mathbf{B}$, $\mathbf{J}$) and fourteen equations. Hence no other equations are needed. This set of equations is quite general, more general than the set of equations used earlier, which assumed that Ohm's law applies (see Section 5.3). These equations assume that the plasma is in thermal equilibrium and is described by a Maxwellian distribution function. Note that the equation of state for a Maxwellian plasma arises "naturally" from the energy conservation equation.

5.7.5 Beta of Plasma

Consider the one-fluid equation of motion (5.178). In equilibrium, $d\mathbf{U}/dt = 0$; hence the momentum equation reduces to

$$\nabla p = \mathbf{J}\times\mathbf{B} \tag{5.183}$$

which is a statement that the particle pressure gradient is balanced by the electromagnetic stress. Now use the relation $\nabla \times \mathbf{B} = \mu_o \mathbf{J}$. The above equation can then be rewritten as

$$\begin{aligned}\nabla p &= \frac{(\nabla\times\mathbf{B})\times\mathbf{B}}{\mu_0}\\ &= \frac{(\mathbf{B}\cdot\nabla)\mathbf{B}}{\mu_0} - \frac{\nabla B^2}{2\mu_0}\end{aligned} \tag{5.184}$$

If the magnetic field does not vary along $\mathbf{B}$ (for a straight homogeneous field), the first term on the right-hand side vanishes. One then concludes that

$$\nabla\left(p+\frac{B^2}{2\mu_0}\right)=0 \tag{5.185}$$

and

$$p+\frac{B^2}{2\mu_0}=\text{constant} \tag{5.186}$$

in space. The sum of the particle pressure (p) and the magnetic pressure ($B^2/2\mu_0$) is a constant for a plasma in equilibrium. The plasma beta (β) parameter is defined as

$$\begin{aligned}\beta &= \frac{\text{particle pressure}}{\text{magnetic field pressure}}\\ &= \frac{p}{B^2/2\mu_0}\end{aligned} \tag{5.187}$$

β measures the relative importance of the particle and magnetic field pressures. A plasma is referred to as a low-beta plasma when $\beta \ll 1$, and a high-beta plasma when $\beta \approx 1$. In space, both high- and low-beta plasmas are encountered.

5.8 Restrictions of MHD Fluids

One of the fundamental properties of ordinary fluids is that they can flow. The physical states of flowing fluids are quantitatively characterized by their velocity, density, and pressure as functions of position and time through the thermodynamic properties and the conservation laws of mass, momentum, and energy. Let L and T be the characteristic distance and time of a fluid system, and consider a fluid element dr so that $dr \ll L$. Since the different parts of a fluid tend to move together, most of the particles in dr at time t will be present at time $t+T$. This persistence of the fluid element is achieved by collisions, as the particles are prevented from leaving the element by collisions with the neighboring particles. Define λ_c as the collision mean free path. Our description of the fluid implies that $dr \gg \lambda_c$. After many collisions, the particles of the fluid will approach some steady-state. Let t_c be the time interval between collisions (in the average sense). Then we have an additional requirement that $T \gg t_c$. Within the characteristic time T, properties of the fluid do not change appreciably.

The requirements of an ordinary fluid are:

$$T \gg t_c \tag{5.188}$$

$$L \gg \lambda_c \tag{5.189}$$

Let us now examine when a charged particle system can be treated as a fluid. An immediate question is whether a system of charged particles can satisfy the conditions of the ordinary fluid. The answer is not obvious because the collisional processes between neutrals (in ordinary fluids) and charged particles are very different. Whereas in the case of neutrals the collisions involve only short-range forces (collisions are felt by particles only upon impact), collisions of charged particles involve long-range electromagnetic forces. A particle's orbit will be affected by another particle even when the two are far apart. Such long-range forces complicate collisional processes considerably when a system contains many particles, since the orbit deflection of one particle has electromagnetic effects on all the particles.

An important property of a charged particle system is that it is anisotropic in the presence of a magnetic field. Motions along and transverse to the magnetic field direction are very different. Hence collisional processes in these directions are expected to be very different. Even if conditions (5.188) and (5.189) were satisfied in one direction, they might not be satisfied in another direction. Another problem is that the particle densities in interplanetary space are so low that no matter how large the system, conditions (5.188) and (5.189) as so prescribed may never be satisfied. The particles in these regions are "collisionless," and t_c and λ_c are very long. Clearly, the requirements of ordinary fluids need to be modified.

Particles in a magnetic field are collisionless if $\lambda_c \gg r_c$ where r_c is the cyclotron radius of the particle. This condition normally implies that the magnetic field must be fairly strong. Introduce now a scale length $L_\perp$ in the transverse direction (transverse to the direction of the magnetic field) and define $T_\perp$ as a transverse characteristic time of the charged particle system. Analogous to ordinary fluids, we then require the charged particle fluid system to satisfy

$$L_\perp \gg r_c \tag{5.190}$$

$$T_\perp \gg T_c \tag{5.191}$$

where T_c is the cyclotron period. A charged particle system that satisfies these conditions can then be considered to behave "similarly" to the ordinary fluid in the direction perpendicular to the magnetic field.

The magnetic field plays the role of collisions, since the field binds the charged particles tightly together on the lines of force. Thus, the particles in the fluid elements are constrained to remain in the elements in flows perpendicular to the magnetic field direction. Conditions (5.190) and (5.191) implicitly states that viscous and heat-transfer effects must be small in the perpendicular direction. This condition requires that the temperature of the fluid be low. In the limit that the temperature approaches zero, the perpendicular flow of a charged particle system is completely analogous to the central mass motion of ordinary fluids discussed earlier. If viscous and heat-transfer effects are small, currents can flow essentially unimpeded, since resistance is small. In the limit of no resistance, we have perfect conductivity and the fluid motion is defined by $\mathbf{E} + \mathbf{U} \times \mathbf{B} = 0$.

In summary, the charged particle fluid flow in the perpendicular direction resembles best the behavior of ordinary fluid flow in the limit of the "cold" plasma approximation where the flow velocity is much larger than the thermal velocity. In view of the importance of the cold plasma approximation, we summarize below the set of one-fluid equations that govern the cold plasma fluid. When a plasma approaches the cold limit ($T \approx 0$), the set of equations is considerably simplified, in part because the thermal effects can be ignored. The set of cold plasma equations includes the continuity equations of mass and charge densities:

$$\frac{\partial \rho_m}{\partial t} + \nabla \cdot \rho_m \mathbf{U} = 0 \tag{5.192}$$

$$\frac{\partial \rho_c}{\partial t} + \nabla \cdot \mathbf{J} = 0 \tag{5.193}$$

the equation of motion:

$$\rho_m \left[\frac{\partial \mathbf{U}}{\partial t} + (\mathbf{U} \cdot \nabla)\mathbf{U} \right] = \rho_c \mathbf{E} + \mathbf{J} \times \mathbf{B} \tag{5.194}$$

Ohm's law:

$$\mathbf{J} = \sigma(\mathbf{E} + \mathbf{U} \times \mathbf{B}) \tag{5.195}$$

and the Maxwell equations:

$$\nabla \times \mathbf{E} = -\frac{\partial \mathbf{B}}{\partial t} \tag{5.196}$$

$$\nabla \times \mathbf{H} = \mathbf{J} + \frac{\partial \mathbf{D}}{\partial t} \tag{5.197}$$

This gives a set of fourteen equations for the fourteen unknowns, ρ_m, ρ_c, $\mathbf{U}$, $\mathbf{J}$, $\mathbf{B}$, and $\mathbf{E}$. The displacement current term is retained now since the inequality (5.41) may no longer be valid. This requires us to retain the term involving the charge density ρ_c in the equation of motion. Hence we include the charge conservation equation. Note also that, to be more precise, the generalized Ohm's law (Chapter 7) should be used instead of the simplified form. Cold plasma equations assume that the thermal energy is small enough that the pressure and the heat tensors can be ignored. Approximating a plasma as cold is highly restrictive but useful information can be obtained, especially in the study of wave propagation in plasmas.

What about fluid flows in the direction parallel to the magnetic field? In this case, the Lorentz force of the magnetic field is identically zero and the fluid will behave as if the magnetic field is not there. For a collisionless fluid, heat transfer and viscous effects may be ignored in the parallel case as well. The flow in the parallel direction may be totally uncoupled from the flow in the perpendicular direction and treated separately. There are two sets of equations, and each set contains the equation of state that describes the state of that system (Chapter 7).

The fluid theory provides a reasonably good picture of the solar wind and the collisionless bow shock. Also, certain large-scale structures in the ionosphere, solar wind, magnetosphere, and the geomagnetic tail regions can be understood within the fluid context. However, there are many observed features that cannot be understood by the simple MHD fluid theory. For example, fine scale features, rapid time variations, and dissipative phenomena observed in auroras and in bow shock regions cannot be explained. In these cases, it is necessary to study the fundamental Lorentz equations of motion that are coupled to Maxwell's equations of electrodynamics.

Bibliography

Alfvén, H. and C.G. Fälthammar, *Cosmical Electrodynamics, Fundamental Principles*, 2nd ed., Oxford University Press, Oxford, England, 1963. Chapter 3 of this book is useful for further reading.

Chandrasekhar, S., *Plasma Physics*, The University of Chicago Press, Chicago, IL, 1960. This paperback book consists of notes compiled by S.K. Trehan of a course given by S. Chandrasekhar at the University of Chicago. We have adopted the procedure outlined by Chandrasekhar in the derivation of the macroscopic fluid equation.

Ferraro, V.C.A. and C. Plumpton, *An Introduction to Magneto-Fluid Mechanics*, 2nd ed., Clarendon Press, Oxford, England, 1966. This book contains the history of how the magneto-fluid concept was developed. MHD theory is also clearly developed. For students at the intermediate level.

Landau, L.D. and E.M. Liftschitz, *Electrodynamics of Continuous Media*, Pergamon Press, Ltd., New York, NY, 1960. Chapter 8 gives a concise but thorough discussion of magnetofluid dynamics.

Longmire, C.L., *Elementary Plasma Physics*, Interscience Publishers, New York, NY, 1963. Chapters 1 and 2 present material that is useful for this chapter.

Newcomb, W.A., Motion of Magnetic Lines of Force, *Annals of Physics*, **3**, 347, 1958. A very detailed piece of work on the concept of magnetic field line motion.

Parker, E.N., *Cosmical Magnetic Fields, Their Origin and Their Activity*, Oxford University Press, New York, NY, 1979. A thorough book on magnetic fields with references to observations. A must for students interested in pursuing research in this field.

Schmidt, G., *Physics of High Temperature Plasmas, An Introduction*, Academic Press, New York, NY, 1966. The properties of the conducting fluids are clearly discussed in Chapters 3 and 4 of this book. The section on the moments of the Boltzmann equation is clearly developed.

Siscoe, G.L., Solar System Magnetohydrodynamics, in *Proceedings of the 1982 Boston College Theory Institute in Solar-Terrestrial Physics*, R.L. Carovillano and J.M. Forbes, eds., D. Reidel Publishing Co., Dordrecht, Holland, 1982. Magnetohydrodynamic theory is developed from first principles. Chapter 1 includes an excellent discussion of the role of thermodynamics in MHD. Also, MHD theory is clearly developed with emphasis on application to space plasmas.

Spitzer, L., *Physics of Fully Ionized Gases*, 2nd ed., Interscience Publishers, New York, NY, 1962. Generally an excellent book for beginning students. Chapters 1 and 2 are appropriate for the discussion of fluids.

Stern, D.P., The Motion of Magnetic Field Lines, *Space Sci. Rev.*, **6**, 147, 1966. This review article presents a formal theory of the concept of magnetic field line motion.

Vasyliunas, V.M., Nonuniqueness of Magnetic Field Line Motion, *J. Geophys. Res.*, **77**, 6271, 1972. The concept of "field line motion" is frequently used in the context of space physics. This concept, which is a consequence of the frozen-in-field theorem, has been a source of confusion for a number of investigators. The article by Vasyliunas clarifies how the confusion arises and presents a formal theory of the field line motion. Examples applied to Earth's magnetospheric media are discussed.

Questions

1. A metal rod of length L is moving in a magnetic field $\mathbf{B}$ with a velocity $\mathbf{U}$. What is the potential difference at the ends of the rod?
2. Consider a metal rod whose length is 1m. Rotate this rod about one of its ends with an angular frequency of 12 radians/sec. Assuming that the Earth's magnetic field is $0.6\ 10^{-4}$T, calculate the voltage difference that will be measured between the ends of the rods.
3. Consider a 100 turn coil compactly wound so that one can assume that all of the loops are in the same plane. Let the average radius of the coil be 0.03m. Now rotate the coil about the diameter with an angular frequency of 15 revolutions/sec. Estimate the average magnetic field of Earth if the average induced EMF is 0.5 millivolts.
4. Consider a sphere of radius R that is rotating about an axis through its center with a constant angular velocity ω. Let there be a surface charge density σ_s rigidly attached to the sphere.
   a. Show that the magnetic field at an external point is a dipole.
   b. Find the equivalent dipole moment.

5. An imbalance of electrical charges in a system of charged particles gives rise to static electric fields. Estimate the strength of the electric field for a 1% deviation from neutrality of a system containing 10^{12} particles $(\text{m})^{-3}$, assuming a spherical geometry of 0.01 m radius.
6. The origin of Earth's magnetic field is not yet known. However, simple considerations indicate that the present magnetic field cannot be the surviving field of the past epoch. Show that the lifetime of Earth's magnetic field is less than 4.6×10^9 years, which is the estimated life of Earth. Assume that the core material is of Cu, whose conductivity is about 10^8 mhos/meter.
7. Consider a magnetic field $\mathbf{B}$ that is uniform in space but is time-dependent. Let $\mathbf{B} = \mathbf{n}B(t)$ where $\mathbf{n}$ is a constant unit vector in the direction of $\mathbf{B}$. In the absence of net space charges, $\mathbf{E}$ satisfies $\nabla \cdot \mathbf{E} = 0$ and $\nabla \times \mathbf{E} = -\partial \mathbf{B}/\partial t$. Show that

$$\mathbf{E} = -\frac{1}{2}(\mathbf{n} \times \mathbf{r})\frac{\partial B}{\partial t}$$

is a solution of the above Maxwell equations.
8. Show that if $\mathbf{E} = -\mathbf{U} \times \mathbf{B}$

$$\nabla \times \mathbf{E} = \mathbf{n}\nabla \cdot (\mathbf{U}B)$$

where $\mathbf{n}$ is a unit vector along $\mathbf{B}$. State what assumptions are required for the above relation to be true.
9. The above result shows that

$$\frac{\partial B}{\partial t} + \nabla \cdot (\mathbf{U}B) = 0$$

which is another way of saying that the magnetic flux is conserved. Use this equation and the equation of continuity of particles to show that

$$\frac{d}{dt}\left(\frac{n}{B}\right) = 0$$

where n is the particle density.
10. Use Ohm's law $\mathbf{J} = \sigma(\mathbf{E} + \mathbf{U} \times \mathbf{B})$ and the equation of motion and show that
    a. The momentum equation for the perpendicular motion can be written as

$$\rho_m \frac{d\mathbf{U}_\perp}{dt} = \mathbf{F}_\perp + \sigma B^2 \left(\frac{\mathbf{E} \times \mathbf{B}}{B^2} - \mathbf{U}_\perp\right)$$

    where $\mathbf{F}_\perp$ is the nonelectromagnetic force perpendicular to $\mathbf{B}$.
    b. The limiting value attained by $\mathbf{U}_\perp$ is

$$\mathbf{U}_\perp = \frac{\mathbf{F}_\perp}{\sigma B^2} + \frac{\mathbf{E} \times \mathbf{B}}{B^2}$$

    Note that in the limit $\sigma \to \infty$, $\mathbf{U}_\perp \to \mathbf{E} \times \mathbf{B}/B^2$.
    c. The time it takes to reach the limiting velocity in (b) is

$$\tau \approx \frac{\rho_m}{\sigma B^2}$$

11. Derive Equation (5.122) from Equation (5.121).

12. The force per unit volume containing charges and currents is $\mathbf{F} = \rho\mathbf{E} + \mathbf{J} \times \mathbf{B}$. Use the Maxwell equations to show that

$$\begin{aligned} \mathbf{F} &= -\epsilon_0 \frac{\partial}{\partial t}(\mathbf{E} \times \mathbf{B}) + \epsilon_0 \mathbf{E}(\nabla \cdot \mathbf{E}) - \frac{1}{2}\epsilon_0 \nabla E^2 + \epsilon_0 (\mathbf{E} \cdot \nabla)\mathbf{E} \\ &+ (\mu_0)^{-1}\mathbf{B}(\nabla \cdot \mathbf{B}) - (2\mu_0)^{-1}\nabla B^2 + (\mu_0)^{-1}(\mathbf{B} \cdot \nabla)\mathbf{B} \end{aligned}$$

Discuss the meaning of each term.

13. The magnetic field data shown below come from two spacecraft separated by about $230R_E$. ISEE-1 was orbiting near Earth and ISEE-3 was at L_1, the gravitation null point between Earth and the Sun. The comparison of the two data sets will show that many features detected on ISEE-3 appear also on ISEE-1. Discuss the meaning of this similarity in the context of the MHD theory. Note that the figures have been aligned by shifting the time and the vertical traces have been offset by 5 nT for clarity. (From Russell *et al.*, *Geophys. Res. Lett.*, **7**, 381, 1984)

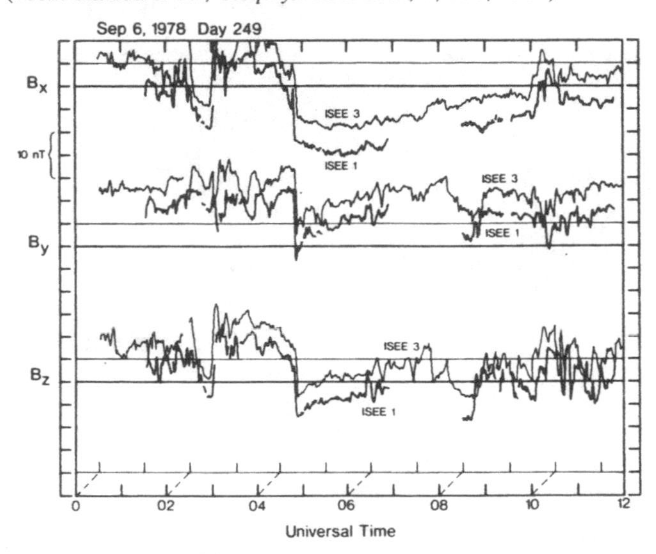

14. In ordinary fluid dynamics, the kth component of the momentum conservation equation can be written in a compact form

$$\frac{\partial}{\partial t}(\rho_m u_k) = -\frac{\partial}{\partial x_i}\pi_{ik}$$

where ρ_m is the mass density, u_k is the kth component of the fluid velocity, and $\pi_{ik} = \rho_m u_i u_k + p\delta_{ik}$, is the momentum transfer tensor. Here p is the pressure and δ_{ik} is the Kronecker delta tensor.

a. Derive the above equation. (Hint: use the continuity equation and the Euler's equation with scalar pressure.)

b. Show that the above representation is equally valid for an MHD fluid if

$$\rho_m u_k = \pi_k = \int m v_k f d^3 v$$

$$\pi_{ik} = \rho_m u_i u_k + p\delta_{ik} - \left[\frac{B_i B_k}{\mu_0} - \left(\frac{B^2}{2\mu_0}\right)\delta_{ik}\right]$$

15. Consider a situation in space in which the current density **J** is parallel to **B**, that is, $\mathbf{J} \times \mathbf{B} = 0$ (the current is field-aligned). In this case, the MHD system is considered force-free and implies that

$$(\nabla \times \mathbf{B}) \times \mathbf{B} = 0$$

This equation is satisfied if

$$\nabla \times \mathbf{B} = \alpha \mathbf{B}$$

where α is a scalar function of position.

a. Show that α must satisfy the equation $(\mathbf{B} \cdot \nabla)\alpha = 0$.

b. Suppose the force-free magnetic field is to decay due to the finite conductivity of the fluid. Show that the magnetic field decays according to

$$\frac{\partial \mathbf{B}}{\partial t} = -\left(\frac{1}{\mu_0 \sigma}\right)[\alpha^2 \mathbf{B} + (\nabla \alpha) \times \mathbf{B}]$$

c. Show that if the force-free field is to decay without distorting the shape of the fluid, α must be a constant and the above equation reduces to

$$\frac{\partial \mathbf{B}}{\partial t} = \left(\frac{\alpha^2}{\mu \sigma}\right) \mathbf{B}$$

16. Energetic solar particles were detected by a spacecraft in the vicinity of Earth following a large solar flare. The plot below shows that protons of various energies, alphas, and medium mass nuclei all arrived at the same time. Explain how this is possible. (From Roeloff, *Physics of Solar Planetary Environments*, American Geophysical Union, Washington D.C., 1976)

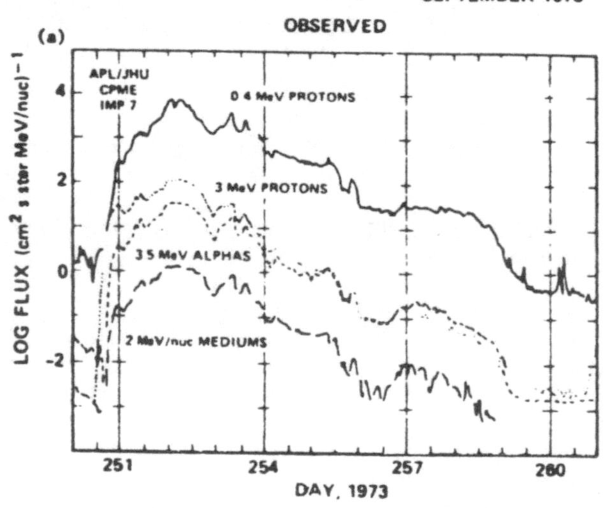

17. For perfect fluids, one can show that

$$\frac{d}{dt}\left(\frac{\mathbf{B}}{\rho_m}\right) = \left(\frac{\mathbf{B}}{\rho_m} \cdot \nabla\right) \mathbf{U}$$

where ρ_m is the fluid density and **U** is the fluid velocity.

a. Start from Equation (5.58) and use the continuity equation to derive this equation.

b. Discuss the meaning of this equation in relation to the line-preservation theorem.

18. The concept of "field line motion" is often used in space physics. This concept comes from application of the frozen-in-field theorem which permits one to visualize the field lines attached to the fluid. Field line motion is not unique and this could lead to misinterpretation of data. Read the articles by Kavanagh, L., *J. Geophys. Res.*, **72**, 6120, 1967; Birmingham, T. and Jones, *J. Geophys. Res.*, **73**, 5505, 1968. Jones and Birmingham, *J. Geophys. Res.*, **76**, 1849, 1971, and Vasyliunas, V., *J. Geophys. Res.*, **77**, 6271, 1972, and summarize briefly the contents and conclusions of each of these articles.

6

Solar Wind, Interplanetary Magnetic Field, and Plasma Convection

6.1 Introduction

The MHD theory has played an important role in advancing our knowledge of space plasma phenomena. The purpose of this chapter is to use the MHD fluid theory and concepts to formulate and discuss the physics of planetary and stellar atmospheres. We are particularly interested in learning about the dynamics of ionized atmospheres, how the solar wind and the interplanetary magnetic field originate, and what the forces are that drive plasma convection in magnetospheres. The solar wind problem is closely tied to the dynamics of the solar corona, the interplanetary magnetic field to the frozen-in-field concept, and the magnetospheric convection to how energy is transported into and circulated around magnetospheres. In regard to the discussion of these models, the reader is reminded that the MHD theory has limitations. Nevertheless, a considerable amount of insight can be gained by studying how MHD theory works in space.

A model of the static atmosphere is first presented to establish the general structure of an atmosphere. An atmosphere supports both neutral and ionized components. Ionization is primarily produced by the solar ultraviolet radiation, and ionized constituents populate the ionosphere and the magnetosphere. To study how ionospheres are formed, we will examine the well-known ionospheric model formulated by Sydney Chapman in 1931. Our formulation of the static atmosphere includes a short discussion of important photochemical interactions that occur in atmospheres to help us understand what factors control the concentration of ionized species in ionospheres.

We next present a dynamic model of atmospheres to explain the solar (stellar) wind and the polar wind phenomena (solar wind describes the escape of particles from the Sun and polar wind the escape of particles from planets). The dynamic model of the solar wind, proposed by E.N. Parker in 1958, treats the solar atmosphere as being in steady-state but includes an outward streaming of particles with a finite velocity at the base of the corona. Parker's model successfully predicted that the coronal particles will arrive at Earth with a radial speed of several hundred kilometers per second. A major feature of Parker's model is that it integrates the corona observed in the solar atmosphere and the continuous flow of solar particles at large distances from the Sun as two aspects of a single coronal dynamic atmospheric system.

The interplanetary magnetic field (IMF) problem is closely tied to the solar wind model. Because the solar coronal constituents are all ionized and are therefore highly conductive, the solar magnetic field is frozen in the solar plasma and is transported out into space with the solar wind. The IMF observed in our solar system far from the Sun therefore represents the extension of the Sun's magnetic field. A model will be constructed to show how the solar magnetic field is convected out with the solar wind. We also discuss qualitatively a model of the ecliptic current sheet to explain the reversal of IMF polarity and the presence of an IMF component perpendicular to the ecliptic plane.

The final problem discussed is plasma convective phenomena observed in magnetospheres. Motion of plasma in the magnetosphere requires the presence of a large-scale electric field for which there are two known sources. The first is the corotational field created by the planetary rotation. The second is the solar wind electric field created by the motion of the solar wind across the IMF. Assuming that a fraction of the solar wind electric field penetrates into the magnetosphere across the magnetopause boundary, the plasma in the magnetosphere is now subject to both types of electric fields. A model will be constructed to show that the corotating ionospheric plasma interacting with a dawn-dusk electric field of solar wind origin will confine the corotation inside a boundary called the plasmapause. The ionosphere is contained inside this boundary.

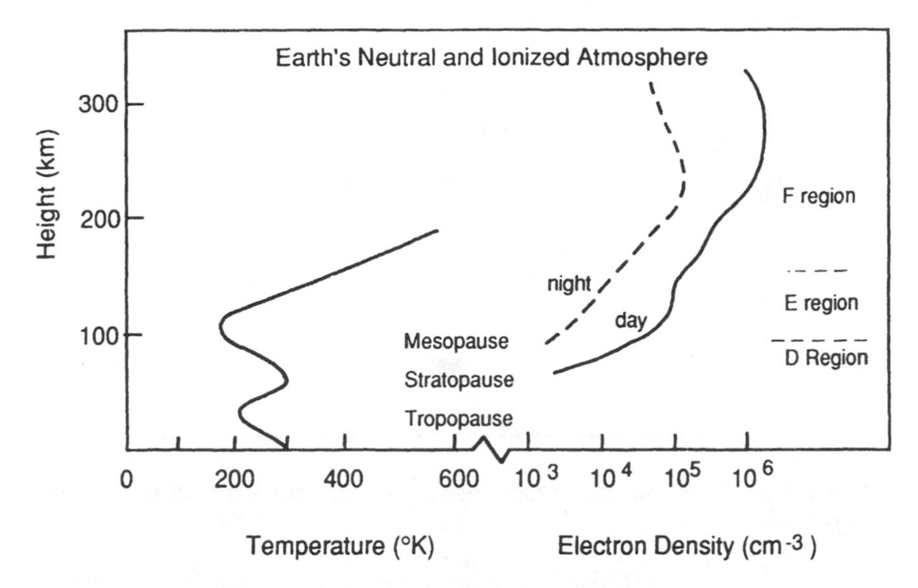

FIGURE 6.1 A schematic diagram of Earth's neutral and ionized atmospheres. Nomenclature for the various regions is due to S. Chapman. Note the various ionospheric layers, designated as D, E, and F regions.

6.2 Static Atmospheres

An atmosphere is characterized by density, temperature, pressure, composition, and motion. Studies of Earth's atmospheric behavior, such as surface wind directions, can be traced back to the ancient Greeks. However, it was not until the invention of barometers and thermometers in the seventeenth century that pressure and temperature were measured. Use of these instruments at various altitudes in the early twentieth century led to the discovery of the troposphere and stratosphere, and later the mesosphere and the thermosphere. The existence of the ionosphere was suggested in the late nineteenth century, but its existence was not firmly established until radiosondes were invented. Figure 6.1 shows a schematic diagram of the distribution of neutral and ionized atmospheres of Earth.

6.2.1 Neutral Atmosphere

One of the most important properties of an atmosphere is the exponential decrease of the density and pressure with height. The equation that describes this decrease is obtained by examining the fluid equation of motion

$$\rho_m \frac{dU}{dt} = -(\nabla p + \rho_m \nabla \Psi) \tag{6.1}$$

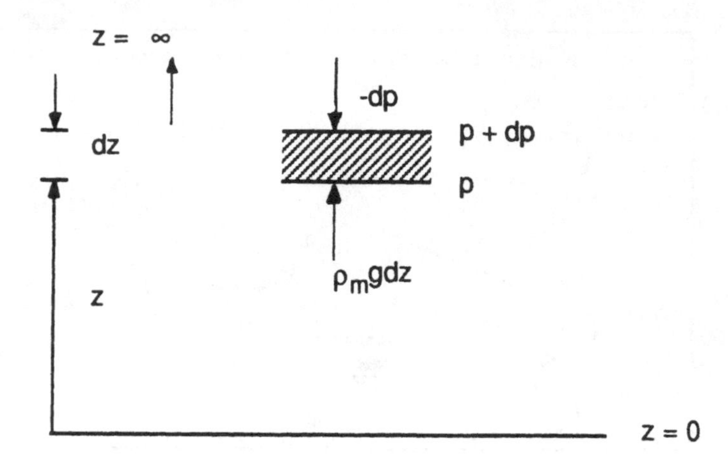

FIGURE 6.2 A schematic diagram showing an atmospheric fluid element in a gravity field.

where ρ_m is the mass density, p is pressure, and Ψ is the gravitational potential. Consider a parcel of fluid at height z from the surface of a planet with vertical incremental dimension dz and unit cross section. For an atmosphere that is in hydrostatic equilibrium, the sum of the forces on a fluid element must vanish and, therefore, the difference of pressure between the top and the bottom must balance the downward force of the weight of the fluid element (Figure 6.2).

Let n be the number density, m the mean mass of the atmospheric constituents of the fluid, and g the acceleration due to gravity. All of these parameters are height-dependent. Equilibrium requires that

$$\begin{aligned} dp &= -\rho_m(z)g(z)dz \\ &= -n(z)m(z)g(z)dz \end{aligned} \tag{6.2}$$

where

$$\rho_m = \sum n_i m_i = nm \tag{6.3}$$

$m = \sum n_i m_i / \sum n_i$ and $n = \sum n_i$. The subscript i represents the individual constituent. If the distribution functions of the atmospheric constituents can be considered Maxwellian (for an atmosphere in equilibrium), we can use the ideal gas law

$$p = nkT \tag{6.4}$$

where k is the Boltzmann constant and T is the kinetic temperature. Then, use of (6.4) and (6.2) yields the relation

$$\begin{aligned} \frac{dp}{p} &= \frac{dn}{n} + \frac{dT}{T} \\ &= -\frac{dz}{H} \end{aligned} \tag{6.5}$$

where $H = kT/mg$ is defined as the scale height. The scale height depends on the temperature, the mean molecular mass, and the gravitational acceleration. Integration of (6.5) yields

$$\begin{aligned} p(z) &= p(z_0)e^{-\int (mg/kT)dz} \\ &= p(z_o)e^{-\int dz/H} \end{aligned} \tag{6.6}$$

where the integration limits are z_0 and z and $p(z_0)$ is an integration constant and represents pressure at z_0. This is the famous barometric equation that describes the exponential decrease of pressure with height. The barometric equations for n and ρ_m can be similarly derived. They are

$$\begin{aligned} n(z)T(z) &= n(z_0)T(z_0)e^{-\int dz/H} \\ \rho_m(z)T(z) &= \rho_m(z_0)T(z_0)e^{-\int dz/H} \end{aligned} \tag{6.7}$$

If the scale height $H = kT/mg$ is independent of z (for example, if a thin layer is considered where H varies slowly enough so that it can be taken out of the integral), integration of (6.6) yields

$$p(z) = p(z_0)e^{-mg(z-z_0)/kT} \tag{6.8}$$

The constancy of the scale height implies that the atmosphere is approximately isothermal and since for isothermal atmospheres $T(z)/T(z_0) \doteq 1$, (6.7) yields

$$\begin{aligned} n(z) &= n(z_0)e^{-mg(z-z_0)/kT} \\ \rho_m(z) &= \rho_m(z_0)e^{-mg(z-z_0)/kT} \end{aligned} \tag{6.9}$$

A more general case is if H depends on z through $g = GM/z^2$, where G is the universal gravitational constant ($G = 6.67 \times 10^{-11}$ Newton-m^2/kg^2), M is the mass of the body to which the atmosphere is bound, and T is constant. In this case, integration of (6.6) yields the solution

$$p(z) = p(z_0)e^{-(GMm/kT)(1/z_0 - 1/z)} \tag{6.10}$$

An important feature to note is the behavior of (6.8) and (6.10) at large distances. For example, let $z \rightarrow \infty$. Then, in the case of (6.8), $p(\infty) \rightarrow 0$, but for (6.10), $p(\infty)$ is finite. As we will show below, the behavior of pressure at large distances is important in atmospheric models. A more general case must also consider the dependence of T on z. If this information is available, (6.6) can be integrated accordingly to obtain an explicit equation for the pressure. A similar set of equations can be obtained for $n(z)$ and $\rho_m(z)$. These equations are quite general and applicable to any atmosphere in hydrostatic equilibrium.

6.2.2 Ionosphere

The existence of Earth's ionosphere was postulated by B. Stewart to explain the geomagnetic field variations recorded at ground level that are due to large currents generated in the atmospheric tidal motions. In 1901, G. Marconi demonstrated trans-Atlantic radio communication and O. Heaviside in 1902 suggested an ionized layer must be present to bend Marconi's radio waves around the curved Earth. The upper regions of Earth's atmosphere were extensively studied after World War II by ground-based radio experiments (ionosondes) and instruments carried on rockets. These experiments firmly established that the ionosphere begins around 60 to 70 km above the surface and that the ionized density increases with height, reaching a maximum around 300 km. Although the various boundaries are not very sharp, the layer between 60–90 km is called the D-layer, 90–120 km is the E-layer, and > 120 km is the F-layer. A midday electron density at 100–300 km heights is typically 10^{11}–10^{13} m^{-3}.

Ionization is produced primarily by solar ultraviolet (UV) radiation, which photoionizes and photodissociates the neutrals (Figure 6.3). For Earth's atmosphere, solar radiation < 2000 Angstroms (Å) is important. The polar regions are also bombarded continually by particles of the outer radiation belt during auroras and by solar flares that produce energetic protons (> 10 MeV). Some events can include particles energetic enough to penetrate deep into the D-region and the enhanced ionization there can cause radio blackouts.

The time rate of change of any ionized constituent in an atmosphere is described by the general continuity equation

$$\frac{\partial n}{\partial t} = q - L - \nabla \cdot (n\mathbf{U}) \tag{6.11}$$

where q represents the source term and L the loss term. (Note that this expression is more general than (5.6).) Consider a steady-state case and ignore all motions. Then the left-hand side and $\nabla \cdot n\mathbf{U}$ vanish and

$$q = L \tag{6.12}$$

In steady-state, the production rate of ionized constituents is exactly balanced by recombination processes that remove ionization. Sydney Chapman in 1931 considered a simple theoretical model to show how q could be

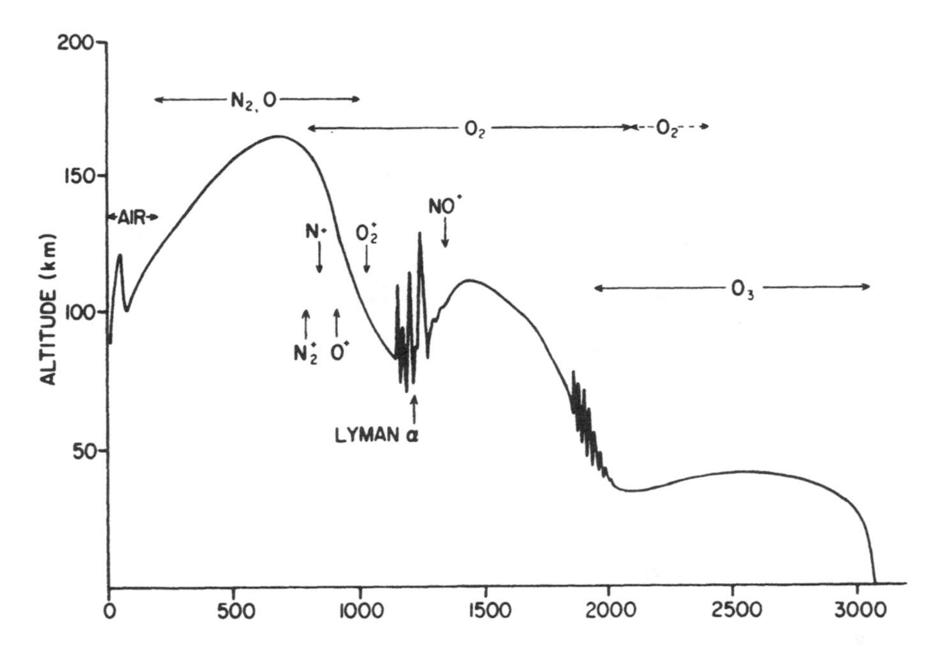

FIGURE 6.3 A schematic diagram to illustrate the absorption of solar radiation at various altitudes of Earth's atmosphere. The curve represents the altitude at which the solar radiation intensity drops to $1/e$ of its value at the top of the atmosphere. Also shown are the various molecular and atomic constituents that absorb the solar UV radiation. (From Herzberg, 1965)

calculated. His model assumes a flat geometry, one-dimensional in height, with a constant temperature, and a monochromatic input solar spectrum (Figure 6.4).

Consider the solar UV radiation impinging on top of an atmosphere. This radiation will be attenuated as it propagates through an atmosphere. The equation describing the attenuation is

$$dI(\lambda, z) = -I(\lambda, z)\mu(\lambda, z) \sec\chi dz \tag{6.13}$$

where $I(\lambda, z)$ is the solar flux spectrum (photons/m^2/sec); ds is the light path, which is related to the vertical height $dz = ds\cos\chi$, where χ is the angle between the incident direction and the zenith; and $\mu(\lambda, z)$ measures the probability of absorption per unit length of path traversed by the radiation. In general, the interaction depends on the wavelength and the molecular species with which the light interacts. We can write

$$\mu(\lambda, z) = n_i(z)\sigma_i(\lambda) \tag{6.14}$$

where $n_i(z)$ is the number density of ith ionizable species at height z and $\sigma_i(\lambda)$ is the interaction cross section measured in units of (meter)^2. Let us

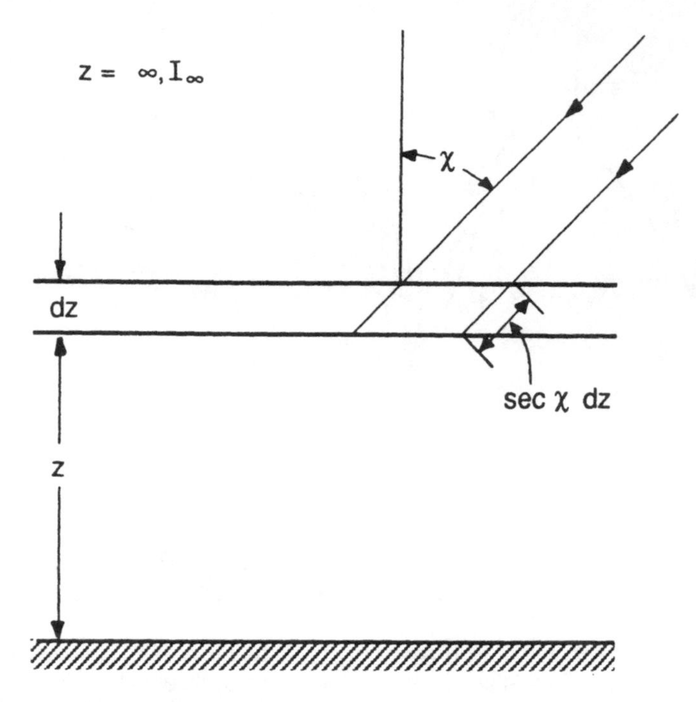

FIGURE 6.4 Slab geometry of Chapman's model.

now assume that there is only one species of neutrals and that the radiation is monochromatic. Then using (6.14) and integrating (6.13) yields

$$I(z) = I_\infty e^{-\int_\infty^z n(z)\sigma \sec\chi dz} \tag{6.15}$$

The (−) sign here comes from the fact that z is measured positively in the upward direction. I_∞ represents the original unattenuated solar flux impinging on the atmosphere for a particular λ.

The interaction of the UV radiation with the neutrals will produce electrons and ions. If the assumption is made that each UV photon produces one electron-ion pair, the rate of ionization produced by this radiation (pairs/m^3/sec) is

$$q(z) = n(z)\sigma I(z) \tag{6.16}$$

where $I(z)$ is given by (6.15) and $n(z)$ represents the density of a specific neutral species at height z that is being ionized. If the atmosphere is isothermal, we can use (6.9)

$$n(z) = n_0 e^{-(z-z_0)/H} \tag{6.17}$$

where n_0 represents the density at a reference height z_0. Now assume that $z_0 = 0$. Then use of (6.15) and (6.17) in (6.16) yields an explicit expression for the ionization production rate

$$q(z) = n_0 e^{-z/H} \sigma I_\infty \exp(-n_0 H \sigma \sec\chi e^{-z/H}) \tag{6.18}$$

To find the height at which maximum ionization is produced, compute $dq(z)/dz = 0$ and, solving for z, obtain

$$z_m = H \ln(n_0 \sigma H \sec\chi) \tag{6.19}$$

Substitution of this expression into (6.18) shows that the maximum production rate is

$$q(z_m) = \frac{I_\infty}{He\sec\chi} \tag{6.20}$$

and the number density of the ionizable constituent there is

$$n_m = (\sigma H \sec\chi)^{-1} \tag{6.21}$$

Another useful way to look at this problem is to reference the variables relative to the overhead direction $\chi = 0$. Let z_0 and q_0 be the value of z_m and q_m taken when $\chi = 0$. Then,

$$z_0 = H \ln(n_0 \sigma H) \tag{6.22}$$

$$q_0 = \frac{I_\infty}{He} \tag{6.23}$$

and substitution of these expressions in (6.18) yields

$$q = q_0 \exp\left(1 + \frac{z_0 - z}{H} - \sec\chi \exp\frac{z_0 - z}{H}\right) \tag{6.24}$$

A plot of (6.24) is shown in Figure 6.5. An interesting property of this function is that as χ is varied, the shape of the curve remains unchanged. However, the peak becomes shifted. This simple model is reasonably good for the F-layer. The model is not adequate for the D and the E layers. Equation (6.24) was first derived by S. Chapman and is known as the Chapman production function. Chapman's original derivation was restricted to monochromatic radiation, but this formulation can be easily extended to include the entire radiation spectrum.

6.2.3 Production and Loss of Ionization

Ionization depends on the input radiation spectrum and the molecular species. The production of ionization in planetary atmospheres is complicated because the solar radiation generated in the solar chromosphere

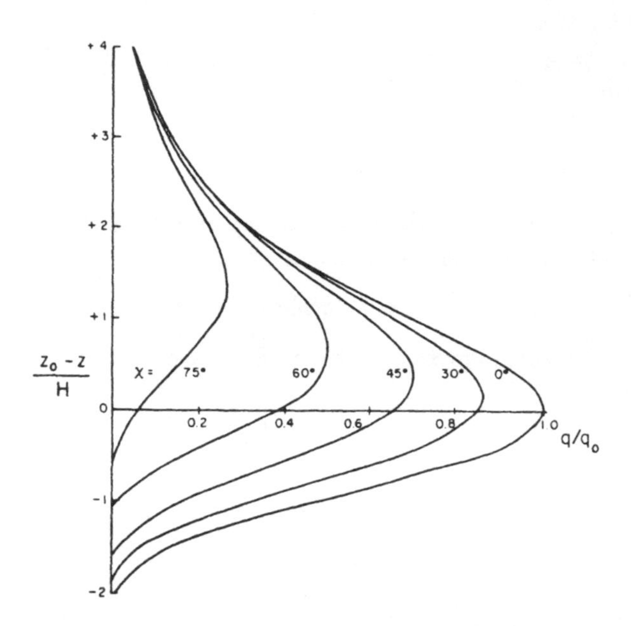

FIGURE 6.5 Plot of normalized Chapman production function for several χ values (Eq. (6.24)).

and the corona covers a broad spectrum. As mentioned above, wavelengths below about 2000 Å are important for the Earth's ionosphere. Note also that the incident radiation will be absorbed over a wide range of altitudes.

The composition of different planetary atmospheres is different and therefore we will not emphasize or identify the participating ions and neutrals. (Reactions specific to a particular planetary atmosphere, for example Earth's, can be found in Herzberg, 1965, cited at the end of the chapter.) Instead, this problem is discussed in a general way to emphasize which photochemical interactions play important roles in the control of ionized constituents.

The concentration of positive and negative ions, electrons, and neutrals in planetary atmospheres depends on six basic physical interactions. Let

$$\begin{aligned} n &= \text{number density of neutral molecules (number } /m^3) \\ n^+ &= \text{number density of positive ions} \\ n^- &= \text{number density of negative ions} \\ n_e &= \text{number density of electrons} \end{aligned} \tag{6.25}$$

Note that a distinction is made between n^- and n_e. Ionized constituents are formed when the radiation interacts photoelectrically with the neutrals.

$$n + h\nu = n^+ + n_e \tag{6.26}$$

where $h\nu$ represents a quantum of energy (h is Planck's constant $= 6.62 \times 10^{-34}$ Joules-sec and ν is the frequency). The volume production rate of a given neutral constituent that absorbs a photon is given by (6.16), $q = n\sigma I$, where σ is the photoelectric cross section. When positive ions and electrons are produced, they can recombine or interact with the neutrals as well as among themselves. We discuss five processes that can occur which will affect the concentration of various ionic species. These are

$$\begin{aligned}
n^+ + n_e &= n^* && \text{recombination } (\alpha_r) \\
n + n_e &= n^- + h\nu && \text{radiative attachment } (a) \\
n^- + h\nu &= n + n_e && \text{photodetachment } (\alpha) \\
n^- + n^+ &= n + n && \text{neutralization } (\alpha_i) \\
n^- + n &= n + n + n_e && \text{collisional detachment } (f)
\end{aligned} \tag{6.27}$$

n^* represents an excited state of a neutral atom (an excited state will eventually transition to a lower-energy state by the emission of a photon $n^* \rightarrow n + h\nu$). The letters in the parentheses are reaction rate coefficients for the various interactions. Note that n^+ or n^- can be a single atom or a more complex molecule or compound. Also, while the equations involve only two-body interactions, in reality three-body interactions may be important.

Certain reactions produce and others remove ionization for a given species. Equation (6.27) permits us to write the differential equations that show how the ion concentrations change in time. These are

$$\begin{aligned}
\frac{dn^+}{dt} &= q - \alpha_r n^+ n_e - \alpha_i n^+ n^- \\
\frac{dn^-}{dt} &= a n_e n - n^-(\alpha + fn + \alpha_i n^+) \\
\frac{dn_e}{dt} &= q - \alpha_r n^+ n_e - a n n_e + n^-(\alpha + fn)
\end{aligned} \tag{6.28}$$

Given the various reaction rates, the distribution of the ionic and the electronic concentration can be obtained for an atmosphere in equilibrium by setting the time derivatives to zero, $dn/dt = 0$. Since plasma is then neutral, the total charge must be balanced and for singly ionized species

$$n^+ = n^- + n_e \tag{6.29}$$

As an example, let us study how electrons are distributed as a function of height. Define the fraction of the negative ions to electrons by

$$\delta = \frac{n^-}{n_e} \tag{6.30}$$

Then (6.29) can be rewritten as

$$n^+ = \delta n_e + n_e = (1 + \delta) n_e \tag{6.31}$$

Using these relations, the last two equations of (6.28) can be rewritten as

$$q + \alpha\delta n_e + f\delta n_e n = \alpha_r(1+\delta)n_e n_e + ann_e \tag{6.32}$$

and

$$ann_e = \alpha\delta n_e + f\delta nn_e + \alpha_i(1+\delta)\delta n_e n_e \tag{6.33}$$

Now add the two equations and obtain

$$q = (1+\delta)(\alpha_r + \alpha_i\delta)n_e^2 \tag{6.34}$$

which, solved for n_e yields

$$n_e^2 = \frac{q}{(1+\delta)(\alpha_r + \alpha_i\delta)} \tag{6.35}$$

Define

$$\alpha_{eff} = (1+\delta)(\alpha_r + \delta\alpha_i) \tag{6.36}$$

as the effective reaction rate and, if this relationship is used, (6.34) reduces to

$$q = \alpha_{eff} n_e^2 \tag{6.37}$$

The electron concentration depends on the square root of q. Given the knowledge about the effective reaction rates (which can be measured in the laboratory) and the electromagnetic radiation spectrum, the behavior of n_e can be studied. All of the quantities in (6.37) are height-dependent.

6.3 Solar Wind Theory

The photosphere of the Sun is about 6000°K and the corona is several million °K. Figure 6.6 shows a schematic diagram of the temperature profile between the photosphere and the lower corona. The minimum just above the photosphere is due to the loss of heat by radiation. This minimum is analogous to the temperature minimum observed in Earth's tropopause and the mesopause at approximately 15 to 80 km. How is the corona heated to such high temperatures?

S. Chapman in 1957 formulated a static coronal model by combining a thermal conduction model with a specific temperature distribution derived for ionized gases in a hydrostatic atmosphere. This model extended the solar coronal atmosphere to large distances, but the ion density and the temperature predicted at the position of Earth were too high. E.N. Parker in 1958 criticized the static model and formulated a dynamic model with streaming of particles at the base of the corona. A unique feature of Parker's model is that the pressure gradient of solar atmospheric constituents continuously accelerates the outward streaming particles and the streaming

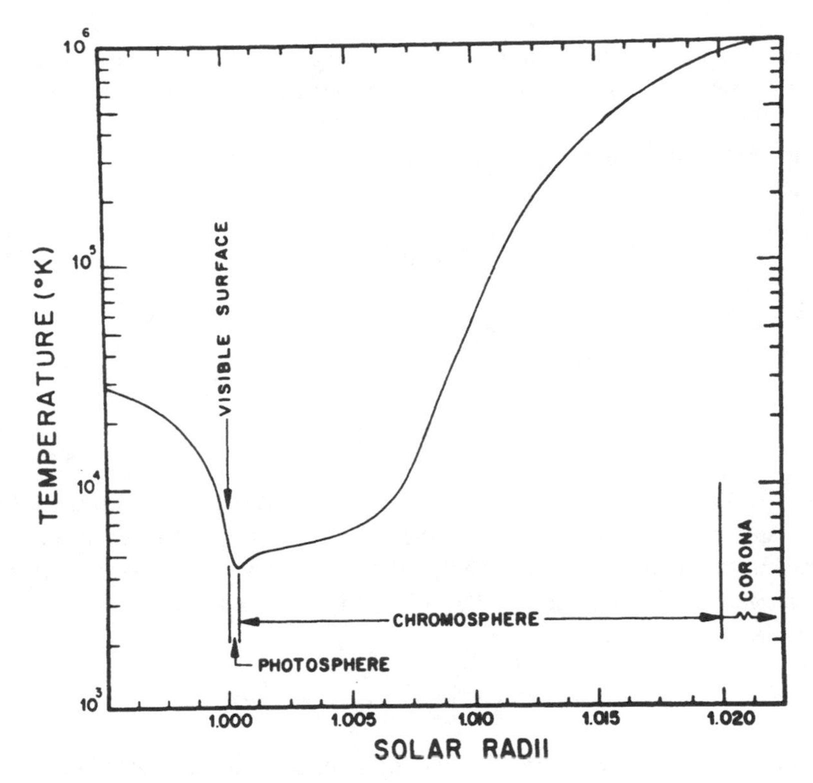

FIGURE 6.6 A sketch of the Sun's temperature profile in the vicinity of the photosphere. The radius of the Sun is about 7×10^8m. (From Dessler, 1967)

velocity increases outward. The streaming undergoes a smooth transition to supersonic speeds and the "solar wind" arrives at Earth with a speed of several hundred kilometers/second. Parker's solar wind model preceded the space age by several years and his predictions were subsequently confirmed by spacecraft observations.

6.3.1 Chapman Model

S. Chapman formulated a hydrostatic model that does not include any external heat source. Heat is therefore transferred by conduction alone. Chapman assumed that an atmosphere consisted of hot plasma and that the thermal conductivity depends on $T^{5/2}$. (We shall not justify why this is so, but interested readers may consult L. Spitzer, given in the reference list.) The thermal conductivity κ can be written in the form

$$\kappa = \kappa_0 T^{5/2} \tag{6.38}$$

where κ_0 is a constant. Chapman's model assumes that the atmosphere is extended to large distances and at large distances the temperature is zero ($T = 0$ at $z = \infty$). Heat is steadily flowing out of the atmosphere by heat conduction, and the flux of heat transported out can be described by the equation

$$\frac{d}{dz}\left(z^2\kappa\frac{dT}{dz}\right) = 0 \tag{6.39}$$

If this differential equation is solved with the expression for heat conduction given above, one obtains

$$T = T_0\left(\frac{z}{z_0}\right)^{-2/7} \tag{6.40}$$

Here, T_0 and z_0 are integration constants. Chapman's model predicts that the temperature falls away from the source as $z^{-2/7}$. Let us insert this temperature dependence into (6.6) to examine how the pressure varies as a function of the distance z. The expression for p with this substitution becomes

$$p = p_0 \exp\left(-\frac{GMm}{2kT_0}\int_{z_0}^{z}\frac{dz}{z^2(z/z_0)^{-\mu}}\right) \tag{6.41}$$

where $\mu = 2/7$. (Note that we use $2kT_0$ instead of kT_0 because a fluid contains electrons and ions. See the discussion of Eq. (6.52).) This equation can be integrated and one obtains

$$p = p_0 \exp\left[-\frac{GMm}{2kT_0}\frac{1}{z_0}\frac{1}{(1-\mu)}\right]\exp\left[\frac{GMm}{2kT_0}\frac{1}{z_0}\frac{1}{(1-\mu)}\left(\frac{z}{z_0}\right)^{-(1-\mu)}\right] \tag{6.42}$$

Note that since $\mu < 1$, the second exponent

$$\exp\left[\frac{GMm}{2kT_0}\frac{1}{z_0}\frac{1}{(1-\mu)}\left(\frac{z}{z_0}\right)^{-(1-\mu)}\right] \rightarrow 1 \tag{6.43}$$

as $z \rightarrow \infty$. Hence (6.42) predicts that the pressure at $z = \infty$ is given by

$$p = p_0 \exp\left[-\left(\frac{GMm}{2kT_0}\right)\frac{1}{z_0}\frac{1}{(1-\mu)}\right] \tag{6.44}$$

This pressure model predicts a finite pressure at $z = \infty$ and it extends the solar coronal atmosphere to large distances. However, the density of particles predicted by Chapman's model at the Earth's position is too high by more than a factor of 50.

The quantity $GMm/2kT_0z_0$ is the ratio of gravitational to thermal binding energy. Applied to the Sun, T_0 is about 10^6 °K and z_0 is about 10^9 meters near solar coronal heights (the radius of the Sun is about

FIGURE 6.7 Halley's comet approaching the Sun, taken on March 8, 9, and 10, 1986. Note the two-tail structure. The diffuse tail can be explained by solar radiation pressure. The sharp tail is explained by solar wind particles and the IMF. It resembles the geomagnetic tail. Note the extreme dynamic nature of the structured tail. (Courtesy of the European Space Agency)

7×10^8 m). Hence, the ratio is about 10 giving a value for $p/p_0 \approx e^{-10}$ which is about 4.5×10^{-5}. This pressure is exerted inward toward the Sun. Chapman's model is static and pressure balance is needed to satisfy the equilibrium condition. Let us compare $p/p_0 = 4.5 \times 10^{-5}$ with experimental data. Observations indicate that the density n is about $10^{14}/\mathrm{m}^3$ near the solar corona, and that T is about 10^6 °K at the solar corona. The observed value of p/p_0 with these numbers is about 10^{-9} or 10^{-10}. This number is much smaller (by five orders of magnitude) than the predicted value of Chapman's model.

6.3.2 Evidence of Solar Wind

The existence of the solar wind was first predicted by Ludwig Biermann in 1951 from studying the shape of cometary tails. When a comet approaches the Sun, the evaporation of cometary material is enhanced. Because of the solar ultraviolet rays, a large fraction of this material is ionized (spectroscopic evidence shows the presence of CO^+, N_2^+, and CO_2^+, for instance). Comets have a nucleus of dense material and a less dense streak of material directed away from the solar direction which resembles a tail. Until Biermann's prediction, comet theories explained the existence of the tail in terms of solar radiation pressure. However, the long tails that Biermann was studying were observed to be always directed radially away from the solar direction. Biermann argued that solar radiation pressure alone cannot account for these tails and suggested that the existence of such tails can only be accounted for by the radial flow of solar corpuscular fluxes. In Figure 6.7, the diffuse tail is due to the solar radiation pressure while the longer tail is due to the solar particle fluxes.

The existence of the solar wind was first reported by the Soviet spacecrafts Lunik 2 and Lunik 3 in 1960. These observations were verified by the observations of Mariner 2 in 1962. Solar wind bulk velocity and proton density observed by Mariner 2 are shown in Figure 6.8. Since then, numerous other spacecraft have performed experiments in the solar wind, and it is now known that the main elemental ionic composition of the solar wind is protons. The second most abundant ionic component is $^4He^{++}$. These two ionic components, together with an equal number of electrons, essentially constitute the main solar wind. Ions such as $^3He^{++}$, $^4He^+$, O^{6+}, and C^{3+} also exist in small amounts. An example of the energy/charge spectrum of the solar wind ion composition is shown in Figure 6.9.

Solar wind is highly variable and depends on solar activity. Figure 6.10 shows a statistical analysis of solar wind parameters. The average temperature is about 10^5 °K and the $^4He^{++}$ content is typically about 5%.

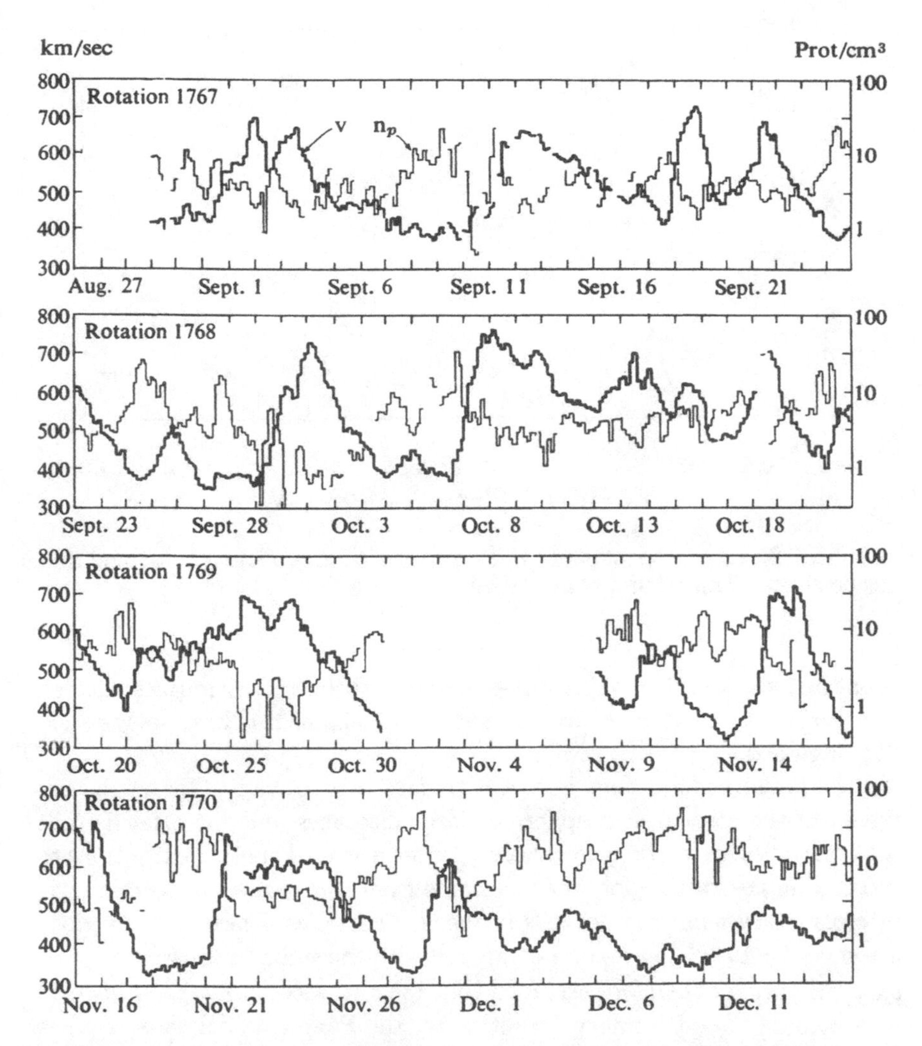

FIGURE 6.8 Solar wind bulk velocity and density observed by the Mariner 2 solar wind experiment. The average solar wind velocity is about 400 km/second and its proton (and electron) density is about $7 \times 10^6 (\mathrm{m})^{-3}$. (From Neugebauer and Snyder, 1966)

6.3.3 Parker's Solar Wind Model

E.N. Parker suggested that the solar coronal heat can be transported by particle streaming. However, in the original model, Parker used a specific temperature distribution tailored to satisfy an expanding solution in the hydrodynamic equations and he did not solve the coronal heat transport

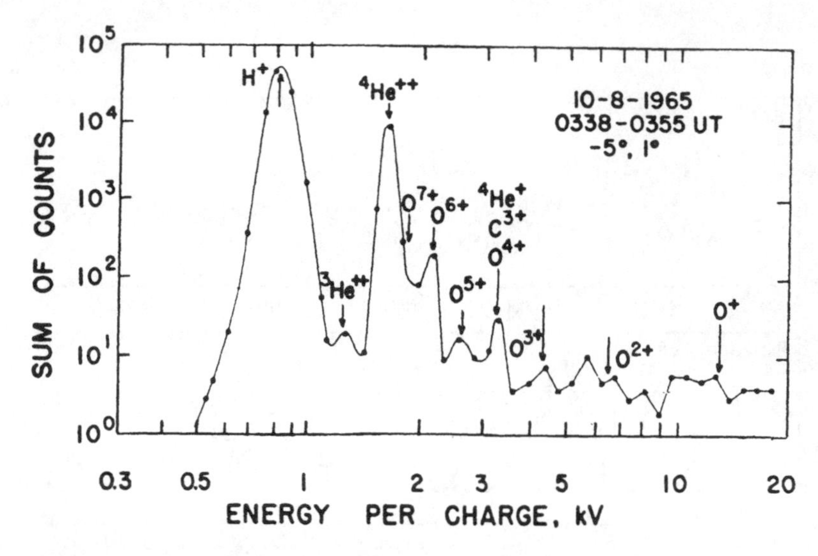

FIGURE 6.9 Energy spectrum of various solar wind ions. Note that H^+ and H^{++} are dominant. (From Bame *et al.*, 1968)

problem (which still remains unresolved). Parker's model implies that an appropriate heat source is present but the details of this heat source were not discussed.

Parker's model differs from Chapman's in two ways. One is the behavior of the coronal atmosphere at large distances and the other is that solar particles are allowed to stream. Parker reasoned that contributions to pressure in the vast region of space come from star radiation, cosmic rays, interplanetary magnetic fields, and thermal particles. These contributions are small ($\approx 10^{-13}$ Joules/m^3) compared with the stellar atmospheric pressure, and Parker therefore assumed that the Sun was essentially immersed in a vacuum. The boundary condition for the Parker model thus requires that $p \to 0$ as $z \to \infty$, contrary to the finite pressure in Chapman's model (see (6.44)). A requisite for this boundary condition is that the integral in (6.6) must diverge as $z \to \infty$. Since p is proportional to $\exp - \int dz/Tz^2$, T must decrease asymptotically as $1/z$ or faster. This condition for T differs from the static atmospheric model and it implies that the atmosphere cannot be in hydrostatic equilibrium. It describes an atmosphere that is steadily expanding into space while new material is fed into it from below.

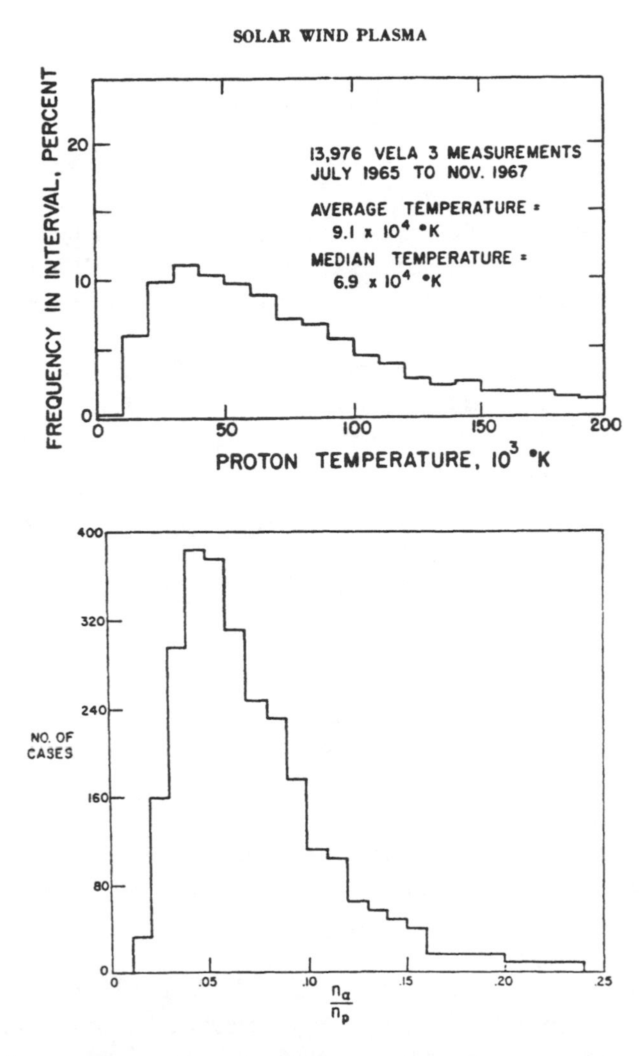

FIGURE 6.10 Average solar wind velocity, density and temperature near Earth's orbit and ratios of two principal constituents H^+/He^{++}. (From Hundhausen *et al.*, 1967)

The complete set of MHD equations needed to study the solar wind problem is (assuming a Maxwellian plasma)

$$\frac{\partial n}{\partial t} + \nabla \cdot n\mathbf{U} = 0$$

$$nm\frac{\partial \mathbf{U}}{\partial t} + nm(\mathbf{U} \cdot \nabla)\mathbf{U} = -\nabla p + \mathbf{J} \times \mathbf{B} + \mathbf{F}$$

and

$$\frac{\partial}{\partial t}\left(\frac{3}{2}nkT + \frac{nmU^2}{2}\right) + \nabla\cdot\left(\frac{5}{2}nkT\mathbf{U} + nmU^2\mathbf{U} - \kappa\nabla T\right) = \mathbf{J}\cdot\mathbf{E} + \mathbf{F}\cdot\mathbf{U}$$

$$p = nkT \tag{6.45}$$

where $\mathbf{F}$ is due to external sources and the MHD equations have been modified to include the thermal contribution. This MHD set of equations must be augmented with the set of Maxwell equations for electromagnetic fields.

These equations are nonlinear and difficult to solve even for steady-state problems. Further simplification is obtained by considering a star with no magnetic field and viscosity (then $\mathbf{J}\times\mathbf{B}$ and $\mathbf{J}\cdot\mathbf{E}$ terms drop). Then the required equations are

$$\nabla\cdot n\mathbf{U} = 0$$

$$nm(\mathbf{U}\cdot\nabla)\mathbf{U} = -\nabla p + nm\mathbf{g}$$

$$p = nkT$$

$$p = p_0\left(\frac{n}{n_0}\right)^{5/3} \tag{6.46}$$

where we assume that only the solar gravitational force contributes to the external force $\mathbf{F}$. Solutions of (6.46) can be obtained once the boundary conditions are specified. Parker's model applies (6.46) in the region of the Sun where the hydrodynamic approximations are valid, and then connects the solutions to the outside regions on the assumption that the fluid behaves as an adiabatic gas.

Table 6.1 shows typical values of n, T, $\mathbf{B}$, and $\mathbf{U}$ as a function of the distance away from the Sun. Here $r_\odot$ represents the solar radius, which is $\approx 7\times10^8$ meters. n near the Sun is obtained from ground observations of the amount of light scattering, which can be translated to the electron density. T near the Sun is determined from atomic spectroscopic results (profiles of line emission) and the magnetic field $|\mathbf{B}|$ from the Zeeman effect. $\mathbf{U}$ near the Sun represents values extrapolated from values obtained by spacecraft near Earth by use of the mass conservation equation. Note that the Sun-Earth distance, also called the astronomical unit (AU), is $\approx 1.5\times10^{11}$ meters.

Table 6.1 shows that while n, T, and $|\mathbf{B}|$ decrease with the distance from the Sun, $\mathbf{U}$ increases with r. Also note that the solar wind flow speed $\mathbf{U}$ at 10 and 215 solar radii does not differ by very much. This means that the solar wind attains high values very close to the Sun. Parker's model must be capable of predicting this behavior.

TABLE 6.1
Typical Parameters of the Solar Wind

$r/r_\odot$	1.03	1.5	10	215
$n(m^{-3})$	2×10^{14}	1.6×10^{13}	4×10^{9}	7×10^{3}
$T(°K)$	1.7×10^{6}	10^{6}	4×10^{5}	4×10^{4}
B(Teslas)	10^{-4}	0.5×10^{-4}	10^{-6}	5×10^{-9}
U(m/sec)	0.6×10^{3}	3×10^{3}	3×10^{5}	4×10^{5}

$r_\odot \simeq 7 \times 10^8$ m

Let us now focus on one unique feature of the solar wind, that is, how the coronal expansion speed becomes supersonic. Parker's model of solar wind expansion is then similar to the way gases flow through a deLaval nozzle (named after a Swiss engineer, P. deLaval). To gain physical insight into the actual solar wind problem, it is therefore instructive to study first how gases flow through a deLaval nozzle.

Consider a compressible gas flowing down a tube of decreasing cross-sectional area, shown in Figure 6.11. In steady-state, the constancy of the mass flow requires that $\rho_m AU$ = constant where ρ_m is the mass density, A is the cross-sectional area, and U the flow velocity. Taking the logarithm of this equation and differentiating yields

$$\frac{d\rho_m}{\rho_m} + \frac{dA}{A} + \frac{dU}{U} = 0 \tag{6.47}$$

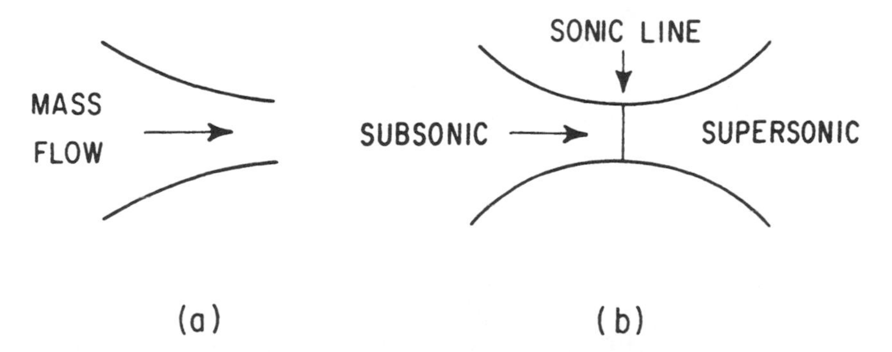

FIGURE 6.11 A schematic diagram of a converging nozzle and a deLaval nozzle.

Euler's equation of motion $\nabla p = -\rho_m(\mathbf{U}\cdot\nabla)\mathbf{U}$ for this problem reduces to $dp = -\rho_m U dU$ which can be rewritten as

$$\frac{dp}{\rho_m} = \frac{dp}{d\rho_m}\frac{d\rho_m}{\rho_m} = -UdU \tag{6.48}$$

Solving for $d\rho_m/\rho_m$ and noting that $dp/d\rho_m$ is equal to the square of the speed of sound in a gas, we obtain

$$\frac{d\rho_m}{\rho_m} = -\frac{U}{C_s^2}dU \tag{6.49}$$

where C_s is the speed of sound of a gas. Now, use this relation in (6.47) and obtain

$$\frac{dA}{A} = \left(\frac{U^2}{C_s^2} - 1\right)\frac{dU}{U} \tag{6.50}$$

The highest flow speed attained for a converging tube is less than C_s because dA/A is negative for a converging tube. The fluid speed U increases as it approaches the narrower region of the tube (dU/U is positive), and this equation is satisfied only if U is less than C_s. When $U = C_s$, the tube most stop converging, $dA/A = 0$. If U is to exceed C_s, dA/A must increase, and this design is incorporated in the deLaval nozzle (Figure 6.11). If the flow speed does not reach C_s at the throat (narrowest part) of the tube, the tube acts as a venturi tube and the flow slows down as it passes through the throat. Whether the tube acts as a venturi tube or a deLaval nozzle depends on the pressure ratio of the upstream-to-downstream section. For example, if the downstream end opens into a vacuum, the flow out of a deLaval nozzle will always be supersonic. The deLaval nozzle is basic for the design of rocket engines.

To achieve supersonic speeds, the flow must first approach a converging section and subsequently expand into a diverging section. Let us now examine whether this condition can be met by flow near the Sun. The velocity streamlines of coronal gas flowing outward diverge as $1/r^2$, opposite to what is required, and it would seem that the expansion cannot become supersonic. However, for the Sun we must modify (6.50) and include the solar gravitational term. The balance of forces then requires that

$$\begin{aligned} dp &= -\rho_m U dU - \rho_m g dr \\ &= -\rho_m U dU - \rho_m \frac{MG}{r^2} dr \end{aligned} \tag{6.51}$$

where $g = GM/r^2$ is the acceleration due to solar gravity, M is the mass of the Sun (2×10^{30} kgm), G is the universal gravitational constant, and r is the distance measured from the center of the Sun (heliocentric distance).

The pressure of a fluid with a Maxwellian distribution function for electrons and ions is

$$\begin{aligned} p &= p_e + p_i \\ &= 2nkT \end{aligned} \tag{6.52}$$

and $C_s^2 = dp/d\rho_m = dp/d(nm) \approx 2kT/m_i$ since for a fluid with electrons and one ion species, $\rho_m = m_i n_i + m_e n_e = n(m_i + m_e) \approx nm_i$. If now the continuity equation $r^2 nU =$ constant and $p = 2nkT$ are used to eliminate n in (6.51), we arrive at

$$\left(\frac{mU^2}{2kT} - 1\right)\frac{dU}{dr} = -\frac{Ur^2}{T}\frac{d}{dr}\left(\frac{T}{r^2}\right) - \frac{GMmU}{2kTr^2} \tag{6.53}$$

Thus far no assumption has been made about T. Let us now assume that T is constant, which means that we are solving an isothermal problem. This is of course incorrect, since an expansion will cool the gas. However, if our interest here is to demonstrate only that the solar wind expansion becomes supersonic, this assumption will not affect the result. Then (6.53) reduces to

$$\left(\frac{U^2}{C_s^2} - 1\right)\frac{dU}{U} = \left(2 - \frac{GM}{C_s^2 r}\right)\frac{dr}{r} \tag{6.54}$$

This equation is valid near the surface of the Sun where the magnetic field is nearly radial and parallel to the expansion direction. This equation is still valid even if the solar wind has a non-radial component, if the expansion is spherically symmetric.

Equation (6.54) possesses a variety of solutions, which we now examine. Consider first a star whose temperature is very high. Consider the limit when $T \to \infty$, which implies $C_s \to \infty$. In this infinite temperature limit, (6.54) reduces to

$$\frac{dU}{dr} \approx -\frac{2U}{r} \tag{6.55}$$

Since U and r are both positive, this equation can only be satisfied if $dU/dr < 0$. This result means that for very hot stars the expansion speed decreases with r and a supersonic solar wind cannot be developed as our Sun does.

To understand why the expansion speed decreases with r in hot stars, consider the mass conservation equation, $nUr^2 =$ constant. Since n and U both depend on r, this equation states that U will increase if and only if n decreases faster than $1/r^2$. A large-density gradient cannot be achieved in hot stars because $p = nkT$ implies that the pressure is also very high and the high pressure pushes the material outward and reduces the density gradient.

Consider now the expansion of an atmosphere whose temperature is lower. Suppose near the surface of this star the flow velocity is finite and $U < C_s$. In this case, $(U^2/C_s^2 - 1) < 0$. If $(2 - GM/C_s^2 r) > 0$, (6.54) requires $dU/dr < 0$, and U decreases with r as before. On the other hand, if $(2 - GM/C_s^2 r) < 0$, then $dU/dr > 0$ and U must increase with r. This solution corresponds to the legitimate solar wind solution.

The solar wind flow speed U equals the speed of sound $C_s (U = C_s)$ when $(2 - GM/C_s^2 r) = 0$. The distance at which the expansion becomes supersonic is called the critical distance and it is given by

$$\begin{aligned} r_c &= \frac{GM}{2C_s^2} \\ &= \frac{GMm_i}{4kT} \end{aligned} \tag{6.56}$$

The critical distance r_c depends on $1/T$.

What happens to the solution at large distances? $GM/C_s^2 r \rightarrow 0$ at large r. Therefore, (6.54) reduces to

$$\left(\frac{U^2}{C_s^2} - 1\right)\frac{dU}{dr} \approx \frac{2U}{r} \tag{6.57}$$

Since $U^2/C_s^2 > 1$ beyond r_c, $dU/dr > 0$ for all $r > r_c$. Hence U continuously grows with r. The behavior of this expanding flow solution is qualitatively sketched in Figure 6.12.

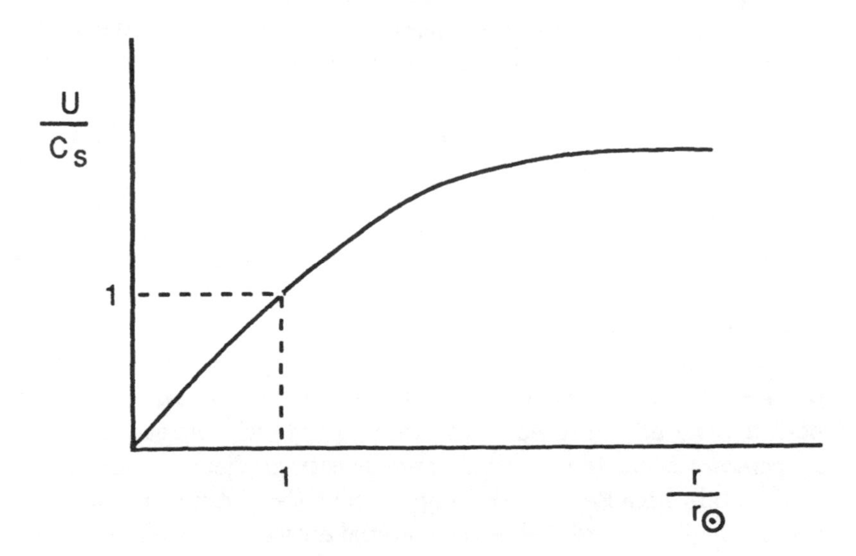

FIGURE 6.12 Sketch of U/C_s as a function of $r/r_\odot$.

The importance of the gravitational term in the supersonic expansion is now examined. Without the GM/r term, (6.54) is

$$\left(\frac{U^2}{C_s^2} - 1\right)\frac{dU}{dr} = \frac{2U}{r} \tag{6.58}$$

This equation states that if U is less than C_s near the Sun, U is always less than C_s since $dU/dr < 0$. Hence, without the gravitational term the expansion will not become supersonic. Physically, gravity helps to produce a large-density gradient that is needed for the supersonic expansion (recall n must decrease as $1/r^2$ or faster).

The analytical solution of the solar wind problem is obtained by integrating (6.54). It yields a solution U that satisfies

$$\frac{1}{v_{th}^2}\frac{3}{8}(U^2 - U_c^2) - \frac{1}{2}\ln\frac{U}{U_c} = \ln\frac{r}{r_c} + \frac{r_c}{r} - 1 \tag{6.59}$$

U_c is an integration constant and represents the solar wind speed at $r = r_c$. Here we have used $v_{th}^2 = 3kT/m$ and $r_c = GMm/4kT$. Note that U_c and v_{th} are free parameters. U_c is also called the critical speed, which, for our model, is just the speed of sound C_s. Recognizing this, we can rewrite (6.59) as

$$\frac{U^2}{U_c^2} - 2\ln\left(\frac{U}{U_c}\right) = 4\ln\left(\frac{r}{r_c}\right) + 4\left(\frac{r_c}{r}\right) - 3 \tag{6.60}$$

A family of solutions can be obtained for different values of U_c and r_c. Equation (6.60) represents one of four possible solutions to the differential equation (6.54). All four solutions are mathematically acceptable. However, physical arguments eliminate three solutions and only one is satisfactory for the solar wind solution. This solution (Figure 6.12) starts with $U < C_s$ near the base of the corona, reaches the speed of sound at the critical distance r_c, and it continues to increase beyond r_c. (A detailed discussion of the solutions of (6.54) is beyond the scope of intention. See Rossi and Olbert, 1970, cited at the end of the chapter.)

How good is the solar wind model? A simple modification of the static atmospheric model to include the effects of steady expansion has resulted in a richer and more complicated equation. It predicts among the many possible mathematical solutions a particular solution that satisfies the physical boundary conditions of an expanding atmosphere, like the solar wind. It is appropriate at this point to re-examine our formulation and state on physical grounds what shortcomings there are in this simple solar wind model.

The first assumption made is that the pressure is isotropic. This assumption is valid near the Sun where isotropy can be maintained by collisions. This approximation becomes less valid as the plasma moves further

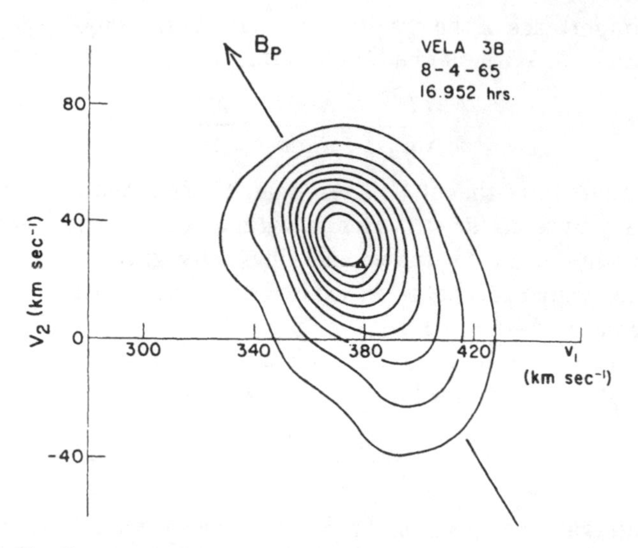

FIGURE 6.13 Contours of a typical velocity distribution function of solar wind protons observed near Earth. The small triangle at $v_1 = 380$ km sec^{-1} and $v_2 = 20$ km sec^{-1} is the mean flow speed. B_p is the projection of the IMF direction. (From Hundhausen, 1970)

away from the Sun. Observations indicate that the pressure perpendicular to $\mathbf{B}$ is less than the pressure parallel to $\mathbf{B}$ (by a factor of two) near Earth (Figure 6.13). Our model starts with a scalar p and the plasma is still isotropic at 1 AU.

Another assumption made is that the solar atmospheric constituent consists of electrons and protons and that the protons carry the inertia. In reality, there is an appreciable amount of He^{++} (5–25%) and these heavier ions must be included in the flow equations. This will add another set of equations. We also assumed that the flow was uniform and radial and that a steady-state situation exists. These assumptions at best describe solar wind flow only during solar quiet periods.

In the equation of state, the electron and the proton temperatures were assumed equal and isotropic. This presupposes that there is thermal equilibrium, which is not generally true for the Sun. A better approach requires solving the energy transport equation which relates the variables n, U, p, and r. This turns out to be exceedingly difficult, since it is not known exactly how the solar energy flows out of the solar atmosphere. One approach is to use the heat conduction equation. However, this approach is incomplete since the outward propagating waves originating from the photosphere and the presence of a magnetic field alters the conventional picture of heat flow.

The model does not include electromagnetic effects, which cannot be ignored in real situations involving magnetized stars such as the Sun. The final comment concerns the external force term **F** (see (6.45)). The expression for **F** has more terms than just the gravitational force we used. The general expression for **F** is

$$\mathbf{F} = -\frac{nGMm}{r^2}\hat{\mathbf{r}} - \frac{4\pi mGM}{r^2}\hat{\mathbf{r}}\int n(r')r'^2dr' + \frac{1}{2}\mathbf{J}\times\mathbf{B}$$
$$+\frac{1}{2}nm\nabla(|\mathbf{\Omega}\times\mathbf{r}|^2) + 2nm\mathbf{U}\times\mathbf{\Omega} + \mathbf{F}_{heat} \qquad (6.61)$$

where the first term represents the gravitational force, the second is the self attraction force, the third is electromagnetic, fourth is centrifugal, fifth is Coriolis, and $\mathbf{F}_{heat}$ is the external heating force. $\mathbf{F}_{heat}$ includes MHD and other waves whose source could be in the stellar interior. Some theories of coronal heating require the presence of these waves.

In spite of the inadequate formulation of the exact solar wind problem, the simple model we studied has been useful. Some of the features predicted by the model are qualitatively correct.

6.4 Interplanetary Magnetic Field

The IMF is a solar magnetic field that is convected out into space by the solar wind. The solar magnetic field was first observed in sunspots in 1908 by G.E. Hale. Scientific studies of solar and stellar magnetic fields began in 1948 with a solar magnetograph built by H.W. Babcock forty years after the sunspot discovery. Observations have since established that the intensity of the solar magnetic field averaged across the solar disk at the surface of the photosphere is around 10^{-4} teslas.

If interplanetary space were a vacuum, the Sun's 10^{-4} tesla (average magnetic field) field strength at the orbit of Earth would be around 10^{-11} teslas, assuming the field is dipolar (the distance from Sun to Earth is about 1.5×10^{11} meters). However, observations show (Figure 6.14) that the IMF intensity is typically 100 times larger than the dipole approximation. Early explanations of this discrepancy included the possibility that the magnetic fields permeating the galaxy may have contributed to the spacecraft measurements. However, the galactic field cannot easily penetrate the heliosphere because of the higher-conducting solar wind. Even with such galactic magnetic field contributions, the observed IMF strength is still larger by a factor of 10 to 50.

Figure 6.14 shows that the largest IMF components are in the ecliptic plane (xy-plane $\theta = 0°$). B_x and B_y can be positive or negative. The direction of the IMF observed during the quiet Sun (Figure 6.15) on the average is about 50 degrees with respect to the Sun-Earth line (x-direction) at 1 AU. The IMF can be directed toward (−) or away (+) from the Sun. During a quiet Sun, in the solar minimum year of 1964, the IMF polarity appeared in a regular pattern for several solar rotations (Figure 6.16).

The structure and dynamics of magnetic fields in interplanetary space pervaded by a conducting solar wind can be described by applying the MHD theory. MHD predicts that the motion of conducting fluid in the presence of a magnetic field induces currents which can then lead to the generation of additional magnetic fields. W.M. Elsasser in 1908 recognized the importance of this dynamo action and suggested that it might be responsible for the persistence of Earth's magnetic field. The dynamo concept is very basic in the studies of the origin of planetary and stellar magnetic fields. The existence of the IMF is a consequence of the extension of the solar magnetic field into space by the expanding solar corona. We now present a simple model of the IMF to illustrate the IMF features observed in space for quiet solar conditions.

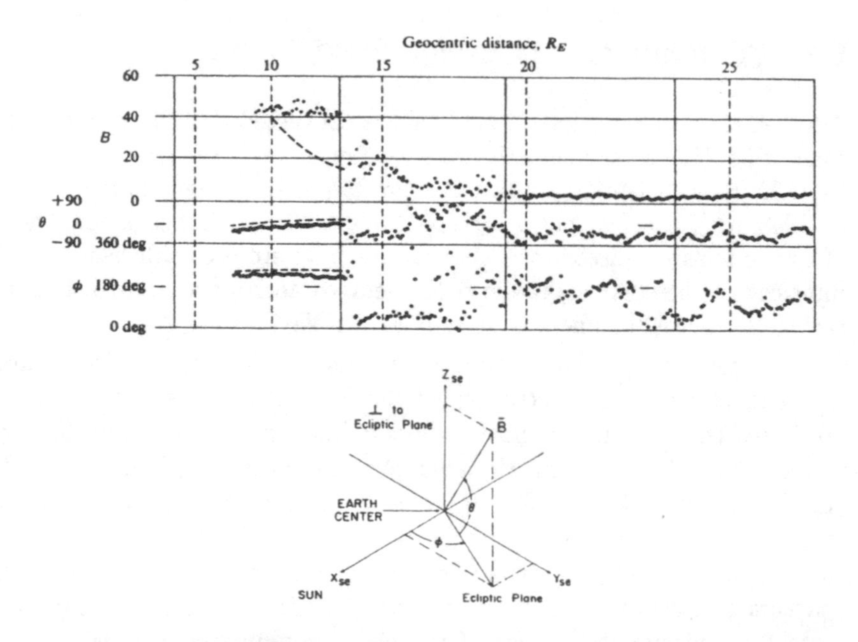

FIGURE 6.14 Observations of the IMF outside Earth's magnetosphere. The magnetopause was encountered at $13.6R_E$, the bow shock at $19.7R_E$. Note that while the total magnetic field intensity beyond the shock remains fairly constant, the direction of the IMF changes often. The data are shown in the geocentric solar ecliptic system. (From Ness, 1965).

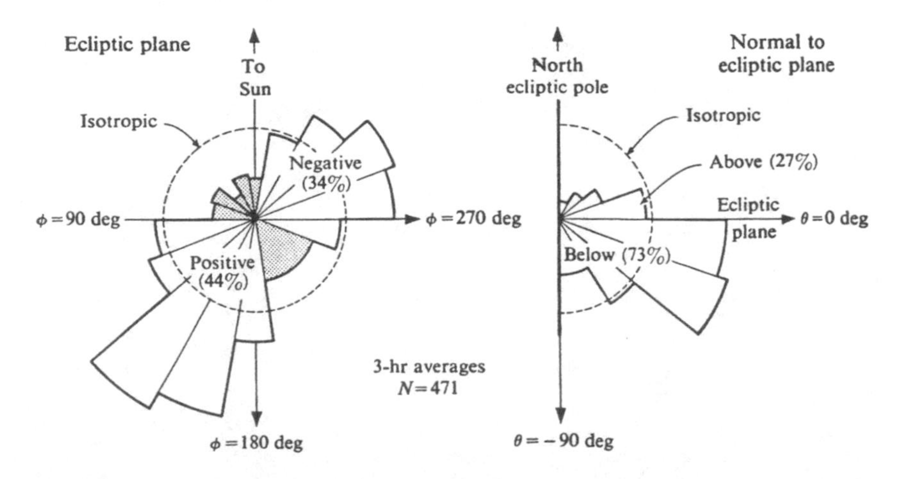

FIGURE 6.15 Direction of the IMF relative to the Earth-Sun line observed in the ecliptic plane near Earth outside the magnetosphere. The observed angle is due to the rotation of the Sun on which the IMF is anchored. This angle is also called Archimede's spiral and the *garden-hose* angle. The IMF can be directed toward (−) or away (+) from the Sun. The IMF at Earth's orbit also shows a component perpendicular to the ecliptic plane. (From Wilcox and Ness, 1965)

6.4.1 Streamline Equation for Solar Wind

Erect two coordinate frames of reference. O is at the center of the Sun, and let this be the rest frame. O′ is fixed at r_c, the critical distance where the solar wind radial flow begins, and let it be the rotating frame (Sun's frame), as shown in Figure 6.17. The relationship of flow velocities in O and O′ is $(r > r_c)$

$$\mathbf{U} = \mathbf{U}' + \boldsymbol{\Omega} \times \mathbf{r}' \tag{6.62}$$

where $\boldsymbol{\Omega}$ is the angular velocity of O′ about O and $\mathbf{r}' = \mathbf{r} - \mathbf{r}_c$. The fluid below $\mathbf{r}_c$ is corotating with the Sun. The component equations of (6.62) in the ecliptic plane are

$$\begin{aligned} U_r &= U_{r'} = U_{sw} \\ U_\theta &= U_{\theta'} = 0 \\ U_\phi &= U_{\phi'} + \Omega(r - r_c) = \Omega(r - r_c) \end{aligned} \tag{6.63}$$

where U_r, U_θ, and U_ϕ are the flow velocities in the directions r, θ, and ϕ.

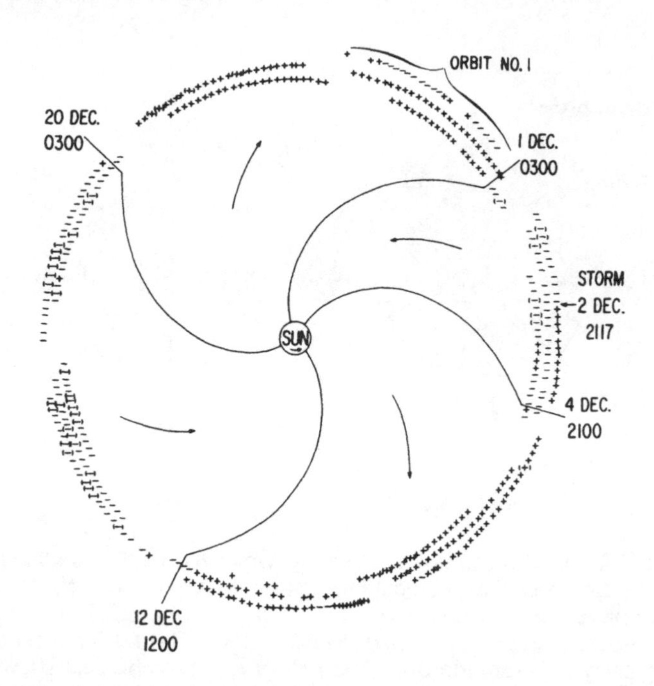

FIGURE 6.16 The (+) and (−) signs indicate the direction of the measured IMF during the successive 3-hour intervals. (+) and (−) designate the IMF direction away and toward the Sun. The inner portion is a schematic representation of the possible magnetic sector structure of the IMF. The sectors are divided into 2/7, 2/7, 2/7, and 1/7 of a solar rotation. (From Wilcox and Ness, 1965)

Define the position of the solar wind parcel in O by the usual spherical coordinates r, θ, and ϕ:

$$\begin{aligned} x &= r\sin\theta\cos\phi \\ y &= r\sin\theta\sin\phi \\ z &= r\cos\theta \end{aligned} \tag{6.64}$$

The differential equation of streamlines obtained from $d\mathbf{l} \times \mathbf{U} = 0$ is

$$\frac{dr}{U_r} = \frac{rd\theta}{U_\theta} = \frac{r\sin\theta d\phi}{U_\phi} \tag{6.65}$$

For flows confined to the ecliptic plane, the streamline equation reduces to

$$\frac{dr}{U_{sw}} = \frac{rd\phi}{\Omega(r - r_c)} \tag{6.66}$$

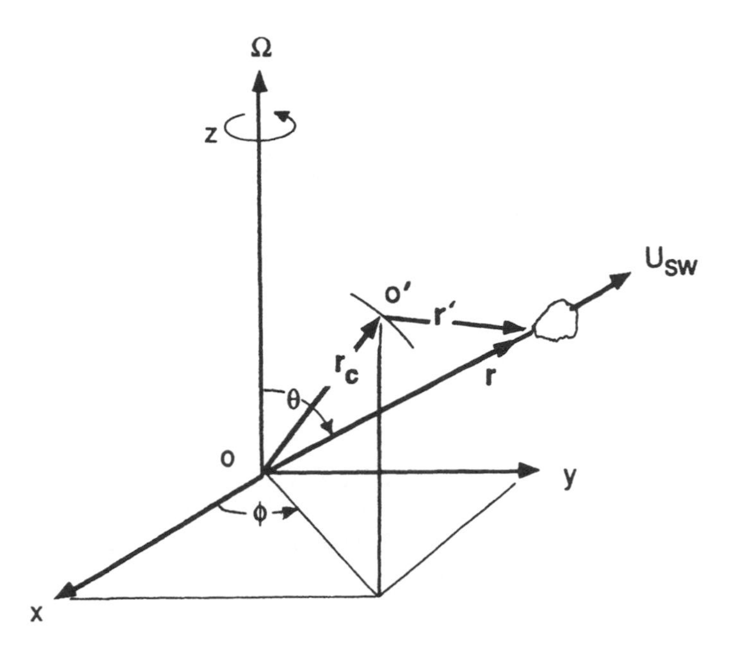

FIGURE 6.17 Coordinate frames of reference to define streamlines and the IMF.

where use was made of the relationships $U_r = U_{sw}$ and $U_\phi = \Omega(r - r_c)$. Integration of this differential equation from r_c to r and ϕ_0 to ϕ yields

$$\frac{r}{r_c} - 1 - \ln\frac{r}{r_c} = \frac{U_{sw}}{\Omega r_c}(\phi - \phi_0) \tag{6.67}$$

which is the equation of streamlines for the solar wind flow.

6.4.2 Equation of the IMF

Information on the IMF in regions $r > r_c$ is now obtained. Our model assumes that the solar magnetic field near the Sun is principally in the radial direction and is corotating with the solar surface. The solar wind is flowing out radially along **B** and **B** is parallel to the flow field **U**. Since the solar wind is a good conductor, we further assume that the frozen-in-field theorem is valid. This implies that if **B** lines are ever parallel to **U**, they will always be parallel to **U**. Hence, the equation of the IMF tangent to the streamline is obtained from the relation $d\mathbf{l} \times \mathbf{B} = 0$. In a spherical coordinate system, this expression is

$$\frac{dr}{B_r} = \frac{rd\theta}{B_\theta} = \frac{r\sin\theta d\phi}{B_\phi} \tag{6.68}$$

We now require $B_\theta = 0$, which reduces the differential equation to be solved as

$$\frac{dr}{B_r} = \frac{r \sin\theta d\phi}{B_\phi} \tag{6.69}$$

Make use of the streamline differential equation (6.65) and obtain the relationship between the flow and the IMF,

$$\frac{U_r}{U_\phi} = \frac{B_r}{B_\phi} \tag{6.70}$$

Information on B_r and B_ϕ is obtained from $\nabla \cdot \mathbf{B} = 0$. In spherical coordinates, this is

$$\frac{1}{r^2}\frac{\partial}{\partial r}(r^2 B_r) + \frac{1}{r\sin\theta}\frac{\partial}{\partial\theta}B_\theta + \frac{1}{r\sin\theta}\frac{\partial}{\partial\phi}B_\phi = 0 \tag{6.71}$$

With the assumptions that $B_\theta = 0$ and that there is axial symmetry, $\partial B_\phi/\partial\phi = 0$, the above equation becomes

$$\frac{1}{r^2}\frac{\partial}{\partial r}(r^2 B_r) = 0 \tag{6.72}$$

The solution of this equation is

$$B_r(r) = B_r(r_c)\frac{r_c^2}{r^2} \tag{6.73}$$

where $B_r(r_c)$ is the magnetic field at $r = r_c$. $B_r(r_c)$ could be the solar dipole field (but this can be more complicated). Substitute this expression into (6.70) and use (6.63) to obtain

$$B_\phi(r) = B_r(r_c)\frac{r_c^2}{r^2}\frac{\Omega(r - r_c)}{U_{sw}} \tag{6.74}$$

This equation defines the IMF at r and ϕ in regions $r > r_c$ in terms of parameters of the Sun near r_c on the ecliptic plane. Equation (6.74) shows that the transverse component of the magnetic field B_ϕ vanishes when $r \leq r_c$ as it must according to our initial assumption. Only the radial component of the magnetic field B_r is non-vanishing. In regions $r > r_c$, B_ϕ spirals according to (6.74) where r is given by (6.67).

The total magnetic field intensity on the ecliptic plane is obtained from (6.73) and (6.74), yielding

$$\begin{aligned} |\mathbf{B}_T| &= (B_r^2 + B_\phi^2)^{1/2} \\ &= B_r(r_c)\left(\frac{r_c}{r}\right)^2\left[1 + \left(\frac{\Omega}{U_{sw}}\right)^2 (r - r_c)^2\right]^{1/2} \end{aligned} \tag{6.75}$$

Note that B_T varies as $1/r^2$ near the Sun ($r \approx r_c$) and as $1/r$ far away from the Sun ($r \gg r_c$).

6.4.3 Northward and Southward IMF Components

The model formulated in the previous section does not include an IMF component perpendicular to the ecliptic plane. This is a requirement that was imposed so that no motional EMF is generated in the Sun's reference frame. However, observations by spacecraft at Earth's orbit show that while the IMF is predominantly in the ecliptic plane, the magnetic component perpendicular to the ecliptic plane does not vanish (Figure 6.15). In the Cartesian coordinate system, this component is represented by B_z and its intensity is typically about 10^{-9} teslas at 1 AU. B_z fluctuates in direction. When B_z is positive, it is referred to as northward and when negative as southward. Physical reasons preclude the possibility of including this perpendicular component in any models dealing with the IMF because convection of this component implies that the Sun's magnetic field must change at an enormous rate, which is not observed. The perpendicular IMF component must be considered as a local phenomenon, produced in transit away from the Sun.

6.4.4 Interplanetary Current Sheet

The model of the solar wind discussed above considered the entire solar atmosphere to be dynamically expanding. However, the fact that the solar surface is inhomogeneous (some regions are hotter than others while other regions have stronger magnetic fields) permits us to ask a more detailed question as to where exactly on the Sun the solar wind originates. Although the answer to this question is still being debated, recent observations suggest that the high-speed solar wind comes from small regions at coronal heights (called coronal holes) where the solar magnetic field is principally in the radial direction.

The question of where the IMF originates is intimately related to the source of the solar wind. A related question is the IMF patterns that can be quite regular and stable during the solar quiet years (Figure 6.16). When the IMF sectors were first observed, it was thought that they mapped directly back to the magnetic active regions on the photosphere of the Sun. The IMF sectors were interpreted simply as photospheric magnetic patterns in space corotating with the Sun. This model, referred to as the "mapping" model, has problems in explaining the IMF features observed in space.

Recently, an ecliptic current sheet model was developed to explain the IMF. This current sheet model is different from the IMF mapping model.

Whereas the mapping model requires the currents to flow in the meridional plane, the current sheet model restricts the currents to the ecliptic plane. As will be shown below, the current sheet model can qualitatively explain many of the observed IMF features in space. (MHD currents will not be discussed until Chapter 7; however, this topic is included here to complete the discussion of the IMF phenomena.)

Observations of magnetic polarity changes indicate that the transition from the positive to negative sectors is very sharp and occurs over a distance of a few Larmor radii, about 10^6 meters. This sharp magnetic field reversal indicates that there exists a fairly intense current sheet across the boundary. The ecliptic current sheet model is based on the existence of a thin current sheet confined to the ecliptic plane.

First consider a static current sheet. It is easily shown that an observer to the north of it will measure a magnetic field directed in the opposite direction from an observer located to the south of it. Thus, the position of the observer relative to the current sheet determines the polarity of the field that will be measured. Suppose the distribution of the current sheet for a given solar cycle is such that the magnetic field above the ecliptic plane is pointing toward the Sun and, below it, away from the Sun (Figure 6.18). Recall that the geometry of the magnetic field in the rest frame of the Sun is a spiral (6.67). Hence the magnetic field above the ecliptic plane is spiraling toward (−) the Sun while the magnetic field below it is spiraling away (+) from the Sun.

How do we account for the sudden changes of the IMF polarity observed in space? Clearly, the static current sheet model cannot explain this feature. In order to account for such changes, we must require the current sheet to be dynamic. Phenomenologically, the polarity change is explained by the up-down motion of the current sheet relative to an observer. The sign change occurs when the observer crosses the current sheet. What produces the up-down motion? One suggestion is that it is produced by waves, for instance Alfvén waves, propagating in the ecliptic plane (waves are discussed in Chapter 9). We now have a wavy ecliptic current sheet (Figure 6.18). What about the effects of solar rotation? If this rotation is included, we have a wavy ecliptic current sheet that resembles the skirt of a ballerina. Note that this model explains the presence of the magnetic component perpendicular to the ecliptic plane as a local feature and avoids the problem of generating large EMF on the Sun.

Observations of the IMF further indicate that during periods of increased solar activity, the IMF patterns are more complicated and changes of the IMF polarity occur more frequently. The wavy current sheet model explains these features by noting that waves are generated more easily and more frequently during disturbed solar wind conditions, which in turn produce more up-down motion of the current sheet.

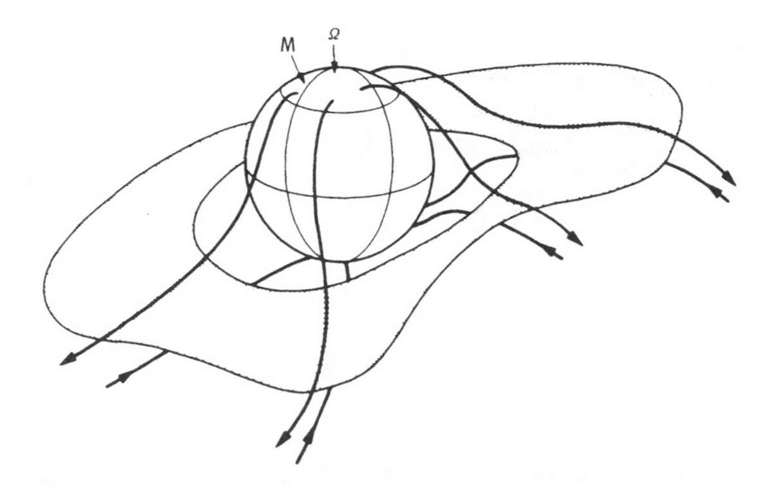

FIGURE 6.18 A model that includes the waves that produce the wavy ecliptic current sheet. The Sun is the center of an extensive and warped disk-like sheet in which electric currents flow azimuthally; that is, around the Sun. The average plane of the disk is approximately the plane of the equator of the Sun's average dipole magnetic field, which may be tilted with respect to its equator of rotation. The sheet separates solar-interplanetary magnetic-field regimes of nearly opposite directions. Solid lines are the magnetic field lines on each side of the current sheet. (From Smith *et al.*, 1978)

Let us now derive a picture of this current. Recall that our model confines the current sheet to a narrow region in the ecliptic plane. Let us refer to the diagram of the IMF shown in Figure 6.18. If we take the curl of the spiraling magnetic fields (ignore the waves and the ballerina effect), we find that there is a current flowing toward the Sun in a logarithmic spiral perpendicular to the IMF lines of force (Figure 6.19). The equation of the spiraling current is identical to (6.67), which describes the spiraling magnetic field in the rest frame. However, since $\mu_0 \mathbf{J} = \nabla \times \mathbf{B}$, the spiraling currents must be perpendicular to the magnetic field lines. The IMF forms equipotential surfaces and the currents are perpendicular to them.

The current closure requires that the same current that flows in must flow out of the Sun. A simple model is to represent this current as flowing out of the polar regions. If I_0 represents the current in the ecliptic plane, the currents flowing out of the northern and southern poles are each $I_0/2$ (Figure 6.20). Note that if the directions of the magnetic field in the ecliptic plane are reversed, the direction of the current must also reverse. Such current reversals are required for solar cycle changes.

The IMF and the currents described above are related through Ampere's circuital law

$$\oint \mathbf{B} \cdot \mathbf{dl} = \mu_0 \int \mathbf{J} \cdot \mathbf{n} da = \mu_0 I_0 \tag{6.76}$$

where I_0 is the total current. In cylindrical coordinates (r, ϕ, z) system, this equation yields

$$\begin{aligned} B_\phi(r) &= \frac{\mu_0 I_0}{4\pi r} \qquad z > 0 \\ &= -\frac{\mu_0 I_0}{4\pi r} \qquad z < 0 \end{aligned} \tag{6.77}$$

where $z = 0$ is the ecliptic plane. This azimuthal magnetic field is confined to the ecliptic plane and is the same field discussed earlier (see (6.74)).

The equation for the radial current density J_r obtained from Ampere's law or from performing the curl of the IMF expressions is

$$J_r = \frac{I_0}{2\pi r} \tag{6.78}$$

Define $J_{r_c} = I_0/2\pi r_c$ and rewrite (6.78) as

$$J_r = J_{r_c} \frac{r_c}{r} \tag{6.79}$$

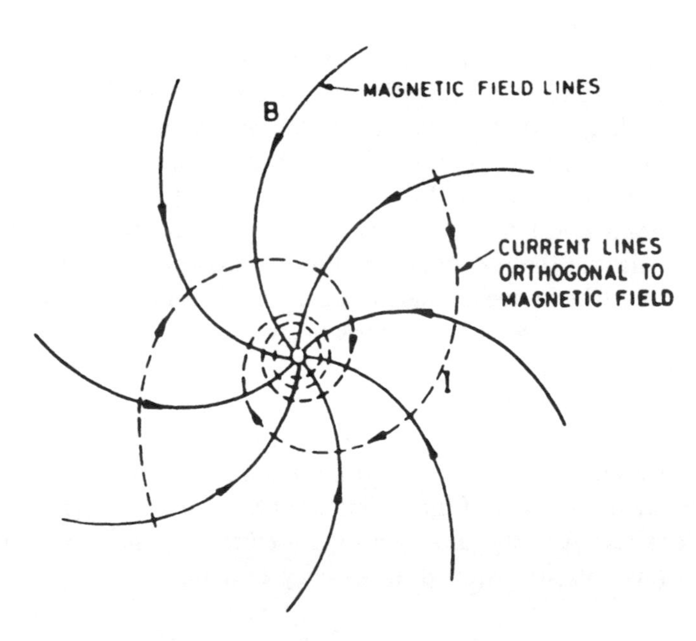

FIGURE 6.19 A sketch of the spiraling magnetic field and current lines in the ecliptic plane. (From Alfvén, 1981)

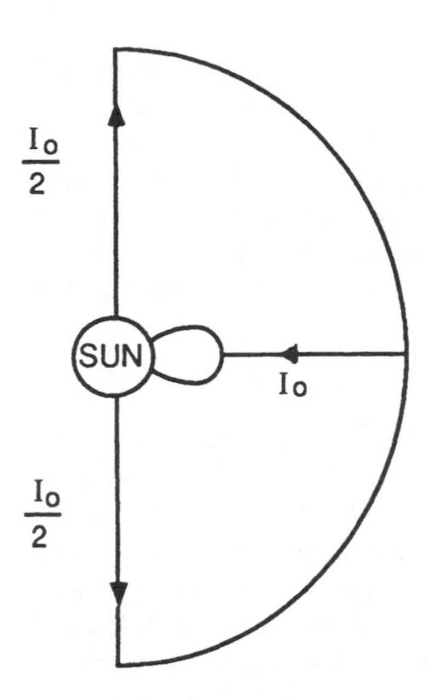

FIGURE 6.20 A schematic diagram of an ecliptic current sheet model. (From Alfvén, 1981)

where J_r is now referenced to the current density at r_c. The azimuthal component of the current density is obtained from

$$J_\phi = \left(\frac{B_\phi}{B_r}\right) J_r \tag{6.80}$$

This equation can be combined with (6.80) and becomes

$$J_\phi = J_{r_c} \left(\frac{B_\phi}{B_r}\right) \frac{r_c}{r} \tag{6.81}$$

B_r and B_ϕ are obtained from (6.73) and (6.74).

6.5 Magnetospheric Convection and Plasmapause

Now that we know that the solar coronal atmosphere and the solar magnetic field pervade interplanetary space, a natural question to ask is, what happens when they encounter a magnetized planet? In Chapter 3 we discussed how the topology of a planetary dipole field is modified by the IMF. This picture, which was studied assuming a vacuum medium, will be modified further since there are now solar wind particles that can produce local currents. These local currents are responsible for boundaries. The general problem of how local currents are produced will be studied in Chapter 7 and boundaries will be discussed in Chapter 8.

Solar wind is magnetized, and therefore the theory of plasma convection can be used to study its motional characteristics. Motions of magnetized plasmas occur in direct response to electric fields and these motions are referred to as convection under certain assumptions (see Section 5.5). The motion of solar wind flowing across an IMF induces an electric field given by

$$\mathbf{E}' = \mathbf{E} + \mathbf{U}_{sw} \times \mathbf{B}_{IMF} \tag{6.82}$$

where $\mathbf{E}'$ and $\mathbf{E}$ are electric fields measured in the moving and rest frames, respectively. Lorentz transformation guarantees that there exists a reference frame in which $\mathbf{E}' = 0$. We choose this frame to be the solar wind frame, which is physically meaningful since in good conductors electric fields vanish. This choice also enables us to employ the frozen-in-field concept and we can "visualize" $\mathbf{B}_{IMF}$ convecting with $\mathbf{U}_{sw}$.

When the solar wind approaches a magnetized planet, it encounters another source of an electric field. This electric field arises from the motion of the ionosphere through the planetary dipole field and is given by

$$\mathbf{E}' = \mathbf{E} + \mathbf{U}_{planet} \times \mathbf{B}_{dipole} \tag{6.83}$$

Let us choose the moving frame of the planet to be the frame in which $\mathbf{E}' = 0$ and also assume that the ionosphere is corotating with the planet. This permits us then to picture $\mathbf{B}_{dipole}$ corotating with the ionosphere, invoking again the frozen-in-field concept.

Our interest here concerns what consequences arise when the two induced electric fields interact. Given these two independent sources of electric fields, how will the individual streamlines modify? Let us answer this question with a specific example of plasma convection that occurs inside magnetospheres.

The action of these electric fields can create large-scale structures in magnetospheres. For example, the plasmapause, which is the outer boundary of the ionosphere, is created by convection. Figure 6.21 shows the positions of the Earth's plasmapause observed at various solar wind disturbance levels. The data are arranged with the quiet solar wind conditions on the top and with increasing disturbance levels toward the bottom. Note that the figures on the left and right panels represent data obtained during inbound and outbound passes, respectively.

The character of the plasmapause differs from one observation to the next. However, examination of data obtained during the inbound passes (inbound passes are close to the equatorial plane) indicates that the structure and the position of the plasmapause depend on the solar wind disturbance level. For example, no boundary was clearly evident when the solar wind was quiet (top left panel). Here the H^+ density decreased monotonically as one moved away from Earth (except for the small notch around $L = 5$, which is interpreted as a temporal effect). The subsequent panels show data obtained when the solar wind was active. Here the plasmapause boundary is well-defined and the boundary is found closer to Earth as the disturbance level increases. Figure 6.22 shows that the position of the plasmapause can be roughly ordered by the planetary disturbance index, K_p. Larger K_p values correspond to higher solar wind disturbance levels.

Let us now construct a simple model of plasma convective phenomena in magnetospheres. Our objectives are to understand how the plasmapause boundary is formed and how the position responds to the varying solar wind disturbance levels. Our model assumes the planetary magnetic field is a

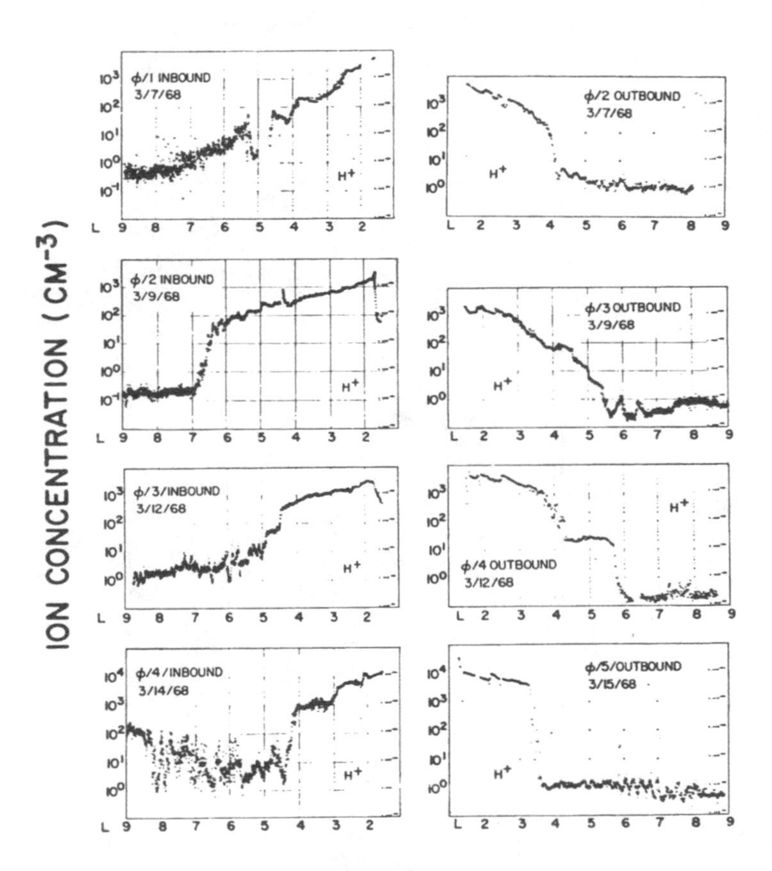

FIGURE 6.21 H^+ density distribution $[n_i(L)]$ profiles from ion mass-spectrometer data obtained on board the OGO-5 satellite. Here L measures the equatorial distance in earth radii $R_E = 6375$ km. (From Harris *et al.*, 1970)

centered dipole and the magnetosphere is in steady-state. The plasma inside this dipole magnetosphere is cold (no thermal energy) and is interacting with a constant corotational electric field and a constant solar wind-induced electric field in the dawn-dusk direction. In the limit of infinite conductivity, an electric field in the rest frame induced by the rotation of the magnetized planet through the conducting plasma is

$$\begin{aligned} \mathbf{E}_c &= -\mathbf{U} \times \mathbf{B} \\ &= -(\boldsymbol{\omega} \times \mathbf{r}) \times \mathbf{B} \end{aligned} \tag{6.84}$$

where $\boldsymbol{\omega}$ is the angular frequency of the planetary rotation, $\mathbf{r}$ is the radial position, and $\mathbf{B}$ is the magnetic field. If we now assume the magnetic field is dipolar

$$\mathbf{B}(r, \lambda) = B_0 \frac{R_p^3}{r^3}(-2\sin\lambda\hat{\mathbf{r}} + \cos\lambda\hat{\boldsymbol{\lambda}}) \tag{6.85}$$

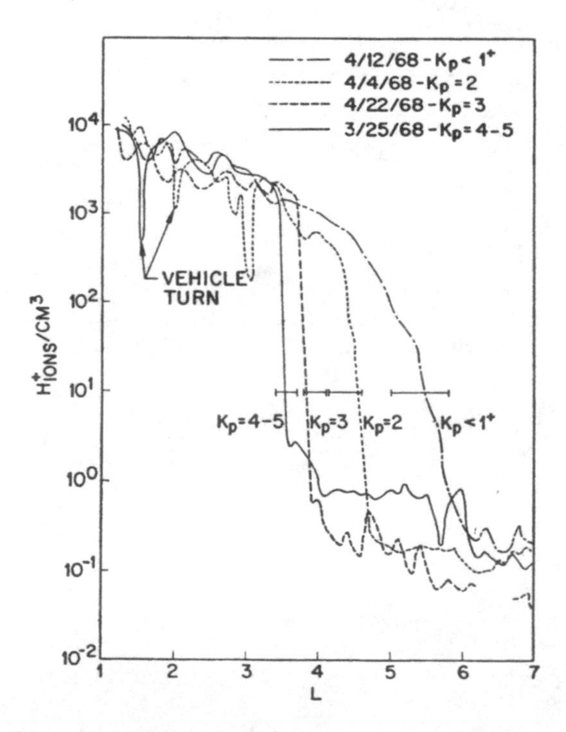

FIGURE 6.22 The plasmapause in H^+ concentration, based on groups of observations from OGO-5 in the range 0000-0400 LT at different levels of magnetic activity. (Chappell *et al.*, 1970)

where $\hat{\mathbf{r}}$ and $\hat{\boldsymbol{\lambda}}$ are unit vectors in the r and λ directions and $B_0 = \mu_0 M/4\pi R_p^3$ is the surface field on the equator of a planet with a radius R_p and magnetic moment M (see Chapter 3). Now let $\boldsymbol{\omega} = \omega\hat{\mathbf{z}}$ and $\mathbf{r} = r\hat{\mathbf{r}}$. Hence

$$\begin{aligned} \boldsymbol{\omega} \times \mathbf{r} &= \omega r \hat{\mathbf{z}} \times \hat{\mathbf{r}} \\ &= \omega r(\sin\lambda\hat{\mathbf{r}} + \cos\lambda\hat{\boldsymbol{\lambda}}) \times \hat{\mathbf{r}} \\ &= \omega r \cos\lambda\hat{\boldsymbol{\phi}} \end{aligned} \tag{6.86}$$

Substitution of (6.85) and (6.86) into (6.84) yields

$$\begin{aligned} \mathbf{E}_c &= -\frac{\omega B_0 R_p^3}{r^2}\cos\lambda\hat{\boldsymbol{\phi}} \times (-2\sin\lambda\hat{\mathbf{r}} + \cos\lambda\hat{\boldsymbol{\lambda}}) \\ &= -\frac{\omega B_0 R_p^3}{r^2}\cos\lambda(2\sin\lambda\hat{\boldsymbol{\lambda}} + \cos\lambda\hat{\mathbf{r}}) \end{aligned} \tag{6.87}$$

Note that $\mathbf{E}_c$ depends on the planetary rotation frequency $\boldsymbol{\omega}$. On the equatorial plane of the magnetosphere ($\lambda = 0$), $\mathbf{E}_c$ is directed radially inward toward the planet. The magnitude of $\mathbf{E}_c$ is

$$E_c = \frac{\omega B_0 R_p^3 \cos\lambda(1 + 3\sin^2\lambda)^{1/2}}{r^2} \tag{6.88}$$

and the corotational electric field falls off as $1/r^2$.

$\mathbf{E}_c$ is a potential field ($\nabla \times \mathbf{E} = 0$, $\mathbf{E} = -\nabla\Psi$). Let us now confine our discussion to the equatorial plane. There, the scalar potential can be represented by

$$\begin{aligned} \Psi_c(L, \lambda = 0, \phi) &= -\frac{\omega B_0 R_p^3}{r} \\ &= -\frac{\omega B_0 R_p^2}{L} \end{aligned} \tag{6.89}$$

where $r = r_0\cos^2\lambda = LR_p\cos^2\lambda = LR_p$ for $\lambda = 0$. In potential flows, particles with zero thermal energy (cold particles) follow contours of constant potential and, therefore, the potential contours are also the streamlines for these particles. Figure 6.23 (top right figure) shows the streamlines for these corotating particles are circles.

Another source of electric fields inside magnetospheres is the solar wind interaction with the planetary magnetic field. This field $\mathbf{E}_{sw}$ is responsible for maintaining, for example, the neutral sheet current which is required to form the tail of the magnetosphere. This field, referred to as the solar wind-induced electric field, will be assumed to be in the dawn-dusk direction on the equatorial plane:

$$\mathbf{E}_{sw}(L, \lambda = 0, \phi) = E_{sw}\hat{\mathbf{y}} \tag{6.90}$$

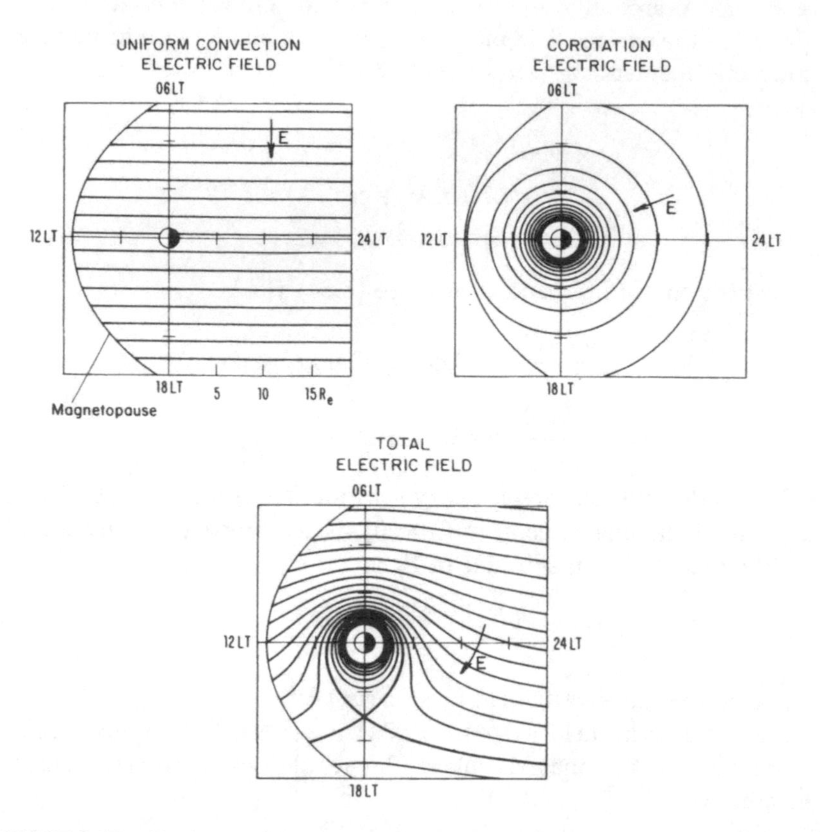

FIGURE 6.23 Equipotential contours for the magnetospheric electric field in the equatorial plane. Upper left: first-order approximation for the convection electric field E_c as uniform. The contours are spaced 3 kV apart for $E_c = 2.5 \times 10^{-4} V$ m^{-1}. Upper right: the corotation electric field, contours spaced 3 kV apart. Lower: sum of convection and corotation electric fields. The heavy contour separates the closed and open convection regions. (From Lyons and Williams, 1984)

where E_{sw} is constant and spatially uniform and directed in the negative y-direction in the geocentric solar ecliptic (GSE) coordinate system, in which x is directed toward the Sun, y toward dusk, and z toward the North Pole. ϕ is measured eastward from local midnight. The potential function for this solar wind electric field is given by

$$\begin{aligned}\Psi_{sw}(L, \lambda = 0, \phi) &= -E_{sw} y \\ &= -E_{sw} r_0 \sin\phi \\ &= -E_{sw} L R_p \sin\phi \end{aligned} \tag{6.91}$$

For an electric field pointing from dawn to dusk, the streamlines are straight lines and are directed toward the Sun (upper left-hand panel of Figure 6.23).

The motion of a plasma inside the magnetosphere is subject to both the corotational and the solar wind-induced electric fields. The streamlines of the zero energy particles, taking into account both of these fields, are obtained from

$$\begin{aligned}\Psi_t(L, \lambda = 0, \phi) &= \Psi_{sw}(L, \lambda = 0, \phi) + \Psi_c(L, \lambda = 0, \phi) \\ &= E_{sw} L R_p \sin\phi - \frac{\omega B_0 R_p^2}{L}\end{aligned} \tag{6.92}$$

where we used (6.89) and (6.91). The equation of the equipotential contours (Ψ_t = constant), obtained by solving the quadratic equation (6.92) for L, is

$$L = \frac{\Psi_t}{A} \pm \left[\left(\frac{\Psi_t}{A}\right)^2 + \frac{2\omega B_0 R_p^2}{A}\right]^{1/2} \tag{6.93}$$

where $A = 2E_{sw} R_p \sin\phi$ and E_{sw} is a free parameter. The bottom panel of Figure 6.23 shows contours derived from (6.93) for the various potential values.

The equipotential contours near the planet are circles and they are closed. These streamlines are dominated by the corotational electric field. The contours further away from the planet are "open" and these streamlines are dominated by the induced solar wind electric field. These two classes of streamlines do not cross each other except on one contour that is located several earth radii from the planet. This contour supports a zero electric field at 1800 local time (singular point) and it separates the closed and open streamlines. This equipotential contour is defined as the plasmapause.

The singular point occurs where the magnitude of the corotation and the induced solar wind electric fields are equal. From (6.89) and (6.91), we obtain

$$E_{sw} = \frac{\omega B_0 R_p}{L^2} \tag{6.94}$$

where $\sin\phi = -1$ at 1800 local time. If this relation for L is used in (6.93), the potential at the plasmapause is

$$\Psi_{pp} = -2R_p (B_0 R_p \omega E_{sw})^{1/2} \tag{6.95}$$

Another useful piece of information is the equation of the plasmapause as a function of the local time. This equation is derived by substituting Ψ_{pp}

into (6.93). We then obtain

$$L_{pp} = \left(\frac{\omega B_0 R_p}{E_{sw}}\right)^{1/2} [(1+\sin\phi)^{1/2} - 1]\csc\phi \tag{6.96}$$

This derivation has ignored the $-(1+\sin\phi)^{1/2}$ case on physical arguments. The coefficient $(\omega B_0 R_p/E_{sw})^{1/2}$ is a dimensionless number and it measures the equatorial geocentric distance to the plasmapause (in units of planetary radii). L_{pp} has a maximum at 1800 local time, minimum at 0600 local time, and has intermediate values at other local times.

The position of the plasmapause depends on $(1/E_{sw})^{1/2}$. A typical value of E_{sw} is about 10^{-4} volts/meter for quiet to moderate solar wind disturbance levels. During large magnetic storms, E_{sw} can be as large as 3×10^{-4} volts/meter, and during such large storms the plasmapause will be found closer to the planet. Evidence that E_{sw} increases during disturbed periods has been obtained by direct observations of the position of the inner boundary of the plasma sheet. Figure 6.24 shows data that compare the relative position of the plasmapause and the inner boundary of the plasma sheet for quiet (left panel) and more disturbed (right panel) times. These data show that the inner plasma sheet boundary moves closer to Earth with increased disturbance level. These observations can be interpreted in terms of enhanced E_{sw} in the magnetosphere.

The simple model we constructed is a reasonably accurate picture of plasma convection close to the planet. In the more distant regions, the details of the streamlines will depend on whether the magnetosphere is open or closed. The left-hand panel of Figure 6.25 shows the streamlines of

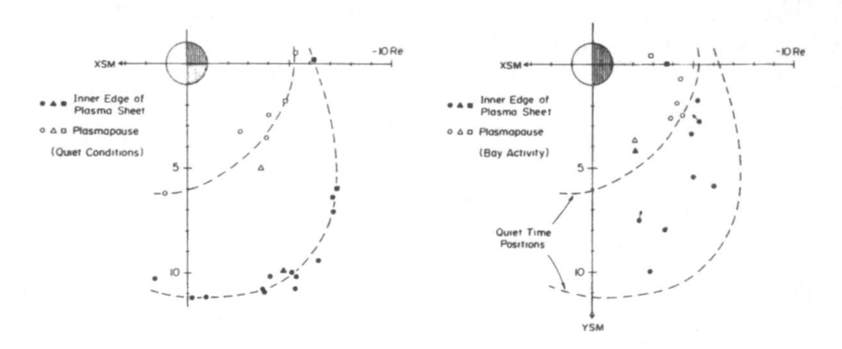

FIGURE 6.24 Observed positions of the inner edge of the plasma sheet (filled-in symbols) and the plasmapause (open symbols) during quiet and disturbed geomagnetic conditions. Circles: OGO-1 measurements (from Vasyliunas, 1968). Triangles: OGO-3 measurements (from Vasyliunas, unpublished). Squares: OGO-3 measurements. (From Schield and Frank, 1970)

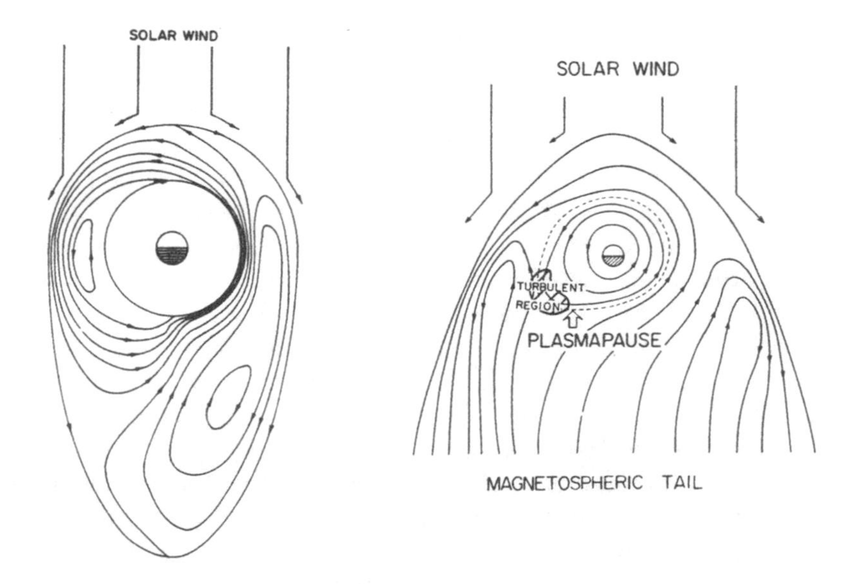

FIGURE 6.25 Streamlines of the solar wind-induced convection on the equatorial plane for open and closed magnetospheric models. The main circulation is not supposed to penetrate closer than about $4.5R_E$. (From Axford and Hines, 1961 and Nishida, 1966)

a closed magnetosphere induced by the solar wind electric field, including the effects of the corotational electric field. Here even the streamlines in the distant magnetospheric regions are closed (the streamlines resemble streamlines of fluid in a closed tank). If the magnetosphere were open, the distant streamlines in the distant regions of the magnetosphere would not be closed. The right-hand panel of Figure 6.25 shows the streamlines of plasma induced by the solar wind electric field superimposed with the corotational electric field. The open streamlines in the distant tail region presumably connect to interplanetary space.

6.6 Concluding Remarks

Our knowledge of the solar wind and the IMF comes mainly from observations made in the vicinity of Earth. Here, in general, Parker's theory has been quite successful. In this regard, it is important to note that very little is known about the evolutional properties of the solar wind and the IMF. How do the solar wind and the IMF evolve with the heliocentric distance? Does Parker's theory explain the characteristics of the solar wind and the IMF at large distances, at tens and hundreds of AU?

Another problem is that essentially all of our observations come from the ecliptic plane. Very little is known about the characteristics of the solar wind and the IMF off the ecliptic plane. The available measurements therefore provide only a two-dimensional picture. These measurements will be improved by observations of the Galileo and Ulysses spacecrafts. There is a plan to divert the trajectory of Galileo (launched on October 17, 1989), after its encounter with Jupiter in 1996, toward the northern ecliptic pole. Ulysses (launched in October 1990) will also be diverted toward the southern pole after its encounter with Jupiter. Knowledge about the solar wind and the IMF off the ecliptic plane must await these observations.

The convection model we constructed assumes that the plasma is cold ($T = 0$). Convective electric fields also affect the warm particles, those particles with thermal energy. The trajectories of warm and cold particles in homogeneous magnetic fields are the same, but they are different in dipole fields because of gradient and curvature drifts that affect only warm particles. In the distant region of the plasma sheet where the fields are homogeneous, the trajectories of warm and cold particles are the same. However, as the particles approach close to the planet, the warm particles will begin to drift and their trajectories will become different. The trajectories of warm particles on the equator can be obtained by conserving their total energy, which is the sum of the kinetic and potential energies, and their magnetic moments. For non-relativistic particles with charge q, we need to conserve

$$\mu B + q\Phi = \text{constant} \tag{6.97}$$

where μ is the magnetic moment. We can then proceed, using (6.92) as the total electrostatic potential and using a dipole for the magnetic field to calculate the trajectories of warm particles.

The final comment concerns the convective electric field in magnetospheres. Although it is assumed that the convective electric field originates in the solar wind, nothing was said about how this electric field appears inside magnetospheres. This is because we do not as yet fully understand the mechanisms by which the solar wind and the planetary magnetopauses interact. The role of the corotational electric field is important for this problem. In this regard, it is worthy of note that while only a portion of Earth's magnetosphere corotates, the entire magnetosphere of Jupiter corotates. On the other hand, none of Mercury's magnetosphere is believed to corotate because Mercury has no ionosphere. These problems are intimately related to how the solar wind mass, momentum, and energy are transported across the magnetopause boundary. Another way to study this problem is to ask how boundary currents are established by the solar wind interacting with planetary magnetic fields. These are topics of Chapters 7 and 8.

Bibliography

Alfvén, H., *Cosmic Plasma*, D. Reidel Publishing Co., Dordrecht, Holland, 1981. This book brings together many of Alfvén's ideas about how space plasmas work in nature. Early in the book he criticizes the misuse of the frozen-in-field concept in space plasma studies. The book, although written for experts, is quite readable. The discussion is intuitive and phenomenological, rather than quantitative.

Axford, W.I. and C.O. Hines, A Unifying Theory of High-Latitude Geophysical Phenomena and Magnetic Storms, *Can. J. Phys.*, **39**, 1433, 1961. Representative convective patterns in the Earth's ionosphere were first presented by these authors.

Bame, S.J., A.J. Hundhausen, J.R. Asbridge and I.B. Strong, Solar Wind Ion Composition, *Phys. Rev. Lett.*, **20**, 393, 1968. One of the first articles reporting on various ionic species in the solar wind.

Chappell, R., K.K. Harris and G.W. Sharp, A Study of the Influence of Magnetic Activity on the Location of the Plasmapause as Measured by OGO 5, *J. Geophys. Res.*, **75**, 50, 1970. This paper shows how the plasmapause moves with the geomagnetic activity.

Dessler, A.J., Solar Wind and Interplanetary Magnetic Fields, *Rev. Geophys.*, **5**, 1, 1967. An excellent review article on solar wind and interplanetary magnetic fields. The article is clearly written and can be read by beginning students.

Harris, K.K., G.W. Sharp, and C.R. Chappell, Observations of the Plasmapause from OGO 5, *J. Geophys. Res.*, **75**, 219, 1970. One of the first articles that studied the details of the plasmapause.

Herzberg, L., Solar Optical Radiation and its Role in Upper Atmospheric Processes, in *Physics of the Earth's Upper Atmosphere*, C.O. Hines, I. Paghis, T.R. Hartz, and J.A. Fejer, eds., Prentice-Hall, Inc., Englewood Cliffs, NJ, 1965. One of the first research monograms on the upper atmosphere. Contributors are experts in their respective fields. For intermediate and advanced students.

Holzer, T.E. and W.I. Axford, The Theory of Stellar Winds and Related Flows, *Annual Rev. of Astronomy and Astrophysics*, **8**, 31, 1970. This article reviews the solar and stellar wind problem in great detail. Mathematical techniques and several classes of solutions are presented. For intermediate and advanced students.

Hundhausen, A.J., Composition and Dynamics of the Solar Wind Plasma, *Rev. of Geophys. and Space Phys.*, **8**, 729, 1970. A good review article on solar wind dynamics. The article covers both observations and theory. For intermediate and advanced students.

Hundhausen, A.J., S. Bame, and N. Ness, Solar Wind Thermal Anisotropies, Vela 3 and IMP 3, *J. Geophys. Res.*, **72**, 5265, 1967. Figure 6.6 is taken from this article.

Landau, L.D. and E.M. Liftshitz, *Fluid Mechanics*, Pergamon Press, Ltd., London, England, 1959. This book is an authoritative reference for the physics of ordinary fluids. However, many concepts of ordinary fluids are valuable in the discussion of MHD fluids.

Lyons, L.R. and D.J. Williams, *Quantitative Aspects of Magnetospheric Physics*, D. Reidel Publishing Co., Boston, MA, 1984. This book brings together data and theory to describe magnetospheric physics. Chapter 4 of this book treats in detail the topic of magnetospheric electric fields. This book is written primarily for researchers in the field (and advanced graduate students).

Neugebauer, M. and C.W. Snyder, Mariner 2 Observations of the Solar Wind, 1. Average Properties, *J. Geophys. Res.*, **71**, 4469, 1966. One of the first results showing long-term solar wind speed and density variations.

Nishida, A., Formation of Plasmapause, or Magnetospheric Plasma Knee, by the Combined Action of Magnetospheric Convection and Plasma Escape from the Tail, *J. Geophys. Res.*, **71**, 5669, 1966.

Parker, E.N., *Interplanetary Dynamical Processes*, Interscience Publishers, New York, NY, 1963. This book brings together in one volume the theory of the solar coronal atmosphere. Chapter 10 discusses the IMF model (the original article on the IMF was published in *Astrophysical Journal*, **128**, 664, 1958). This book includes a discussion of relevant observations that motivated the development of the theory.

Rossi, B.B. and S. Olbert, *Introduction to the Physics of Space*, McGraw Hill Book Co., New York, NY, 1970. This book is one of the first textbooks on the subject of space physics and is written primarily for graduate and advanced undergraduate students. Mathematical solutions of the solar wind equation are discussed in Chapter 14.

Schield, M.A. and L.A. Frank, Electron Observations Between the Inner Edge of the Plasma Sheet and the Plasmasphere, *J. Geophys. Res.*, **75**, 5401, 1970. This paper discusses the measurements of electrons near $\approx$ 100 eV to 50 keV between 3 and 10 R_E near local midnight.

Schulz, M., Interplanetary Sector Structure and the Heliomagnetic Equator, *Astrophys. and Space Sci.*, **24**, 371, 1973. The conventional interpretation of interplanetary sector structures was that the magnetic field mapped back to the photosphere. This article presents an interpretation of the sector structure based on the ecliptic current sheet model, which has been proven to be correct.

Smith, E.J., B.T. Tsurutani and R.L. Rosenberg, Observations of the Interplanetary Sector Structure up to Heliographic Latitudes of 16°: Pioneer 11, *J. Geophys. Res.*, **83**, 717, 1978. Data from off the ecliptic plane was used to construct a model of the heliospheric current sheet.

Spitzer, L., *Physics of Fully Ionized Gases*, 2nd ed., Interscience Publishers, New York, NY, 1962. A good plasma physics book. Can be read by beginning students.

Stern, D.P., The Motion of Magnetic Field Lines, *Space Sci. Rev.*, **6**, 147, 1966. This article reviews the concept of the motion of magnetic lines of force as an aid in describing velocity fields in highly conducting fluids. It provides a mathematical foundation for the flux conservation theorem, which is required for the discussion of the concept of magnetic field line motion.

Vasyliunas, V.M., A Survey of Low Energy Electrons in the Evening Sector of the Magnetosphere with OGO 1 and OGO 3, *J. Geophys. Res.*, **73**, 2839, 1968. One of the first articles that reported on the systematic properties of the plasma sheet electrons.

Vasyliunas, V.M., Nonuniqueness of Magnetic Field Line Motion, *J. Geophys. Res.*, **77**, 6271, 1972. This short article presents a detailed theory of the motion of magnetic lines of force and stresses that the solutions are not unique. Caution must be therefore exercised in interpretations of observations in space that require the concept of the motion of lines of force.

Wilcox, J., and N. Ness, Quasi-Stationary Corotating Structure in the Interplanetary Medium, *J. Geophys. Res.*, **70**, 1233, 1965. These authors noted a pattern in IMF polarity for quiet years. See the reference of M. Schulz for an explanation of the sector structures.

Questions

1. If the solar coronal atmosphere is 10^{6}°K, what is the proton thermal energy in eV?
2. What is the kinetic energy (in eV) of the solar wind protons at 1 AU?
3. Start with Equation (6.16) and fill in the missing steps to arrive at Equation (6.24).
4. Use Chapman's static model of the solar corona to estimate the coronal temperature and the density at 1 AU. Assume $T_0 = 10^6$ °K, and $n_0 = 10^{14}(\mathrm{m}^{-3})$.
5. Plot the critical distance r_c as a function of the solar coronal temperature for temperatures from 10^5 °K to 10^7 °K. Express r_c in terms of the solar radius $r_\odot$.
6. Compute and then plot the solar wind velocity U_{sw} as a function of the normalized heliocentric distance $r/r_\odot$, where $r_\odot$ is the radius of the Sun, from $r/r_\odot = 1$ to $r/r_\odot = 215$ assuming the solar atmosphere is isothermal and for coronal temperatures of 0.5×10^6 °K, 10^6 °K, and 2×10^6 °K. From these plots, state how long it takes the solar wind to reach Earth. Assume that $v_c = c_s$.
7. Compute and then plot the number density n of the solar wind as a function of the normalized heliocentric distance $r/r_\odot$ for the three temperatures considered in problem 6 from $r/r_\odot = 1$ to $r/r_\odot = 215$. State how the number density measured at 1 AU depends on the coronal temperature.
8. How much mass is lost by the Sun per unit time owing to the solar wind? Assume that $n = 10^7$ protons/m^3 at 1 AU. How long will it take to deplete the Sun?
9. Estimate the intensity of the IMF at 1 AU if the Sun were immersed in vacuum and the solar field could be approximated as a dipole. Compare this result with the result obtained assuming the model adopted in the chapter. Use the fact that a coronal magnetic field intensity is 10^{-4} teslas and the solar wind velocity beyond r_c is 400 km/sec. The average angular period of rotation of the Sun is 27 days.
10. Compute the spiral angle of the IMF relative to the Earth-Sun line (solar wind flow direction) at 1 AU. Assume the solar wind speed = 400 km/sec.
11. Compute the strength and the direction of the IMF for an observer at rest with respect to the solar wind flow. Assume the solar wind flow is 400 km/sec and the IMF of 10 nanoteslas is confined to the ecliptic plane.
12. The solar wind flow was assumed to be strictly radial. However, because of the rotation of the Sun, the solar wind must have a component of flow in the transverse direction. Show from conservation of angular momentum that the transverse flow component is negligible as compared to the speed of radial flow.
13. Compute the total EMF generated in one solar rotation if the solar wind convects out a 1-nanotesla magnetic field component perpendicular to the ecliptic plane. Take the line integral of the electric field $\mathbf{E}$ around the Sun in the ecliptic plane inside 1 AU.
14. The total torque exerted on the Sun about the z-axis across the surface of a sphere of radius r is $T = \int \mathbf{r} \times (\mathbf{J} \times \mathbf{B}) dV$. Now $\mathbf{J} = (\nabla \times \mathbf{B})/\mu_0$ and $((\nabla \times \mathbf{B}) \times \mathbf{B})/\mu_0 = \nabla B^2/2\mu_0 + (\mathbf{B} \cdot \nabla)\mathbf{B}/\mu_0$.
   a. Show that

$$T = r^2 \int_0^{\pi} \sin\theta d\theta \int_0^{2\pi} r \sin\theta \frac{B_r B_\theta}{\mu_0} d\phi$$

b. Assume the Sun's field is a dipole so that $B_r(r_c) = B_p \cos\theta$ at the poles. Show that

$$T(r) = \frac{8\pi}{15\mu_0} r_c^4 B_p^2 \frac{\Omega}{U_{sw}} \left(1 - \frac{r_c}{r}\right)$$

7

Currents in Space

7.1 Introduction

Currents are produced by charges in motion. Large-scale currents in space are produced by the charged particles of the solar wind, magnetospheres, and ionospheres. These currents are sources of magnetic fields in the distant regions of space. For example, the heliospheric current sheet produced by the dynamics of the solar wind is responsible for the IMF. Large-scale currents are also associated with planetary magnetospheres. A convenient way to discuss currents in planetary magnetospheres is to classify them as boundary currents, ring currents, ionospheric currents, field-aligned currents, and magnetotail currents. The main purpose of this chapter is to study how these currents are produced in planetary ionospheres and magnetospheres (Figure 7.1).

Currents responsible for the magnetopause boundary were first studied by S. Chapman and V.A. Ferraro in 1931. They were interested in magnetic storms, which usually begin a few days after a major disturbance (flares) on the Sun. Magnetic storms on Earth produce large deviations of the surface

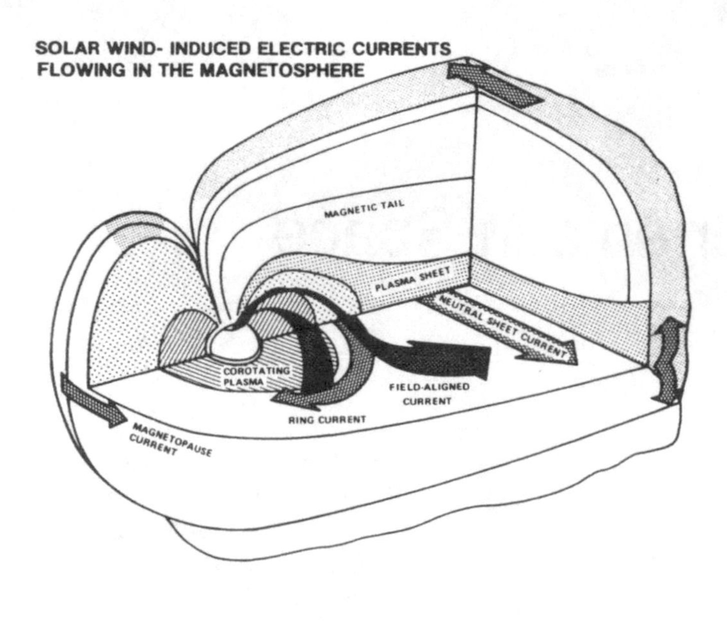

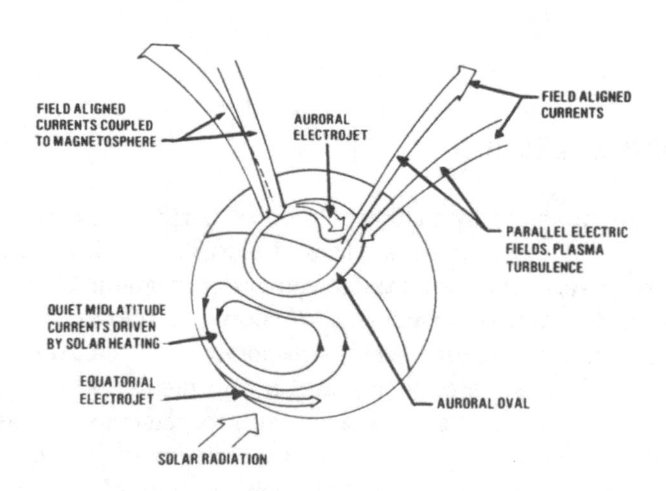

FIGURE 7.1 A schematic diagram that shows where currents run in the magnetosphere and the ionosphere.

magnetic field, which require the presence of currents external to the solid planet. Chapman and Ferraro developed a model in which the geomagnetic field, originally in vacuum, becomes enveloped in solar particles (the solar wind was not yet discovered). This model shows that a current will be produced at the interface that separates the geomagnetic field and the solar particle medium. This boundary current, also called the Chapman-Ferraro current, is produced by the solar protons and electrons that penetrate the geomagnetic field. The original Chapman-Ferraro model was a transient model, since the planetary magnetic field is immersed in the solar particle medium only after a solar flare. However, with the discovery of the solar wind, we now know that magnetopause currents are a permanent feature of the interface that separates the magnetic field of the planet from the solar wind.

Currents produced at plasma boundaries are diamagnetic and the Chapman-Ferraro current at the magnetopause is an example of a diamagnetic current produced by space plasmas. Another example of a diamagnetic current produced in space is the ring current, which resides inside magnetospheres. This ring current, produced by the motion of trapped particles in an inhomogeneous planetary magnetic field as the particles undergo gradient and curvature drifts, is greatly enhanced during magnetic storms. Major deviation of the surface magnetic field is attributed to the ring current (which is closer to the surface) rather than the magnetopause current, as originally thought by Chapman and Ferraro. Diamagnetic currents are normally associated with fluid properties of particles. However, since fluids consist of many individual particles, it can be shown that the collection of individual particle motions adds up to yield the same diamagnetic current. We show that this is the case for both isotropic and anisotropic MHD fluids.

The density of particles in ionospheres is high enough that collisions cannot be ignored. Collisions give rise to momentum transfer and the concept of electrical conductivity is meaningful. Current flows in ionospheres can therefore be described by Ohm's law. The simple Ohm's law must be modified, however, to describe how currents flow in a fluid system consisting of electrons and ions. A generalized Ohm's law is required and this will be derived for an MHD fluid. The ionospheric plasma is also anisotropic if the planetary magnetic field is strong. In this case, currents do not necessarily flow parallel to the direction of the electric field. It is necessary then to replace the scalar conductivity by a tensor. Use of the tensor conductivity in Ohm's law for Earth's ionosphere leads to Hall, Pedersen, and Cowling conductivities and currents, named after the scientists who performed the original studies.

This chapter concludes with a short discussion of field-aligned and magnetotail currents. Field-aligned currents flow along the magnetic field direction and they connect the ionosphere and the more distant regions of the magnetosphere. Magnetotail currents are responsible for the long magnetic tails of planetary magnetospheres. Understanding the origin of magnetotail currents is important because they are tied to mechanisms that transfer the solar wind mass, momentum, and energy into planetary magnetic fields. The magnetotail currents are also a major source of energy for auroral currents, since tail currents are diverted into the ionosphere along the magnetic field during an aurora.

7.2 Currents in Plasmas

Let us first define and establish some general concepts about currents. The fundamental law governing the behavior of currents is given by one of the Maxwell equations,

$$\nabla \times \mathbf{H} = \mathbf{J} + \frac{\partial \mathbf{D}}{\partial t} \tag{7.1}$$

However in MHD approximation, we can ignore the displacement current. We also note that currents in MHD fluids are intimately coupled to the motion of fluids,

$$\rho_m \frac{d\mathbf{U}}{dt} = -\nabla \cdot \mathbf{p} + \mathbf{J} \times \mathbf{B} \tag{7.2}$$

Equation (7.2) is the vectorial form of (5.153), which is more general than (5.35) because the pressure $\mathbf{p}$ is a tensor rather than a scalar.

Consider the charges of an arbitrary volume bounded by a surface S. Now suppose the charges of species α are moving across an incremental area dS in S with unit normal $\mathbf{n}$ pointing outward of the surface. Let $<\mathbf{v}>^\alpha$ be the average velocity of particle species α flowing across dS. The infinitesimal amount of charge dq^α crossing dS in time dt in the direction of $\mathbf{n}$ is

$$dq^\alpha = n^\alpha q^\alpha <\mathbf{v}>^\alpha \cdot \mathbf{n}\, dt\, dS \tag{7.3}$$

where n^α is the density of particle species α. A current is defined as the rate at which these charges move across the surface S. Hence

$$dI^\alpha = \frac{dq^\alpha}{dt} = n^\alpha q^\alpha <\mathbf{v}>^\alpha \cdot \mathbf{n}\, dS \tag{7.4}$$

It is convenient to define the current density

$$\mathbf{J}^\alpha = n^\alpha q^\alpha <\mathbf{v}>^\alpha \tag{7.5}$$

Equation (7.4) then becomes

$$dI^\alpha = \mathbf{J}^\alpha \cdot \mathbf{n}\, dS \tag{7.6}$$

Integration of (7.6) yields the total current crossing the surface S,

$$I^\alpha = \int \mathbf{J}^\alpha \cdot \mathbf{n}\, dS \tag{7.7}$$

by the particle species α. If there is more than one particle species, we must sum up the contributions from all of the species. Then

$$\mathbf{J} = \sum_{\alpha} \mathbf{J}^{\alpha} = \sum_{\alpha} n^{\alpha} q^{\alpha} < \mathbf{v} >^{\alpha} \tag{7.8}$$

and

$$I = \sum_{\alpha} I^{\alpha} = \int \mathbf{J} \cdot \mathbf{n} dS \tag{7.9}$$

Equation (7.8) involves computing the average velocity $< \mathbf{v} >^{\alpha}$, which is defined as

$$< \mathbf{v} >^{\alpha} = \int \mathbf{v} f^{\alpha}(\mathbf{r}, \mathbf{v}, t) d^3 v \tag{7.10}$$

where $f^{\alpha}(\mathbf{r}, \mathbf{v}, t)$ is the distribution function of particle species α. These equations show that currents exist wherever there is a plasma.

The total current I is the rate at which the charges are flowing out of the volume V across S and since the charges in V are being lost at a rate $-dq/dt$, the conservation of charges requires that

$$\int \mathbf{J} \cdot \mathbf{n}\, dS + \frac{dq}{dt} = 0 \tag{7.11}$$

The total charge q may be expressed in terms of the charge density ρ

$$q = \int \rho\, dV \tag{7.12}$$

and if the volume in space is assumed fixed, (7.11) reduces to

$$\int \mathbf{J} \cdot \mathbf{n}\, dS + \int \frac{\partial \rho}{\partial t} dV = 0 \tag{7.13}$$

Now use the divergence theorem, which permits (7.13) to be rewritten as

$$\int \nabla \cdot \mathbf{J}\, dV + \int \frac{\partial \rho}{\partial t}\, dV = 0 \tag{7.14}$$

Since this equation is valid for any arbitrary volume V bounded by S, we can write the charge conservation equation in differential form as

$$\nabla \cdot \mathbf{J} + \frac{\partial \rho}{\partial t} = 0 \tag{7.15}$$

For problems in which $\partial \rho / \partial t$ vanishes, (7.15) yields

$$\nabla \cdot \mathbf{J} = 0 \tag{7.16}$$

$\mathbf{J}$ is in this case solenoidal. Note that (7.16) can also be obtained by taking the divergence of $\nabla \times \mathbf{H} = \mathbf{J}$, since $\nabla \cdot \nabla \times \mathbf{H} = 0$.

In principle, the current density in space can be directly measured by particle detectors (Chapter 2). However, (7.8) requires as an input complete information on the particles that are carrying the current. This requires detecting all of the different particle species over all energies and pitch-angles. Such measurements are extremely difficult to make in part because of the limitations of detection techniques. The three-dimensional distribution function over all velocities must be measured. One practical problem concerns the spacecraft charging, which will distort the measurement of low-energy particles. The spacecraft is usually several volts positive relative to the ambient plasma because of the photoelectrons that are produced by the solar ultraviolet radiation interacting with the spacecraft. The photoelectron contribution must be unambiguously separated from the naturally occurring current carriers, whose energies may be similar to the energies of the photoelectrons.

For steady-state currents, we can ignore the contribution of the displacement current. Then (7.1) relates the current density $\mathbf{J}$ to the curl of the magnetic induction $\mathbf{B}$, which can be measured by magnetometers ($\mathbf{B} = \mu_0 \mathbf{H}$). Thus one can in principle obtain information on $\mathbf{J}$ by measuring $\nabla \times \mathbf{B}$. However, single-point measurements in space do not give information on the gradients of the various $\mathbf{B}$ components that are required to determine $\mathbf{J}$. Certain assumptions need to be made about the structure of currents before the magnetic field measurements can be interpreted in terms of currents.

7.3 Boundary Current (Chapman-Ferraro Current)

To study the Chapman-Ferraro current requires that we understand how currents are produced in MHD fluids. Currents in MHD fluids are coupled to the magnetic field and they play an important role in the dynamics of the momentum transport. The characteristics of MHD currents can be studied by deriving an explicit expression for this current. Let us derive this expression for a fluid whose pressure is isotropic and constant throughout.

Consider a conducting fluid consisting of one species of ions and electrons. Ignoring gravity and collisions, the momentum equation for each of the species is

$$\begin{aligned} m_i n_i \frac{d\mathbf{U}_i}{dt} &= -\nabla p_i + n_i q_i (\mathbf{E} + \mathbf{U}_i \times \mathbf{B}) \\ m_e n_e \frac{d\mathbf{U}_e}{dt} &= -\nabla p_e + n_e q_e (\mathbf{E} + \mathbf{U}_e \times \mathbf{B}) \end{aligned} \tag{7.17}$$

As usual, we assume $q_i = -q_e = |q|$ and $n_i = n_e = n$. We have shown in Chapter 5 that the addition of the above equations leads to a one-fluid

MHD momentum equation

$$\rho_m \frac{d\mathbf{U}}{dt} = -\nabla p + \mathbf{J} \times \mathbf{B} \tag{7.18}$$

where the various quantities have already been defined. Cross this momentum equation with $\mathbf{B}$, and expand the vector triple product,

$$\begin{aligned} \rho_m \frac{d\mathbf{U}}{dt} \times \mathbf{B} &= -\nabla p \times \mathbf{B} + (\mathbf{J} \times \mathbf{B}) \times \mathbf{B} \\ &= -\nabla p \times \mathbf{B} + [(\mathbf{J} \cdot \mathbf{B})\mathbf{B} - B^2 \mathbf{J}] \end{aligned} \tag{7.19}$$

and solving for $\mathbf{J}_\perp$, obtain

$$\mathbf{J}_\perp = \frac{\mathbf{B} \times \nabla p}{B^2} + \rho_m \frac{\mathbf{B}}{B^2} \times \frac{d\mathbf{U}_\perp}{dt} \tag{7.20}$$

The first term on the right side comes from the coupling of the pressure gradient (found usually at a boundary) and the local magnetic field $\mathbf{B}$. This term is called the diamagnetic current. This current is diamagnetic because the current always flows in such a direction as to reduce the magnetic field intensity in the fluid. This means the gyromotion plays an important role. The second term represents the contribution to the current resulting from the coupling of the acceleration of the perpendicular fluid flow to the magnetic field. This term is referred to as an inertial current and it gives rise to the polarization currents discussed in Chapter 4.

7.3.1 Diamagnetic Current

An interesting feature about the diamagnetic current is that it exists even if the fluid is not flowing. Under static conditions, (7.20) reduces to (omit the subscript $\perp$)

$$\mathbf{J} = \frac{\mathbf{B} \times \nabla p}{B^2} \tag{7.21}$$

which is also equivalent to

$$\nabla p = \mathbf{J} \times \mathbf{B} \tag{7.22}$$

Equation (7.22) is a statement that the stresses (fluid and electromagnetic) are balanced in steady state. It implies that

$$\begin{aligned} \mathbf{J} \cdot \nabla p &= 0 \\ \mathbf{B} \cdot \nabla p &= 0 \end{aligned} \tag{7.23}$$

$\mathbf{J}$ and $\mathbf{B}$ are perpendicular to ∇p and they must lie on surfaces of constant pressure. Constant-pressure surfaces are called isobaric surfaces. Regardless of the MHD geometry, which can be complicated, $\mathbf{J}$ and $\mathbf{B}$ are parallel to the isobaric surfaces and cannot cross these constant-p surfaces. In this sense, the isobaric surfaces are also magnetic and current surfaces. They are magnetic surfaces because no magnetic lines cross the isobaric surface. Note that although the isobaric surfaces are magnetic surfaces, it does not mean that the magnetic pressure $B^2/2\mu_0$ is constant on an isobaric surface.

∇p is associated with both the gradient and curvature of the magnetic field and hence the diamagnetic current has contributions from both gradient and curvature drift currents of the particles (Chapter 4). To see this, expand the right-hand side of (7.22), using one of Maxwell's equations,

$$\begin{aligned} \nabla p &= \mathbf{J} \times \mathbf{B} \\ &= (\nabla \times \mathbf{B}) \times \frac{\mathbf{B}}{\mu_0} \\ &= -\nabla \frac{B^2}{2\mu_0} + \frac{(\mathbf{B} \cdot \nabla)\mathbf{B}}{\mu_0} \end{aligned} \tag{7.24}$$

where the first term on the right represents the gradient of the magnetic pressure and the second term is the stress (tension) along the direction of $\mathbf{B}$. When the lines of force are curved, the second term includes the centrifugal force. Equation (7.24) is often written as

$$\nabla \left(p + \frac{B^2}{2\mu_0} \right) = (\mathbf{B} \cdot \nabla) \frac{\mathbf{B}}{\mu_0} \tag{7.25}$$

which emphasizes that the gradient of the total pressure (fluid plus magnetic) is balanced by the tension along $\mathbf{B}$. If $\mathbf{n}$ is a unit vector normal to the isobaric surface, then

$$\mathbf{n} \cdot \nabla p = \mathbf{n} \cdot \left[-\nabla \frac{B^2}{2\mu_0} + (\mathbf{B} \cdot \nabla) \frac{\mathbf{B}}{\mu_0} \right] = 0 \tag{7.26}$$

and we see that on an isobaric surface,

$$\nabla \frac{B^2}{2\mu_0} = (\mathbf{B} \cdot \nabla) \frac{\mathbf{B}}{\mu_0} \tag{7.27}$$

Hence the magnetic stresses are balanced on the isobaric surfaces. For the special case when the magnetic lines of force are straight, for example $\mathbf{B} = B\hat{\mathbf{z}}$, then from the requirement that $\nabla \cdot \mathbf{B} = 0$, $B^2/2\mu_0 =$ constant everywhere on the isobaric surfaces. Equation (7.25) then indicates that the total pressure, $p + B^2/2\mu_0$, is constant. However, this is not a general statement, since the lines of force may be curved.

7.3.2 Relation Between Fluids and Particles

Insight as to how diamagnetism arises in plasmas can be obtained by studying the motions of the individual particles. Since plasma fluids are made up of many single particles, one can ask how individual particle motions add up to yield the same current derived from the fluid equations.

We showed in Chapter 4 that a charged particle in a magnetic field executes gyration motion and, in inhomogeneous magnetic fields, it also drifts. Let there be many particles in a volume each having a dimension of the order of one Larmor radius. Each gyrating orbit produces a current loop i whose magnetic moment is

$$\mu = imr_c^2 \tag{7.28}$$

If the phases of these particles in the orbits are random, the average current produced by the particles acting collectively can be computed.

Consider an element of area S bounded by a closed curve C (Figure 7.2). We are interested in computing the net current crossing S due to the orbits that penetrate it. Orbits that penetrate the area twice do not produce a net current. Orbits that encircle the bounding curve C penetrate the area only once and they produce a net current. Let $d\mathbf{l}$ be the element of length along C and if there are N current loops per unit volume, each carrying a current i with an area $A = \pi r_c^2$, the number of current loops

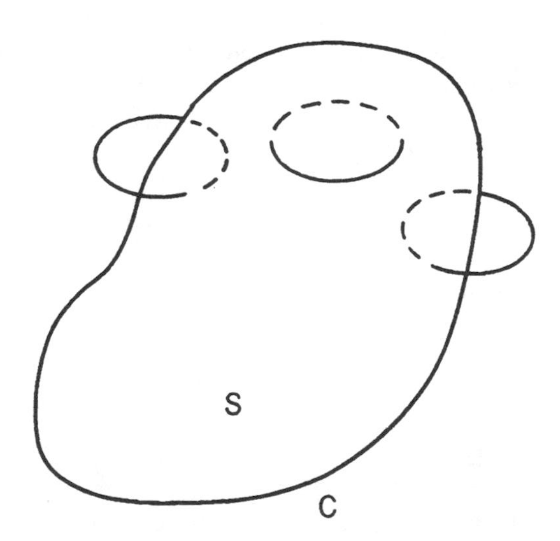

FIGURE 7.2 Only orbits that encircle the bounding surface S contribute to the current.

penetrated by $d\mathbf{l}$ is $N\mathbf{A}\cdot d\mathbf{l}$. The total net current penetrating S is

$$\begin{aligned} I &= \oint Ni\mathbf{A}\cdot d\mathbf{l} \\ &= \oint \mathbf{M}\cdot d\mathbf{l} \\ &= \int (\nabla\times\mathbf{M})\cdot d\mathbf{S} \\ &= \int \mathbf{J}_M\cdot d\mathbf{S} \end{aligned} \tag{7.29}$$

where $\mathbf{M}$ is the magnetization (magnetic moment/volume), $d\mathbf{S}$ denotes an area element on S, $\mathbf{J}_M$ is the average magnetization current density, and we have used Stoke's theorem.

Consider now a straight magnetic field with a spatial gradient, $\mathbf{B}(x,y)\hat{\mathbf{z}}$. In this field geometry, particles execute the usual gyration motion but they also gradient drift. The magnetic moment/volume is, from (4.19)

$$\mathbf{M} = -\frac{N\varepsilon_\perp}{B^2}\mathbf{B} \tag{7.30}$$

Hence the magnetization current density is

$$\begin{aligned} \mathbf{J}_M &= \nabla\times\mathbf{M} \\ &= -\nabla\times\left(\frac{N\varepsilon_\perp\mathbf{B}}{B^2}\right) \end{aligned} \tag{7.31}$$

where $\varepsilon_\perp = mv_\perp^2/2$, and $v_\perp$ is the magnitude of the velocity in the direction perpendicular to $\mathbf{B}$. In writing this equation, we have assumed that the individual particles act together to produce a resultant magnetic moment.

The gradient drift of a particle in an inhomogeneous magnetic field gives rise to the gradient drift current, which is from (4.84)

$$\mathbf{J}_{\nabla B} = N\varepsilon_\perp\frac{\mathbf{B}\times\nabla B}{B^3} \tag{7.32}$$

The total current of a particle in an inhomogeneous magnetic field is

$$\begin{aligned} \mathbf{J} &= \mathbf{J}_M + \mathbf{J}_{\nabla B} \\ &= -\nabla\times\left(\frac{N\varepsilon_\perp\mathbf{B}}{B^2}\right) + N\varepsilon_\perp\frac{\mathbf{B}\times\nabla B}{B^3} \\ &= \frac{\mathbf{B}}{B^2}\times\nabla N\varepsilon_\perp - N\varepsilon_\perp\frac{\mathbf{B}\times\nabla B}{B^3} + N\varepsilon_\perp\frac{\mathbf{B}\times\nabla B}{B^3} \\ &= \frac{\mathbf{B}}{B^2}\times\nabla N\varepsilon_\perp \end{aligned} \tag{7.33}$$

It is worthy of note that contribution to the magnetization current comes from gradients of the total energy density and the inhomogeneous magnetic field. The gradient term is exactly equal and opposite to the current produced by the gradient drift of the guiding center. Hence they cancel and the total currents are given by (7.33).

When many particles act together, the resultant magnetic moment may be large enough to alter the applied magnetic field. To tie the particle results back to the diamagnetic current derived from the MHD fluid equation, note that the two expressions (7.21) and (7.33) are equal if

$$p = N\varepsilon_\perp \tag{7.34}$$

This relation sheds light on how individual particles acting together behave as a fluid. If an ensemble of particles is defined in terms of a distribution function, (7.34) is precisely the expression that one would obtain for the momentum transfer tensor were the tensor evaluated for a plasma distribution with $\varepsilon_\perp$ as its average perpendicular energy. In an isotropic plasma we can use a Maxwellian distribution function, which yields

$$\begin{aligned} p &= \int \frac{mv^2}{2} f(r, v, t) d^3v \\ &= N \left\langle \frac{mv^2}{2} \right\rangle \\ &= N\varepsilon \end{aligned} \tag{7.35}$$

where $\varepsilon = \langle mv^2/2 \rangle$ (more on this below).

A picture of how a diamagnetic current rises can be seen from the following considerations. Suppose a fluid obeys the ideal gas law, $p = nkT$, where k is the Boltzmann constant and T is the temperature. If an assumption is made that T is constant,

$$\frac{\nabla p}{p} = \frac{\nabla n}{n} \tag{7.36}$$

Now suppose $\mathbf{B} = B\hat{\mathbf{z}}$ where $\hat{\mathbf{z}}$ is a unit vector along $\mathbf{B}$. Then the diamagnetic drift velocity is

$$\begin{aligned} \mathbf{U}_\perp &= \frac{\mathbf{B} \times \nabla p}{nqB^2} \\ &= \frac{kT}{qB} \left(\hat{\mathbf{z}} \times \frac{\nabla n}{n} \right) \end{aligned} \tag{7.37}$$

For the purpose of illustration, let ∇n be in the $+y$ direction (Figure 7.3). The dots represent the density of points and the loops represent the individual gyration orbits (we show only ion orbits). Because of the density gradient in the $+y$ direction, there will be more ions moving in the $-x$ direction than in the $+x$ direction. It is this differential gyromotion that gives rise to the diamagnetic currents. The guiding centers of the individual par-

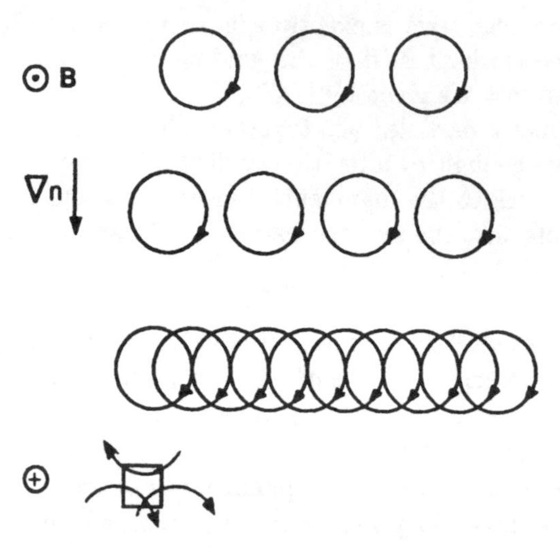

FIGURE 7.3 A schematic diagram showing how a diamagnetic current is produced.

ticles themselves need not be moving physically. The diamagnetic current exists as long as there is a pressure gradient regardless of the shape of the magnetic field. This current exists even if the magnetic field is straight. The diamagnetic drift does not depend on the mass of the particle, and electrons and ions of the same temperature and density gradient contribute equally to the diamagnetic current. The diamagnetic drift and the diamagnetic current are peculiar to conducting fluids.

7.3.3 Chapman-Ferraro Current at the Magnetopause

The above results can now be applied to the magnetopause to help us understand how a boundary is produced. Recall that planetary magnetic fields are constrained to exist inside a magnetic cavity (a closed magnetospheric model) by the continuous flow of the solar wind. The fluid pressure across a magnetopause is therefore discontinuous, and magnetopause boundaries must support diamagnetic currents.

Chapman and Ferraro's model considers the planetary magnetic field to butt against a sheet of solar particles (which we can consider as the solar wind) and that the boundary is a plane surface. Let the planetary magnetic field at the interface be straight and the boundary be stationary. There are no particles in the planetary magnetic field and there is no magnetic field in the solar wind medium. A schematic diagram of this boundary on the equatorial plane is shown in Figure 7.4.

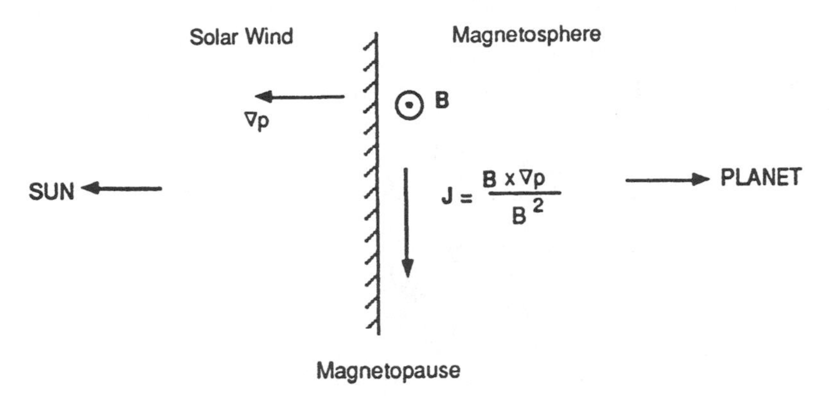

FIGURE 7.4 A simple diagram showing how a Chapman-Ferraro current, a diamagnetic current is produced at the interface between the solar wind and a magnetopause.

The balance of forces requires that at the boundary

$$\nabla p_{sw} = \mathbf{J}_{mp} \times \mathbf{B}_p \tag{7.38}$$

where p_{sw} is the solar wind pressure, $\mathbf{J}_{mp}$ is the magnetopause current, and $\mathbf{B}_p$ is the planetary magnetic field at the boundary. The magnetopause current is

$$\mathbf{J}_{mp} = \mathbf{B}_p \times \frac{\nabla p_{sw}}{B^2} \tag{7.39}$$

and since on the equator the magnetic field is pointing outward (for Earth), $\mathbf{J}_{mp}$ runs from dawn to dusk as shown.

The only particles that exist in this simple model are solar wind particles. Hence $\mathbf{J}_{mp}$ must be created by the solar wind. While the fluid theory does not give a clear picture of how $\mathbf{J}_{mp}$ is created, the single-particle physics shows that the current must be created by the temporary penetration of the solar wind particles (Figure 7.5). These particles must return back into the solar wind because of the $\mathbf{v} \times \mathbf{B}$ force. However, the partial penetration of these particles is sufficient to create a current in a way similar to the production of the magnetization current as discussed above. Hence the magnetopause current is a diamagnetic current. Note that this configuration is only possible if the β of the plasma is exactly equal to unity ($\beta = 1$).

To model real magnetopauses, the model must deal with the actual magnetic field geometry and must also include the magnetic field in the solar wind and particles in the planetary field medium. Not only can the problem become quite complicated because of the curved magnetic geometry, but there are also several populations of particles in magnetospheres

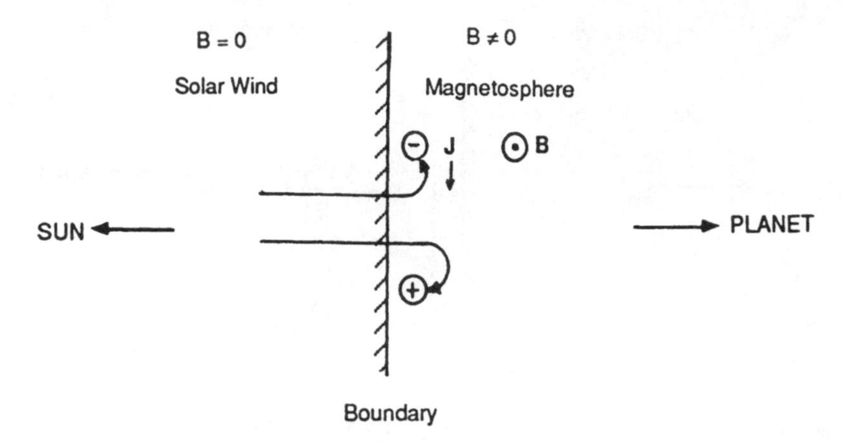

FIGURE 7.5 A schematic showing charged particle penetration into a magnetic field. Note that the particles return to the field-free region.

that must be considered. The Chapman-Ferraro model is a closed model and may only be applicable under very quiet solar wind conditions when the IMF is in the northward direction (see Chapter 3). The question of what currents run at the boundary and what kind of boundary one should consider for realistic magnetospheres is not completely answered. This is central to the problem of how the solar wind interacts with magnetospheres (further discussion of the magnetopause boundary is given in Chapter 8).

7.4 Ring Current in Magnetospheres

The above results are based on the assumption that the fluid pressure is a scalar. This result is valid only if there is equipartition of energy in all directions. In general, this assumption is incorrect, especially if the density is low and the plasma is magnetized and collisionless, as is the case in space. Although in some situations equipartition of energy may be assumed for motions parallel to $\mathbf{B}$, this cannot be assumed for the particle motion in the perpendicular direction, since in a tenuous plasma the particles gyrate around a magnetic line of force many times between collisions. Hence the perpendicular and parallel pressures are expected to be different.

To study how ring currents are produced in magnetospheres, we must treat the pressure as a tensor. It was mentioned earlier that when the fluid pressure is anisotropic, the fluid motion is described by (5.147). For conducting fluids, free charges vanish and the equation of motion for electron or ion fluids is given by (7.2):

$$mn\frac{d\mathbf{U}}{dt} = -\nabla \cdot \mathbf{p} + \mathbf{J} \times \mathbf{B} \tag{7.2}$$

This equation is identical in form to the one-fluid MHD equation (7.18) except that the pressure is now a tensor and the force involves the divergence of this tensor (instead of the gradient of a scalar pressure).

7.4.1 Pressure Tensor

The general formula for the pressure tensor was defined as (see Chapter 5)

$$p_{ik} = \pi_{ik} - U_i U_k \sum_\alpha m^\alpha n^\alpha \tag{7.40}$$

where π_{ik} is the momentum transfer tensor defined by

$$\pi_{ik} = \sum_\alpha m^\alpha \int v_i v_k f^\alpha(\mathbf{r}, \mathbf{v}, t) d^3 v \tag{7.41}$$

and U_i and U_k are the ith and k^{th} components of the center of mass velocity $\mathbf{U}$ given by

$$\mathbf{U} = \frac{\sum_\alpha m^\alpha \int \mathbf{v} f^\alpha(\mathbf{r}, \mathbf{v}, t) d^3 v}{\sum_\alpha m^\alpha n^\alpha} \tag{7.42}$$

$\mathbf{p}$ in general has nine components,

$$\mathbf{p} = \begin{pmatrix} p_{xx} & p_{xy} & p_{xz} \\ p_{yx} & p_{yy} & p_{yz} \\ p_{zx} & p_{zy} & p_{zz} \end{pmatrix} \tag{7.43}$$

$\mathbf{p}$ is, however, symmetric, $p_{ik} = p_{ki}$, and the number of variables is reduced to six. The computation of p_{ik} depends explicitly on the form of the distribution function.

$\mathbf{p}$ can be simplified under certain conditions. Consider the off-diagonal terms of $\mathbf{p}$, for example of a particular species α. In the fluid frame of reference

$$p_{xy} = m \int v_x v_y f(\mathbf{r}, \mathbf{v}, t) dv_x dv_y dv_z \tag{7.44}$$

which involves the average $< v_x v_y >$. In order to get an idea what these off-diagonal terms mean, consider a geometry where $\mathbf{B}$ is along the z-direction. Then, from the solutions of the Lorentz equation, we can write $v_x = v_\perp \cos(\omega_c t + \alpha)$ and $v_y = v_\perp \sin(\omega_c t + \beta)$ where ω_c is the gyrofrequency and α and β are the phases of the particles, which depend on the initial conditions. $< v_x v_y >$ therefore depends on the phases of the particles averaged over the distribution function.

A distribution function is defined as gyrotropic when $<v_x v_y>= 0$ and non-gyrotropic when $<v_x v_y>\neq 0$. The phases of the particles are random and not correlated in gyrotropic distributions, whereas the phases of the particles are correlated in non-gyrotropic distributions. As an example, consider a distribution of the particles that is isotropic in velocity, that is,

$$f(\mathbf{r}, \mathbf{v}, t) = f(\mathbf{r}, v^2, t) \tag{7.45}$$

Equation (7.45) is a gyrotropic distribution function, since with this type of distribution $<v_x v_y>= 0$. Hence the pressure tensor of the fluid with the type of distribution (7.45) has only diagonal elements.

Define a coordinate system in which **B** is along the **z**-axis and let $\mathbf{v}_\perp$ and $\mathbf{v}_\parallel$ represent the motions of the particles perpendicular and parallel to **B**. In this frame of reference, we can write (7.43) for a gyrotropic distribution as

$$p = \begin{pmatrix} p_\perp & 0 & 0 \\ 0 & p_\perp & 0 \\ 0 & 0 & p_\parallel \end{pmatrix} \tag{7.46}$$

since the directions x and y are perpendicular to **B**, which is in the z-direction. The scalar pressures $p_\perp$ and $p_\parallel$ are obtained from

$$p_\perp = m \int v_\perp^2 f(\mathbf{r}, v^2, t) d^3v \tag{7.47}$$

$$p_\parallel = m \int v_\parallel^2 f(\mathbf{r}, v^2, t) d^3v \tag{7.48}$$

Here $p_\perp$ and $p_\parallel$ are the average kinetic energy per unit volume of a distribution of particles in the directions perpendicular and parallel to the magnetic field **B**. Equation (7.46) can be written as a combination of two tensors

$$\mathbf{p} = \begin{pmatrix} p_\perp & 0 & 0 \\ 0 & p_\perp & 0 \\ 0 & 0 & p_\perp \end{pmatrix} + \begin{pmatrix} 0 & 0 & 0 \\ 0 & 0 & 0 \\ 0 & 0 & p_\parallel - p_\perp \end{pmatrix} \tag{7.49}$$

Thus, **p** has the form

$$\mathbf{p} = p_\perp \mathbf{I} + (p_\parallel - p_\perp)\mathbf{bb} \tag{7.50}$$

where **I** is the unit tensor (the Kronecker tensor) and **bb** is a second-order tensor (a dyadic) formed from the unit vectors **b** along the magnetic field direction.

If the magnetic field is curved, the orientation of **B** changes and this change must be taken into account. Writing (7.50) in tensor notation

$$p_{ij} = p_\perp I_{ij} + (p_\parallel - p_\perp) b_i b_j \tag{7.51}$$

we find that in an arbitrary frame of reference, the pressure takes the form

$$\mathbf{p} = \begin{pmatrix} p_\perp + (p_\parallel - p_\perp)b_x b_x & (p_\parallel - p_\perp)b_x b_y & (p_\parallel - p_\perp)b_x b_y \\ (p_\parallel - p_\perp)b_y b_x & p_\perp + (p_\parallel - p_\perp)b_y b_y & (p_\parallel - p_\perp)b_y b_z \\ (p_\parallel - p_\perp)b_z b_x & (p_\parallel - p_\perp)b_x b_y & p_\perp + (p_\parallel - p_\perp)b_z b_z \end{pmatrix} \tag{7.52}$$

The expressions in (7.52) are complex, since each term requires information on how the unit vector **b** changes from point to point in space.

7.4.2 Divergence of the Pressure Tensor

Consider **p** as given by (7.50). We now compute the divergence of **p** as required by the equation of motion (7.2). First, note that the divergence of the unit tensor **I** vanishes because **I** is invariant under a transformation of axes and has the same form in any Cartesian coordinate system. Hence

$$\begin{aligned} \nabla \cdot \mathbf{p} &= \nabla \cdot p_\perp \mathbf{I} + \nabla \cdot (p_\parallel - p_\perp)\mathbf{bb} \\ &= \nabla p_\perp + (p_\parallel - p_\perp)\nabla \cdot \mathbf{bb} + \mathbf{b}(\mathbf{b} \cdot \nabla)(p_\parallel - p_\perp) \\ &= \nabla p_\perp + (p_\parallel - p_\perp)[(\mathbf{b} \cdot \nabla)\mathbf{b} + \mathbf{b}(\nabla \cdot \mathbf{b})] \\ &\quad + \mathbf{b}(\mathbf{b} \cdot \nabla)(p_\parallel - p_\perp) \end{aligned} \tag{7.53}$$

Define now $\nabla = \nabla_\parallel - \nabla_\perp$. Noting that $\nabla_\parallel = \mathbf{b}(\mathbf{b} \cdot \nabla)$ and that the left side of (7.53) can then be written as $\nabla \cdot \mathbf{p} = \nabla_\parallel \cdot \mathbf{p} + \nabla_\perp \cdot \mathbf{p}$, (7.53) can be separated into

$$\nabla_\parallel \cdot \mathbf{p} = \nabla_\parallel p_\parallel + (p_\parallel - p_\perp)\mathbf{b}(\nabla \cdot \mathbf{b}) \tag{7.54}$$

and

$$\nabla_\perp \cdot \mathbf{p} = \nabla_\perp p_\perp + (p_\parallel - p_\perp)(\mathbf{b} \cdot \nabla)\mathbf{b} \tag{7.55}$$

It is important to note that while $\nabla \cdot \mathbf{B} = 0$, $\nabla \cdot \mathbf{b} \neq 0$.

Substitution of (7.54) and (7.55) into (7.2) splits the momentum equation into two equations, one parallel and the other perpendicular to **B**,

$$\begin{aligned} mn\frac{d\mathbf{U}_\parallel}{dt} &= -\nabla_\parallel p_\parallel - (p_\parallel - p_\perp)(\nabla \cdot \mathbf{b})\mathbf{b} \\ mn\frac{d\mathbf{U}_\perp}{dt} &= -\nabla_\perp p_\perp - (p_\parallel - p_\perp)(\mathbf{b} \cdot \nabla)\mathbf{b} + \mathbf{J} \times \mathbf{B} \end{aligned} \tag{7.56}$$

An expression of the current density in the direction perpendicular to **B** is obtained by taking the cross-product of the second equation of (7.56) with **B**. We find that

$$\mathbf{J}_\perp = \frac{\mathbf{B}}{B^2} \times \nabla_\perp p_\perp + \frac{(p_\parallel - p_\perp)}{B^2}[\mathbf{B} \times (\mathbf{b} \cdot \nabla)\mathbf{b}] + mn\frac{\mathbf{B}}{B^2} \times \frac{d\mathbf{U}_\perp}{dt} \tag{7.57}$$

where $\mathbf{B} = B\mathbf{b}$. Equation (7.57) replaces the expression of the current density (7.20) when the pressure is anisotropic. Note that if $p_\parallel = p_\perp$, (7.20) is recovered. The successive terms in (7.57) come from the gyromotion (first term), curvature and gradient drifts (second term), and inertial effects (last term). This expression for the current density is quite general and must satisfy the Maxwell equation $\nabla \times \mathbf{B} = \mu_0 \mathbf{J}$.

7.4.3 Relation Between Anisotropic Fluids and Particles

To continue discussion of the relationship between particles and fluids, we now derive the expression of the current density (7.57) directly from the first-order orbit theory. We showed in Chapter 4 that the drift velocity of a single particle in an inhomogeneous, static, and time-independent magnetic field is

$$\mathbf{W}_D = \frac{mv_\parallel^2}{qB^4}[\mathbf{B} \times (\mathbf{B} \cdot \nabla)\mathbf{B}] + \frac{mv_\perp^2}{2qB^3}\mathbf{B} \times \nabla B \tag{7.58}$$

where the first term is the curvature drift (4.77) and the second the gradient drift (4.84). Since these drifts are charge-dependent, they give rise to currents.

Consider now a particle system in which there are only electrons and one ion species. Let n^- and n^+ denote their number densities. Then the expression for the drift current is

$$\mathbf{J}_D = n^+ q^+ \mathbf{W}_D^+ + n^- q^- \mathbf{W}_D^- \tag{7.59}$$

We have already seen for the case of isotropic pressure that the particles act together to produce particle pressure. This concept can be generalized to the case of the anisotropic pressure. We construct pressures in the directions parallel and perpendicular to the direction of the magnetic field by using the single-particle parameters,

$$\begin{aligned} p_\parallel &= n^+ m^+ v_\parallel^{+2} + n^- m^- (v_\parallel^-)^2 \\ 2p_\perp &= n^+ m^+ v_\perp^{+2} + n^- m^- (v_\perp^-)^2 \end{aligned} \tag{7.60}$$

where contributions from both species of particles have been included. The factor 2 appears in $p_\perp$ and not in $p_\parallel$ because there are two components of motion in the xy direction ($v_\perp^2 = v_x^2 + v_y^2$). Also, the velocities in (7.60) should be looked at as velocities averaged over the ensemble of these particles making up a particle system in the manner discussed in Chapter 2. In terms of these pressures, (7.59) can be written as

$$\mathbf{J}_D = p_\parallel \frac{\mathbf{B} \times (\mathbf{B} \cdot \nabla)\mathbf{B}}{B^4} + p_\perp \frac{\mathbf{B} \times \nabla B}{B^3} \tag{7.61}$$

This drift current must now be added to the magnetization current to obtain the expression for the total current. The magnetization current is

$$\begin{aligned}\mathbf{J}_M &= \nabla \times (n^+\mu^+ + n^-\mu^-) \\ &= -\nabla \times p_\perp \left(\frac{\mathbf{B}}{B^2}\right) \\ &= \frac{\mathbf{B}}{B^2} \times \nabla p_\perp - 2p_\perp \frac{\mathbf{B} \times \nabla B}{B^3} - p_\perp \frac{\nabla \times \mathbf{B}}{B^2} \end{aligned} \tag{7.62}$$

where μ^+ and μ^- are the magnetic moments of the individual ions and electrons. In going from the first to the second line, we used the relation that the magnetic moment/volume $M = M^+ + M^-$ and (7.60).

Add (7.61) and (7.62) and obtain the total current perpendicular to $\mathbf{B}$, as

$$\begin{aligned}\mathbf{J}_\perp &= \mathbf{J}_D + \mathbf{J}_M \\ &= p_\parallel \frac{\mathbf{B} \times (\mathbf{B} \cdot \nabla)\mathbf{B}}{B^4} + \frac{\mathbf{B}}{B^2} \times \nabla p_\perp \\ &\quad - p_\perp \frac{\mathbf{B} \times \nabla B}{B^3} - p_\perp \frac{\nabla \times \mathbf{B}}{B^2} \end{aligned} \tag{7.63}$$

Note that as in the case of the isotropic pressure, a portion of $\mathbf{J}_M$ cancels a portion of $\mathbf{J}_D$. Now use the vector identities

$$\begin{aligned}\frac{\mathbf{B}}{B} \times \left(\frac{\mathbf{B}}{B} \cdot \nabla\right) \frac{\mathbf{B}}{B} &= \frac{\mathbf{B}}{B} \times \frac{(\mathbf{B} \cdot \nabla)\mathbf{B}}{B^2} \\ \frac{\mathbf{B}}{B^2} \times \frac{(\mathbf{B} \cdot \nabla)\mathbf{B}}{B^2} &= \frac{\mathbf{B} \times \nabla B}{B^3} + \frac{\nabla \times \mathbf{B}}{B^2} \end{aligned} \tag{7.64}$$

in (7.63) and arrive at

$$\mathbf{J}_\perp = \frac{\mathbf{B}}{B^2} \times \nabla p_\perp + (p_\parallel - p_\perp) \frac{\mathbf{B} \times (\mathbf{B} \cdot \nabla)\mathbf{B}}{B^4} \tag{7.65}$$

which is identical to (7.57) except for the inertial term (the model used to calculate (7.65) is static and time-independent). The fluid model agrees with the single-particle model to first order. The reader should note however, that the single-particle calculation assumes that $\mathbf{B}$ is given whereas this assumption is not required for the fluid calculation.

For a more general situation in which there are forces other than the static magnetic field, (7.65) must be modified. For example, if we include electric field and gravitational drifts and allow the electric field to be time-dependent, (7.65) is modified to

$$\mathbf{J}_\perp = \frac{\mathbf{B}}{B^2} \times \left[\nabla p_\perp + (p_\parallel - p_\perp) \frac{(\mathbf{B} \cdot \nabla)\mathbf{B}}{B^2} - \rho_m \frac{d}{dt}(\mathbf{W}_D - \mathbf{g})\right] \tag{7.66}$$

The new terms are current contributions from inertial and gravitational drifts. The inertial term is related to the time-dependent electric field and (as stated earlier) is equal to the polarization current. The static electric field contributes another term, $\rho_c(\mathbf{E}\times\mathbf{B}/B^2)$, where ρ_c is the charge density. This term is defined as a convective current but it can be ignored since free charge density vanishes in a neutral plasma.

7.4.4 Ring Current and Magnetic Storms

We now apply the above results to the problem of ring current in the Earth's magnetosphere. Disturbances in Earth's magnetic field were discovered around 1722 in the slight changes of compass direction in London. Systematic studies of Earth's magnetic disturbance did not begin until around 1847 when K.F. Gauss established the first magnetic observatory. A magnetic storm is accompanied by a worldwide decrease of the horizontal component of the geomagnetic field (Figure 7.6). A magnetic storm usually starts with a sudden commencement and includes the initial phase, the main phase, and the recovery phase.

The sudden commencement is very rapid and the increase of the field intensity is accomplished in a few minutes. The field intensity remains increased during the initial phase and this phase lasts typically from tens of minutes to hours. The main phase is accompanied by a decrease in the

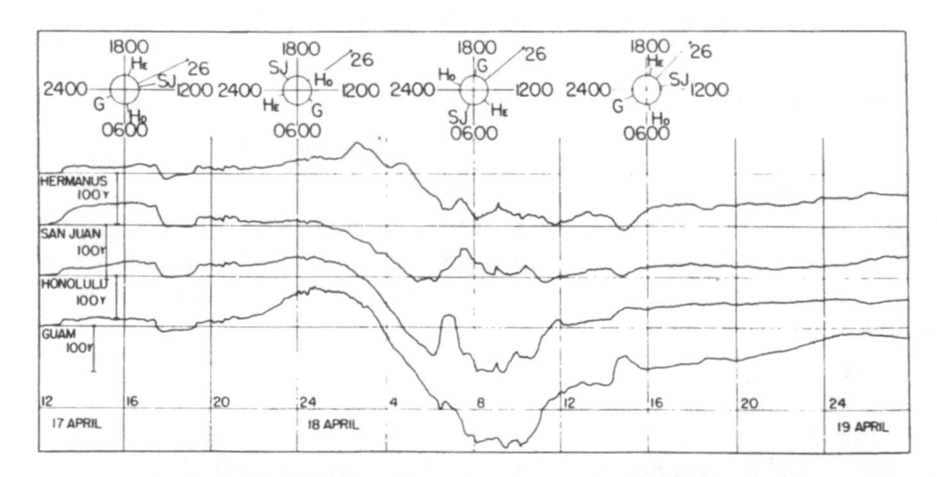

FIGURE 7.6 Horizontal component of Earth's magnetogram for a magnetic storm plotted as a function of Universal Time. Records from four low-latitude stations are shown. Top insets show the local times of the stations at different Universal Times. The onset of the sudden commencement occurred around 13 UT, 17 April. The main phase began around 00 UT, 18 April. (From Cahill, 1968)

geomagnetic field and this phase lasts typically 6 to 48 hours. The decrease in the magnetic field is greatest at the geomagnetic equator and less at higher latitudes. The recovery of the field to the normal undisturbed value can take several days.

The sudden commencement and the initial phases can be explained in terms of compression of the planetary magnetic field by increased solar wind pressure, and the model of Chapman-Ferraro is applicable. However, the main phase requires an alternative mechanism. A global ring current above the equatorial ionosphere flowing in the westward direction is needed to account for the behavior of the main phase (a westward current decreases the magnetic field inside the ring). The recovery phase corresponds to the dissipation of this ring current.

The ring current is now known to be formed by a large number of trapped energetic particles drifting in the magnetosphere. Ring current particles were identified by satellite-borne particle detectors in 1967. Figure 7.7 shows an example in which the intensity of trapped particles in Earth's magnetosphere built up and decayed away during a magnetic storm. Note that the time of maximum particle intensity coincided with the largest decrease of the surface geomagnetic field. Ring currents on Earth can carry several million amperes of current during a magnetic storm.

The ring current problem can be studied quantitatively by evaluating the Biot-Savart's equation, which relates the current and the magnetic field. Ignoring the time dependence (see (3.3)),

$$\mathbf{B}(\mathbf{r}) = \frac{\mu_0}{4\pi} \int \frac{\mathbf{J}(\mathbf{r}') \times (\mathbf{r} - \mathbf{r}')}{|\mathbf{r} - \mathbf{r}'|^3} d^3r' \tag{7.67}$$

where the magnetic field is evaluated at $\mathbf{r}$ due to the presence of a current at $\mathbf{r}'$ (for example, $\mathbf{r}$ could be the position of a magnetometer on the surface of a planet and $\mathbf{r}'$ is in the magnetosphere). To evaluate this integral equation requires knowledge about $\mathbf{J}$, which, for a plasma, comes from magnetization and drift currents.

A simpler problem to solve is to evaluate the magnetic field at the center of a planet rather than at the surface. For a circular loop, the Biot-Savart equation yields

$$\mathbf{B} = \frac{\mu_o I}{2r} \hat{\mathbf{z}} \tag{7.68}$$

for the magnetic field at the origin (see any standard textbook on electricity and magnetism). The current I can be computed from the first-order particle orbit theory or from the equivalent fluid theory.

We start from the drift velocity equation of a charged particle trapped in a magnetic field (see (4.78) and (4.84)),

$$\mathbf{W}_D = \frac{mv_\parallel^2}{qB^4}[\mathbf{B} \times (\mathbf{B} \cdot \nabla)\mathbf{B}] + \frac{mv_\perp^2}{2qB^3}\mathbf{B} \times \nabla B \tag{7.69}$$

Now substitute $v_\parallel = v\cos\alpha$, $v_\perp = v\sin\alpha$ where α is the pitch-angle of the particle, and $(\mathbf{B} \cdot \nabla)\mathbf{B} = \nabla B^2/2 = B\nabla B$ since $\nabla \times \mathbf{B} = 0$, and rewrite the above equation as

$$\mathbf{W}_D = \frac{mv^2}{2qB^3}(1 + \cos^2\alpha)\mathbf{B} \times \nabla B \tag{7.70}$$

To compute the $\mathbf{B} \times \nabla B$ term of a dipole field on the equator, first note that $B_r = -2\mu_0 M \cos\theta/4\pi r^3$ vanishes (see (3.26)) and the magnetic field has only a θ-component,

$$B_\theta = B_s \left(\frac{R_p}{r}\right)^3 \tag{7.71}$$

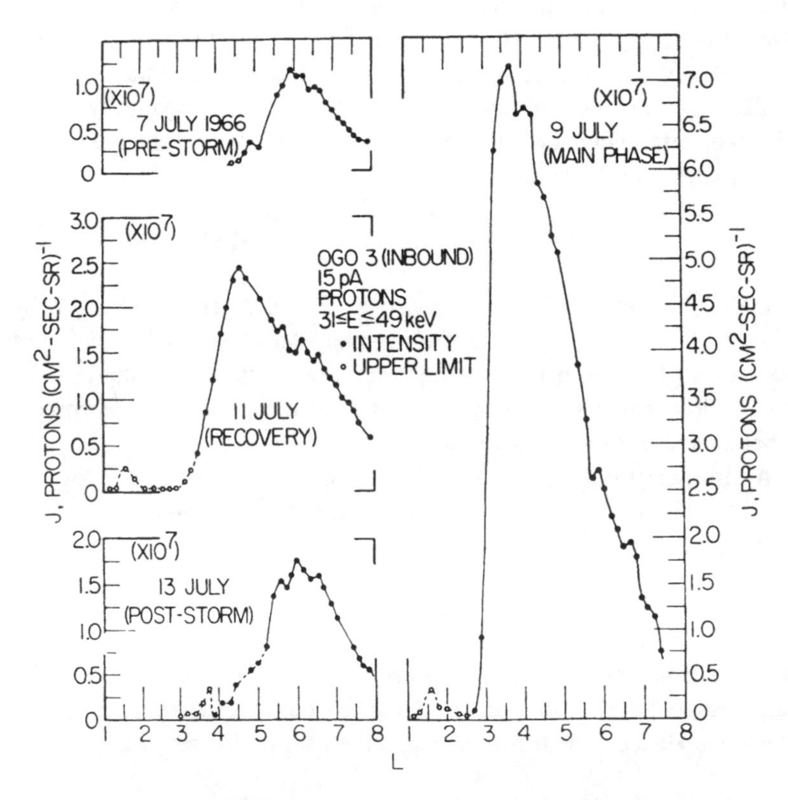

FIGURE 7.7 Variation of particle fluxes in Earth's magnetosphere during a magnetic storm. (From Frank, 1967)

where $B_s = \mu_0 M/4\pi R_p^3$ is the field at the surface of a dipole planet of radius R_p and M is the magnetic moment. For a dipole on the equator, one finds

$$\mathbf{B} \times \nabla B = -\left(\frac{3B^2}{r}\right)\hat{\boldsymbol{\phi}} \tag{7.72}$$

Use this result in the drift equation (7.70) and, for simplicity, consider only particles with $\alpha = 90°$ (this applies only to a ring current confined to the equator). This yields

$$\mathbf{W}_D = -\frac{3mv^2}{2qBr}\hat{\boldsymbol{\phi}} \tag{7.73}$$

for the drift velocity on a dipole equator. Here $\hat{\boldsymbol{\phi}}$ is positive in the eastward direction, and positive and negative particles drift in the westward and eastward directions, creating a ring current flowing in the westward direction.

The current density is

$$\begin{aligned}\mathbf{J} &= nq\mathbf{W}_D \\ &= -\frac{3nmv^2}{2Br}\hat{\boldsymbol{\phi}} \\ &= -\frac{3\varepsilon}{Br}\hat{\boldsymbol{\phi}}\end{aligned} \tag{7.74}$$

where n is the number density and $\varepsilon = nmv^2/2$ is the kinetic energy density of the drifting particles. The relationship between the current density $\mathbf{J}$ and the total current I is

$$Id\mathbf{l} = \mathbf{J}dV \tag{7.75}$$

Define $E = \int \varepsilon dV$ as the total energy of the particles and, noting that $\int d\mathbf{l} = 2\pi r\hat{\boldsymbol{\phi}}$, we can write the total current as

$$I = -\frac{3E}{2\pi r^2 B} \tag{7.76}$$

Substitution of this result in (7.68) leads to

$$\mathbf{B}_D = -\frac{3\mu_0 E}{4\pi r^3 B}\hat{\mathbf{z}} \tag{7.77}$$

as the magnetic field perturbation at the origin due to the drifting particles. The $(-)$ sign here denotes that the perturbation field is in the opposite direction of the planetary magnetic field.

The total perturbation of the planetary field by the ring current must also include the diamagnetic contribution (due to the cyclotron motion). The diamagnetic current can be calculated by solving the Biot-Savart's equation with $\mathbf{J}_\mu = \nabla \times \boldsymbol{\mu}$ where $\boldsymbol{\mu}$ is the total magnetic moment of the particles ($\boldsymbol{\mu}$ is used instead of $\mathbf{M}$ so as not to confuse the magnetic moment of the particles with the magnetic moment of a planet, which is designated as $\mathbf{M}$ here).

This problem can be solved by recognizing that the diamagnetic contribution can be physically modelled by a ring of dipoles of radius r and total magnetic moment $\boldsymbol{\mu}$. Noting that $r \gg r_c$, the cyclotron radius, we can estimate the magnetic field at the center of a dipole ring by

$$\begin{aligned} \mathbf{B}_{diamag} &= -\frac{\mu_0}{4\pi r^3}\boldsymbol{\mu} \\ &= \frac{\mu_0 E}{4\pi r^3 B}\hat{\mathbf{z}} \end{aligned} \tag{7.78}$$

where $\boldsymbol{\mu} = -mnv^2\mathbf{B}/2B^2 = -E\mathbf{B}/B^2 = -(E/B)\hat{\mathbf{z}}$ is the total magnetic moment of the ring current particles on the equator. Note that the individual dipoles are aligned along the planetary magnetic field direction. In contrast to the drift current, whose magnetic field is in the opposite direction of the planetary field (7.77), the magnetic field of a ring of dipoles adds to the main planetary magnetic field.

The perturbation of the planetary magnetic field at the origin due to the ring current is obtained by adding the two contributions (7.77) and (7.78)

$$\begin{aligned} \Delta\mathbf{B}_T &= \mathbf{B}_D + \mathbf{B}_{diamag} = -\frac{2\mu_0 E}{4\pi r^3 B}\hat{\mathbf{z}} \\ &= -\frac{2E}{M}\hat{\mathbf{z}} \end{aligned} \tag{7.79}$$

where use was made of $B = \mu_0 M/4\pi r^3$. This simple result shows that the total magnetic perturbation at the center of a planet is related to the magnetic moment M of the planet and the total particle energy E of the ring current particles.

It is useful to express the perturbation field in terms of the surface field B_s of a planet,

$$\begin{aligned} \frac{\Delta\mathbf{B}_T}{B_s} &= -\frac{2E/M}{\mu_0 M/4\pi R_p^3}\hat{\mathbf{z}} \\ &= \frac{-2E\mu_0}{4\pi B_s^2 R_p^3}\hat{\mathbf{z}} \end{aligned} \tag{7.80}$$

where $M = 4\pi B_s R_p^3/\mu_0$. This equation can be rewritten since the total magnetic field energy of a dipole above the planetary surface is

$$\begin{aligned} E_M &= \int \frac{B^2}{\mu_0} dV \\ &= \frac{4\pi}{\mu_0} \frac{B_s^2 R_p^3}{3} \end{aligned} \tag{7.81}$$

where the integration limits are R_p to ∞. Use of this result in the above equation yields

$$\Delta \mathbf{B}_T = -\frac{2}{3} \frac{E}{E_M} \hat{\mathbf{z}} \tag{7.82}$$

as the total decrease of the magnetic field at the center of a planet due to a ring current.

This equation was derived with many restrictions (for example, the pitch-angle $\alpha = 90°$, β of plasma is small and negligible), but it turns out that this relationship is generally valid. The same relationship is obtained with an arbitrary distribution of particles. This result can be derived from the virial theorem for steady-state plasmas (see Carovillano and Maguire, 1968, cited at the end of the chapter).

Two important points are noted if one were to apply the above results to real observations. The first is that the magnetic field is measured on the surface of a planet rather than at the center. This will result in an error. Another error comes because the above calculations assume a fixed dipole geometry, which is not exactly correct. Intense currents in the magnetosphere will elongate the dipole geometry, making it more parabolic (see Chapter 3). This in turn will change the form of the current. In practice, the calculation must be iterated several times for the theory and observations to agree. Figure 7.8 shows an example of a magnetic storm that occurred on Earth where the magnetic field perturbation was compared to the particle energy of the ring current. The peak of the ring current was located between 3 and $4R_E$. Observation and theory were made to fit fairly well after three iterations.

7.4.5 Double Adiabatic Equations

We conclude this section with a discussion of the closure problem for the anisotropic fluid. There are twelve unknowns in this case, $\mathbf{E}$, $\mathbf{B}$, ρ_m, $\mathbf{U}$, $p_\parallel$, and $p_\perp$. The equations we have thus far are the equation of continuity (which is the same for both isotropic and anisotropic cases), the two equations of motion (7.56), and the assumption that $\mathbf{E}+\mathbf{U}\times\mathbf{B} = 0$. This gives a total of ten equations. Thus, two more equations are needed. As in the case of isotropic pressure (treated in Section 5.7.3), under certain assumptions the energy equation leads to an equation of state. In our case, we need to obtain separate equations of state for the two directions.

We start from the collisionless Boltzmann-Vlasov equation

$$\frac{\partial f}{\partial t} + v_k \frac{\partial f}{\partial x_k} + a_k \frac{\partial f}{\partial v_k} = 0 \tag{7.83}$$

whose meaning has been discussed in Chapter 2. The second moment is obtained by multiplying the above equation with $v_i v_j$ and integrating over the velocity space. Let us also use the definition of the thermal velocity given by $w_i = v_i - u_i$. The first term then becomes

$$\begin{aligned} \int v_i v_j \frac{\partial f}{\partial t} d^3 v &= \frac{\partial}{\partial t} \int (w_i w_j + u_i u_j) f d^3 v \\ &= \frac{\partial}{\partial t} p_{ij} + u_i u_j \frac{\partial \rho_m}{\partial t} \end{aligned} \tag{7.84}$$

where $< w_i > = < w_j > = 0$, $p_{ij} = \int w_i w_j f d^3 v$ is the pressure tensor (see (5.144) and (7.40)), and $\rho_m = \int f d^3 v$ is the mass density.

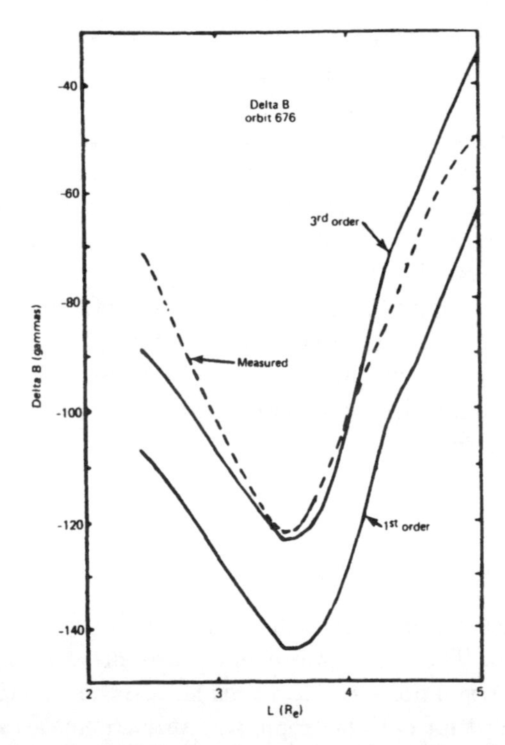

FIGURE 7.8 Comparison of theory and observations for a ring current. (From Berko *et al.*, 1975)

The second term will contain eight terms, but since three terms involve $<w>= 0$, we will be left with

$$\int v_i v_j v_k \frac{\partial f}{\partial x_k} d^3v = \frac{\partial}{\partial x_k} \int (w_i w_j w_k + w_i w_k u_j + w_j w_k u_i + w_i w_j u_k + u_i u_j u_k) f d^3v \tag{7.85}$$

We now use the definitions of the pressure tensor $p_{ij} = \int w_i w_j f d^3v$ and the heat tensor $q_{ijk} = \int w_i w_j w_k f d^3v$ and rewrite the above equation as

$$\begin{aligned} \int v_i v_j v_k \frac{\partial f}{\partial x_k} d^3v &= \frac{\partial q_{ijk}}{\partial x_k} + \frac{\partial}{\partial x_k}(u_j p_{jk} + u_j p_{ik} + \partial u_k p_{ij}) + u_i u_j u_k \frac{\partial \rho_m}{\partial x_k} \\ &= \frac{\partial q_{ijk}}{\partial x_k} + p_{jk}\frac{\partial u_i}{\partial x_k} + p_{ik}\frac{\partial u_j}{\partial x_k} + p_{ij}\frac{\partial u_k}{\partial x_k} + u_i \frac{\partial p_{jk}}{\partial x_k} \\ &\quad + u_j \frac{\partial p_{ik}}{\partial x_k} + u_k \frac{\partial p_{ij}}{\partial x_k} + u_i u_j u_k \frac{\partial \rho_m}{\partial x_k} \end{aligned} \tag{7.86}$$

The average of the last term in (7.83) involves $<v_i v_j a_k>$ and for the electromagnetic force this reduces to $(eB_k/m)(\epsilon_{ink} p_{jn} + \epsilon_{jnk} p_{in})$ where ϵ is the permutation tensor, equal to +1 if the indices are cyclic, −1 if the cyclic order is mixed, and 0 if any indices repeat (the term involving the electric field does not appear because the charge density ρ_c averages out to zero). This term involves the cyclotron frequency eB_k/m and, since MHD deals with frequencies much smaller than the ion cyclotron frequency, we can ignore this term to first order.

Now, combining (7.84) and (7.86) and recognizing that

$$u_i u_j \left(\frac{\partial \rho_m}{\partial t} + u_k \frac{\partial \rho_m}{\partial x_k} \right) = 0 \tag{7.87}$$

$$\frac{dp_{ij}}{dt} = \frac{\partial p_{ij}}{\partial t} + u_i \frac{\partial p_{jk}}{\partial x_k} + u_j \frac{\partial p_{ik}}{\partial x_k} + u_k \frac{\partial p_{ij}}{\partial x_k} \tag{7.88}$$

we obtain

$$\frac{dp_{ij}}{dt} + \frac{\partial q_{ijk}}{\partial x_k} + p_{jk}\frac{\partial u_i}{\partial x_k} + p_{ik}\frac{\partial u_j}{\partial x_k} + p_{ij}\frac{\partial u_k}{\partial x_k} = 0 \tag{7.89}$$

This general equation defines how energy is transported in an anisotropic collisionless MHD fluid.

To learn about the transport of energy parallel and perpendicular to the direction of the magnetic field, (7.89) must be solved for $p_{\parallel}$ and $p_{\perp}$. In general, this cannot be done unless we have more information about the heat transport term, $\partial q_{ijk}/\partial x_k$. Heat transfer for a collisionless plasma in the perpendicular direction can be ignored however, if the magnetic field is strong enough to inhibit motion in this direction. Heat transfer in the parallel direction can also be ignored if the plasma approaches a low-temperature limit. We therefore assume the plasma is cold enough that the heat transfer term can be ignored, but the pressure $p_{\|}$ term will be retained because the temperature is finite. Then (7.89) simplifies to

$$\frac{dp_{ij}}{dt} + p_{jk}\frac{\partial u_i}{\partial x_k} + p_{ik}\frac{\partial u_j}{\partial x_k} + p_{ij}\frac{\partial u_k}{\partial x_k} = 0 \tag{7.90}$$

Evaluate this equation by using $p_{ik} = p_{\|}b_i b_k + p_{\perp}(\delta_{ik} - b_i b_k)$. The last three terms of (7.90) can then be written as

$$\begin{aligned} p_{jk}\frac{\partial u_i}{\partial x_k} &= [p_{\|}b_j b_k + p_{\perp}(\delta_{jk} - b_j b_k)]\frac{\partial u_i}{\partial x_k} \\ p_{ik}\frac{\partial u_j}{\partial x_k} &= [p_{\|}b_i b_k + p_{\perp}(\delta_{ik} - b_i b_k)]\frac{\partial u_j}{\partial x_k} \\ p_{ij}\frac{\partial u_k}{\partial x_k} &= p_{ij}\nabla \cdot \mathbf{u} \end{aligned} \tag{7.91}$$

By using these relationships and contracting (7.90) with δ_{ij} and $\hat{b}_i\,\hat{b}_j$, (7.90) becomes (for $i = j$)

$$\frac{dp_{ii}}{dt} + [2p_{\|}b_i b_k + 2p_{\perp}(\delta_{ik} - b_i b_k)]\frac{\partial u_i}{\partial x_k} + p_{ii}\nabla \cdot \mathbf{u} \tag{7.92}$$

In Cartesian coordinates,

$$\begin{aligned} p_{ii} &= p_{xx} + p_{yy} + p_{zz} \\ &= p_{\perp} + (p_{\|} - p_{\perp})b_x b_x + p_{\perp} + (p_{\|} - p_{\perp})b_y b_y + p_{\perp} + (p_{\|} - p_{\perp})b_z b_z \\ &= p_{\|} + 2p_{\perp} \end{aligned} \tag{7.93}$$

Then (7.92) yields

$$\frac{d}{dt}(p_{\|}+2p_{\perp})+2(p_{\|}-p_{\perp})b_i b_k\frac{\partial u_i}{\partial x_k}+2p_{\perp}\frac{\partial u_i}{\partial x_k}\delta_{ik}+(p_{\|}+2p_{\perp})\nabla\cdot\mathbf{u}=0$$

$$\frac{dp_{\|}}{dt} + 2p_{\|}b_i b_k\frac{\partial u_i}{\partial x_k} + p_{\|}\nabla \cdot \mathbf{u} = 0 \tag{7.94}$$

From these, we obtain

$$\frac{dp_{\perp}}{dt} - p_{\perp}b_i b_k\frac{\partial u_i}{\partial x_k} + 2p_{\perp}\nabla \cdot \mathbf{u} = 0 \tag{7.95}$$

Multiply the second equation of (7.94) by $p_{\perp}$ and (7.95) by $2p_{\|}$ and add to eliminate the middle term. The result is

$$p_{\perp}\frac{dp_{\|}}{dt} + 2p_{\|}\frac{dp_{\perp}}{dt} + 5p_{\|}p_{\perp}\nabla \cdot \mathbf{u} = 0 \tag{7.96}$$

Use the continuity relation $-\rho_m(d\rho_m)/dt = \nabla \cdot \mathbf{u}$ and eliminate the $\nabla \cdot \mathbf{u}$ term. This leads to

$$\frac{d}{dt}\left(\frac{p_{\parallel}p_{\perp}^2}{\rho_m^5}\right) = 0 \tag{7.97}$$

which is one of the two equations of state needed. The other equation of state is obtained by assuming that the first adiabatic invariant is valid. Then, $<mv_{\perp}^2>/B = p_{\perp}/\rho_m B =$ constant, and we can write

$$\frac{d}{dt}\left(\frac{p_{\perp}}{\rho_m B}\right) = 0 \tag{7.98}$$

We can also use (7.98) to rewrite (7.97) as

$$\frac{d}{dt}\left(\frac{p_{\parallel}B^2}{\rho_m^3}\right) = 0 \tag{7.99}$$

The two equations of state (7.97) and (7.98) are referred to as double adiabatic equations of state. They were first derived by G.F. Chew, M.L. Goldberger, and F.E. Low in 1956.

We summarize the above results as follows. The double adiabatic theory (also known as CGL theory) describes a plasma fluid in which the magnetic field is strong enough that the pressure tensor is diagonal. The assumption that no heat flows in this model allows us to decouple motions in the perpendicular and parallel directions of the magnetic field. This results in two equations of state, one parallel and the other perpendicular to the direction of the magnetic field. The equations describing the CGL theory form a closed set for a plasma that moves across the magnetic field with a velocity $\mathbf{U} = \mathbf{E} \times \mathbf{B}/B^2$. This theory is more suitable for the description of collisionless fluids than the theory that assumes the fluid is isotropic.

7.5 Currents in Plasmas Including Collisions

Plasmas in ionospheres and in the lower atmosphere of the Sun (photosphere and chromosphere) undergo collisions. Collisions change the velocities of the individual particles and momentum will be exchanged among the colliding particles. Collisions affect the transport of charged particles and therefore how currents are produced. Before we discuss how currents flow in a medium that includes collisions, let us introduce some well-known terminologies concerning the physics of collisions.

The concept of the mean free path λ is useful in describing particles undergoing collisions. The mean free path represents the average distance a particle travels before it collides with another particle. Given a system containing n particles/volume, the mean free path is

$$\lambda = \frac{1}{n\sigma} \tag{7.100}$$

where σ is the collision cross-section. Cross-section measures an effective target area for scattering (discussed further below). Another useful parameter is the collision frequency

$$\nu = n\sigma v \tag{7.101}$$

which represents the rate at which collisions occur. Here v is the average speed of the particles.

Both λ and ν depend on the types of particles that are undergoing collisions. For example, collisions can occur between an electron and another electron, an ion and another ion, an electron and an ion, an electron and a neutral, and an ion and a neutral. These collisions have different cross-sections because, for example, collisions between two charged particles involve the Coulomb force whereas this force does not play a role in collisions involving a charged particle and a neutral particle. It is important to specify which species are involved in collisions.

7.5.1 Electrical Conductivity of Plasmas

To study how currents are generated in plasmas that include collisions, consider first a simple plasma model consisting of an electron fluid moving under the action of an applied electric field $\mathbf{E}$ relative to the ions that are at rest. Assume the plasma is cold ($T = 0$) so that we can ignore the pressure term. Then the equation of motion including collisions is

$$mn\frac{d\mathbf{U}}{dt} = -en\mathbf{E} - mn\nu\mathbf{U} \tag{7.102}$$

where ν represents the collision frequency. The collisional term is sometimes referred to as the "frictional" term because it impedes motion. Collisions occur between the electrons and ions and these collisions can involve significant momentum and energy transfers. Assume now the plasma is in steady state and that $\mathbf{U}$ is uniform. Then

$$\mathbf{U} = \frac{-e\mathbf{E}}{m\nu} \tag{7.103}$$

The electrical current associated with this motion is given by

$$\begin{aligned} \mathbf{J} &= -ne\mathbf{U} \\ &= \frac{ne^2}{m\nu}\mathbf{E} \\ &= \Sigma\mathbf{E} \end{aligned} \tag{7.104}$$

where the proportionality constant

$$\Sigma = \frac{ne^2}{m\nu} \tag{7.105}$$

is called the electrical conductivity of the medium (we use Σ instead of σ to avoid confusion since σ is already used to define the cross-section). Note that Σ depends on $(1/\nu)$. The inverse of Σ is electrical resistivity,

$$\eta = \frac{m\nu}{ne^2} \tag{7.106}$$

Equation (7.104) is the familiar Ohm's law and it states that the current density and the total electric field are linearly related if the conductivity is a constant. **E** is then parallel to **J**. This occurs only if the medium is isotropic. However, plasmas in nature tend to be anisotropic. Currents running in one direction will be different from those running in another direction. The electrical conductivity is then not a scalar but becomes a tensor (the general relationship between the current density and the electric field for an anisotropic plasma medium will be studied in a later section).

To study the details of how currents are conducted in plasmas, let us now consider a fully ionized plasma. The equations of motion for a magnetized fluid consisting of electrons and one ion species (including the gravitational force) are

$$\begin{aligned} m_i n \frac{d\mathbf{U}_i}{dt} &= Zen(\mathbf{E} + \mathbf{U}_i \times \mathbf{B}) - \nabla p_i + m_i n\mathbf{g} - \mathbf{P}_{ie} \\ m_e n \frac{d\mathbf{U}_e}{dt} &= -en(\mathbf{E} + \mathbf{U}_e \times \mathbf{B}) - \nabla p_e + m_e n\mathbf{g} - \mathbf{P}_{ei} \end{aligned} \tag{7.107}$$

All of the terms in each equation are familiar to us except the last term. $\mathbf{P}_{ie}$ represents the total momentum transferred to the ions colliding with the electrons per unit volume and per unit time and it is equal to

$$\mathbf{P}_{ie} = m_i n(\mathbf{U}_i - \mathbf{U}_e)\nu_{ie} \tag{7.108}$$

where ν_{ie} is the collision frequency of ions with electrons. Similarly, $\mathbf{P}_{ei}$ represents the total momentum transferred to the electrons colliding with the ions and is equal to

$$\mathbf{P}_{ei} = m_e n(\mathbf{U}_e - \mathbf{U}_i)\nu_{ei} \tag{7.109}$$

where ν_{ei} represents the collision frequency of electrons with ions. For a two-component plasma $\nu_{ei} = \nu_{ie}$, and momentum conservation requires that $\mathbf{P}_{ie} = -\mathbf{P}_{ei}$.

Consider now a single electron whose momentum is $p = mv$. The momentum lost by this electron interacting with the ion is

$$\Delta p = p(1 - \cos\theta) \tag{7.110}$$

where θ is the scattering angle and we assume elastic scattering (Figure 7.9). The average value of momentum loss multiplied by the total number of collisions per unit distance, $n\sigma$, averaged over all angles is $n\sigma p < 1 - \cos\theta >$, and this is equal to $m\nu$,

$$\begin{aligned} m\nu &= n\sigma p < 1 - \cos\theta > \\ &\approx n\sigma p \frac{< \theta^2 >}{2} \end{aligned} \tag{7.111}$$

where we now assume small-angle scattering. Equation (7.111) involves computation of $< \theta^2 >$.

To learn about electron scattering that occurs in Coulomb collisions, consider an electron that is approaching a positive ion of charge $+Ze$, as shown in Figure 7.9. Since the ion is much more massive than the electron, we can assume that the ion is essentially at rest. The Coulomb force

$$\mathbf{F} = -\frac{Ze^2}{4\pi\epsilon_0 r^2}\hat{\mathbf{r}} \tag{7.112}$$

where $\hat{\mathbf{r}}$ is a unit vector in the direction of $\mathbf{r}$, deflects the trajectory of the electron and the component of the force that causes the deflection is

$$\begin{aligned} F_\perp &= F \sin\theta \\ &= \frac{Ze^2}{4\pi\epsilon_0 b^2}\sin^3\theta \end{aligned} \tag{7.113}$$

where we define the impact parameter b as the closest distance of approach of the electron to the positive ion (impact parameter is the undeflected distance of the electron relative to the ion). θ is the angle to the particle

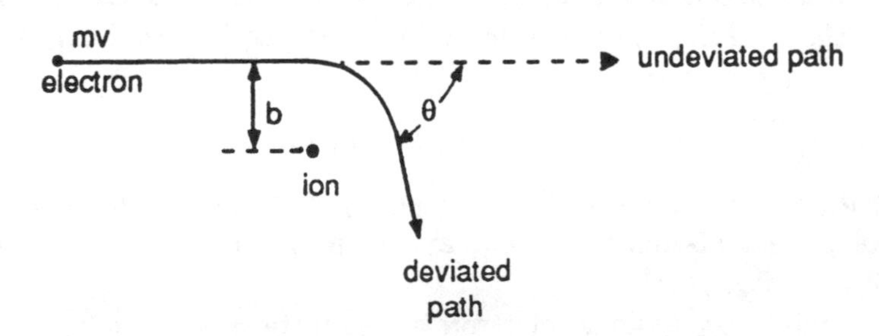

FIGURE 7.9 A schematic diagram to show the various parameters involved in a collision of a charged particle with another charged particle.

measured from $x = -\infty$ in a clockwise sense. Then the change of momentum is

$$\begin{aligned} \Delta p_\perp &= \int F_\perp dt \\ &= \frac{1}{v}\int F_\perp dx \\ &= \frac{Ze^2}{4\pi\epsilon_0 vb^2}\int \sin^3\theta dx \end{aligned} \tag{7.114}$$

where the limits of integration are from $-\infty$ to $+\infty$. Now note that $\tan(\pi/2-\theta) = \text{ctn}\theta = x/b$, hence $dx = -(b\csc^2\theta)d\theta$. By using this relation, the above equation can be rewritten as

$$\begin{aligned} \Delta p_\perp &= -\frac{Ze^2}{4\pi\epsilon_0 vb}\int \sin\theta d\theta \\ &= \frac{2Ze^2}{4\pi\epsilon_o vb} \end{aligned} \tag{7.115}$$

where the integral has been evaluated between 0 and π. This equation represents the change of the momentum arising from the Coulomb collision of an electron with the ion.

The deflection angle of the electrons by the Coulomb force is obtained from

$$\begin{aligned} \tan\theta &= \frac{\Delta p_\perp}{p} \\ &= \frac{2Ze^2}{4\pi\epsilon_0 mv^2 b} \end{aligned} \tag{7.116}$$

since $p = mv$. For small deflection angles, $\tan\theta \approx \theta$. Small (large) θ corresponds to a large (small) impact parameter since then the Coulomb force will be weaker (stronger). For small deflection angles,

$$\theta \approx \frac{2Ze^2}{4\pi\epsilon_0 mv^2 b} \tag{7.117}$$

Equation (7.117) is relativistically correct if m is replaced by γm, where $\gamma = (1 - v^2/c^2)^{1/2}$. The deflection angle θ depends on the inverse of the impact parameter $(1/b)$ and computing $<\theta^2>$ means we must compute $<1/b^2>$. Thus we have not made any progress in computing $<\theta^2>$, since all we have done is to replace one unknown by another.

To proceed further, we need a probability density function so that for an arbitrary target particle, the incident particle will be uniformly distributed in a circle of some maximum radius, b_{max} (Figure 7.10). Since we are considering only small angle deflection, there is also a b_{min}. A density function that is uniform in area throughout the annular region can be

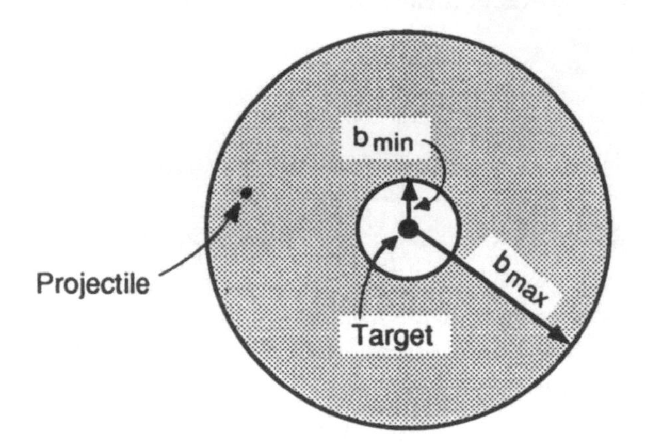

FIGURE 7.10 A schematic diagram that shows the behavior of the uniform probability density function.

represented by [1]

$$p(b)db = \frac{2b\,db}{b_{max}^2 - b_{min}^2} \tag{7.118}$$

where $b_{min} < b < b_{max}$. Note that

$$\int p(b)db = 1 \tag{7.119}$$

as required if the integration is carried out between b_{min} and b_{max}. Now change variables using (7.117), $b = 2Ze^2/4\pi\epsilon_0 mv^2\theta$. Then, the density function becomes

$$f(\theta)d\theta = \frac{2}{1/\theta_{min}^2 - 1/\theta_{max}^2}\frac{d\theta}{\theta^3} \tag{7.120}$$

where $\theta_{min} < \theta < \theta_{max}$. Here

$$\begin{aligned} \theta_{min} &= \frac{2Ze^2}{4\pi\epsilon_0 mv^2 b_{max}} \\ \theta_{max} &= \frac{2Ze^2}{4\pi\epsilon_0 mv^2 b_{min}} \end{aligned} \tag{7.121}$$

Note that

$$\int f(\theta)d\theta = 1 \tag{7.122}$$

where the integration limits are θ_{min} and b_{max}.

[1]This formulation was suggested by Dr. Michael McCarthy.

We can compute $< \theta^2 >$ using (7.120). This yields

$$\begin{aligned} <\theta^2> &= \int f(\theta)\theta^2 d\theta \\ &= \int \frac{2}{1/\theta^2_{min} - 1/\theta^2_{max}} \frac{\theta^2 d\theta}{\theta^3} \\ &= \frac{2}{1/\theta^2_{max} - 1/\theta^2_{min}} \ln \left| \frac{\theta_{max}}{\theta_{min}} \right| \end{aligned} \tag{7.123}$$

Now, $\theta_{max} \gg \theta_{min}$ and $\theta_{min} > 0$. Hence, ignoring $1/\theta^2_{max}$ in the denominator yields

$$<\theta^2> \approx 2\theta^2_{min} \ln \Lambda \tag{7.124}$$

where $\Lambda = \theta_{max}/\theta_{min}$.

We can now obtain an equation of electrical conductivity for a two-component plasma. We use (7.124) in (7.111). Also note that the total cross-section, using (7.117), is

$$\begin{aligned} \sigma &= \pi b^2_{max} \\ &= \pi \left(\frac{2Ze^2}{4\pi\epsilon_0 m v^2 \theta_{min}} \right) \end{aligned} \tag{7.125}$$

The electrical conductivity for a fully ionized two-component plasma is

$$\begin{aligned} \Sigma &= \frac{nZe^2}{m\nu} \\ &= \frac{Ze^2}{p\sigma < \theta^2 >} \\ &= \frac{(4\pi\epsilon_0)^2 m v^3}{4\pi Z e^2 \ln(\theta_{max}/\theta_{min})} \\ &\approx \frac{(4\pi\epsilon_0)^2 m v^3}{4\pi Z e^2 \ln \Lambda} \end{aligned} \tag{7.126}$$

Λ involves the kinetic properties of the plasma.

If this result is applied to a plasma, (7.126) must be averaged over the plasma distribution function because of the v^3 factor. We use as an example a Maxwellian distribution function and assume that Λ is not velocity-dependent (which is incorrect). Then it can be shown that

$$\Sigma \approx \frac{(4\pi\epsilon_0)^2 m}{(Ze^2) \ln \Lambda} \left(\frac{2kT}{\pi m} \right)^{3/2} \tag{7.127}$$

This conductivity equation was first derived by L. Spitzer and is known as the Spitzer conductivity. This result agrees within a factor of two with the results obtained from the more rigorous Boltzmann equation.

7.5.2 Generalized Ohm's Law

Now that we have gained some understanding of how electrical conductivity is produced by plasmas, we continue the discussion of currents in plasmas. The electric field $\mathbf{E}$ in Ohm's law (7.104) is the total field. Since conducting material flowing across a magnetic field induces a $\mathbf{U} \times \mathbf{B}$ field, Ohm's law is generally written as

$$\mathbf{J} = \sigma(\mathbf{E} + \mathbf{U} \times \mathbf{B}) \tag{7.128}$$

Consider now a magnetized plasma defined by the two equations of motion given by (7.107). If the two are added, one obtains

$$\rho_m \frac{\partial \mathbf{U}}{\partial t} = -\nabla p + \mathbf{J} \times \mathbf{B} + \rho_m \mathbf{g} \tag{7.129}$$

which is the same as (5.35) except for the gravitational term. The derivation of the one-fluid equation (7.129) in Chapter 5 ignored collisional and gravitational effects. Comparison of (7.129) with the one-fluid equation of Chapter 5 shows that the one-fluid motion is affected by the presence of the gravitational force but unaffected by the plasma collision.

If the top equation of (7.107) is multiplied by m_e and the bottom by m_i and the latter subtracted from the former, one obtains

$$\begin{aligned} m_i m_e n \frac{\partial}{\partial t}(\mathbf{U}_i - \mathbf{U}_e) &= qn(m_e + m_i)\mathbf{E} + qn(m_e \mathbf{U}_i + m_i \mathbf{U}_e) \times \mathbf{B} \\ &\quad - m_e \nabla p_i + m_i \nabla p_e - (m_e + m_i)\mathbf{P}_{ie} \end{aligned} \tag{7.130}$$

Note that $\mathbf{E}$ and $\mathbf{P}_{ie}$ are retained in (7.130) whereas they were "subtracted out" in (7.129). On the other hand, (7.130) does not contain the gravitational term, while this term is retained in (7.129).

With the help of the definition of the mass density $\rho_m = n(m_e + m_i)$, the current density $\mathbf{J} = nq(\mathbf{U}_i - \mathbf{U}_e)$, and (7.108) for $\mathbf{P}_{ie}$, (7.130) can be transformed (after some algebra) to

$$\begin{aligned} \frac{m_i m_e n}{q} \frac{\partial}{\partial t}\left(\frac{\mathbf{J}}{n}\right) &= q\rho_m \mathbf{E} + qn(m_e \mathbf{U}_i + m_i \mathbf{U}_e) \times \mathbf{B} \\ &\quad - m_e \nabla p_i + m_i \nabla p_e - \rho_m q \eta \mathbf{J} \end{aligned} \tag{7.131}$$

Now the second term on the right side can be rewritten as

$$\begin{aligned} qn(m_e \mathbf{U}_i + m_i \mathbf{U}_e) \times \mathbf{B} &= qn[m_e \mathbf{U}_e + m_i \mathbf{U}_i + m_i(\mathbf{U}_e - \mathbf{U}_i) \\ &\quad + m_e(\mathbf{U}_i - \mathbf{U}_e)] \times \mathbf{B} \\ &= qn\left[(m_e + m_i)\mathbf{U} - (m_i - m_e)\frac{\mathbf{J}}{qn}\right] \times \mathbf{B} \\ &= qn\left[\frac{\rho_m}{n}\mathbf{U} - (m_i - m_e)\frac{\mathbf{J}}{qn}\right] \times \mathbf{B} \end{aligned} \tag{7.132}$$

where $\mathbf{U}$ is given by $\mathbf{U} = (n_i m_i \mathbf{U}_i + n_e m_e \mathbf{U}_e)/(m_i n_i + m_e n_e)$. Substituting (7.132) into (7.131) and rearranging the various terms yields

$$\mathbf{E} + \mathbf{U} \times \mathbf{B} - \eta \mathbf{J} = \frac{1}{q\rho_m}\left[\frac{m_i m_e n}{q}\frac{\partial}{\partial t}\frac{\mathbf{J}}{n} + (m_i - m_e)\mathbf{J} \times \mathbf{B} + m_e \nabla p_i - m_i \nabla p_e\right] \tag{7.133}$$

where η, the specific resistivity of the conducting fluid, is

$$\eta = \frac{m_e m_i}{m_e + m_i}\frac{1}{q^2 n}\nu \approx \frac{m_e}{q^2 n}\nu \tag{7.134}$$

valid for $m_e/m_i \ll 1$. In the limit $m_e/m_i \ll 1$, and in steady state, (7.133) reduces to

$$\mathbf{E} + \mathbf{U} \times \mathbf{B} = \eta \mathbf{J} + \frac{1}{qn}(\mathbf{J} \times \mathbf{B} - \nabla p_e) \tag{7.135}$$

Equation (7.135) is known as the generalized Ohm's law. Define now $\mathbf{E}' = \mathbf{E} + \mathbf{U} \times \mathbf{B}$ and $\mathbf{E}'' = (1/qn)\nabla p_e$. Then (7.135) can be written as

$$\mathbf{J} + \left(\frac{\omega_{ce}}{\nu}\right)\mathbf{J} \times \frac{\mathbf{B}}{B} = \sigma(\mathbf{E}' + \mathbf{E}'') = \sigma \mathbf{E}_T \tag{7.136}$$

Here ω_{ce} is the cyclotron frequency of the electron, qB/m_e, and $\mathbf{E}_T$ is the total electric field $\mathbf{E}' + \mathbf{E}''$. When we write it in this form, we see that the pressure gradient of electrons gives rise to an equivalent electric field $\mathbf{E}''$. The ∇p_e term leads to ambipolar diffusion. $\mathbf{E}'$ is an electric field in the moving frame of reference, the center of mass frame. The $\mathbf{J} \times \mathbf{B}$ term is referred to as the Hall current term. If the situation permits us to ignore the Hall term (for example if $\omega_{ce} \ll \nu$) and the second term on the right-hand side ($\mathbf{E}''$), the simple form of Ohm's law is recovered. In general, (7.133) or (7.135) should be used instead of (7.128).

7.6 Currents in Ionospheres

The transient variations of Earth's geomagnetic field were correctly interpreted in terms of ionospheric currents in the late nineteenth century. Ionospheric currents flow in a narrow horizontal layer at an altitude between 100 and 150 km concentric with Earth's surface. Ionospheric currents are observed during both quiet and disturbed solar wind conditions. The quiet ionospheric currents, designated as S_q currents, are produced by the motion of the ionized ionospheric particles across the planetary magnetic field. This motion, driven by the daily heating of the ionosphere by the Sun and by the lunar and solar tidal forces, induces an electromotive force that produces an equivalent current pattern that is fixed with respect to the Sun. A diurnal variation of the geomagnetic field is observed by a magnetic station fixed on Earth and rotating through this current system.

The disturbed ionospheric currents, designated as S_D, are observed in conjunction with the auroral activity at northern magnetic latitudes. During an aurora, an excess of 10^{19} ergs of particle energy is deposited into the auroral ionosphere. The auroral ionospheric conductivity thus becomes greatly enhanced and ionospheric currents flow in both eastward and westward directions. These currents are referred to as eastward and westward electrojets. The intensity of the electrojets is several million amperes for a moderately sized aurora. These currents cause a deviation of several hundred nanoteslas of Earth's main magnetic field on the surface of the planet (Figure 7.11).

Ionospheric plasma is anisotropic and therefore currents in general do not flow parallel to the direction of the electric field. The ratio of the current density and the electric field (conductivity) is thus not a constant but varies from one direction to another. The conductivity is now a tensor and Ohm's law can be written as $\mathbf{J} = \boldsymbol{\sigma} \cdot \mathbf{E}$ where $\boldsymbol{\sigma}$ is the conductivity tensor. We now derive the expression of $\boldsymbol{\sigma}$ for a plasma consisting of electrons and one species of ions.

It is instructive to examine this problem starting from the Lorentz equation. Consider a charged particle in a magnetic field $\mathbf{B} = (0, 0, B)$ and let this particle be subject to an additional constant force $\mathbf{F} = (F_x, 0, F_z)$. The components of the Lorentz equation of motion for the ions are,

$$\begin{aligned} m_i \frac{dv_x}{dt} &= F_x + qBv_y \\ m_i \frac{dv_y}{dt} &= -qBv_x \\ m_i \frac{dv_z}{dt} &= F_z \end{aligned} \tag{7.137}$$

The solutions to this set of differential equations are

$$\begin{aligned} v_x &= v_{xo} \cos \omega_c t + \frac{v_{yo} + F_x}{qB} \sin \omega_c t \\ v_y &= \frac{v_{yo} + F_x}{qB} \cos \omega_c t - v_{xo} \sin \omega_c t - \frac{F_x}{qB} \\ v_z &= v_{zo} + \frac{F_z}{m_i} t \end{aligned} \tag{7.138}$$

where v_{xo}, v_{yo} and v_{zo} are the initial velocities of the particle at $t = 0$, and ω_c is the gyrofrequency of the ions, qB/m_i. Equation (7.138) describes a

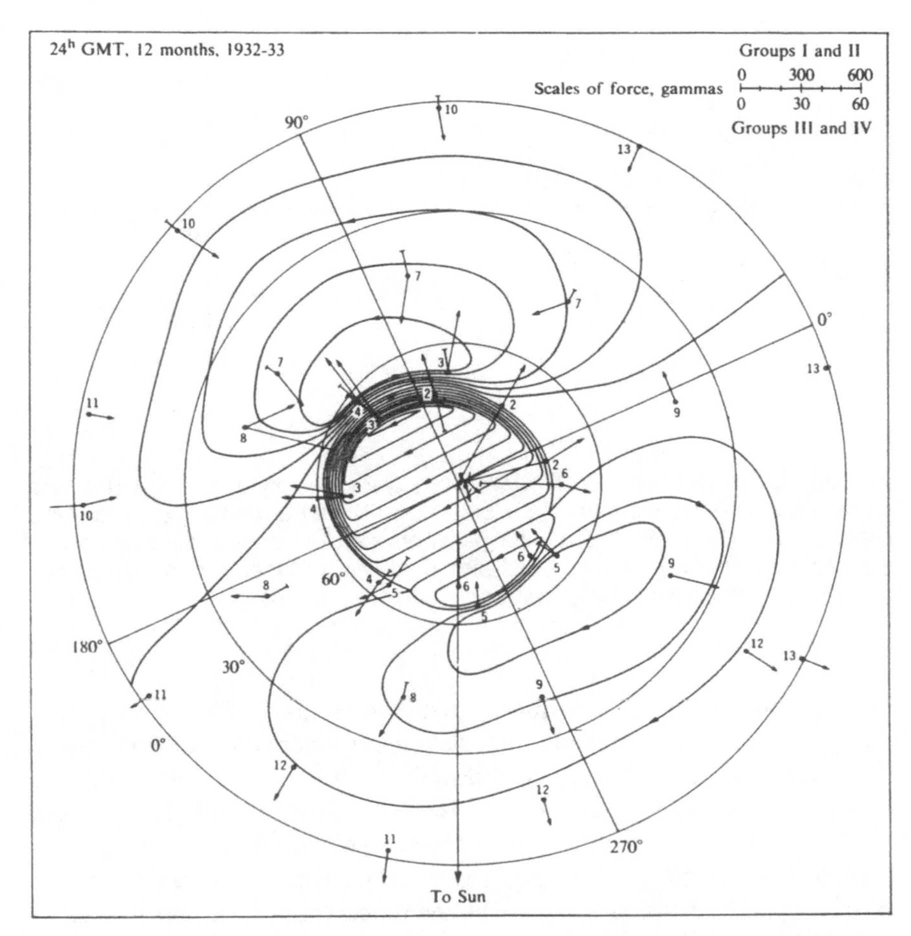

FIGURE 7.11 S_D ionospheric current systems. A current of 50,000 amperes flows between successive contours. Note that the Sun is toward the bottom. Current density is maximum near local midnight. (From Silsbee and Vestine, 1942)

gyration motion of the particle and the drift of the center of gyration due to the presence of the constant force **F**.

Suppose this particle now undergoes collisions. Collisions will disrupt the ordered motion described by (7.138). For the sake of simplicity, assume that after each collision the particle starts from rest in a random direction. In between each collision, the motion is still validly described by (7.138). Hence the main effect of the collisions is to introduce a drift in the direction of **F**, which modifies the motion perpendicular to **F**. A schematic diagram illustrating the collisional modification of the motions is shown in Figure 7.12.

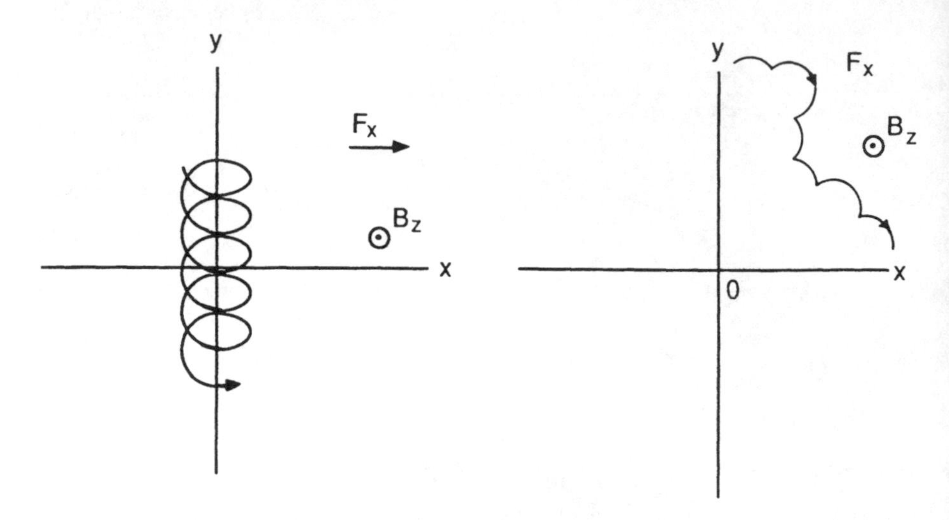

FIGURE 7.12 The panel on the left shows the trajectory in the xy-plane of a positive particle in a magnetic field $\mathbf{B} = (0, 0, B)$ that is under the action of a constant force $\mathbf{F} = (F_x, 0, F_z)$. The panel on the right shows the trajectory of the same particle if it suffers a collision randomly in time. It is assumed that immediately after each collision the particle velocity is zero.

Let ν be the collision frequency and assume that collisions occur randomly in time. The probability distribution of random collisional processes is proportional to $e^{-\nu t}$ and the probability that a particle will escape a collision in the time t and $t + dt$ is then given by $\nu e^{-\nu t} dt$. During such collision-free times, the particle drifts because of the constant force $\mathbf{F}$. The average drift velocities of the particles in between collisions are calculated by folding in the collision-free distribution function with the velocities given in (7.138).

$$
\begin{aligned}
< v_x > &= \frac{\int \nu e^{-\nu t}(F_x/qB)\sin\omega_c t dt}{\int \nu e^{-\nu t} dt} = \frac{F_x}{qB}\frac{\omega_c \nu}{(\nu^2 + \omega_c^2)} \\
< v_y > &= \frac{\int \nu e^{-\nu t}(F_x/qB)(\cos\omega_c t - 1) dt}{\int \nu e^{-\nu t} dt} = -\frac{F_x}{qB}\frac{\omega_c^2}{(\nu^2 + \omega_c^2)} \\
< v_z > &= \frac{\int \nu e^{-\nu t}(F_z/m)t dt}{\int \nu e^{-\nu t} dt} = \frac{F_z}{m_i \nu}
\end{aligned} \tag{7.139}
$$

Note that because we assumed that the particle starts from rest immediately after a collision, the initial velocities in (7.138) do not contribute. Equation (7.139) shows that the average drift velocities in x and y directions are only affected by F_x, as we stated earlier.

A similar set of equations for electron motions can be obtained by the same procedure. The Lorentz equation of motion for the electron is identical to the ion equation except that $q_i = -q_e$ and $m_i = m_e$. Then in (7.139) let $q = -q$ and $\omega_c = -\omega_c$. Now the gyrofrequency is that of the electrons, $\omega_{ce} = -qB/m_e$. The components of the average electron drift velocities are

$$
\begin{aligned}
<v_x> &= \frac{F_x}{qB} \frac{\omega_{ce}\nu}{(\nu^2 + \omega_{ce}^2)} \\
<v_y> &= \frac{F_x}{qB} \frac{\omega_{ce}^2}{(\nu^2 + \omega_{ce}^2)} \\
<v_z> &= \frac{F_z}{m_e \nu}
\end{aligned} \qquad (7.140)
$$

where ω_{ce} is the cyclotron frequency of electrons. For a plasma consisting of electrons and one species of singly charged ions, the current density is $\mathbf{J} = nq(<\mathbf{v}_i> - <\mathbf{v}_e>)$ where $<\mathbf{v}_i>$ and $<\mathbf{v}_e>$ are obtained from (7.139) and (7.140) and n is the number density of electrons and ions, $n_e = n_i = n$. Let the constant external force $\mathbf{F}$ be the force of the electric field $q\mathbf{E}$ and $\omega_{ci} = qB/m_i$, the cyclotron frequency of ions. Then, for $\mathbf{E} = (E_x, 0, E_z)$,

$$
\begin{aligned}
J_x &= \frac{nq}{B}\left[\frac{\omega_{ci}\nu}{(\nu^2 + \omega_{ci}^2)} - \frac{\omega_{ce}\nu}{(\nu^2 + \omega_{ce}^2)}\right] E_x \\
J_y &= \frac{nq}{B}\left[\frac{-\omega_{ci}^2}{(\nu^2 + \omega_{ci}^2)} + \frac{\omega_{ce}^2}{(\nu^2 + \omega_{ce}^2)}\right] E_x \\
J_z &= \frac{nq}{B}\left(\frac{\omega_{ci}}{\nu} - \frac{\omega_{ce}}{\nu}\right) E_z
\end{aligned} \qquad (7.141)
$$

The cyclotron frequency here is always defined positive whereas, in an earlier usage, the cyclotron frequencies of electrons and ions differed in sign. The coefficients preceding the components of $\mathbf{E}$ in (7.141) represent the conductivities in the three orthogonal directions.

The conductivity tensor of a plasma with electrons and one species of ions, obtained from comparison of (7.141) and $\mathbf{J} = \boldsymbol{\sigma} \cdot \mathbf{E}$, is

$$
\sigma = \begin{pmatrix} \sigma_1 & -\sigma_2 & 0 \\ \sigma_2 & \sigma_1 & 0 \\ 0 & 0 & \sigma_0 \end{pmatrix} \qquad (7.142)
$$

where

$$
\begin{aligned}
\sigma_1 &= \frac{nq}{B}\left[\frac{\omega_{ci}\nu}{(\nu^2 + \omega_{ci}^2)} - \frac{\omega_{ce}\nu}{(\nu^2 + \omega_{ce}^2)}\right] \\
\sigma_2 &= \frac{nq}{B}\left[\frac{-\omega_{ci}^2}{(\nu^2 + \omega_{ci}^2)} + \frac{\omega_{ce}^2}{(\nu^2 + \omega_{ce}^2)}\right] \\
\sigma_0 &= \frac{nq}{B}\left[\frac{\omega_{ci}}{\nu} - \frac{\omega_{ce}}{\nu}\right]
\end{aligned} \qquad (7.143)
$$

σ_1 conducts currents in the direction of the external electric field **E** and perpendicular to the magnetic field **B**. σ_1 is referred to as the Pedersen conductivity and the component of the current associated with this conductivity is called the Pedersen current. The Pedersen current is dissipative, since $\mathbf{J} \cdot \mathbf{E} \neq 0$. Pedersen currents are an important part of the auroral current system.

σ_2 conducts the current perpendicular to both electric **E** and magnetic **B** fields. σ_2 is called the Hall conductivity and the component of the current associated with this conductivity is called the Hall current. Since the Hall current runs in the direction perpendicular to the electric field, this current is not dissipative (that is, $\mathbf{E} \cdot \mathbf{v} = 0$). Hall currents also play an important role in the auroral current system.

If an electric field is applied along the direction of the magnetic field, currents will run along **B** (field-aligned current). σ_0 conducts this field-aligned current. As is evident, the current along **B** is unaffected by **B**. Hence σ_0 is the same as the conductivity in the absence of the magnetic field. σ_0 is called the ordinary conductivity (it is the same conductivity that was defined in (7.105)). Particles of the field-aligned currents are a source of auroral luminosity. Field-aligned currents are important and fundamental to the dynamics of MHD and, in general, to MHD large-scale configurations.

Consider the magnetic equatorial ionosphere in which exists an eastward electric field E_x (this can be developed, for instance, by the horizontal neutral wind system, which transfers the momentum to the charged particles). Let the magnetic field be in the y-direction. Currents will flow in the x-direction due to the Pedersen conductivity σ_1 and in the z-direction due to the Hall conductivity σ_2. Assume now that the ionosphere is a thin shell and the currents are confined to flow in this narrow vertical region. The inhibition of the flow of the vertical current gives rise to a polarization electric field E_z (due to charges piling up at the edge). We can therefore write

$$\begin{aligned} J_x &= \sigma_1 E_x + \sigma_2 E_z \\ J_z &= -\sigma_2 E_x + \sigma_1 E_z = 0 \end{aligned} \tag{7.144}$$

which gives

$$J_x = \left(\sigma_1 + \frac{\sigma_2^2}{\sigma_1}\right) E_x \tag{7.145}$$

The Cowling conductivity is defined as

$$\sigma_3 = \sigma_1 + \frac{\sigma_2^2}{\sigma_1} \tag{7.146}$$

This conductivity relates the current flow in the direction of the electric field. The Cowling current is dissipative and power dissipation by Joule heating can be significant in the direction perpendicular to **B**.

Ohm's law for a planetary ionosphere can be written in terms of the various conductivities as

$$\mathbf{J} = \sigma_p(\mathbf{E}_\perp + \mathbf{U} \times \mathbf{B}) + \sigma_H \frac{\mathbf{B}}{B} \times (\mathbf{E}_\perp + \mathbf{U} \times \mathbf{B}) + \sigma_\| \mathbf{E}_\| \qquad (7.147)$$

where σ_p, σ_H, and $\sigma_\|$ are the Pedersen, Hall, and parallel conductivities.

The calculation of conductivity requires knowledge about the density and the temperature of both neutral and ionized species. S. Chapman first computed Earth's ionospheric conductivities in 1956. A more recent calculation of night-time conductivities as a function of height (> 80 km) is shown in Figure 7.13. Note that the conductivity along the magnetic field direction $\sigma_\|$ is much larger than σ_p or σ_H. This means that $\mathbf{E}_\|$ will be generally smaller than $\mathbf{E}_\perp$. Note also that the Cowling conductivity (7.146) can be larger than σ_p or σ_H. The Cowling conductivity is the dominant conductivity at the 100-km altitude range and it is responsible for the equatorial electrojet phenomenon.

7.7 Field-Aligned Currents

Kristian Birkeland, a Norwegian physicist studying the dynamics of Earth's aurora at the turn of the century, hypothesized that the horizontal currents

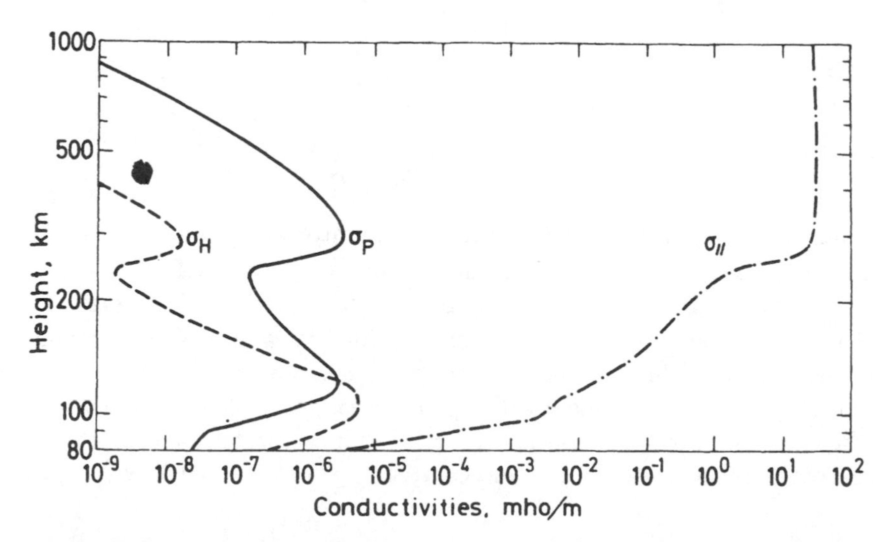

FIGURE 7.13 Various conductivities as a function of height for Earth's ionosphere. (From Boström, 1967)

in the auroral ionosphere may be coupled to vertical currents. Currents that flow vertically along the magnetic field direction are called field-aligned currents. (The field-aligned current on Earth is also referred to as a Birkeland current.) Magnetic fields associated with field-aligned currents were first detected in 1966 by a satellite-borne magnetometer flown through an aurora.

Field-aligned currents $\mathbf{J}_{\|}$ do not contribute to the electromagnetic stress because

$$\mathbf{J} \times \mathbf{B} = 0 \tag{7.148}$$

Therefore, these currents are associated with a "force-free" magnetic configuration. Field-aligned currents provide a means for coupling the magnetosphere and the ionosphere and they are also a source of visual auroras. The current density of Earth's field-aligned currents is typically a few tens of micro-amperes/m^2 during a moderate-size auroral event.

An equation of field-aligned current $\mathbf{J}_{\|}$ cannot be obtained from the momentum equation as was done for the perpendicular current because $\mathbf{J}_{\|}$ does not appear explicitly in the MHD momentum equation. Instead, let us examine what can be learned about $\mathbf{J}_{\|}$ from one of Maxwell's equations $\nabla \times \mathbf{B} = \mathbf{J}$. If we take the divergence of this equation, we obtain the continuity equation for the current density $\nabla \cdot \mathbf{J} = 0$. Now let $\mathbf{J} = \mathbf{J}_{\|} + \mathbf{J}_{\perp}$. Then the continuity equation can be written as

$$\nabla \cdot (\mathbf{J}_{\|} + \mathbf{J}_{\perp}) = 0 \tag{7.149}$$

and the divergence of $\mathbf{J}_{\|}$ can be related to the divergence of $\mathbf{J}_{\perp}$.

$$\begin{aligned} -\nabla \cdot \mathbf{J}_{\perp} &= \nabla \cdot \mathbf{J}_{\|} \\ &= \nabla \cdot J_{\|} \left(\frac{\mathbf{B}}{B} \right) \\ &= B \frac{\partial}{\partial s} \left(\frac{J_{\|}}{B} \right) \end{aligned} \tag{7.150}$$

where $\partial / \partial s = (\mathbf{B}/B) \cdot \nabla$ is the gradient operator along the direction of $\mathbf{B}$. Given $\mathbf{J}_{\perp}$, $\mathbf{J}_{\|}$ can be obtained by integrating $\nabla \cdot \mathbf{J}_{\perp}$ along $\mathbf{B}$.

Consider $\mathbf{J}_{\perp}$ to be generated by the drift motion of trapped particles in the magnetosphere (ring current). Suppose a part of this current is allowed to flow along $\mathbf{B}$ and let it close in the ionosphere (how this current closes in the ionosphere will depend on the ionospheric conductivities, whose details need not concern us here). Since the formula for the ring current is known, we can use the continuity equation for the current and obtain information on $J_{\|}$. Recall that $\mathbf{J}_{\perp}$ obtained from the MHD momentum equation (7.2) is

$$\mathbf{J}_{\perp} = \left(\frac{\mathbf{B}}{B^2}\right) \times \left(\rho_m \frac{d\mathbf{U}}{dt} + \nabla \cdot \mathbf{p}\right) \tag{7.151}$$

where $\mathbf{p}$ is the pressure tensor. Let $\mathbf{F} = \rho_m d\mathbf{U}/dt + \nabla \cdot \mathbf{p}$. The divergence of (7.151) is

$$\begin{aligned}
\nabla \cdot \mathbf{J}_{\perp} &= \nabla \cdot \left(\frac{\mathbf{B}}{B^2} \times \mathbf{F}\right) \\
&= \mathbf{F} \cdot \left(\nabla \times \frac{\mathbf{B}}{B^2}\right) - \frac{\mathbf{B}}{B^2} \cdot (\nabla \times \mathbf{F}) \\
&= \mathbf{F} \cdot \left(\frac{1}{B^2} \nabla \times \mathbf{B} - \frac{2}{B^3} \nabla B \times \mathbf{B}\right) - \frac{\mathbf{B}}{B^2} \cdot (\nabla \times \mathbf{F}) \\
&= \frac{\mathbf{F}}{B^2} \cdot \nabla \times \mathbf{B} - 2\mathbf{J}_{\perp} \cdot \frac{\nabla B}{B} - \frac{\mathbf{B}}{B^2} \cdot (\nabla \times \mathbf{F}) \\
&= -2\mathbf{J}_{\perp} \cdot \frac{\nabla B}{B} - \left(\frac{\mathbf{B}}{B^2}\right) \cdot \nabla \times \mathbf{F}
\end{aligned} \tag{7.152}$$

where $\mathbf{F} = \rho_m d\mathbf{U}/dt + \nabla \cdot \mathbf{p}$, and $(\nabla \times \mathbf{B}/B^2) \cdot \mathbf{F} = 0$. Inserting (7.152) into (7.150) yields

$$B \frac{\partial}{\partial s}\left(\frac{J_{\|}}{B}\right) = 2\mathbf{J}_{\perp} \cdot \frac{\nabla B}{B} + \left(\frac{\mathbf{B}}{B^2}\right) \cdot \nabla \times \mathbf{F} \tag{7.153}$$

and upon integration along the length of the magnetic field line $J_{\|}$ is obtained. Note that $J_{\|}$ depends on the combination of the magnetic topology and the thermal properties of plasmas (through the pressure tensor). If the pressure is a scalar, the perpendicular current reduces to (7.20), which has a simpler form than (7.151). If the plasma pressure is anisotropic, (7.151) leads to (7.57) for $\mathbf{J}_{\perp}$, assuming the anisotropic pressure is diagonal and given by (7.46).

Another useful piece of information on field-aligned currents is that, since $\mathbf{J}_{\|}$ is parallel to $\mathbf{B}$, one can write

$$\begin{aligned}
\nabla \times \mathbf{B} &= \mu_0 \mathbf{J} \\
&= \alpha \mathbf{B}
\end{aligned} \tag{7.154}$$

α is in general a function of position and time. However, the choice of α is restricted. Taking the divergence of (7.151), one deduces that

$$\begin{aligned}
0 &= \nabla \cdot (\nabla \times \mathbf{B}) \\
0 &= \nabla \cdot (\alpha \mathbf{B}) \\
0 &= (\mathbf{B} \cdot \nabla)\alpha
\end{aligned} \tag{7.155}$$

which shows that the gradient of α must vanish along $\mathbf{B}$. Hence α is constant along the lines of force on which the currents are flowing. Note that the field-aligned current is intimately tied to the field-aligned plasma flow (Chapter 5).

7.7.1 Mininum Energy Configuration

An MHD system that supports field-aligned currents is referred to as a force-free configuration, since $\mathbf{J} \times \mathbf{B} = 0$. An important theorem derived by L. Woltjer in 1958 shows that the field-aligned current with $\alpha =$ constant represents the state of lowest magnetic energy for an ideal MHD system in which the magnetic helicity (to be defined below) is a constant. This theorem has important implications about the large-scale current structures in space and about the equilibrium MHD configurations.

For ideal fluids with $\sigma = \infty$,

$$\frac{\partial \mathbf{B}}{\partial t} = \nabla \times (\mathbf{U} \times \mathbf{B}) \tag{7.156}$$

Since $\mathbf{B} = \nabla \times \mathbf{A}$ where $\mathbf{A}$ is the vector potential, we can rewrite the above equation as

$$\nabla \times \frac{\partial \mathbf{A}}{\partial t} = \nabla \times [\mathbf{U} \times (\nabla \times \mathbf{A})] \tag{7.157}$$

This equation implies that

$$\frac{\partial \mathbf{A}}{\partial t} = \mathbf{U} \times (\nabla \times \mathbf{A}) + \nabla \Phi \tag{7.158}$$

where Φ is some scalar. Now choose the gauge such that $\nabla \Phi = 0$. Then (7.158) shows that

$$\frac{\partial \mathbf{A}}{\partial t} \cdot (\nabla \times \mathbf{A}) = 0 \tag{7.159}$$

The helicity of the magnetic vector potential $\mathbf{A}$ in a volume of fluid V is defined as

$$\int \mathbf{A} \cdot (\nabla \times \mathbf{A}) dV = \int \mathbf{A} \cdot \mathbf{B} dV \tag{7.160}$$

Under certain conditions, helicity is a constant in time. To see this, differentiate the helicity with respect to time,

$$\begin{aligned} \frac{\partial}{\partial t} \int \mathbf{A} \cdot (\nabla \times \mathbf{A}) dV &= \int \left[\frac{\partial \mathbf{A}}{\partial t} \cdot (\nabla \times \mathbf{A}) + \mathbf{A} \cdot \nabla \times \left(\frac{\partial \mathbf{A}}{\partial t} \right) \right] dV \\ &= \int \left[\mathbf{A} \cdot \nabla \times \left(\frac{\partial \mathbf{A}}{\partial t} \right) \right] dV \end{aligned} \tag{7.161}$$

since the first term vanishes. We now use the vector identity $\nabla \cdot (\mathbf{A} \times \mathbf{B}) = \mathbf{B} \cdot (\nabla \times \mathbf{A}) - \mathbf{A} \cdot (\nabla \times \mathbf{B})$ to transform the above equation. The result is

$$\begin{aligned} \frac{\partial}{\partial t} \int \mathbf{A} \cdot (\nabla \times \mathbf{A}) dV &= \int \left[\frac{\partial \mathbf{A}}{\partial t} \cdot (\nabla \times \mathbf{A}) - \nabla \cdot \left(\mathbf{A} \times \frac{\partial \mathbf{A}}{\partial t} \right) \right] dV \\ &= \int \left[\frac{\partial \mathbf{A}}{\partial t} \cdot (\nabla \times \mathbf{A}) \right] dV - \int \left[\left(\mathbf{A} \times \frac{\partial \mathbf{A}}{\partial t} \right) \right] \cdot d\mathbf{S} \\ &= - \int \mathbf{A} \times (\mathbf{U} \times \mathbf{B}) \cdot d\mathbf{S} \end{aligned} \tag{7.162}$$

where, to arrive at the second line, we have used the divergence theorem. The first term of the second line again vanishes. To arrive at the third line, we used (7.156). The condition for the helicity of the magnetic potential $\mathbf{A}$ to be constant time is for $\int \mathbf{A} \times (\mathbf{U} \times \mathbf{B}) \cdot dS$ to vanish. Then $\int \mathbf{A} \cdot (\nabla \times \mathbf{A}) dV$ is constant for all time in an MHD fluid with $\sigma = \infty$ (more on this later).

We now examine under what condition the total magnetic energy $\int (B^2/2\mu_0) dV$ is a minimum. Let $\mathbf{B} = |\nabla \times \mathbf{A}|$ and use the Lagrange method of undetermined multipliers to enforce helicity conservation,

$$\begin{aligned} \delta \int |\nabla \times \mathbf{A}|^2 dV &= 0 \\ &= \int \{ (\nabla \times \mathbf{A}) \cdot (\nabla \times \delta \mathbf{A}) - \alpha [(\delta \mathbf{A}) \cdot (\nabla \times \mathbf{A}) \\ &\qquad + \mathbf{A} \cdot (\nabla \times \delta \mathbf{A})] \} dV \end{aligned} \tag{7.163}$$

where α is a Lagrange multiplier and is a constant. Now use the same vector identity used above and rewrite (7.163) as

$$\begin{aligned} &= \int [\nabla \times (\nabla \times \mathbf{A}) - \alpha \nabla \times \mathbf{A}] \cdot \delta \mathbf{A} dV \\ &\quad - \int \nabla \cdot [(\nabla \times \mathbf{A}) \times \delta \mathbf{A} + \alpha \mathbf{A} \times \delta \mathbf{A}] dV \\ &= \int [\nabla \times (\nabla \times \mathbf{A}) - \alpha \nabla \times \mathbf{A}] \cdot \delta \mathbf{A} dV \\ &\quad - \int (\nabla \times \mathbf{A} + \alpha \mathbf{A}) \times \delta \mathbf{A} \cdot dS \end{aligned} \tag{7.164}$$

where use is made again of the divergence theorem. We require $\delta\mathbf{A}$ to vanish at the surface for reasons already given above. Hence the surface contribution vanishes. Since $\delta\mathbf{A}$ is arbitrary, the integrand over the volume must vanish. Hence

$$\nabla \times (\nabla \times \mathbf{A}) - \alpha \nabla \times \mathbf{A} = 0 \tag{7.165}$$

Recalling that $\mathbf{B} = \nabla \times \mathbf{A}$, we arrive at the important result

$$\nabla \times \mathbf{B} = \alpha \mathbf{B} \tag{7.166}$$

The results just derived are summarized as follows. When the electromagnetic stress vanishes ($\mathbf{J} \times \mathbf{B} = 0$) and the current density in the force-free state is defined by $\mathbf{J} = \alpha \mathbf{B}$, the case with $\alpha =$ constant represents a state of the lowest magnetic energy in a system containing an MHD fluid with $\sigma = \infty$ and in which magnetic helicity is an invariant. In this system, a force-free field with a constant α is an equilibrium configuration. Other systems supporting field configurations different from the force-free configuration have energies higher than the force-free state and one expects that, if left alone, they would relax to a force-free configuration.

7.7.2 Field-Aligned Current and Parallel Electric Field

The equation of the field-aligned current for a two-component plasma can be written as

$$\mathbf{J}_{\parallel} = n^+ q^+ < \mathbf{v}_{\parallel}^+ > + n^- q^- < \mathbf{v}_{\parallel}^- > \tag{7.167}$$

where $< \mathbf{v}_{\parallel}^+ >$ and $< \mathbf{v}_{\parallel}^- >$ are velocities of ions and electrons averaged over the distribution function in the direction of the magnetic field $\mathbf{B}$. These velocities will change if there are electric fields in the direction parallel to $\mathbf{B}$.

Observations in Earth's auroral ionosphere suggest that electric fields exist in the direction parallel to the geomagnetic field $\mathbf{B}$. Evidence of $\mathbf{E}_{\parallel}$ was first deduced in 1967 from observations of precipitated energetic electron distributions measured at auroral altitudes. This has since been verified by numerous particle experiments on rockets and polar orbiting satellites.

A convenient way to display the experimental data is to show the behavior of the distribution of electrons in the phase space. Figure 7.14 shows an example of the distribution function of electrons measured in an aurora by a sounding rocket experiment (top panel). The five panels represent data obtained at five different times (time after launch of the rocket is shown on the upper right-hand corner). Contours of constant-phase space densities are displayed in the $(v_{\parallel}, v_{\perp})$ plane. The field-aligned electrons appear as the fingerlike extensions along the $v_{\parallel}$ axis.

The bottom panels show cuts through 0° and 65° pitch-angles for the distributions shown in the panels above. These figures show the contrast between the field-aligned and the more isotropic components of precipitation. A prominent feature of these spectra is the presence of peaks centered around a few keV. Such "monoenergetic" peaks have been interpreted to be the result of electrons falling through a field-aligned electrostatic potential located above the altitude where measurements are made. This potential accelerates the primary electrons travelling toward Earth (upward current) and the peak represents the value of the potential drop. The large fluxes observed below the monoenergy peak require another source since all electrons falling through the potential gain the same amount of energy. An interpretation given to the fluxes below the peak is that they are the energy-degraded electrons (produced by the interaction of the primary electrons colliding with the atmosphere) and the low-energy ionospheric electrons that are trapped between the potential and the mirror altitudes. These electrons have nearly isotropic pitch-angle distributions and model calculations have shown that this explanation is reasonable.

The electric field that accelerates electrons downward will accelerate ions upward. An example showing this effect is shown in Figure 7.15. These data come from a particle experiment flown on a polar orbiting satellite at an altitude of $\approx$ 7000 kilometers. The ion distribution function shows a peak in the upward direction with a $v_{\parallel}$ around -600 km/s (this corresponds to a proton energy of about 2 keV). The electrons were moving downward with a $v_{\parallel} \approx 2 \times 10^4$ km/s (this corresponds to an energy of about 1 keV). The peak energies here are different and this suggests that the potential occupies

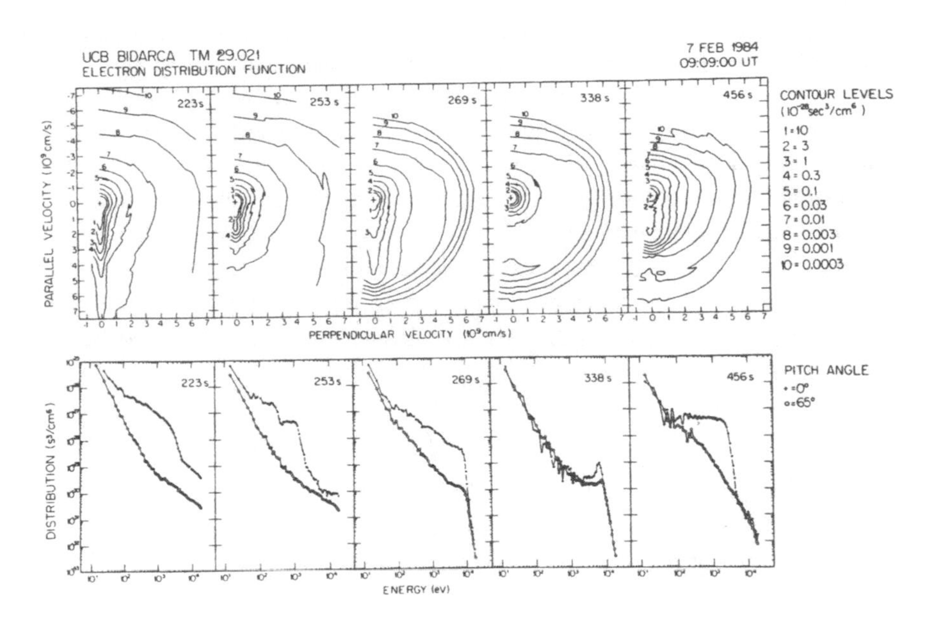

FIGURE 7.14 The top panel shows contour plots of electron distribution functions measured in an aurora. Positive $v_{\parallel}$ is in the downward direction. The bottom panel shows cuts through the five distributions at $0°$ and $65°$ pitch-angles. (From McFadden *et al.*, 1987)

an extended region along **B**. For a fixed-potential structure, data here can be interpreted to mean that electrons and ions entered the potential at different heights and resulted in different amounts of acceleration.

The existence of a field-aligned electric field $\mathbf{E}_{\|}$ requires that space charges be maintained along **B**. Space charges, however, are difficult to maintain because the highly mobile electrons along the magnetic field will neutralize any excess space charges. For example, consider a plasma immersed in a uniform magnetic field. Suppose a potential is now applied along the magnetic field. Electrons and ions will respond to this potential and they will move in such a way to cancel the $\mathbf{E}_{\|}$ (Figure 7.16).

Consider now a plasma in an inhomogeneous magnetic field, for example a dipole field. Suppose a potential is applied between the equator and the ionosphere. In this case, the parallel motion of the particles will be affected by the combination of the electric $qE_{\|}$ and the magnetic "mirror" $-\mu\partial B/\partial s$ force. The mirror force could reinforce or counteract the electrical force and the dependence of the particle's speed on position will be different for positive ions and electrons. The net result is that the plasma charge distribution in an inhomogeneous magnetic field under the action of parallel electric fields is much more complex than the case of a homogeneous magnetic field.

A proper way to study this problem is to show that a plasma in an inhomogeneous magnetic field supporting $\mathbf{E}_{\|}$ can still satisfy the plasma charge neutrality requirement (actually quasineutrality). This problem can be solved but is quite complicated and will not be treated here (see Stern, 1981, cited at the end of the chapter). Instead, we will show how $\mathbf{E}_{\|}$ arises in magnetospheres by use of the familiar "single" particle concepts.

Consider the particles trapped in a planetary dipole magnetic field and assume that these particles are electrons and protons. In Chapter 4, it was shown that particles travelling along an inhomogeneous magnetic field experience a force (see (4.72)). In the presence of a parallel electric field, this force is

$$F_{\|} = -\mu\frac{\partial B}{\partial s} + |q|E_{\|} \tag{7.168}$$

where μ is the magnetic moment of the particle and the distance s is measured along the magnetic field. Now $F_{\|} = mdv_{\|}/dt = mv_{\|}dv_{\|}/ds$, so for a plasma consisting of electrons and one species of ions (protons), we can write

$$\begin{aligned} m_e v_{\|e}\frac{dv_{\|e}}{ds} &= -\mu_e\frac{\partial B}{\partial s} + eE_{\|} \\ m_i v_{\|i}\frac{dv_{\|i}}{ds} &= -\mu_i\frac{\partial B}{\partial s} - eE_{\|} \end{aligned} \tag{7.169}$$

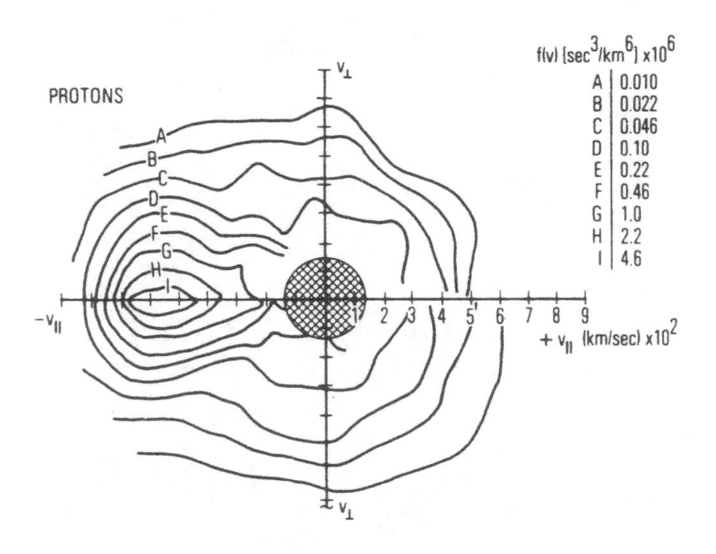

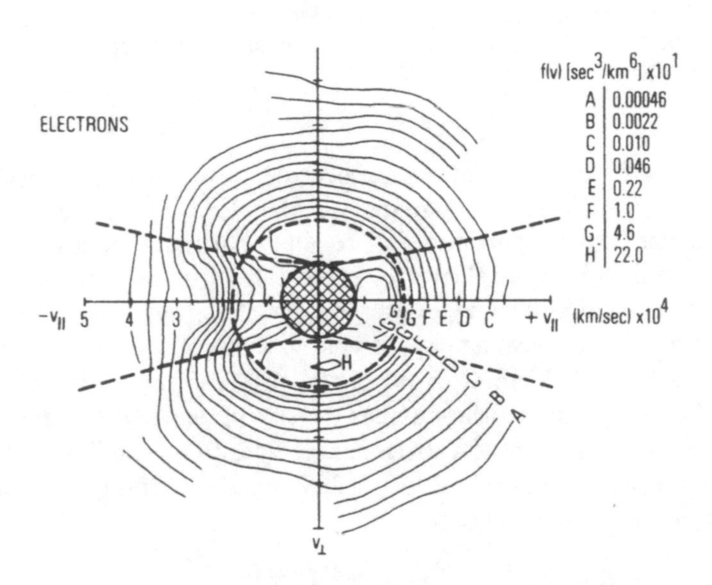

FIGURE 7.15 Contour plots of ion and electron distribution functions obtained by a polar orbiting satellite-borne instrument. The dashed ellipse represents the boundary between accelerated electrons and those electrons that cannot penetrate the electrostatic barrier. The dashed hyperbola represents the boundary of the loss cone in the atmosphere at the satellite altitudes. (From Mizera and Fennel, 1977)

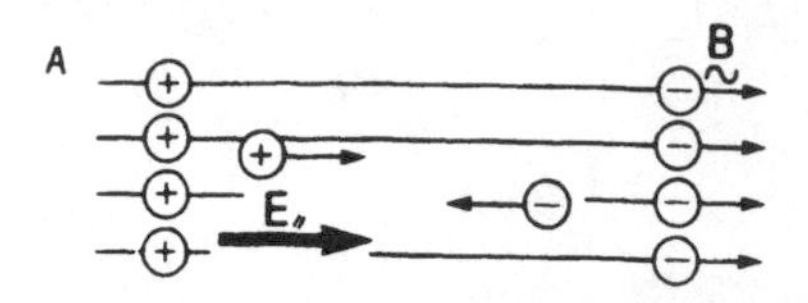

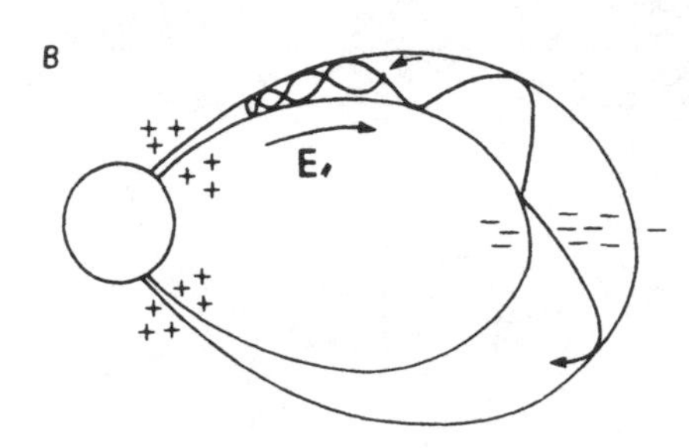

FIGURE 7.16 A schematic diagram showing how electrons and ions respond to electric fields. (A) homogeneous magnetic field, (B) inhomogeneous magnetic field. (From Stern, 1981)

where $|q| = e$ for protons and $-e$ for electrons ($e = 1.6 \times 10^{-19}$ coulombs) and the subscripts i and e designate ions and electrons, respectively.

Consider now the particles that oscillate back and forth between the points s_1 and s_2 with a velocity $v_\parallel$ (note that $v_\parallel$ varies with position in the mirroring field geometry). A particle spends a time $dt = ds/v_\parallel$ on the line element ds. The average space charge in ds is then $dQ = q\,ds/\tau v_\parallel$, where τ is the half bounce-period of the particles travelling between s_1 and s_2. Let n_i and n_e be the number density of ions and electrons. Then the explicit expression of the space charge density on the line element ds is $n_i dQ_i + n_e dQ_e$. Since charge neutrality (quasineutrality) is required at every point in space we must set

$$n_i dQ_i + n_e dQ_e = 0 \tag{7.170}$$

(Note that charge neutrality is also required for the entire plasma and thus the plasma must also satisfy $n_i q_i + n_e q_e = 0$.) This neutrality condition implies that $t_i v_{\parallel i} = t_e v_{\parallel e}$ and if we let $\alpha = t_i/t_e$, then

$$v_{\parallel e} = \alpha v_{\parallel i} \tag{7.171}$$

In general $\alpha \neq 1$ because electrons and ions (protons) in a magnetosphere do not necessarily have the same velocity or energy distributions.

Using (7.171), we can rewrite (7.169) as

$$\begin{aligned}
-\mu_e \frac{\partial B}{\partial s} - eE_{\parallel} &= m_e v_{\parallel e} \frac{dv_{\parallel e}}{ds} \\
&= \alpha^2 m_e v_{\parallel i} \frac{dv_{\parallel i}}{ds} \\
&= \alpha^2 \left(\frac{m_e}{m_i}\right) F_{\parallel i} \\
&= \alpha^2 \left(\frac{m_e}{m_i}\right) \left(-\mu_i \frac{\partial B}{\partial s} + eE_{\parallel}\right) \qquad (7.172)
\end{aligned}$$

Note that the magnetic moment μ and the velocities $v_{\parallel}$ are averaged quantities. Solving for $E_{\parallel}$ yields

$$E_{\parallel} = -K \frac{\partial B}{\partial s} \qquad (7.173)$$

where

$$\begin{aligned}
K &= \frac{(\mu_e/m_e - \alpha^2 \mu_i/m_i)}{e(1/m_e + \alpha^2/m_i)} \\
&= \frac{W_{\parallel i} W_{\perp e} - W_{\parallel e} W_{\perp i}}{eB(W_{\parallel e} + W_{\parallel i})} \qquad (7.174)
\end{aligned}$$

Equation (7.174) made use of the relations $\mu = W_{\perp}/B$ and $W_{\parallel} = mv_{\parallel}^2/2$ and $W_{\perp} = mv_{\perp}^2/2$. Equation (7.173) shows that $E_{\parallel}$ vanishes if and only if the magnetic field is homogeneous $\partial B/\partial s = 0$ or if

$$\begin{aligned}
K &= W_{\parallel i} W_{\perp e} - W_{\parallel e} W_{\perp i} \\
&= 0 \qquad (7.175)
\end{aligned}$$

Here

$$\begin{aligned}
W_{\parallel} &= \int \left(\frac{mv_{\parallel}^2}{2}\right) f(v_{\parallel}, v_{\perp}) d^3v \\
W_{\perp} &= \int \left(\frac{mv_{\perp}^2}{2}\right) f(v_{\parallel}, v_{\perp}) d^3v \qquad (7.176)
\end{aligned}$$

are energies associated with particles moving in the parallel and perpendicular directions of the magnetic field averaged over the distribution function. There is no requirement that the electrons and protons in magnetospheres have the same distribution function. Hence it is unlikely that $K = 0$.

The results can be summarized as follows. Unless the distributions of electrons and ions are equal, plasmas in inhomogeneous magnetic fields will support parallel electric fields. For example, suppose the pitch-angle distributions of electrons and ions in the magnetosphere are different so that the electrons mirror closer to the planet than the ions. This will create a charge separation along **B** and the electric field will accelerate the ions downward and electrons upward (a downward current). If the source is static, the plasma will eventually neutralize itself. However, magnetospheric sources are dynamic and charge separation along **B** will be continuously created. This simple picture qualitatively explains why one could detect parallel electric fields in magnetospheres at auroral altitudes.

Let us now return to the discussion of the two-dimensional contour plots shown in Figures 7.14 and 7.15. Suppose the electron distribution function can be described by a Maxwellian function (we caution the reader that the actual distribution is *not* a simple Maxwellian).

$$f(v) = f_0 \exp\left(\frac{-mv^2}{2kT}\right) \tag{7.177}$$

where $mv^2/2$ is the kinetic energy of the electron and kT its characteristic thermal energy. An isotropic distribution has circular iso-intensity contours in the velocity space centered about $v = 0$. Now let this distribution pass through a potential ψ along a constant magnetic field. After acceleration, most of the phase space contours will remain circular (because of conservation of particle energy and phase space density; Liouville's theorem), except that the contours for $v_\parallel < (2e\psi/m)^{1/2}$ vanish. Assuming the first invariant is conserved, this cutoff surface can be expressed as a maximum pitch-angle for any given contour,

$$\sin^2 \alpha_{max} = \frac{W}{W + e\psi} \tag{7.178}$$

where W is the energy of the corresponding contour in the original distribution.

The pitch-angle distribution becomes narrower as $e\psi$ increases. However, note that if the acceleration takes place at high altitudes, the pitch-angles will increase as the electrons move into stronger magnetic fields at lower altitudes. If B_0 and B are the magnetic field strength at the height where the potential is encountered and where the electrons are detected, (7.178) will be modified to

$$\sin^2 \alpha_{max} = \frac{B_0}{B}\frac{W}{W + e\psi} \tag{7.179}$$

Thus we see that the shape of the pitch-angle distribution function of the parallel electrons is determined by the combination of electrostatic potential and the magnetic mirror forces. Detailed studies of these distributions thus provide important information on the nature and structure of the potential and where the particle acceleration occurs.

7.8 Magnetic Tail Currents

The topic of magnetic tails was introduced in Chapter 3. There it was mentioned that magnetic tails often are a part of large magnetic structures in space. For example, planetary magnetospheres, comets, and magnetic structures of active regions on the Sun all support magnetic tails. Magnetic tails are of great interest to space physicists because they include neutral sheets whose dynamics are involved in many flare and auroral theories.

Magnetic tails are created by a tail current and it is customary to place this current in the neutral sheet. The configuration of this current must be such that when the magnetic field of this current and the principal magnetic field (a dipole, for instance) are superposed, a magnetic tail geometry ensues. For Earth's magnetotail, the current flows from dawn to dusk. A considerable amount of effort has been given to modelling the magnetic tail of Earth. These models are generally considered functional or empirical (or semi-empirical) or both. A functional model of a distorted geomagnetic field considers an infinitesimally thin (z-direction) current sheet embedded in the neutral sheet (Figure 7.17). This current sheet has a finite dimension in the x-direction and extends from $-\infty$ to $+\infty$ in the y-direction. The magnetic field of this current sheet is defined by

$$\begin{aligned} B_x &= -\frac{\mu_0}{2\pi} J(\theta_2 - \theta_1) \\ B_z &= \frac{\mu_0}{2\pi} J \ln\left(\frac{r_2}{r_1}\right) \end{aligned} \tag{7.180}$$

where J, the linear current density (current/length), is a free parameter chosen by observations. For example, J can be chosen so that its magnitude at $x = 10R_E$ in the tail direction will produce a magnetic field intensity around 10–20 nanoteslas. When this field of a sheet current is superposed with the dipole field, a magnetic tail configuration will be obtained. This

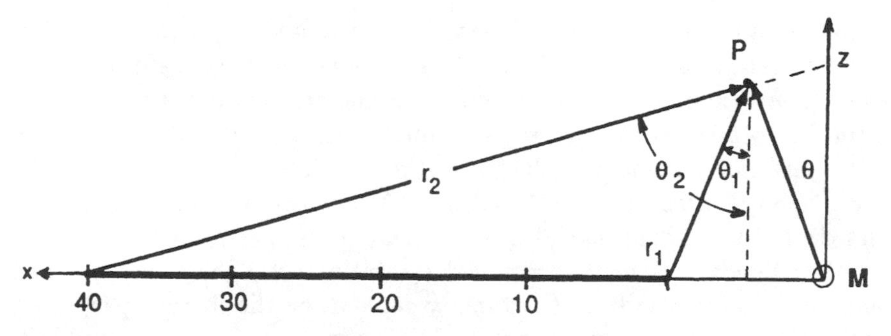

FIGURE 7.17 Geometry of the tail current sheet.

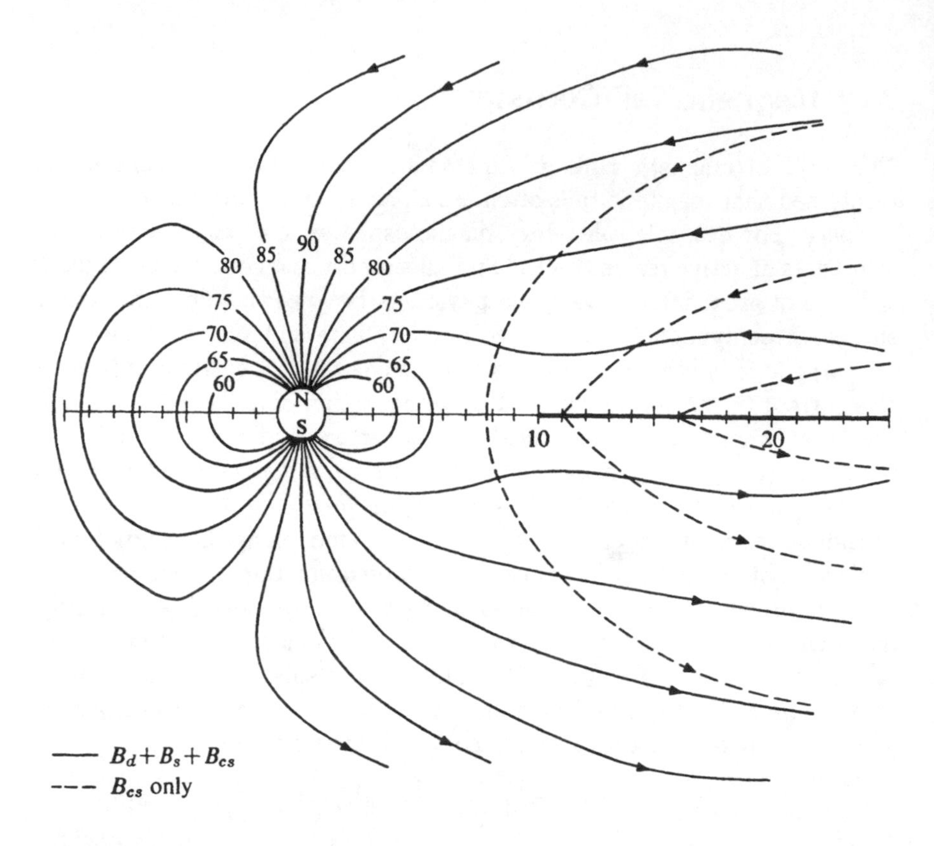

FIGURE 7.18 Geomagnetic field model used to explain the behavior of trapped particle distributions. The dashed lines are the magnetic field of the current sheet B_{cs}. The solid lines represent the superposition of the sheet field and the dipole B_d and a surface magnetopause field B_s. (From Williams and Mead, 1965)

functional model is useful and it was formulated in 1966 to explain qualitatively the behavior of trapped particle fluxes during moderately disturbed geomagnetic times (Figure 7.18).

From a theoretical point of view, functional and empirical models are not completely satisfactory. While these models may be useful and may even agree with observations, it seems that the problem ought to be solved from first principles. A theoretical approach that is currently considered most advanced in modelling large-scale magnetic structures relies on the MHD theory. However, these MHD models can become quite complicated and the full three-dimensional problem has not yet been solved.

MHD models take into account the stability factor that functional and empirical models ignore. MHD stability depends on the thermal properties of the plasmas, and Maxwell and the MHD equations must be solved self-consistently. The simplest MHD models assume that the magnetic structure is in hydrostatic equilibrium. Thus the equations governing static models are

$$\mathbf{J} \times \mathbf{B} = \nabla \cdot \mathbf{p}$$

$$\nabla \times \mathbf{B} = \mu_0 \mathbf{J}$$

$$\nabla \cdot \mathbf{B} = 0 \tag{7.181}$$

Here $\mathbf{p}$ is the pressure tensor and if it can be assumed to be isotropic, $\nabla \cdot \mathbf{p}$ reduces to ∇p. The solutions of (7.181) must ensure that the equilibrium condition is always satisfied (the topic of MHD equilibrium will be introduced in Chapter 11).

A simple MHD model of the magnetotail is that of the Harris sheet discussed in Chapter 3. There we showed that the magnetic field is given by

$$B_x = B_0 \tanh\left(\frac{z}{L}\right)$$

$$B_y = B_z = 0 \tag{7.182}$$

where B_0 is a constant. The current density, derived from $\mu_o \mathbf{J} = \nabla \times \mathbf{B}$, is $J_x = J_z = 0$, and

$$J_y = \left(\frac{B_0}{\mu_0 L}\right) \operatorname{sech}^2\left(\frac{z}{L}\right) \tag{7.183}$$

This current has a finite dimension in the z-direction. The plasma density (mass density) and temperature used in this model are

$$\rho_m(x, y, z) = \rho_0 \tag{7.184}$$

$$T(z) = T_0 \operatorname{sech}^2\left(\frac{z}{L}\right) \tag{7.185}$$

where ρ_0 and T_0 are constants. A magnetic tail described by the above equations is a stable MHD configuration.

Unfortunately, the stability configuration for the general MHD problem remains unsolved. Even for special cases, for example a two-dimensional problem, stability is not achieved for any arbitrary density and temperature functions. It has also been shown that if the pressure is anisotropic, instabilities arise (further information is found in Voigt, 1986, cited at the end of the chapter).

A phenomenological description pictures magnetic tails to form by "stretching" planetary fields in the solar wind flow direction. Stretching here means that the solar wind momentum is transferred to the planetary magnetic fields. To stretch a dipole field to yield a tail-like field requires a considerable amount of momentum transfer. Just how this is achieved in collisionless plasmas remains unknown. Another description pictures mag-

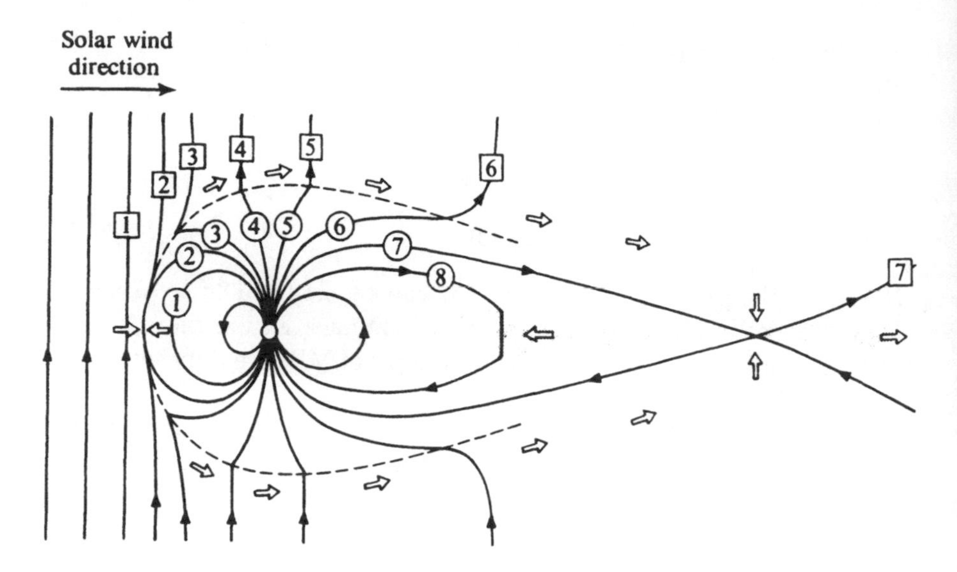

FIGURE 7.19 Evolution of merged magnetic field lines being swept back to form a tail (originally proposed by J. Dungey in 1961).

netotails as forming by "merging" a planetary magnetic field with the interplanetary magnetic field. Merged field lines have one end attached to the planet and the other in interplanetary space. Since the portion of the field line in interplanetary space is frozen in the solar wind and flows with the solar wind, it will stretch into a magnetotail (Figure 7.19).

Merging magnetic fields as indicated above relies on the idea that magnetic neutral points or sheets can be created. However, we must note that the appearance of neutral points or sheets in plasmas is not universally accepted. This is because the processes that create neutral points are still under much theoretical discussion and are not yet completely understood.

Figure 7.20 shows a more detailed schematic diagram of magnetic field configurations occurring in a merging process. Diagram (a) represents $t < 0$ configuration when two plasmas carrying oppositely directed magnetic fields move toward each other. Diagram (b) shows $t = 0$ configuration when the two plasmas collide and their magnetic fields merge. Diagram (c) shows $t > 0$ configuration when merged field lines have been created. Note that merging produces a change of magnetic field topology.

Currents associated with these configurations are as follows: for the $t < 0$ configuration, the two independent plasma systems include the diamagnetic currents $\mathbf{J} = \mathbf{B} \times \nabla p / qnB^2$ and these currents are in the same direction (directed out of the page). To produce the $t = 0$ configuration, the two plasmas must be allowed to "coalesce" or merge. Let this occur

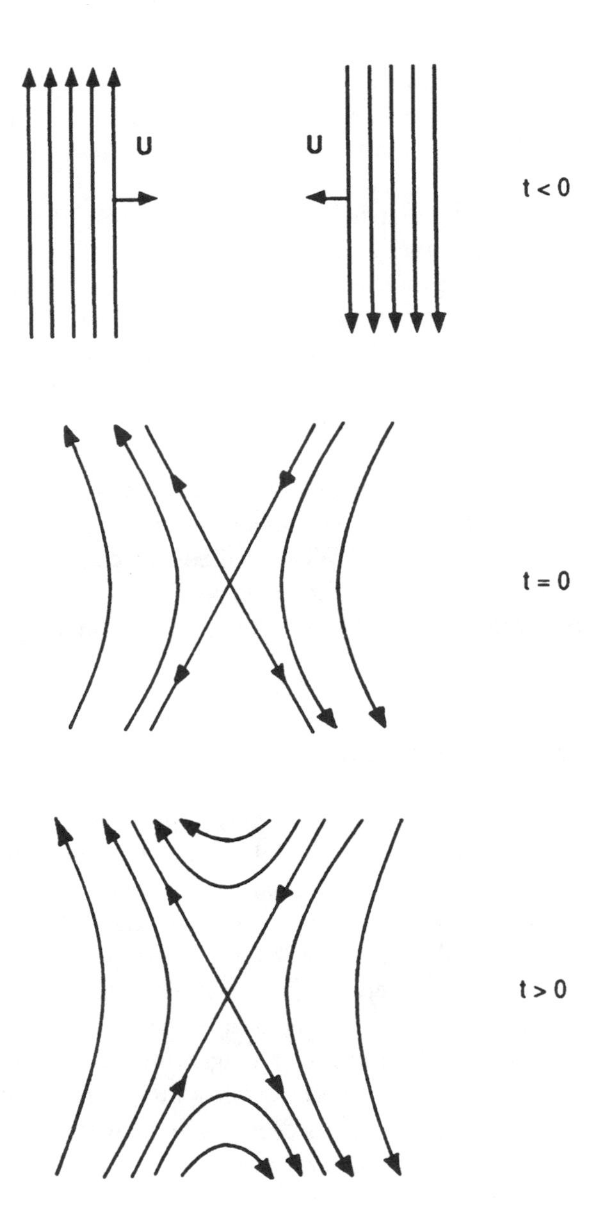

FIGURE 7.20 A schematic diagram to show how merging of the magnetic field works.

at an x-point where $\mathbf{B} = 0$. Since $\mathbf{B} = 0$ at the x-point, the diamagnetic current vanishes there. It does not mean, however, that x-points do not support currents. Since the electric field $\mathbf{E} = \nabla\Phi - \partial\mathbf{A}/\partial t = -\nabla\Phi$, $\mathbf{E}$ need not vanish when $\mathbf{B} = 0$ and currents can be driven by this electric field. The existence of this electric field is a requirement of the merging process.

Note that the current driven by the electric field at the x-point is directed into the paper, in the opposite direction from the original diamagnetic currents in the two plasma systems (before merging occurs the two systems are considered separate).

Each merging of a pair of field lines creates a pair of new field lines in steady-state merging process. Hence the $t > 0$ configuration requires two current systems, one responsible for the field lines of incoming plasmas and one for the newly created field lines. The newly created current is dissipative since $\mathbf{E} \cdot \mathbf{J} \neq 0$ and Joule heating is expected to occur at the x-points. This also implies that at the x-point where the two fluids coalesce and merging of the field lines occurs, the fluid conductivity must be finite, $\sigma \neq \infty$.

As the two plasmas are brought together, the fluid pressure between them will increase. How much pressure can be built up between the two fluids pushing against each other? What confines the plasma into the small region to form the x-points? This discussion will be continued in Chapter 8 where a model is presented to show how fluids flow in the neighborhood of neutral points.

7.9 Concluding Remarks

Understanding how electrical currents are produced and flow in space is important because currents are intimately associated with the dynamics of large-scale solar, galactic, and universal magnetic structures. For convenience, our presentation organized the currents in terms of where they are flowing. Thus we discussed currents at magnetic boundaries, in the ring, in magnetic tails, and in ionospheres. We also briefly discussed field-aligned currents associated with force-free magnetic structures. In a quiet state, these currents determine the shapes of magnetospheres. However, magnetic disturbances (for example, auroras and flares), are associated with topological changes of structures caused by the dissipation of some of these currents while others are being redistributed.

Magnetospheric currents are produced by the solar wind interaction with planetary magnetic fields, motions of trapped particles in radiation belts, convection of plasma sheet particles in the tail, and ionospheric particles. These currents contribute to the momentum flow and, in MHD, an equation of current that flows in the direction perpendicular to the magnetic field can be obtained directly from the momentum equation. However, an equation of current that runs along the magnetic field direction cannot be obtained directly from the momentum equation because $\mathbf{J} \times \mathbf{B} = 0$.

Instead, information on these field-aligned currents is obtained from the integration of the perpendicular current equation along the magnetic field, which invokes current continuity.

Field-aligned currents play an important role in coupling magnetospheres and ionospheres. Although the distant magnetospheric regions and the ionospheres are already connected by the magnetic field, these regions become dynamically coupled because of the field-aligned current. Understanding the dynamic coupling requires studying the processes by which electric fields and currents are generated and transmitted in magnetized plasmas. For example, understanding is required into how convective electric fields drive fluid flows and how convective electric fields are transmitted into ionospheres for dynamic situations.

The processes by which Earth's ionosphere-magnetosphere system is coupled have received much attention and there exist several models. A common theme of these models is that magnetospheres are driven by the solar wind. This induces electric fields to drive flows in magnetospheres and currents in ionospheres. The starting point is the diamagnetic boundary or the ring current and one calculates the currents that flow into and out of the ionosphere along the magnetic field directions. The electric fields that drive the resistive currents are then computed by invoking current closure through the ionosphere and applying the necessary boundary conditions, such as current termination in the low- or high-altitude boundaries. These calculations must be performed in a self-consistent way by taking into account for example, modifications of magnetospheric flows due to ionospheric electric fields that are mapped back into the magnetosphere. In these models, magnetospheres act as a current source. The ionosphere is a load and for a constant current, the ionospheric feedback effects are strong (weak) if the ionospheric conductivities are low (high). Similar physics can be applied to study the interaction between the I_0 torus and Jupiter's atmosphere.

A topic not discussed is the dynamo theory. This topic is very specialized and it can be developed by expanding the concepts discussed in this chapter. Dynamo theories are intimately associated with how currents are generated and amplified in nature and these theories are important in the studies of origins of solar, planetary, and cosmic magnetic fields.

Bibliography

Alfvén, H. and C.-G. Fälthammar, *Cosmical Electrodynamics, Fundamental Principles*, 2nd ed., Oxford University Press, Oxford, England, 1963. Chapter 5 discusses the concept of plasma conductivity and currents with emphasis on naturally occurring plasmas in space (natural plasmas as distinguished from laboratory plasmas). They also show how parallel electric fields arise in plasmas with unequal electron and ion distributions in inhomogeneous magnetic fields.

Berko, F.W., L.J. Cahill, Jr. and T.A. Fritz, Protons as the Prime Contributors to Storm Time Ring Current, *J. Geophys. Res.*, **80**, 3549, 1975. One of the first articles that used observational data to test the ring current theory.

Boström, R., Electrodynamics of the Ionosphere, in *Cosmical Electrodynamics*, A. Egeland, Ø. Holter, and A. Omholt, eds., Scandinavian University Books, Universitetsfor-Laget, Oslo, Sweden, 1973. This article presents a concise and clear formulation of ionospheric currents. It complements the topics treated in this chapter.

Cahill, L.J. Jr., Inflation of the Inner Magnetosphere, in *Physics of the Magnetosphere*, R.L. Carovillano, J.F. McClay and H.R. Radoski, eds., D. Reidel Publishing Co., Dordrecht, Holland, 1968. This article discusses the search for the terrestrial ring current, using satellite- and ground-based magnetic field data.

Carovillano, R.L. and J.J. Maguire, Magnetic Energy Relationships in the Magnetosphere, in *Physics of the Magnetosphere*, R.L. Carovillano, J.F. McClay and H.R. Radoski, eds., D. Reidel Publishing Co., Dordrecht, Holland, 1968. This paper presents methods of calculating magnetic field energy changes and formulates several fundamental theorems that can be applied to planetary magnetospheric processes.

Carovillano, R.L. and G.L. Siscoe, Energy and Momentum Theorems in Magnetospheric Processes, *Rev. Geophys. Space Phys.*, **11**, 289, 1973. This article reviews basic energy and momentum theorems applicable to systems containing electromagnetic fields and charged particles. Emphasis is given toward understanding magnetospheric processes.

Chen, F.F., *Introduction to Plasma Physics*, Plenum Press, New York, NY, 1974. Chapters 3 and 5 are relevant to the topics of currents and plasma in a collision-dominated medium.

Dungey, J.W., Interplanetary Magnetic Field and the Auroral Zones, *Phys. Rev. Letters*, **6**, 47, 1961. Dungey first suggested the importance of the southward interplanetary magnetic field on the topology of Earth's magnetopause.

Evans, D.S., Precipitating Electron Fluxes Formed by a Magnetic Field Aligned Potential Difference, *J. Geophys. Res.*, **79**, 2853, 1974. Evans is one of the first who interpreted the particle data in terms of the parallel electric field.

Fairfield, D.H., Solar Wind Control of the Magnetotail, in *Solar-Wind Magnetospheric Coupling*, Y. Kamide and J.A. Slavin, eds., Terra Scientific Publishing Co., Tokyo, Japan, 1986. This article reviews the current status of observations of the magnetotail of Earth.

Frank, L.A., On the Extraterrestrial Ring Current During Geomagnetic Storms, *J. Geophys. Res.*, **72**, 3753, 1967. The first article that showed the buildup and decay of proton fluxes in the magnetosphere during a geomagnetic storm.

Hasegawa, A. and T. Sato, Generation of Field Aligned Current During Substorm, in *Dynamics of the Magnetosphere*, S.-I. Akasofu, ed., D. Reidel Publishing Co., Dordrecht, Holland, 1980. This article shows how one can obtain an explicit expression of field-aligned currents from the equation of motion.

Jackson, J.D., *Classical Electrodynamics*, 2nd ed., John Wiley and Sons, Inc., New York, NY, 1975. Chapter 13 discusses the physics of charged particle collisions.

Kamide, Y., *Electrodynamic Processes in the Earth's Ionosphere and Magnetosphere*, Kyoto Sangyo University Press, Kyoto, Japan, 1988. This book contains a considerable amount of observational results and is mainly written for researchers in the field. Various models are discussed with emphasis on understanding the polar aurora.

Kamide, Y. and G. Rostoker, The Spatial Relationship of Field-Aligned Currents and Auroral Electrojets to the Distribution of Nightside Auroras, *J. Geophys. Res.*, **82**, 5589, 1977. This article shows examples of field-aligned currents that map to discrete auroral structures.

Krall, N.A. and A.W. Trivelpiece, *Principles of Plasma Physics*, McGraw-Hill Book Co., Inc., New York, NY, 1973. Chapter 3 of this book contains material relevant to the physics of MHD fluids. Both this and the book by Longmire emphasize the behavior of laboratory plasmas (fusion plasmas).

Longmire, C.L., *Elementary Plasma Physics*, Interscience Publishers, New York, NY, 1963. This book includes an excellent discussion of currents in plasmas resulting from individual particle motions. The basic concepts are clearly presented in Chapters 2 and 3.

McFadden, J., C.W. Carlson, M.H. Boehm and T.J. Hallinan, Field-Aligned Electron Flux Oscillations that Produce Flickering Aurora, *J. Geophys. Res.*, **92**, 11133, 1987. Recent observations of parallel electron fluxes interpreted in terms of parallel electric fields.

Mizera, P.F. and J.F. Fennell, Signatures of Electric Fields from High and Low Altitude Particle Distributions, *Geophys. Res. Lett.*, **4**, 311, 1977. This paper reports observations of features in particle distributions that can be interpreted in terms of parallel electric fields.

Peck, E.R., *Electricity and Magnetism*, McGraw-Hill Book Co., Inc., New York, NY, 1953. This old undergraduate textbook discusses clearly some of the basic concepts of currents in electrodynamics. The emphasis is on currents in wires rather than plasmas.

Potemra, T., ed., *Magnetospheric Currents*, Geophysics Monograph, American Geophysical Union, Washington, D.C., 1984. This monograph is a conference proceeding on magnetospheric currents and includes many papers on current research. Several articles that are quite interesting and illuminating deal with the history of field-aligned currents. The papers include observational results as well as models of currents in space. The paper by V. Vasyliunas is an excellent tutorial and points out clearly the conceptual differences between currents running on wires and currents in tenuous plasmas that permeate space. Vasyliunas also discusses the origin of several representations of currents commonly encountered in different plasma physics texts.

Salingaros, N., On Solutions of the Equation $\nabla \times \mathbf{a} = k\mathbf{a}$, *J. Physics A*, **19**, L101, 1986.

Schindler, K. and J. Birn, Magnetotail Theory, *Space Science Rev.*, **44**, 307, 1986. This article reviews the current theory of magnetotail structure and complements the article by G. Voigt. Emphasis is given to magnetic merging as the dominant process.

Silsbee, H. and E. Vestine, Geomagnetic Bays, Their Frequency, and Current Systems, *Terr. Mag. and Atmos. Elect.*, **47**, 195, 1942. These authors constructed the disturbance ionospheric current system responsible for the magnetic variations by studying hundreds of polar magnetic records obtained from many stations.

Stern, D.P., One Dimensional Models of Quasi-Neutral Parallel Electric Fields, *J. Geophys. Res.*, **86**, 5839, 1981. A good tutorial article on parallel electric fields. Readers interested in pursuing further the physics of parallel electric fields should read this article. The references are fairly complete.

Voigt, G.H., Magnetospheric Equilibrium Configurations and Slow Adiabatic Convection, in *Solar Wind-Magnetospheric Coupling*, Y. Kamide and J.A. Slavin, eds., Terra Scientific

Publishing Co., Tokyo, Japan, 1986. This review article discusses the global magnetic field configuration of magnetospheres in terms of MHD equilibrium theory.

Williams, D.J. and G.D. Mead, Nightside Magnetospheric Configuration as Obtained from Trapped Electrons at 1100 Kilometers, *J. Geophys. Res.*, **70**, 3017, 1965. A functional model of the geomagnetic tail was developed to explain the behavior of energetic electrons trapped in Earth's magnetosphere.

Woltjer, L., A Theorem on Force-Free Magnetic Fields, *Proc. Nat. Acad. Sci.*, **44**, 489, 1958.

Zumuda, A., J.H. Martin and F.T. Heuring, Transverse Magnetic Disturbances at 1100 Kilometers in the Auroral Region, *J. Geophys. Res.*, **71**, 5033, 1966. The first article that reported the measurement of magnetic signatures arising from field-aligned currents.

Questions

1. Currents in excess of a million amperes are observed in the Earth's ionosphere and the magnetosphere (ring current) during storms. The average energy of the ring current particles is about 1 keV. Compute the total energy that is available in these currents.
2. Suppose you wish to harness this energy and use it in New York. Assuming that the population is ten million, how much average power is available to each person during a magnetic storm?
3. Assume the auroral belt consists of a strip of circle between 60° and 65° latitude. Suppose the ionospheric current is distributed uniformly in this auroral belt. Compute the current density.
4. Discuss a situation in which the plasma has a mean velocity but the current vanishes.
5. It is sometimes overheard in the corridors of a meeting place that currents running along **B** (field-aligned currents) are analogous to currents running along wires. Discuss and criticize this statement.
6. The dipole field has a gradient and therefore a force which is proportional to the gradient of the magnetic energy density. Discuss why the field lines don't "blow" away.
7. Show that for a centered dipole field, the expression of the ring current is

$$\mathbf{J}_\perp = \frac{r^2}{M(1+3\sin^2\lambda)}\left[2\sin\lambda\frac{\partial P_\perp}{\partial\lambda} + r\cos\lambda\frac{\partial P_\perp}{\partial r}\right.$$

$$\left. + 3(P_\parallel - P_\perp)\cos\lambda\frac{1+\sin^2\lambda}{1+3\sin^2\lambda}\right]\hat{\boldsymbol{\phi}}$$

where $P_\perp$ and $P_\parallel$ are pressures perpendicular and parallel to the magnetic field direction, r is the radial distance, and λ is the magnetic latitude.
8. Compute the expressions of the parallel and perpendicular pressures, $p_\parallel$ and $p_\perp$ for a bi-Maxwellian distribution in which the temperatures in the directions parallel and perpendicular to the magnetic field are not equal, $T_\parallel \neq T_\perp$.

9. $\mathbf{J}_\perp$ given by Equation (7.65) satisfies the Maxwell equation $\nabla \times \mathbf{B} = \mu_0 \mathbf{J}$. Show that

$$\frac{\mathbf{B} \times (\nabla \times \mathbf{B})}{\mu_0} = -\nabla_\perp p_\perp - (p_\parallel - p_\perp)(\mathbf{b} \cdot \nabla)\mathbf{b}$$

10. For uniform plasma flow in steady state, the equation of motion along B is given by Equation (7.54). Show that this equation is equivalent to

$$B \frac{\partial}{\partial s} \frac{p_\parallel}{B} + \frac{p_\perp}{B} \frac{\partial B}{\partial s} = 0$$

where $\partial / \partial s = (\mathbf{b} \cdot \nabla)$ is the gradient operator along the magnetic field direction.

11. Consider the expression of the current density in the x-direction

$$J_x = q \int f v_x d^3 v$$

Now introduce the cylindrical coordinates in velocity space

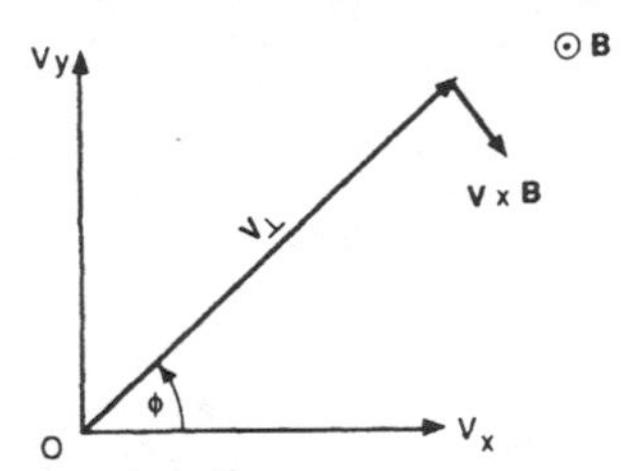

a. Transform the above equation and show that

$$J_x = q \int f v_\perp^2 \cos \phi dv_\perp d\nu_\parallel d\phi$$

b. Compute J_x for a bi-Maxwellian distribution function

$$f(v_\parallel, v_\perp) = \left(\frac{m}{2\pi k T_\perp}\right) \left(\frac{m}{2\pi k T_\parallel}\right)^{1/2} e^{-(mv_\perp^2/2kT_\perp + mv_\parallel^2/2kT_\parallel)}$$

12. An explicit expression for the field-aligned current derived by Hasegawa and Sato (1980) is

$$J_\parallel = B \int \left(\frac{P}{B} \frac{d}{dt} \frac{\Omega}{B} + \frac{2}{B^2} \mathbf{J}_\perp \cdot \nabla B + \frac{1}{\rho B} \mathbf{J}_{in} \cdot \nabla \rho \right) dl_\parallel$$

where $\mathbf{J}_{in}$ is the inertial current, Ω is the vorticity, ρ is the mass density, and the other terms have their usual meaning. Derive this expression starting from

$$\mathbf{J}_\perp = \frac{\mathbf{B} \times \nabla p}{B^2} - \frac{\rho}{B^2} \frac{d\mathbf{u}}{dt} \times \mathbf{B}$$

13. A publication by Salingaros (1986) states that the solutions of the differential equation $\nabla \times \mathbf{a} = k\mathbf{a}$ where k is a constant that "exhibits several inconsistencies."

a. Read and summarize the main points of this paper.

b. State whether you agree or disagree with the conclusions of this paper. Your reasons should include a discussion of the gauge dependency of the solutions and the behavior of the solutions under parity transformation.

8

Boundaries in Space

8.1 Introduction

A boundary is an interface that separates the different regions of space. A plasma boundary is defined through discontinuous fluid parameters such as density and temperature in the same way ordinary fluid boundaries are defined. However, plasmas are coupled to electric and magnetic fields, and therefore the requirements must also include the behavior of discontinuous electromagnetic fields. The complete set of discontinuous variables in collisionless plasma fluids includes the mass density ρ_m, flow velocity $\mathbf{U}$, pressure tensor $\mathbf{p}$, current $\mathbf{J}$, and the magnetic and electric fields $\mathbf{B}$ and $\mathbf{E}$.

The solar wind encounters both magnetized and non-magnetized objects in our solar system. The interactions that occur in these two cases are quite different; one involves electromagnetic forces while the other does not, and they produce different kinds of boundaries. The main purpose of this chapter is to develop an understanding of how boundaries are produced by flowing plasma interacting with planetary magnetic fields. We will use the magnetohydrodynamic (MHD) continuum theory of fluids to obtain information on large-scale properties of plasma boundaries. MHD theory is valid when the dimensions of the boundaries are much larger than the ion Larmor radius.

MHD fluids can develop more types of discontinuities than can ordinary fluids (see *Electrodynamics of Continuous Media*, by L.D. Landau and E.M. Lifshitz). These discontinuities, classified as tangential, contact, rotational, and shock, define how mass, momentum, and energy are transported across them. Shocks are created by supersonic flows. Contact discontinuities separate two regions of space at rest with different densities and temperatures and may play an important role in describing non-magnetized ionospheres. Tangential and rotational discontinuities define closed and open boundary configurations and they play an important role in the solar wind interaction with planetary magnetic fields.

Fluid theory and the first-order orbit theory are not applicable if a boundary is thin and the dimensions are of the order of one to several Larmor radii, as observations of the Earth's magnetopause boundary sometimes show. A more exact Lorentz theory must be used to describe the orbits of particles in the neighborhood of a thin boundary. As mentioned in Chapter 7, "micro" structures of Earth's magnetopause for a closed boundary configuration were first studied by S. Chapman and V.A. Ferraro in 1931. Although the model of Chapman and Ferraro is idealistic, their theory provides a basic picture of how a closed boundary is formed.

An open boundary is based on the concept that magnetic neutral points may form when two oppositely directed magnetic fields are superposed. "Merging" of magnetic fields provides an efficient mechanism for converting magnetic field energy into particle energy and the concept is often invoked to explain how particles are accelerated in solar flares and auroras. Transport of plasma also occurs much more efficiently in boundaries that include neutral points (transport in closed boundaries is achieved mainly by stochastic diffusion). Open-boundary models are still in infancy, and we discuss the model of E.N. Parker and P.A. Sweet.

This chapter includes a short discussion of the status of observations in relation to the theoretical predictions of the various fluid models. The last section of this chapter presents a simple model of boundary layers to illustrate how boundary layers arise in closed models.

8.2 Basic Equations and Assumptions

Observations by experiments carried on unmanned spacecraft have led us to conclude that there are four basic types of boundaries in space. Figure 8.1 shows a schematic diagram of the boundaries the solar wind creates in the neighborhood of (a) a solid object, (b) an unmagnetized planet with an ionosphere, (c) a magnetized planet with a closed boundary, and (d) a magnetized planet with an open boundary. The different types of boundaries can be characterized by the behavior of electromagnetic fields and the mass, momentum, and energy of fluids across these boundaries.

Assume that a boundary is stable and in equilibrium with the surrounding plasma medium. The coupled Maxwell and fluid equations then reduce to

$$\nabla \cdot \mathbf{B} = 0 \tag{8.1}$$

$$\nabla \cdot \mathbf{D} = 0 \tag{8.2}$$

$$\nabla \times \mathbf{H} = \mathbf{J} \tag{8.3}$$

$$\nabla \times \mathbf{E} = -\frac{\partial \mathbf{B}}{\partial t} \tag{8.4}$$

$$\frac{\partial \rho_m}{\partial t} + \nabla \cdot \rho_m \mathbf{U} = 0 \tag{8.5}$$

$$\frac{\partial(\rho_m U_i)}{\partial t} + \frac{\partial \pi_{ik}}{\partial x_k} = 0 \tag{8.6}$$

$$\frac{\partial}{\partial t}\left(\frac{\rho_m U^2}{2} + p_{ik} + \frac{B^2}{2\mu_0}\right) + \frac{\partial q_i}{\partial x_i} = 0 \tag{8.7}$$

where the mass, momentum, and energy equations have been written in conservational form. Here

$$\pi_{ik} = \rho_m U_i U_j + p_{ik}\delta_{ik} - \frac{(2B_i B_k - B^2 \delta_{ik})}{2\mu_o} \tag{8.8}$$

is the momentum transfer tensor,

$$p_{ik} = p_\perp \delta_{ik} + (p_\| - p_\perp) b_i b_k \tag{8.9}$$

is the pressure tensor and δ_{ik} is the Kronecker delta, and

$$q_i = \rho_m U_i \left(\frac{U^2}{2} + p_{ik}\delta_{ik}\right) + \frac{(\mathbf{E} \times \mathbf{B})_i}{\mu_0} \tag{8.10}$$

is the heat flux density. The energy conservation equation is not normally required for ideal fluids and we have included this equation here for completeness.

These equations are non-linear and to date they have not been solved analytically or numerically. They can be simplified if we can assume that the fluid can be treated as ideal, which is valid when dissipation effects arising from viscous and thermal properties can be ignored.

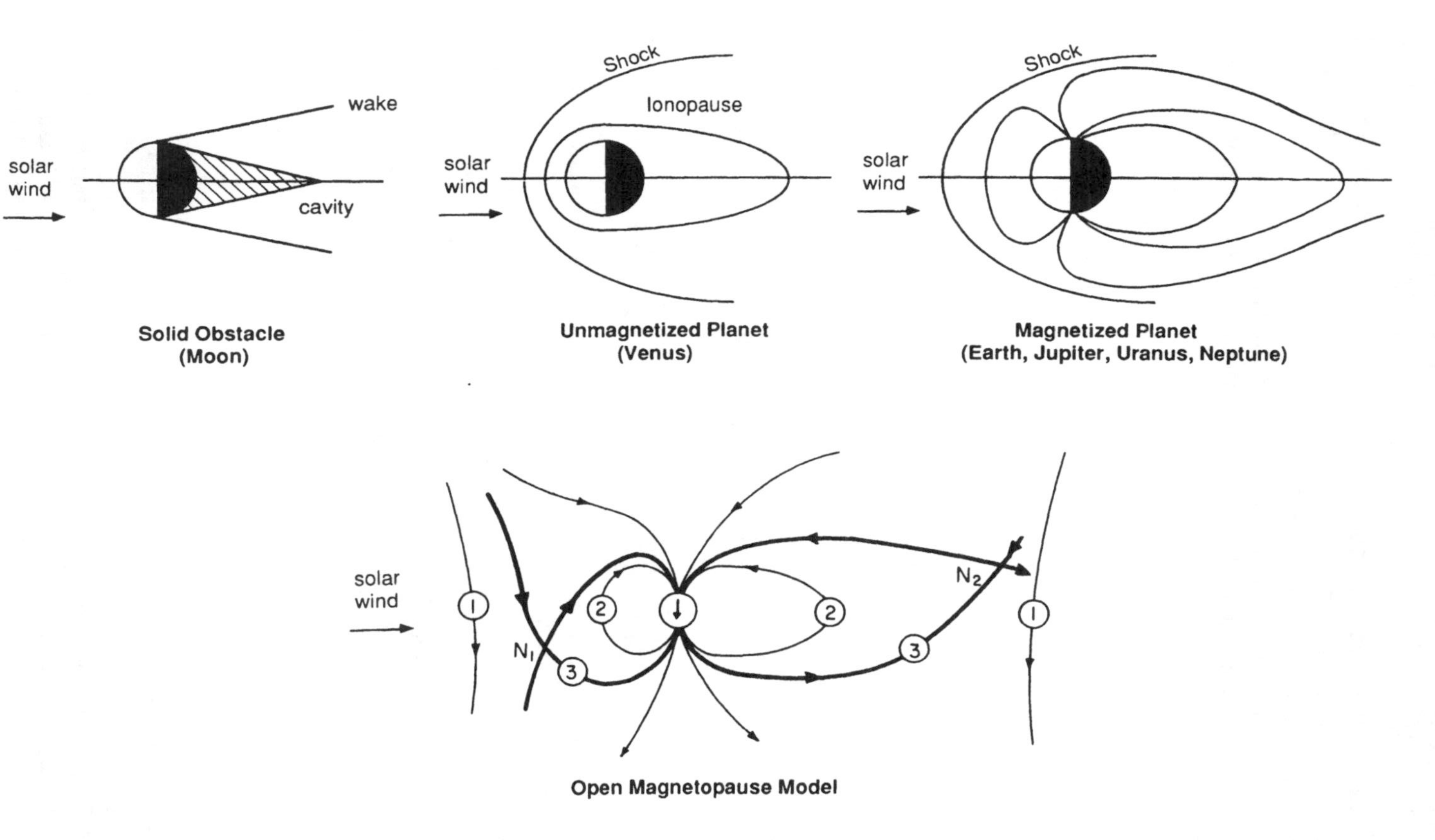

FIGURE 8.1 A schematic diagram to illustrate the different types of boundaries created by the solar wind interacting with an obstacle placed in its flow.

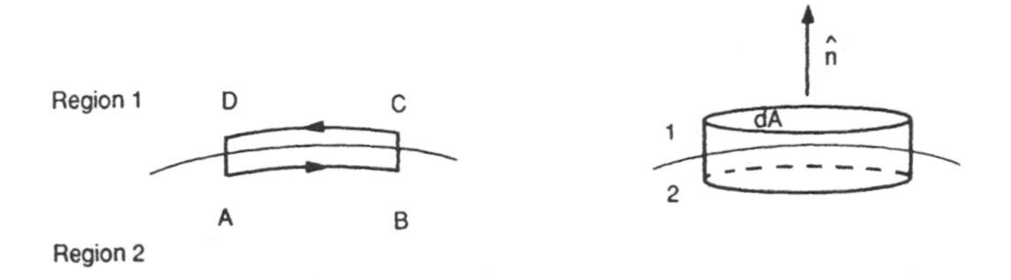

FIGURE 8.2 A schematic diagram illustrating how the boundary conditions for the field vectors are derived.

8.2.1 Boundary Conditions for Electric and Magnetic Fields

We begin our discussion with the well-known requirements of electric and magnetic fields. Suppose $\mathbf{E}$ is different in two regions separated by a boundary. Choose a coordinate frame in which the boundary is at rest and consider only stationary phenomena. The boundary condition for $\mathbf{E}$ is obtained by integrating (8.4) across the interface and using Stoke's law. In Figure 8.2, let the segments AB and CD be represented by $\Delta\mathbf{l}$ and $-\Delta\mathbf{l}$ and let the segments AD and BC be negligibly small. Then

$$\mathbf{E}_2 \cdot \Delta\mathbf{l} + \mathbf{E}_1 \cdot (-\Delta\mathbf{l}) = 0$$

$$(\mathbf{E}_2 - \mathbf{E}_1) \cdot \Delta\mathbf{l} = 0 \tag{8.11}$$

where the subscripts 1 and 2 refer to the two sides of the boundary. Equation (8.11) states that the component of the electric field parallel (tangent) to the interface is continuous. In terms of the unit normal vector $\mathbf{n}$, (8.11) is

$$(\mathbf{E}_2 - \mathbf{E}_1) \times \mathbf{n} = 0 \tag{8.12}$$

Let us denote the difference between the values of any electrodynamic quantity on the two sides of the discontinuity by enclosing the quantity in the square brackets [ ]. Thus, the continuity of the tangential component of the electric field $\mathbf{E}_t$ is written as

$$[\mathbf{E}_t] = 0 \tag{8.13}$$

To obtain the boundary condition for $\mathbf{B}$, use (8.1). The procedure here is similar to the method used in obtaining the boundary conditions for $\mathbf{E}$. We construct a small pill box at the interface and shrink the side to a negligibly small height. Let $\mathbf{n}$ be the unit normal vector and, using Gauss' law, we obtain

$$(\mathbf{B}_2 - \mathbf{B}_1) \cdot \mathbf{n} = 0 \tag{8.14}$$

$$[\mathbf{B}_n] = 0 \tag{8.15}$$

The normal component of **B** is continuous across a boundary. The boundary condition for **H** is obtained by integrating $\Delta \times \mathbf{H} = \mathbf{J}$ along the path ABCD as was done for **E**. We then obtain

$$\mathbf{H}_2 \cdot \Delta \mathbf{l} + \mathbf{H}_1 \cdot (-\Delta \mathbf{l}) = |\mathbf{j}_s \times \Delta \mathbf{l}| \tag{8.16}$$

where $\mathbf{j}_s$ is the surface current density flowing at the interface. Equation (8.16) can be rewritten in terms of the normal **n** as

$$\mathbf{n} \times (\mathbf{H}_2 - \mathbf{H}_1) = \mathbf{j}_s \tag{8.17}$$

$$[\mathbf{H}_t] = |\mathbf{j}_s| \tag{8.18}$$

The discontinuity of the tangential component of **H** is related to the presence of a surface current $\mathbf{j}_s$. Surface currents can only exist in boundaries if the electrical conductivity is very large. This is because in a finite conductivity medium, the magnetic field will diffuse (see (5.55)) and this diffusion will destroy the surface current. Space plasmas do have very high electrical conductivity and surface currents are meaningful.

We can use $\nabla \times \mathbf{D} = \rho$ and by similar procedure derive

$$[\mathbf{D}_n] = \sum \tag{8.19}$$

where $\sum$ is the surface charge density. The normal component of **D** is discontinuous at the surface density of free charges. Note, however, that plasmas in steady-state do not support free charges.

8.2.2 Boundary Condition for Mass Flux

At the surface of discontinuity, the mass flux is required to be continuous. Ignoring time dependence, $\nabla \cdot \rho_m \mathbf{U} = 0$ where ρ_m is the mass density and $\rho_m \mathbf{U}$ is the mass (since there is no confusion between ρ_m and the charge density ρ, the subscript m will be omitted for convenience). Mass flux conservation requires

$$\begin{aligned} \rho_1 \mathbf{U}_1 \cdot \mathbf{n} &= \rho_2 \mathbf{U}_2 \cdot \mathbf{n} \\ &= G \\ [\rho \mathbf{U} \cdot \mathbf{n}] &= 0 \end{aligned} \tag{8.20}$$

The mass flux on either side is represented for convenience by the letter G. In steady-state, mass flux entering side 1 in the normal direction of the boundary is equal to the mass flux leaving side 2 in the normal direction. This boundary requirement is the same for both isotropic and anisotropic MHD fluids.

For ideal MHD fluids ($\sigma = \infty$), the electric field in the rest frame is related to the bulk velocity of the plasma, $\mathbf{E} = -\mathbf{U} \times \mathbf{B}$. The requirement that the tangential electric field is continuous across boundaries is obtained from $\mathbf{n} \times \mathbf{E}$. This yields

$$\mathbf{n} \times \mathbf{E} = (\mathbf{n} \cdot \mathbf{B})\mathbf{U} - (\mathbf{n} \cdot \mathbf{U})\mathbf{B} \tag{8.21}$$

Let $(\mathbf{n} \cdot \mathbf{B}) = B_n$ and $(\mathbf{n} \cdot \mathbf{U}) = U_n$. Then

$$\mathbf{n} \times \mathbf{E} = B_n \mathbf{U}_t - U_n \mathbf{B}_t \tag{8.22}$$

The continuity of the tangential electric field across a boundary requires $[\mathbf{n} \times \mathbf{E}] = 0$. Hence

$$B_n \mathbf{U}_{t1} - B_n \mathbf{U}_{t2} = U_{n1} \mathbf{B}_{t1} - U_{n2} \mathbf{B}_{t2} \tag{8.23}$$

Now use the mass conservation equation across the boundary. Then

$$\begin{aligned} B_n[\mathbf{U}_t] &= G\left(\frac{\mathbf{B}_{t1}}{\rho_1} - \frac{\mathbf{B}_{t2}}{\rho_2}\right) \\ &= G\left[\frac{\mathbf{B}_t}{\rho}\right] \end{aligned} \tag{8.24}$$

If $\mathbf{n}(\mathbf{n} \cdot \mathbf{B})G(1/\rho_1 - 1/\rho_2)$ is added to both sides of (8.24), this equation becomes

$$B_n(\mathbf{U}_1 - \mathbf{U}_2) = G\left(\frac{\mathbf{B}_1}{\rho_1} - \frac{\mathbf{B}_2}{\rho_2}\right) \tag{8.25}$$

which is more general than (8.24). Given $\mathbf{U}$ and $\mathbf{B}$ on both sides of a boundary, this equation admits several classes of solutions, depending on B_n and G. Suppose $B_n \neq 0$ and $G \neq 0$. Then the vectors $\mathbf{B}_1$, $\mathbf{B}_2$, and $(\mathbf{U}_1 - \mathbf{U}_2)$ are coplanar and

$$(\mathbf{U}_1 - \mathbf{U}_2) \cdot (\mathbf{B}_1 \times \mathbf{B}_2) = 0 \tag{8.26}$$

If $B_n = 0$ but $G \neq 0$, $\mathbf{B}_1$ and $\mathbf{B}_2$ are parallel. If $B_n \neq 0$ and $G = 0$, then $\mathbf{U}_1 = \mathbf{U}_2$. If $B_n = 0$ and $G = 0$, then $\mathbf{B}_1$, $\mathbf{B}_2$, and $\mathbf{U}_1 - \mathbf{U}_2$ can have any value, but they are required to all lie in a plane whose normal is $\mathbf{n}$. The various combinations of B_n and G determine the type of discontinuity.

8.2.3 Classification of Boundaries

Our objective is to apply these equations to the different types of boundaries. Let us first study a simple classification scheme originally proposed by L.D. Landau and E.M. Lifshitz. They organized the MHD boundaries according to whether a fluid crosses a boundary or not and whether a boundary supports a magnetic field normal to it. Since in MHD mass flow is related to convective electric fields, mass flow across a boundary can be related to whether a boundary supports an electric field tangent to it.

The classification scheme is based on the behavior G and B_n at the boundaries. There are four combinations of G and B_n and a special case of G and B_t, which results in the classification of the discontinuities as follows:

$$
\begin{array}{lll}
\text{Tangential Discontinuity} & G = 0 & B_n = 0 \\
\text{Contact Discontinuity} & G = 0 & B_n \neq 0 \\
\text{Perpendicular Shock} & G \neq 0 & B_n = 0 \\
\text{Parallel Shock} & G \neq 0 & B_t = 0 \\
\text{Oblique Shock} & G \neq 0 & B_n \neq 0 \\
\text{Rotational Discontinuity} & G \neq 0 & B_n \neq 0
\end{array}
\tag{8.27}
$$

Note that $G \neq 0$, $B_n \neq 0$ for both oblique shocks and rotational discontinuities. The two are distinguished by $[U_n] \neq 0$ or $[\rho] \neq 0$ for oblique shocks and $[U_n] = 0$ or $[\rho] = 0$ for rotational discontinuities. A special geometry of shocks concerns the case when the magnetic field $\mathbf{B}$ is completely parallel to the surface normal direction $\mathbf{n}$. This discontinuity is classified as a parallel shock. We now focus on tangential and rotational discontinuities, with emphasis on understanding closed (tangential) and open (rotational) boundaries. Shocks will be discussed in Chapter 10.

8.3 Closed and Open Boundary Models

Although we do not as yet fully understand the ways the solar wind interacts with various objects in the flow, observations suggest that boundaries in space may be classified as closed or open. An example of a closed boundary is the model of the Chapman-Ferraro boundary already discussed in Chapter 7. Later, the Chapman-Ferraro model will be further developed to illustrate other properties of closed boundaries. An example of an open boundary is the Parker-Sweet model developed in 1958. These models contrast the ways the fluid mass, momentum, and energy are transferred across the boundaries.

8.3.1 Flow Around a Solid Object

The basic concepts of fluid flows can be obtained from the behavior of fluid flows when they encounter a solid object. Consider a two-dimensional problem that can be modeled by placing an infinitely long cylinder in a uniform incompressible flow field (unless stated otherwise, fluids will be assumed incompressible in the models to be discussed below). The Eulerian description of this problem requires information on the velocity of the particles at every point in the flow field (Figure 8.3). A flow field is conveniently represented by streamlines. In steady flows, the velocity of the particles (which is tangent to the streamlines) does not change, and the streamlines remain constant. Note that crowding of streamlines implies higher velocities, since the amount of fluid flow per unit time across any section of two streamlines is the same.

Let U_∞ be the flow velocity far away from the cylinder. The streamlines far away from the cylinder are undisturbed and therefore parallel. As the fluid approaches the cylinder, all of the streamlines go around it except for one which collides with the cylinder at point A. Since the boundary is impenetrable and thus the normal component of the fluid velocity vanishes at the boundary, the termination of the streamline indicates that the flow velocity vanishes at A. This point is called the stagnation point.

Now apply Bernoulli's equation (energy conservation equation) to two points on the streamline that intersect at A. This yields

$$\rho\frac{U_\infty^2}{2} + p_\infty = p_A \tag{8.28}$$

where p_∞ and p_A are respectively, the fluid pressure far away in the undisturbed region and the fluid pressure at point A. Bernoulli's equation shows that the pressure at the stagnation point is higher than the free stream pressure p_∞ by $\rho U_\infty^2/2$, which is the kinetic energy per unit volume of undisturbed flow. p_A is the total pressure, which is the sum of the dynamic pressure $\rho U_\infty^2/2$ and the static pressure p_∞.

Consider now a neighboring streamline that goes around the obstacle. Bernoulli's equation on this streamline is

$$\rho\frac{U_\phi^2}{2} + p_\phi = p_\infty \tag{8.29}$$

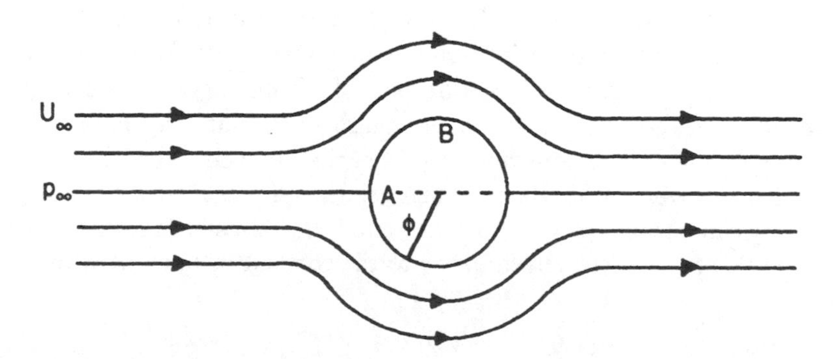

FIGURE 8.3 Flow of fluid around an infinitely long cylinder.

where U_ϕ and p_ϕ are flow velocity and pressure at angle ϕ. Our objective is to learn about the distribution of U_ϕ and p_ϕ on the surface in terms of undisturbed flow quantities.

Information on the pressure p_ϕ is obtained by considering the particles colliding with the obstacle to be elastic. Define the initial momentum of the particles before collision as $\mathbf{p} = m\mathbf{v}_\infty$. The momentum after collision is $-m\mathbf{v}_\infty$ and thus the change of momentum is $\Delta\mathbf{p} = 2m\mathbf{v}_\infty$. We now average this over the distribution of the particles. Then the averaged momentum change is $2\rho < \mathbf{v}_\infty >= 2\rho\mathbf{u}_\infty$ where $\rho = nm$. For a two-component plasma fluid consisting of ions and electrons, note that the mass density $n^+m^+ + n^-m^- \approx nm^+$. Also, if $u^+ \approx u^-$, which we will assume, the fluid velocity $\mathbf{U} \approx \mathbf{u}$.

Let $\mathbf{n}$ be the unit normal vector positive in the outward direction. Then $\mathbf{U} \cdot \mathbf{n} = U \cos\phi$ and the change of the magnitude of the momentum in the normal direction is $\Delta p = 2mU_\infty \cos\phi$. The number of particles striking the obstacle per unit area per unit time is $nU_\infty \cos\phi$. The total pressure imparted to the obstacle in the normal direction is the product of these two terms and is equal to

$$p_\phi = 2\rho U_\infty^2 \cos^2\phi \tag{8.30}$$

Bernoulli's equation (8.29) then becomes

$$2U_\infty^2 \cos^2\phi + \frac{U_\phi^2}{2} = 2U_\infty^2 \tag{8.31}$$

This equation yields

$$U_\phi = 2U_\infty \sin\phi \tag{8.32}$$

The pressure is highest at the stagnation point ($\phi = 0$) and is lowest at point $B(\phi = \pi/2)$. The fluid is accelerated from a place of higher pressure to a point of lower pressure. Hence U_ϕ is maximum at point B and minimum at point A.

In treating this simple problem, we have ignored frictional effects. This is not a bad estimate for air or liquid with low viscosity. However, even for these fluids, one must take caution near the boundary. Since the normal component of the flow velocity must vanish at the surface, there is a large velocity gradient as one moves away from the obstacle. The Navier-Stokes equation involves the gradient of the velocity and the term involving the viscosity can become very important even if the viscosity is small. Inclusion of the velocity gradient effects leads to the concept of boundary layers.

8.3.2 Chapman-Ferraro Boundary

Solar wind interaction with planetary magnetic fields, modelled on MHD theory, produces several features of interest for the sunward portion of the magnetosphere. Adopt the Chapman-Ferraro model to describe the magnetopause boundary of a closed boundary. This model incorporates solar wind particles on one side with no IMF and, on the magnetospheric side, a planetary magnetic field with no particles (see Figures 7.4 and 7.5).

Assume now that the solar wind particles that strike the boundary are elastically reflected (Figure 8.4). The interaction of the solar wind particles with the magnetopause boundary here is identical to the interaction that occurs at hard boundaries. Hence the pressure normal to the magnetopause is given by (8.30). The magnetopause boundary is formed by the solar wind as it pushes against the planetary magnetic field B_p until the magnetic energy density equals the solar wind pressure. The balance of forces then requires

$$2mnU_{SW}^2 \cos\phi = \frac{B_p^2}{2\mu_0} \tag{8.33}$$

The magnetopause is supported by a diamagnetic current created by the partial penetration of the solar wind particles (Chapter 7). The Chapman-Ferraro current is "infinitely thin" and flows from dawn to dusk for a plane boundary infinite in extent (Figure 8.5). The magnetic field of this current is tangent to the boundary and there is no magnetic field in

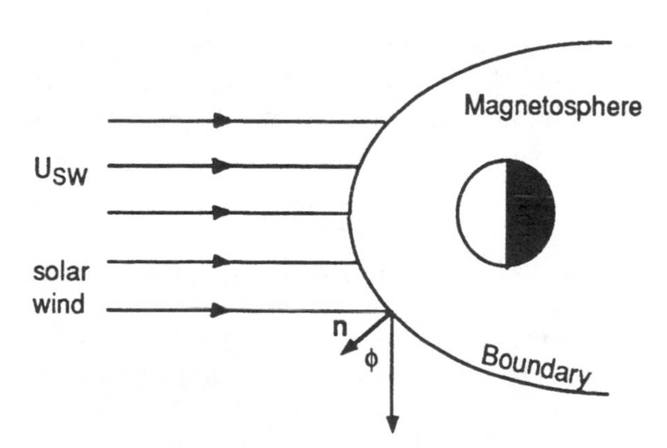

FIGURE 8.4 A schematic diagram to illustrate specular reflection of solar wind off a closed magnetopause boundary.

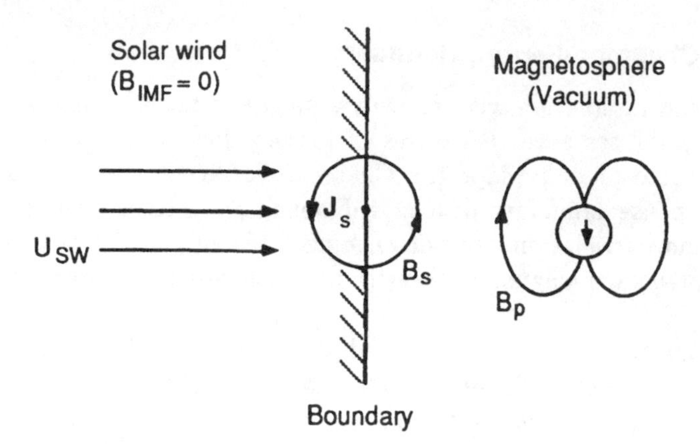

FIGURE 8.5 Model of a Chapman-Ferraro boundary to illustrate basic features of a closed magnetopause.

the normal direction. This field is in the same direction as the planetary field on the magnetospheric side and is in the opposite direction on the solar wind side.

Figure 8.5 shows that the total magnetic field intensity just inside the boundary is

$$B_{in} = B_p + B_s \tag{8.34}$$

where B_p is the planetary field at the boundary and B_s is the magnetic field created by the surface current. The magnetic field intensity just outside the boundary is

$$B_{out} = B_p - B_s \tag{8.35}$$

The Chapman-Ferraro model requires the boundary current to produce a surface magnetic field that will completely cancel the planetary magnetic field inside the solar wind. Hence if we let $B_{out} = 0$, the above equations yield

$$B_{in} = 2B_p \tag{8.36}$$

This shows that the field intensity just inside the boundary is twice that of the planetary field at the boundary.

The Chapman-Ferraro model produces a perfect diamagnetic cavity, since the solar wind completely confines the planetary magnetic field. This ideal boundary can only be created by a $\beta = 2\mu_0 nkT/B^2 = 1$ plasma. The first observation of Earth's magnetopause agreed with this prediction (see Figure 3.2). However, this agreement was fortuitous and subsequent observations show no such simple relationship exists.

The plane model can be improved by letting the boundary become curved. Inclusion of curvature will yield

$$B_{in} = 2(B_p + B_c) \tag{8.37}$$

where B_c represents the magnetic field at the surface whose contribution comes from the curvature current. The total field just inside a curved magnetopause boundary is larger than the field of a plane boundary. Note, however, that solving this problem requires knowledge about the total field B_{in}, which includes B_c. Hence this problem cannot be solved exactly except by approximation. An approach that is reasonable to take is to first let $B_c = 0$, solve for the shape, then use this shape to calculate the values for B_c and, by successive iteration, obtain a more realistic picture.

Although the problem of a closed magnetopause boundary shape can be solved in principle, this problem can become quite complicated. In addition to the curved magnetic field geometry, in realistic models one must include the IMF and several populations of particles inside and outside of the boundary. More advanced models of boundaries must also examine whether the boundary is stable. Instability alters the shape of a boundary. Magnetopause boundaries are very likely to be unstable because the solar wind streaming creates a velocity shear (which is a source of free energy) across the boundary.

The Chapman-Ferraro boundary model assumes the solar wind is a perfect conductor. This implies that the boundary is closed from the magnetic field point of view because magnetic fields cannot diffuse into perfect conductors. Hence the solar wind will always remain free of planetary magnetic fields. It is a closed boundary from an electric field point of view because, in conductors, all free charges from inside the conductor migrate to the surface and on the surface these charges are distributed in such a way as to give a zero electric field inside the conductor. Since the electric field inside the conductor is zero, the tangential electric field boundary condition (8.13) cannot be satisfied and any electric field at the boundary must be normal to it.

In the Chapman-Ferraro model, $B_n = 0$ and $B_t \neq 0$ at the boundary. Since the tangential electric field must vanish at the boundary, we see from (8.22), $\mathbf{n} \times \mathbf{E} = B_n \mathbf{U}_t - B_t \mathbf{U}_n$, that $U_n = 0$ at the boundary. Hence the solar wind cannot penetrate the boundary and the magnetosphere will always remain free of solar wind plasma. $U_t \neq 0$ and the solar wind, which was originally in the $-x$-direction, must divert and flow parallel to the boundary.

8.3.3 Parker-Sweet Model

Space plasma systems may support magnetic field geometries that include neutral points created by oppositely directed magnetic fields. R.G. Giovanelli in 1947 and F. Hoyle in 1949 suggested that energetic solar flare and auroral particles could be accelerated at these magnetic neutral points. This idea was further developed by J. Dungey in 1953, who studied the behavior of plasma in the neighborhood of an x-type neutral point. Then, in 1958, P.A. Sweet examined whether magnetic energy converted into plasma energy at neutral points could be a viable energy source for flares. In 1963, E.N. Parker improved Sweet's model and formulated the problem, using MHD theory.

The magnetic field configuration considered by Parker is shown in Figure 8.6. This diagram shows the configuration for $t > 0$ after the oppositely directed field lines have merged. The model assumes that the fluid approaches the boundary with the convection velocity and the fluid is subsequently ejected along the boundary. Let the boundary that contains the neutral point be a region where the magnetic field is reduced. The dimensions of the boundary layer are 2δ for the width (in the x-direction) and $2L$ for the length (in the z-direction).

Let U_x be the fluid velocity toward the boundary and U_z the velocity along the boundary. The continuity equation then yields

$$U_x L = U_z \delta \tag{8.38}$$

Assume now that the boundary is maintained by a pressure balance between the fluid and the magnetic field. The balance of pressure requires that

$$p = p_0 + \frac{B_z^2}{2\mu_0} \tag{8.39}$$

where B_z is the magnetic field just outside the boundary, p is the pressure in the middle of the boundary (neutral point), and p_0 is the pressure far from the boundary. Note that the pressure in the neutral point is higher than the pressure in the surrounding medium. This pressure gradient will eject the fluid and accelerate it along the boundary. The velocity U_z along the boundary can be obtained from Bernoulli's equation

$$p - p_0 = \rho \frac{U_z^2}{2} \tag{8.40}$$

where ρ is the mass density of the fluid.

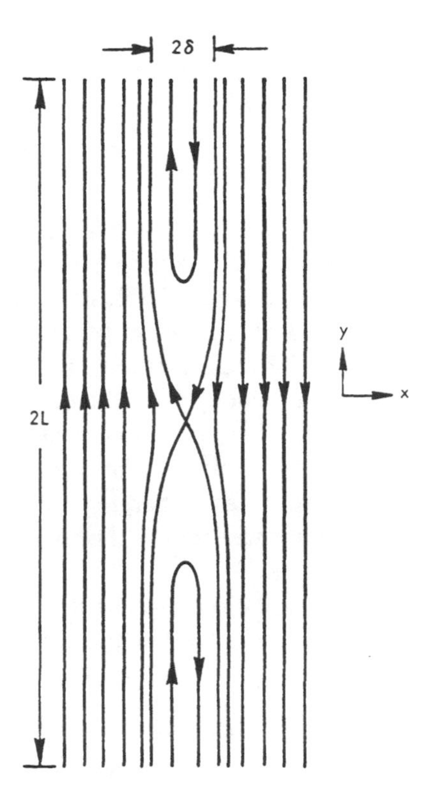

FIGURE 8.6 Magnetic geometry used in the Parker-Sweet model.

The fluid approaches the boundary with a convection velocity $\mathbf{U} = \mathbf{E} \times \mathbf{B}/B^2$ and, for the simple geometry being considered, this reduces to

$$U_x = -\frac{E_y}{B_z} \tag{8.41}$$

where E_y is the electric field. Note that in steady-state, $\nabla \times \mathbf{E} = 0$ and E_y is constant and uniform in the flow. This conclusion also follows from the fact that our model assumes the current density outside the boundary layer vanishes. Since E_y is continuous across the boundary, it can be related to the current in the boundary layer. We use Ohm's law $J_y = \sigma E_y$ where σ is the electrical conductivity and J_y is the current density. J_y can be obtained from $\nabla \times \mathbf{B} = \mu_0 \mathbf{J}$, yielding

$$\begin{aligned} \mu_0 J_y &= \frac{\partial B_z}{\partial x} \\ &\approx \frac{B_z}{\delta} \end{aligned} \tag{8.42}$$

where the change of the magnetic field across the boundary width 2δ is approximated to be equal to $2B_z$. Eliminating E_y and J_y by using these equations and solving for U_x yields

$$U_x = -\frac{1}{\mu_0 \sigma \delta} \tag{8.43}$$

The expression on the right-hand side of this equation is simply the expression of the diffusion velocity of the magnetic field relative to the fluid. This can be seen from the expression of diffusion time $t_D = \mu_0 \sigma \delta^2$ and by letting $\delta = v_D t_D$ where v_D is the diffusion velocity of the magnetic field. This yields $v_D = (\mu_0 \sigma \delta)^{-1}$. The main result of the Parker-Sweet steady-state model is that the fluid velocity into the boundary layer is equal to the rate at which the magnetic field diffuses out of the plasma. Since U_x is the rate at which magnetic fields are brought together to form the neutral point, U_x can thus be looked at as a merging rate (also referred to as a reconnection and annihilation rate) that is the conversion rate of magnetic field energy into plasma energy.

Now use the continuity equation and the Bernoulli equation to eliminate δ and U_z. Then the fluid velocity can be rewritten as

$$U_x^2 = \frac{B_z}{\sqrt{\mu_0 \rho}} \left(\frac{1}{\mu_0 \sigma L} \right) \tag{8.44}$$

$V_A = B_z / \sqrt{\mu_0 \rho}$ is the Alfvén velocity. If one now defines the magnetic Reynolds number using the Alfvén velocity,

$$R_m = \mu_o \sigma L V_A \tag{8.45}$$

we can rewrite the flow velocity as

$$U_x = \frac{V_A}{R_m} \tag{8.46}$$

In general, R_m for space plasmas is very large and therefore the rate of conversion of magnetic into plasma energy is rather slow. Hence this model cannot explain solar flare and auroral phenomena, which require much faster conversion rates (by a factor of 10 to 100).

We conclude this section with a short discussion of how one can improve the above model. In 1964, H. Petschek argued that the Parker-Sweet model overlooked an important mechanism for the dissipation of magnetic energy. Noting that an x-geometry in the region of a neutral point necessarily includes a magnetic field component normal to the boundary, Petschek suggested that magnetic energy can also be dissipated by the propagation of Alfvén waves. Petschek's boundary geometry differs from the Parker-Sweet boundary because it includes a normal component of the magnetic field and the diffusion region has been confined to a small region about the neutral point. The net effect of shrinking the diffusion region is to enhance the energy conversion rate (see (8.44)), since U_x is proportional to $(L)^{-1/2}$ (see Petschek, 1963, cited at the end of the chapter).

The diffusion velocity due to finite conductivity decreases as the square root of time; therefore the diffusion mechanism will be dominant at early times. Assuming that merging takes place, fluid entering the boundary is subsequently ejected at the Alfvén speed. Alfvén velocity is independent of conductivity and consequently the rate at which magnetic energy is converted into plasma energy is also independent of conductivity. Alfvén propagation velocity is much faster and its effect could dominate at later times (after merging has occurred).

Petschek's boundary has a normal component of the magnetic field and a tangential component of the electric field. Hence this boundary can be considered to possess a rotational discontinuity. As shown below, the flow rate across a rotational discontinuity occurs at the Alfvén speed. This implies that the "merging rate" can occur much more rapidly than the diffusion rate given by (8.46) and can phenomenologically explain that magnetic field energy converted into plasma energy is responsible for auroral and solar flares.

The merging theory discussed above relies on the MHD theory and, while it offers useful information, the MHD approach has often been criticized. First of all, the theory assumes that the fluid can be characterized by a simple form of Ohm's law that has not been verified for collisionless plasmas. The theory also relies on the concept of magnetic field reconnection whose physical basis has not been unambiguously tested and no acceptable mechanism has been offered. Some of these objections can be avoided if this problem is studied using the Lorentz equation of motion. This requires solving many coupled equations and evaluating self-consistently the trajectories of the individual particles in electromagnetic field geometries that include magnetic neutral points.

8.4 Continuity of Momentum and Energy

Section 8.2 needs to be expanded so that all of the requirements of fluids are included across boundaries. As before, consider a stable boundary configuration and ignore all time dependence. Let the fluid possess infinite conductivity $\sigma = \infty$ and $\mathbf{U}$ be the bulk velocity of the fluid. Allow the boundary to move with a velocity $\mathbf{V}$ (which is unknown) relative to a spacecraft where measurements are made. (Assume that the spacecraft is essentially at rest because its velocity is much less than $\mathbf{U}$ or $\mathbf{V}$.) Now erect a coordinate frame of reference in which the boundary is at rest, $\mathbf{V} = 0$ (this is also called a boundary coordinate system). Let us now examine the boundary conditions of momentum and energy in isotropic and anisotropic fluids.

8.4.1 Momentum Flux

Equation (8.6) shows that in steady-state, the divergence of the momentum transfer tensor must vanish. This immediately tells us that the normal component of π_{ik} must be continuous across a boundary. Assume now $p_{ik} = p\delta_{ik}$ where p is a scalar. Then the continuity of the normal component is expressed by

$$\begin{aligned}[(\pi_{ik} + T_{ik})n_i] &= \left[\rho U_i U_k + p\delta_{ik} - \left(\frac{B_i B_k}{\mu_0} - \frac{B^2}{2\mu_0}\delta_{ik}\right)\right] n_i \\ &= 0 \end{aligned} \tag{8.47}$$

In vectorial form, the right-hand side of (8.47) is

$$\left[\rho \mathbf{U}(\mathbf{U}\cdot\mathbf{n}) + p\mathbf{n} - \left(\mathbf{B}\frac{\mathbf{B}\cdot\mathbf{n}}{\mu_0} - \frac{B^2}{2\mu_0}\mathbf{n}\right)\right] = 0 \tag{8.48}$$

Isotropic Fluid Decompose the vectors into components parallel and perpendicular to the boundary. Then the continuity of the momentum across the interface is deduced from (8.48) by scalar multiplying with $\mathbf{n}$ and $\mathbf{t}$. This yields

$$\left[p + \rho U_n^2 + \frac{B_t^2 - B_n^2}{2\mu_0}\right] = 0 \tag{8.49}$$

and

$$\left[\rho U_n \mathbf{U}_t - \frac{B_n \mathbf{B}_t}{\mu_0}\right] = 0 \tag{8.50}$$

where $B^2 = B_t^2 + B_n^2$ has been used. These two equations can be cast into a form more suitable for classifying boundaries. Use (8.20) and, noting that $[B_n] = 0$, rewrite (8.49) and (8.50) as

$$\begin{aligned}[p] + G^2\left[\frac{1}{\rho}\right] + \frac{[B_t^2]}{2\mu_0} &= 0 \\ G[\mathbf{U}_t] - B_n\frac{[\mathbf{B}_t]}{\mu_0} &= 0 \end{aligned} \tag{8.51}$$

Anisotropic Fluid The assumption of isotropic pressure is only valid in regions where collisions dominate. In most regions of space, the plasma is tenuous and, in the presence of magnetic fields, the plasma as a rule is distributed anisotropically. This is clearly the case of the solar wind near Earth's orbit. We let the pressure in (8.47) become a tensor. Then the boundary condition becomes

$$\left[p_{ik} + \rho U_i U_k + \frac{B^2}{2\mu_0}\delta_{ik} - \frac{(B_i B_k)}{\mu_0}\right] n_i = 0 \tag{8.52}$$

which is identical to (8.47) except the pressure p_{ik} is a tensor given by

$$p_{ik} = p_\perp \delta_{ik} + (p_\parallel - p_\perp)\frac{B_i B_k}{B^2} \tag{8.53}$$

The vectorial form of (8.52) is

$$\left[\rho \mathbf{U}(\mathbf{U}\cdot\mathbf{n}) + \left(p_\perp + \frac{B^2}{2\mu_0}\right)\mathbf{n} + (p_\parallel - p_\perp)\mathbf{B}\frac{(\mathbf{B}\cdot\mathbf{n})}{B^2} - \mathbf{B}\frac{(\mathbf{B}\cdot\mathbf{n})}{\mu_0}\right] = 0 \tag{8.54}$$

from which we can immediately deduce that

$$\left[\rho U_n^2 + p_\perp + \frac{B^2}{2\mu_0} + (p_\parallel - p_\perp)\frac{B_n^2}{B^2} - \frac{B_n^2}{\mu_0}\right] = 0 \tag{8.55}$$

as the stress balance required of the normal momentum flux, and

$$\left[\rho U_n \mathbf{U}_t + (p_\parallel - p_\perp)\frac{B_n \mathbf{B}_t}{B^2} - \frac{B_n \mathbf{B}_t}{\mu_0}\right] = 0 \tag{8.56}$$

as the stress balance of the tangential momentum flux. Equations (8.55) and (8.56) can be rewritten as

$$[p_\perp] + \frac{[B_t^2]}{2\mu_0} + G^2\left[\frac{1}{\rho}\right] - B_n^2\left[\frac{(p_\parallel - p_\perp)}{B^2}\right] = 0$$

$$G[\mathbf{U}_t] - B_n\left[\frac{\mathbf{B}_t}{\mu_0} - \frac{(p_\parallel - p_\perp)\mathbf{B}_t}{B^2}\right] = 0 \tag{8.57}$$

Note that these equations reduce to (8.51) if the pressure anisotropy vanishes.

8.4.2 Energy Flux

The energy conservation equation (5.52) for an ideal MHD fluid was derived by assuming an adiabatic equation of state. Use now the relation $\mathbf{E} = -\mathbf{U} \times \mathbf{B}$. Therefore, the continuity requirement for the energy becomes

$$\left[\frac{\rho U^2}{2}\mathbf{U} + \frac{\gamma}{\gamma - 1}p\mathbf{U} - \frac{(\mathbf{U} \times \mathbf{B}) \times \mathbf{B}}{\mu_0}\right] = 0 \qquad (8.58)$$

The requirement for the continuity of the normal component of the energy flux is

$$\left[\frac{\rho U^2}{2}U_n + \frac{\gamma}{\gamma - 1}pU_n - \frac{(\mathbf{U} \cdot \mathbf{B})}{\mu_0}B_n + \frac{B^2 U_n}{\mu_0}\right] = 0 \qquad (8.59)$$

The adiabatic factor $\gamma/(\gamma - 1)$, which appears throughout the energy equations, is related to the internal energy per unit mass of the fluid, $w = [\gamma/(\gamma-1)]p/\rho$. Also, we will state without justification that for an adiabatic MHD fluid, $\gamma = 2$. Equation (8.59) then can be rewritten as

$$\left[\rho U_n \left(\frac{U^2}{2} + w\right) + U_n \frac{B^2}{\mu_0} - B_n \frac{(\mathbf{U} \cdot \mathbf{B})}{\mu_0}\right] = 0 \qquad (8.60)$$

The first term is the flow of kinetic and the internal energy, the second term represents the flow of the magnetic energy, and the last term represents the coupling between the magnetic field and the flow. Use $G = \rho U_n$ and, since $[B_n] = 0$, (8.60) can be written as

$$G\left[\frac{G^2}{2\rho_m^2} + w + \frac{U_t^2}{2} + \frac{B_t^2}{\mu_0 \rho}\right] = B_n \frac{[\mathbf{U}_t \cdot \mathbf{B}_t]}{\mu_0} \qquad (8.61)$$

If the anisotropic pressure tensor p_{ik} given by (8.53) is used, the above equation is modified to

$$G\left[\epsilon + \frac{U^2}{2} + \frac{p_\perp}{\rho} + \frac{B_t^2}{\mu_0 \rho}\right] + GB_n^2\left[\frac{(p_\parallel - p_\perp)}{B^2 \rho}\right]$$
$$- B_n\left[\mathbf{U}_t \cdot \mathbf{B}_t \left(\frac{1}{\mu_0} - \frac{(p_\parallel - p_\perp)}{B^2}\right)\right] = 0 \qquad (8.62)$$

as the boundary condition for the conservation of the energy flux across a boundary. Here $\epsilon = w - p/\rho$ is the enthalpy, which is also defined as the heat function per unit mass of the fluid. This equation reduces to (8.61) if $p_\parallel - p_\perp = 0$.

8.5 Equations of MHD Discontinuities

Section 8.2 discussed the boundary conditions for electric and magnetic fields and the mass flux. This discussion is now expanded to include the requirements of momentum and energy fluxes across the various types of boundaries. We use the same classification scheme shown in (8.27).

8.5.1 Tangential Discontinuity ($B_n = 0, G = 0$)

A tangential discontinuity describes a boundary across which there is no mass flow and which does not support a normal component of the magnetic field. At such discontinuities, the velocity and the magnetic field are both tangential and they can have any values in both direction and magnitude. The normal $\mathbf{n}$ to the boundary is then defined by

$$\mathbf{n} = \frac{\mathbf{B}_1 \times \mathbf{B}_2}{|\mathbf{B}_1 \times \mathbf{B}_2|} \tag{8.63}$$

where $\mathbf{B}_1$ and $\mathbf{B}_2$ are magnetic fields on the two sides of the boundary. The density can also have any value (8.25). However, the pressure p is related to B_t according to (8.49). The equations governing the tangential discontinuity are

$$\begin{aligned} \rho U_n &= G = 0 \\ B_n &= 0 \\ [\mathbf{U}_t] &\neq 0 \\ [\mathbf{B}_t] &\neq 0 \\ [\rho] &\neq 0 \\ \left[p + \frac{1}{2\mu_0} B_t^2\right] &= 0 \end{aligned} \tag{8.64}$$

If the pressure is anisotropic, the last equation of (8.64) must be modified. From the first equation of (8.57) we obtain

$$[p_\perp] + \left[\frac{B_t^2}{2\mu_0}\right] = 0 \tag{8.65}$$

which is the same as for the isotropic case. As in the case of the isotropic fluid, the bulk velocity (U), the fluid density (ρ), the parallel pressure p, the internal energy density (w), and the magnetic field are allowed to change in any way across a tangential boundary and the rest of the equations from (8.58)–(8.62) are valid.

8.5.2 Contact Discontinuity ($B_n \neq 0$, $G = 0$)

A contact discontinuity describes a boundary across which the mass flow ρU_n is zero but there is a component of the magnetic field through the surface. Hence $U_n = 0$ but $B_n \neq 0$. Since $G = 0$ for the contact discontinuity, (8.25) shows that $\mathbf{U}_1 = \mathbf{U}_2$. The plasmas on the two sides of a contact discontinuity are at rest relative to each other.

A contact discontinuity permits the density (and other thermodynamic quantities, such as the temperature) to have any value. Since $B_n \neq 0$, the second equation of (8.51) requires that $[B_t] = 0$. Then the pressure p must be continuous, according to the first equation of (8.51). The requirements of a contact discontinuity for an isotropic fluid are

$$\begin{aligned} G &= 0 \\ B_n &\neq 0 \\ [\mathbf{U}] &= 0 \\ [\rho] &\neq 0 \\ [\mathbf{B}_t] &= 0 \\ [p] &= 0 \end{aligned} \tag{8.66}$$

A contact discontinuity is a boundary between two plasma media at rest with different densities and temperatures. This type of boundary in principle may describe the different layers of static ionospheric boundaries. However, the plasma in general is free to flow along the magnetic field and a contact discontinuity is expected to be destroyed rapidly. A contact discontinuity is therefore difficult to maintain and these boundaries may not be observed in magnetized plasmas.

The requirements of the contact discontinuity for an anisotropic pressure can be examined from the two equations of (8.57). For example, if $\mathbf{B}_1 = \mathbf{B}_2$, then we see that $p_{\perp 1} = p_{\perp 2}$ and $p_{\parallel 1} = p_{\parallel 2}$. Hence pressure is continuous across the surface as in the case of isotropic pressure. If $\mathbf{B}_1 \neq \mathbf{B}_2$, (8.57) shows $\mathbf{B}_{t1}$ is parallel to $\mathbf{B}_{t2}$ and therefore $\mathbf{B}_1$, $\mathbf{B}_2$, and $\mathbf{n}$ are coplanar. Hence

$$\mathbf{n} = \frac{(\mathbf{B}_2 - \mathbf{B}_1) \times (\mathbf{B}_1 \times \mathbf{B}_2)}{|\mathbf{B}_1 - \mathbf{B}_2||\mathbf{B}_1 \times \mathbf{B}_2|} \tag{8.67}$$

The first equation of (8.57), written out explicitly for this case, is

$$(p_{\perp 1} - p_{\perp 2}) + \frac{(B_1^2 - B_2^2)}{2\mu_0} = B_n^2 \left(\frac{p_{\parallel 2} - p_{\perp 2}}{B_2^2} - \frac{p_{\parallel 1} - p_{\perp 1}}{B_1^2} \right) \tag{8.68}$$

Use was made of the fact that $[B_n] = 0$ and $[U_n] = 0$. The right side of (8.68) cannot vanish, since otherwise (8.57) would imply that $\mathbf{B}_1 = \mathbf{B}_2$.

8.5.3 Rotational Discontinuity ($G \neq 0$, $B_n \neq 0$)

This discontinuity is different from contact and tangential discontinuities, since the fluid is permitted to flow across the boundary ($G \neq 0$). A rotational discontinuity also supports a normal component of the magnetic field, $B_n \neq 0$. From (8.25) we see that $\mathbf{B}_1$, $\mathbf{B}_2$ and the velocity jump $\mathbf{U}_1 - \mathbf{U}_2$ are coplanar; hence (8.63) again applies. A necessary condition for the rotational discontinuity is that the flow discontinuity ($\mathbf{U}_1 - \mathbf{U}_2$) be parallel to $[(\mathbf{B}_1)/(\rho_1) - (\mathbf{B}_2)/(\rho_2)]$.

Combine (8.49) and (8.25) and obtain the relation

$$G^2 \left[\frac{\mathbf{B}_t}{\rho} \right] = \frac{B_n^2}{\mu_0} [\mathbf{B}_t] \tag{8.69}$$

An MHD fluid with isotropic pressure supporting a rotational discontinuity requires that the fluid density be continuous across the boundary. Hence ρ in (8.69) can be taken out of the square bracket, and this equation can then be written as

$$\left(\frac{G^2}{\rho} - \frac{B_n^2}{\mu_0} \right) [\mathbf{B}_t] = 0 \tag{8.70}$$

The solution of (8.70) is

$$G^2 = \frac{B_n^2 \rho}{\mu_0} \tag{8.71}$$

which states that

$$U_n = \frac{B_n}{(\rho \mu_0)^{1/2}} \tag{8.72}$$

This is another condition for a rotational discontinuity of an MHD fluid with isotropic scalar pressure. Equation (8.72) represents the speed of Alfvén waves, which is a characteristic speed of an MHD fluid (Alfvén waves will be discussed in Chapter 9). Note also that since the mass flux is always continuous across a boundary $[G] = 0$, the normal velocity component must also be continuous in rotational discontinuities, $[U_n] = 0$. Equation (8.50) then becomes

$$[\mathbf{U}_t] = \frac{[\mathbf{B}_t]}{(\mu_0 \rho)^{1/2}} \tag{8.73}$$

Equation (8.61) can be rewritten as

$$G[w] + \frac{G}{\rho} \left[p + \frac{\mathbf{B}_t^2}{2\mu_0} \right] + \frac{G}{2} \left[\left(\mathbf{U}_t - \frac{\mathbf{B}_t}{\mu_0 \rho} \right)^2 \right] = 0 \tag{8.74}$$

where (8.71) has been used. Equations (8.49) and (8.51) show that the second and third terms vanish in (8.74). Hence we find that the internal energy is continuous, $[w] = 0$. Since thermodynamic quantities are determined if w and ρ are given, all thermodynamic quantities are continuous in rotational discontinuities. Thus $[p] = 0$, and from (8.61) B_t^2 is also continuous, $[B_t] = 0$. Since both B_n and B_t are continuous, the magnitude of the total field $\mathbf{B}$ must also be continuous. Thus we have a peculiar situation that the magnetic field across a rotational discontinuity may be rotated about the normal but the total intensity remains unchanged. The vectors $\mathbf{B}_t$ and $\mathbf{U}_t$ are discontinuous according to (8.73) but U_n is continuous (8.72). The requirements of a rotational discontinuity are

$$\begin{aligned} G &\neq 0 \\ B_n &\neq 0 \\ [\mathbf{U}_t] &\neq 0 \\ [\rho] &= 0 \\ [p] &= 0 \\ [U_n] &= 0 \\ U_n &= \frac{B_n}{(\mu_0 \rho)^{1/2}} \end{aligned} \tag{8.75}$$

A rotational discontinuity permits mass flow across the boundary and therefore this type of discontinuity is important in solar wind-magnetopause interactions. It should be emphasized that the requirements of the fluid discontinuities derived above are for the case of isotropic pressure only. Anisotropic pressure yields an additional degree of freedom and therefore the continuity requirements are expected to be quite different. We now examine what additional requirements arise for anisotropic pressure.

If (8.24) and (8.56) are combined, we find that

$$\begin{aligned} B_n G[\mathbf{U}_t] &= G^2 \left[\frac{\mathbf{B}_t}{\rho}\right] \\ &= B_n^2 \left[\frac{\mathbf{B}_t}{\mu_0} - \mathbf{B}_t \frac{(p_\parallel - p_\perp)}{B^2}\right] \end{aligned} \tag{8.76}$$

which is more complex than the case of isotropic pressure (8.69). Write out (8.76) with parameters of the two sides and obtain

$$\begin{aligned} \mathbf{B}_{t1} \left\{ \frac{G^2}{\rho_1} - B_{n1}^2 \left(\frac{1}{\mu_0} - \frac{p_{\parallel 1} - p_{\perp 1}}{B_1^2} \right) \right\} &= \\ \mathbf{B}_{t2} \left\{ \frac{G^2}{\rho_2} - B_{n2}^2 \left(\frac{1}{\mu_0} - \frac{p_{\parallel 2} - p_{\perp 2}}{B_2^2} \right) \right\} & \end{aligned} \tag{8.77}$$

Let us now define

$$\mathbf{V}_{A1} = \frac{\mathbf{B}_1}{\sqrt{\mu_0 \rho_1}} \left(1 - \frac{p_{\parallel 1} - p_{\perp 1}}{B_1^2/\mu_0}\right)^{1/2}$$
$$\mathbf{V}_{A2} = \frac{\mathbf{B}_2}{\sqrt{\mu_0 \rho_2}} \left(1 - \frac{p_{\parallel 2} - p_{\perp 2}}{B_2^2/\mu_0}\right)^{1/2} \quad (8.78)$$

as modified Alfvén velocities. A rotational discontinuity for an anisotropic fluid requires that

$$\mathbf{U}_1 \cdot \mathbf{n} = \mathbf{V}_{A1} \cdot \mathbf{n}$$
$$\mathbf{U}_2 \cdot \mathbf{n} = \mathbf{V}_{A2} \cdot \mathbf{n} \quad (8.79)$$

in which case (8.77) is identically satisfied. Hence in rotational discontinuity, $\mathbf{B}_1$, $\mathbf{B}_2$, and $\mathbf{n}$ do not all lie in the same plane. Alfvén waves are modified in the sense that the anisotropic pressure introduces additional factors to the conventional definition of the Alfvén wave (8.72). When the pressure is anisotropic, the fluid velocity on either side of the boundary is the modified Alfvén velocity $\mathbf{V}_A$. If the anisotropy vanishes, we recover (8.72).

The fluid velocities on the two sides are linked by $\rho_1 \mathbf{V}_{A1} \cdot \mathbf{n} = \rho_2 \mathbf{V}_{A2} \cdot \mathbf{n}$ and if the Alfvén velocities are known, this equation also determines the density ratio of the two sides. Note that this requirement is different from the case of the isotropic pressure, where the rotational discontinuity was defined as a boundary across which the density was continuous.

8.5.4 Summary

The set of fundamental equations of discontinuities in an ideal MHD fluid with isotropic scalar pressure that were derived above are collected here for convenience.

$$[B_n] = 0 \quad (8.15)$$

$$[\rho U_n] = 0 \quad (8.20)$$

$$B_n[\mathbf{U}_t] = G\left[\frac{\mathbf{B}_t}{\rho}\right] \quad (8.24)$$

$$B_n[\mathbf{U}] = G\left[\frac{\mathbf{B}}{\rho}\right] \quad (8.25)$$

$$[p] + G^2\left[\frac{1}{\rho}\right] + \frac{[B_t^2]}{2\mu_0} = 0 \tag{8.52}$$

$$G[\mathbf{U}_t] - B_n\frac{[\mathbf{B}_t]}{\mu_0} = 0 \tag{8.52}$$

$$G\left[\frac{G^2}{2\rho^2} + w + \frac{\mathbf{U}_t^2}{2} + \frac{\mathbf{B}_t^2}{\rho\mu_0}\right] = B_n\frac{[\mathbf{U}_t \cdot \mathbf{B}_t]}{\mu_0} \tag{8.62}$$

For the anisotropic fluids, the first four relations given above are still valid, since they are general and apply to both isotropic and anisotropic fluids. However, the momentum and energy conservation equations need to be modified. The equations of discontinuities for the anisotropic fluids are

$$[B_n] = 0 \tag{8.15}$$

$$[\rho U_n] = 0 \tag{8.20}$$

$$B_n[\mathbf{U}_t] = G\left[\frac{\mathbf{B}_t}{\rho}\right] \tag{8.24}$$

$$B_n[\mathbf{U}] = G\left[\frac{\mathbf{B}}{\rho}\right] \tag{8.25}$$

$$[p_\perp] + G^2\left[\frac{1}{\rho}\right] + \frac{[B_t^2]}{2\mu_0} - B_n^2\left[\frac{(p_\parallel - p_\perp)}{B^2}\right] = 0 \tag{8.58}$$

$$G[\mathbf{U}_t] - B_n\left[\frac{\mathbf{B}_t}{\mu_0} - \frac{(p_\parallel - p_\perp)\mathbf{B}_t}{B^2}\right] = 0 \tag{8.58}$$

$$G\left[\epsilon + \frac{U^2}{2} + \frac{p_\perp}{\rho}\frac{B_t^2}{\mu_0 p}\right] + GB_n^2\left[\frac{(p_\parallel - p_\perp)}{B^2\rho}\right]$$

$$-B_n\left[\mathbf{U}_t \cdot \mathbf{B}_t\left(\frac{1}{\mu_0} - \frac{(p_\parallel - p_\perp)}{B^2}\right)\right] = 0 \tag{8.63}$$

8.6 Test of Theory

The purpose of this section is to attempt to test the validity of the fluid theory. Our objective is to establish whether tangential and rotational discontinuities occur in space. The theory will be tested by examining the details of data obtained by spacecraft experiments as the spacecraft crosses the magnetopause boundary. Since tangential and rotational discontinuities correspond to closed and open boundaries, these two types of boundaries are most likely to occur when the IMF points in the northward and southward directions, respectively.

8.6.1 Tangential Discontinuity

Figure 8.7 shows an example of magnetic field data obtained by a recent spacecraft experiment in the vicinity of the magnetopause when the solar wind was very quiet. The data were obtained by a magnetometer on a spacecraft inbound toward Earth on the sunward hemisphere near local noon. Data displayed here are the three components of the magnetic field, the total field intensity, and the fluxes of energetic electrons. Note that the IMF in the magnetosheath included a northward (positive B_z) magnetic field component.

This magnetopause boundary was not accompanied by a major discontinuity in the magnetic field and at first glance, it is difficult to identify that a boundary exists here. However, use of the energetic particles as tracers of magnetospheric trapped particles helps to identify the magnetopause boundary. The quiet magnetopause boundary is identified by the change of sign of the B_y component at approximately 1714:10 UT. The magnetic field on the solar wind side was intense and the field slowly increased as the spacecraft approached the magnetopause. The total field intensity varied smoothly across the magnetopause.

The observed behavior of the magnetic field can be compared more easily against the transitional properties of the magnetic field defined by (8.15) by transforming the data into a boundary coordinate system. This boundary is defined by the unit vectors **L**, **M**, and **N** (Figure 8.8). **N** is normal to the boundary, **M** is tangent to the boundary and positive pointing toward dawn, and $\mathbf{L} = \mathbf{M} \times \mathbf{N}$. In this coordinate frame, the normal magnetic field component for this magnetopause vanishes, $B_N = 0$. The boundary here is totally defined by the tangential components B_L and B_M. The absence of B_N is one of the two parameters that would classify this boundary as a tangential discontinuity.

Figure 8.9 shows macroscopic plasma parameters in the vicinity of the magnetopause: density (top panel), temperature (second panel), and flow velocity (third and fourth panels). The bottom panel shows pressure (nkT) and the magnetic energy density $B^2/2\mu_0$. These macroscopic plasma parameters were deduced from an experiment that measured the electron and ion distribution functions in the energy range of a few eV to about 40 keV. The vertical line marks 1714 UT, the time when $B_M(B_y)$ changed sign. Table 8.1 summarizes the relevant plasma parameters observed for the magnetospheric and magnetosheath sides of the boundary.

The solar wind flow was originally along the Sun-Earth direction ($\Phi_p \approx 180°$) and as it approached the boundary, it was diverted toward dawn

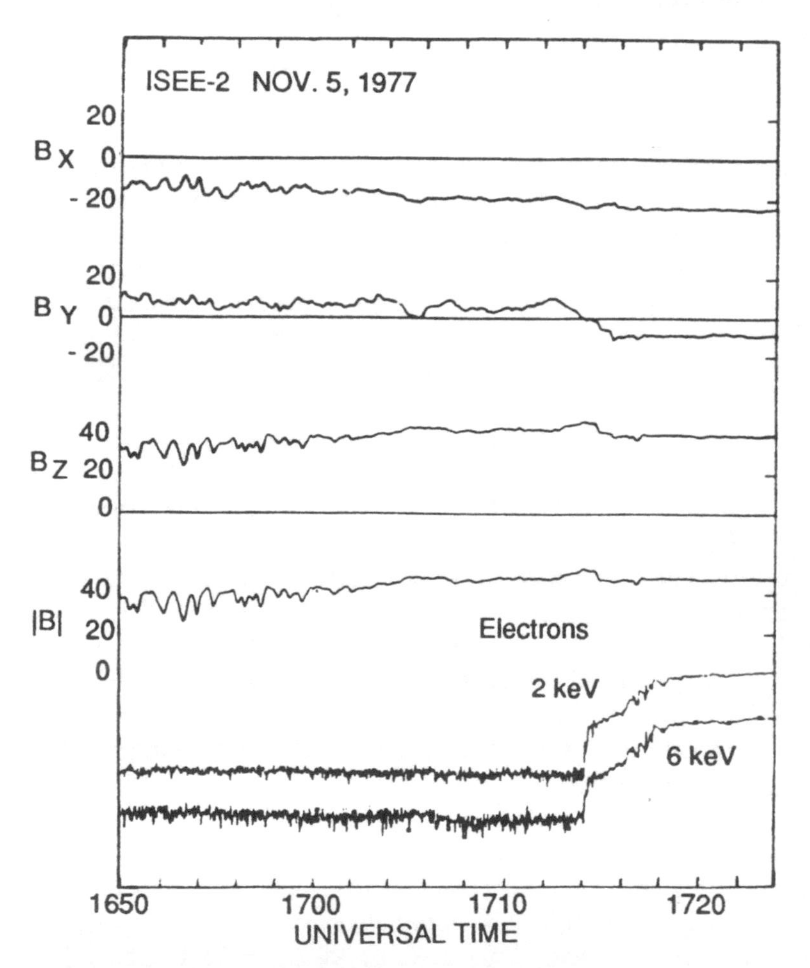

FIGURE 8.7 An example of a magnetopause boundary crossing during a very quiet solar wind condition. Shown are the three components of the magnetic field and 2 and 6 keV electrons measured on the spacecraft ISEE-2. The magnetopause boundary is marked by the change of the sign B_y and an increase of particle fluxes identified as trapped magnetospheric fluxes.

($\Phi \approx 270°$). Thus, at the boundary, the solar wind was flowing tangent to it and $U_n \approx 0$. (This conclusion is consistent with the data from the electric field experiment, which showed that this magnetopause did not support a tangential electric field. See Mozer, 1984, cited at the end of the chapter.) Since the solar wind flowed tangent to the boundary, it did not cross the magnetopause. Hence $G = \rho U_n = 0$.

An interesting feature of this magnetopause is the behavior of the density profile across the magnetopause. The solar wind density decreased gradually as the magnetopause was approached, and the region between the two dashed lines in Figure 8.9 contains both magnetosheath and magnetospheric plasmas. This region is called a boundary layer. Note that magnetospheric and magnetosheath particles apparently come from different

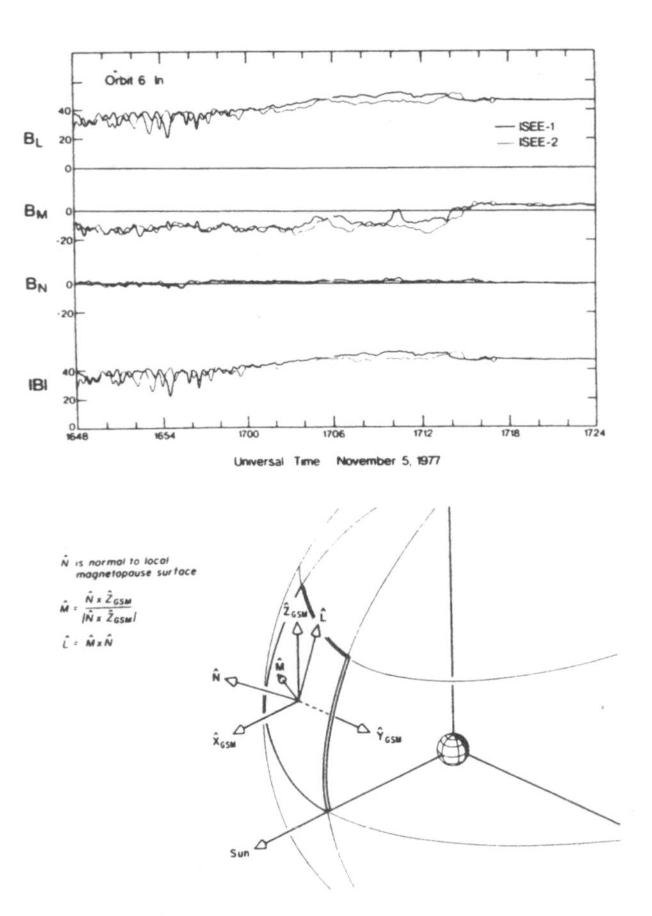

FIGURE 8.8 Magnetic field data transformed into a boundary coordinate system. (From Russell and Elphic, 1979)

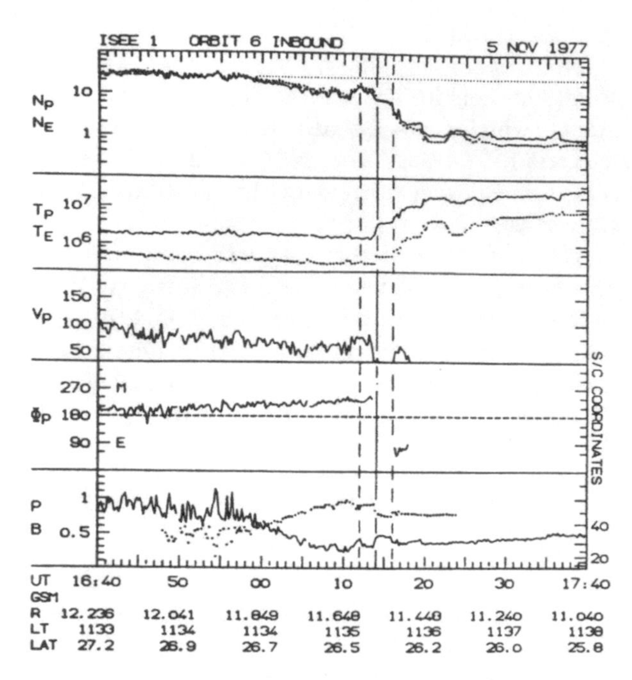

FIGURE 8.9 Macroscopic plasma data for the magnetopause boundary crossing shown in Figure 8.7. (From Paschmann *et al.*, 1978)

populations. The plasma inside the magnetosphere was much hotter than the plasma in the magnetosheath. It was not flowing, hence the convective electric field was absent inside the magnetosphere.

The diversion of solar wind at the magnetopause and the absence of a tangential electric field across the magnetopause (and therefore no occurrence of mass exchange) suggest that this magnetopause boundary supported a "closed" boundary configuration. This magnetopause boundary supports the Chapman-Ferraro model we discussed earlier. $B_N = 0$ as required by the model and $B_L \neq 0$ on the magnetosheath side of the boundary. However, note that this magnetopause includes a boundary layer with a non-vanishing B_M component (the source of B_M could be a field-aligned current at the surface of the boundary).

The total momentum transferred from the solar wind to the Earth's magnetic field can be calculated from the momentum transfer tensor, which includes contributions from both thermal and flowing particles. From (8.52) we have $\pi_{ik} = p_{ik} + U_i U_k \Sigma n^\alpha m^\alpha$. The pressure tensor $p_{ik} = \int w_i w_j f d^3v$, where $\mathbf{w} = \mathbf{v} - \mathbf{U}$, $\mathbf{v}$ is the velocity of the particle, $\mathbf{U} = \Sigma m^\alpha n^\alpha < \mathbf{v} >^\alpha / \Sigma n^\alpha m^\alpha$, and $< \mathbf{v} > = \int v f d^3v$, reduces to a scalar pressure and for a Maxwellian distribution, Table 8.1 shows that $p = nkT \approx 4.5 \times 10^{-10}$

TABLE 8.1
Plasma Parameters

	Magnetosheath	*Magnetosphere*
n_p(m^{-3})	1.5×10^7	8×10^6
n_e(m^{-3})	1.5×10^7	8×10^6
T_p(°K)	2×10^6	4×10^6
T_e(°K)	2×10^5	4×10^5
U_{sw}(km/sec)	80	0
B(10^{-9} Teslas)	49	47
nkT (Joules/m^3)	5.9×10^{-10}	4.5×10^{-10}
$B^2/2\mu_0$ (Joules/m^3)	9.6×10^{-10}	8.8×10^{-10}

Joules/m^3. The second term represents the energy of the flow momentum. For a two-component plasma flowing uniformly in space, this term is $n(m_e + m_i)U^2 \approx nm_iU^2$. The flow energy is $\approx 1.4 \times 10^{-10}$ J/m^3. The total momentum density of the solar wind impinging on the magnetopause for this event was about 5.9×10^{-10} J/m^3.

Pressure balance (8.51) is required if the boundary is stationary. Since $B_n \approx 0$, the total pressure to be balanced is $p + nmU^2 + B_t^2/2\mu_0$. We have already computed the first two terms for the magnetosheath. The magnetic energy density is $B_t^2/2\mu_0 \approx 9.6 \times 10^{-10}$ J/m^3, which is roughly twice the particle energy density. The total pressure on the magnetosheath side was $\approx 1.5 \times 10^{-9}$ J/m^3. The total pressure on the magnetospheric side is estimated to be $p = nkT \approx 4.5 \times 10^{-10}$ J/m^3, $B^2/2\mu_0 \approx 8.8 \times 10^{-10}$ J/m^3, and $nmU^2 \approx 0$. Thus, the total energy density on the magnetospheric side is $\approx 1.3 \times 10^{-9}$ J/m^3. This shows that the energy densities on the two sides of the boundary are nearly equal.

8.6.2 Rotational Discontinuity

Figure 8.10 shows an example of the magnetopause when the solar wind was more disturbed and when the magnetic field in the magnetosheath included a southward B_z component. Use again the energetic trapped particle fluxes as a marker for when the spacecraft encountered the magnetopause. It then becomes apparent that the spacecraft crossed (or partially crossed) the magnetopause many times before its final entry at $\approx$ 0313 UT into the magnetosphere. For example, the intermittent encounter of trapped fluxes

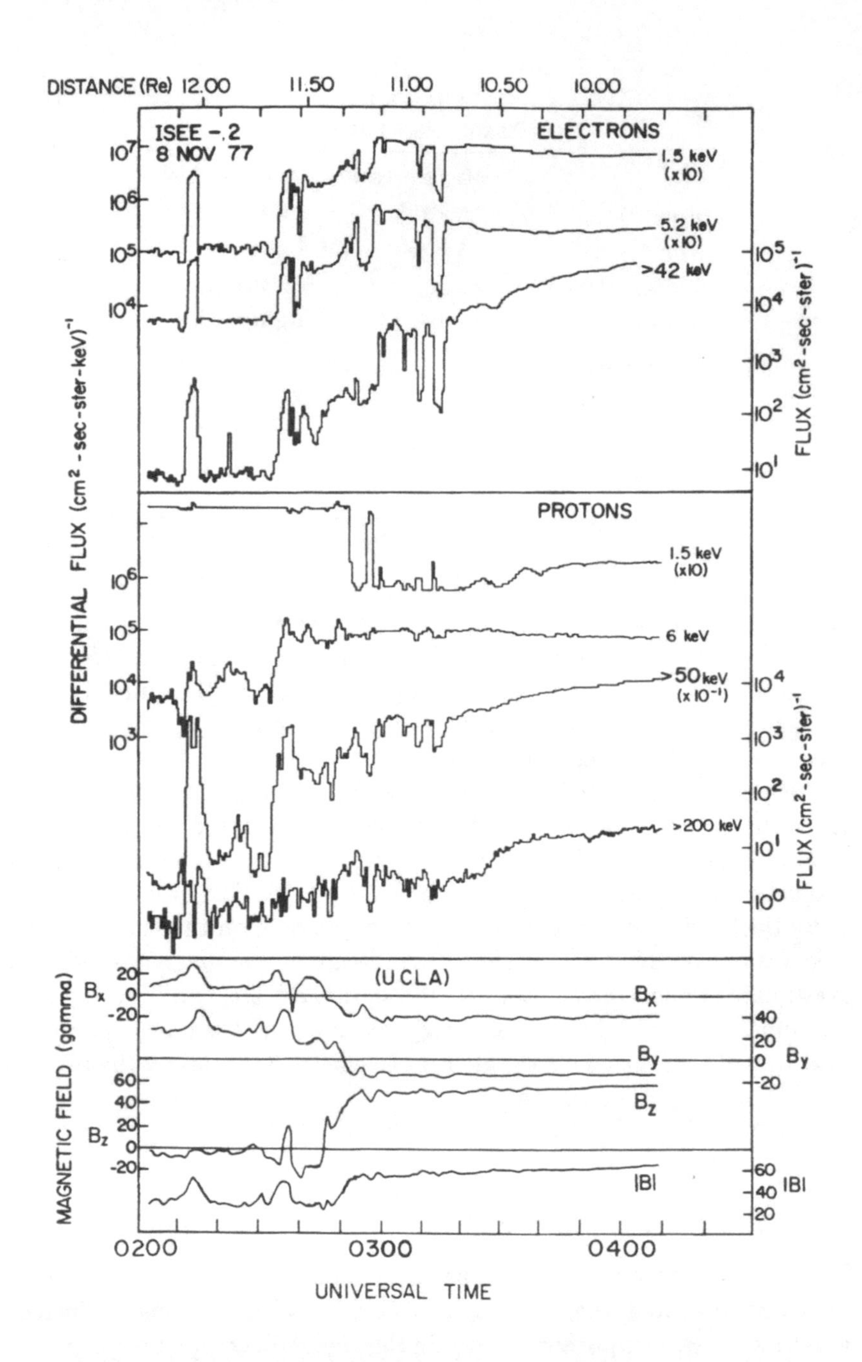

FIGURE 8.10 Magnetopause crossing on a more disturbed solar wind condition.

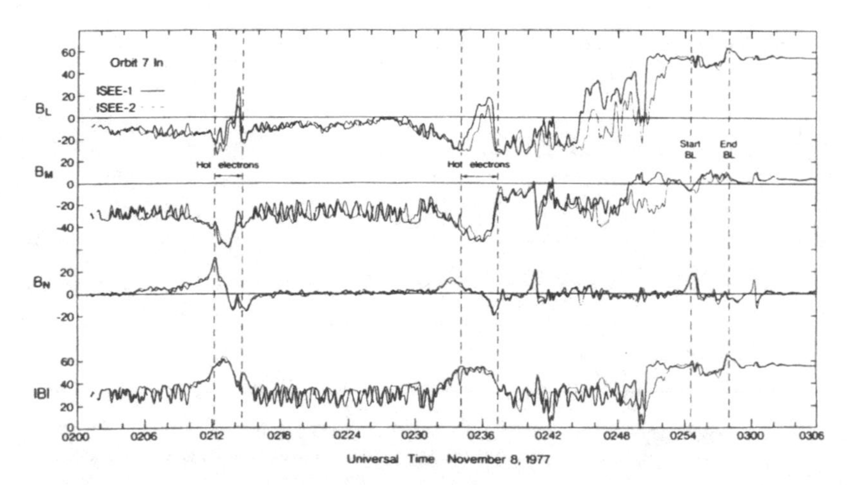

FIGURE 8.11 Magnetic field data transformed into a boundary coordinate system. Light and heavier lines come from two spacecraft separated by a few hundred kilometers. (From Elphic and Russell, 1979)

for intervals of up to a few minutes occurred at $\approx$ 0212 UT, 0236 UT and 0250 UT.

When the solar wind is disturbed, the magnetopause boundary is dynamic and moves in and out very rapidly. The multiple encounters of the magnetopause occur as the magnetopause sweeps by the spacecraft at speeds greater than the spacecraft speed ($\approx$ 10 km/sec vs. $\approx$ 2 km/sec). Figure 8.11 shows the magnetic field data transformed to the boundary coordinate system. The magnetic signature of a disturbed magnetopause is different from that of the quiet magnetopause shown earlier. Here, we see variations in all three components. A new feature is that the magnetic field orientation at the magnetopause boundary becomes more horizontal as indicated by the B_N component, which first turns northward, then southward, as the spacecraft returns to the magnetosheath. This behavior of the B_N component is characteristic of only the disturbed magnetopause boundary and is not found when the solar wind is quiet. The non-vanishing B_N component is one of the requirements of a rotational discontinuity.

Figure 8.12 shows the plasma data in the vicinity of this magnetopause. The geocentric distance in Earth radii (R_E), local time (LT), and magnetic latitude are indicated at the bottom of the figure. This figure shows the density of protons (top panel), the temperature of protons (second panel), the bulk velocity (third panel), and the flow directions given in (L, M, N) system. One notable difference between this magnetopause and the quiet one shown earlier is that the solar wind flow does not seem to come to an

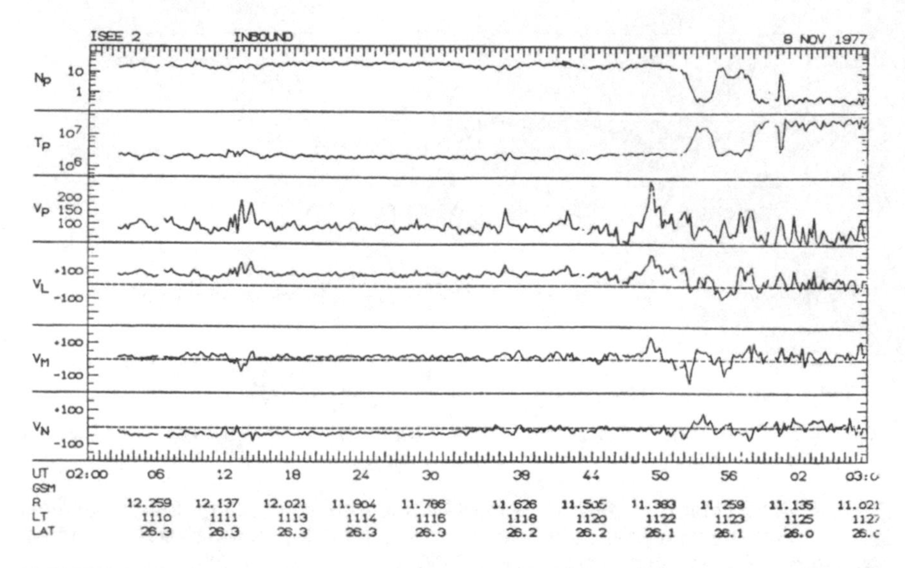

FIGURE 8.12 Plasma data from the same spacecraft that detected the magnetopause boundary shown in Figure 8.10. (Courtesy of G. Paschmann)

abrupt stop at the magnetopause (see, for example, $\approx$ 0250:30 UT). This suggests that $U_n \neq 0$ and solar wind mass may have crossed the boundary. Other features that are different include the presence of flowing plasma inside the magnetosphere, a change of direction in V_M (continuity of the tangential electric field would require this), and the very high plasma flows observed in conjunction with the bipolar B_N signatures at $\approx$ 0214 UT and 0249 UT.

If the interpretation that $G \neq 0$ is correct, then this magnetopause can be considered to possess a rotational discontinuity. However, the behavior of plasma during disturbed solar wind is extremely complicated and not all of the features are completely understood. Rotational discontinuities permit the solar wind mass, momentum, and energy to be transported across the magnetopause boundary. Clearly it is important to ascertain with more certainty that nature supports rotational discontinuities (for other works on this topic, see Paschmann *et al.*, 1978 and Sonnerup *et al.*, 1981, cited at the end of the chapter).

8.7 Boundary Layer of Closed Boundaries

A boundary layer is defined as a region adjacent to a boundary in which particles and fields from both sides of the boundary contribute. Earth's

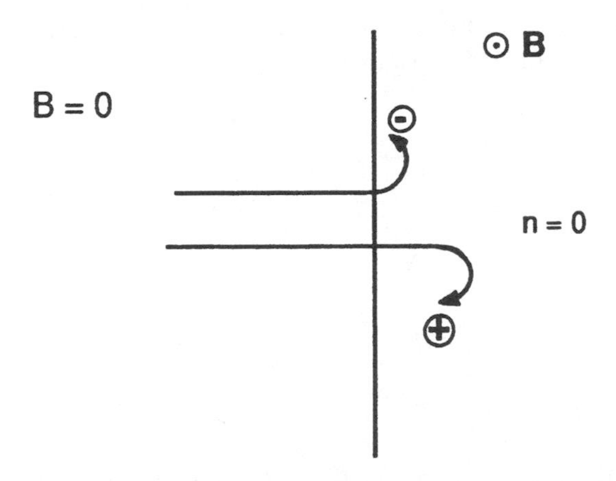

FIGURE 8.13 Orbits of particles in the neighborhood of a magnetic boundary. Note that because the particles partially penetrate the boundary before they are deflected back, a boundary layer is formed.

magnetopause boundary has a boundary layer (even for very quiet periods see Figures 8.7 and 8.8). This boundary layer is created, for instance, by solar wind particles that penetrate the boundary by some distance before they are turned around (by the $\mathbf{v} \times \mathbf{B}$ force). The boundary dimension depends on how far the particles can penetrate into a boundary. This problem is now examined by studying the Lorentz equation of motion, since a plasma boundary may be very thin and the assumptions of the MHD fluid and the first-order orbit theory may not be satisfied.

Consider a steady-state situation in which the solar wind particles are incident on a plane boundary. Let the magnetic field $\mathbf{B}$ be in the z-direction, the plane boundary in the y-direction, and let the solar wind particles approach the boundary from the $-x$-direction. A surface current will be created by the deflection of the solar wind particles so the boundary is not "infinitely" sharp and $\mathbf{B}$ falls off "slowly" in the solar wind, vanishing far away from the boundary.

The partial penetration of these particles creates a boundary layer that is a new feature not predicted by MHD (Figure 8.13). To examine the effects of these penetrating particles, let us examine first the properties of a boundary, assuming that the penetrating ions and electrons have the same mass. In this simple case, there is no separation of charges and thus no electric field is created (later, the formulation will be expanded to allow for the different masses).

The particles move under the Lorentz force, and the component equations for the two-dimensional problem we are considering are

$$\begin{aligned} \frac{dv_x}{dt} &= \frac{q}{m} v_y B \\ \frac{dv_y}{dt} &= -\frac{q}{m} v_x B \end{aligned} \tag{8.80}$$

Now (8.80) can be rewritten as

$$\begin{aligned} v_x \frac{dv_x}{dx} &= \frac{q}{m} v_y B \\ \frac{dv_y}{dx} &= -\frac{q}{m} B \end{aligned} \tag{8.81}$$

using the relation $d/dt = v_x d/dx$, which is permitted since all variables are a function of x only. Multiplying the bottom equation of (8.81) with v_y and adding the two equations yields

$$\frac{d}{dx}(v_x^2 + v_y^2) = 0 \tag{8.82}$$

and, upon integration, one obtains

$$(v_x^2 + v_y^2) = v_0^2 \tag{8.83}$$

where v_0 is an integration constant equal to the initial velocity of the particles, far away from the boundary. The total energy of the particles is conserved as expected.

Introduce now a vector potential $A(x)$ to describe the magnetic field

$$B = \frac{\partial A}{\partial x} = \frac{dA}{dx} \tag{8.84}$$

If this equation is combined with the bottom equation in (8.81), it yields

$$\frac{d}{dx}(mv_y + qA) = 0 \tag{8.85}$$

and, upon integration, one obtains

$$mv_y + qA = 0 \tag{8.86}$$

Our model assumes that $A = 0$ far from the boundary inside the plasma. Note that (8.86) is the canonical momentum in the y-direction and it is a constant of motion because the Hamiltonian is not dependent on the y-coordinate.

Let us now assume a fluid flow with streamlines that follow the particle trajectories. Then the total particle number density n at x, which includes both ions and electrons impinging on the boundary and those that are reflected, must satisfy the mass conservation equation $\nabla \cdot n\mathbf{v} = 0$. The solution of this equation is

$$nv_x = n_0 v_0 \tag{8.87}$$

where n_0 is the density inside the plasma far away from the boundary.

Now one of the Maxwell equations (8.3) can be rewritten as

$$\begin{aligned} -\frac{\partial B}{\partial x} &= \mu_0 J_y \\ &= \mu_0 n q v_y \end{aligned} \tag{8.88}$$

where n is given by (8.87). Using (8.87) to eliminate n and the energy conservation equation (8.83) to eliminate v_x, (8.88) can be transformed to

$$\begin{aligned} -\frac{\partial B}{\partial x} &= \frac{\mu_0 n_0 q v_0 v_y}{v_x} \\ &= \frac{\mu_0 n_0 q v_0 v_y}{(v_0^2 - v_y^2)^{1/2}} \end{aligned} \tag{8.89}$$

From (8.86), $v_y = -(q/m)A$. Using this, (8.89) becomes

$$\frac{\partial B}{\partial x} = \frac{\mu_0 n_0 q^2}{m} \frac{A}{[1 - (qA/mv_0)^2]^{1/2}} \tag{8.90}$$

Now, $dB/dx = (dB/dA)(dA/dx) = BdB/dA = d(B^2/2)/dA$. Hence we can write (8.90) as

$$\frac{d}{dA}\frac{B^2}{2\mu_0} = \frac{n_0 q^2}{m} \frac{A}{[1 - (qA/mv_0)^2]^{1/2}} \tag{8.91}$$

This differential equation can be directly integrated to yield

$$\frac{B^2}{2\mu_0} = n_0 m v_0^2 \left\{1 - \left[1 - (qA/mv_0)^2\right]^{1/2}\right\} \tag{8.92}$$

We have chosen the integration constant so that B goes to zero when A goes to zero far away from the boundary ($B \to 0$, $A \to 0$, as $x \to \infty$). Now note from the energy conservation equation (8.83) that at the point where the particles turn, $v_y = -v_0$, and the value of A is a maximum equal to mv_0/q. If we now designate B_0 as the magnetic field beyond the turning point of the particle, the quantity in the square root of (8.92) vanishes, yielding

$$\frac{B_0^2}{2\mu_0} = n_0 m v_0^2 \tag{8.93}$$

which shows that the total plasma energy density inside the plasma is equal to the magnetic pressure outside the plasma (same as the MHD result).

Equation (8.92) gives the relation of B as a function of the vector potential A. Since $B = dA/dx$, (8.92) can be rewritten as a function of x,

$$B_0 dx = \frac{dA}{\{1 - [1 - (qA/mv_0)^2]^{1/2}\}^{1/2}} \tag{8.94}$$

This equation can be simplified, by introducing new variables,

$$s = \frac{qA}{mv_0} \qquad 0 \leq s \leq 1 \tag{8.95}$$

$ds = (q/mv_0)dA$ and substitution of these in (8.94) yields

$$dx = r_c \left\{ \frac{ds}{[1 - (1 - s^2)^{1/2}]^{1/2}} \right\} \tag{8.96}$$

where $r_c = mv_0/qB_0$ is the cyclotron radius. Note that near the boundary, A is mv_o/q and hence $s \approx 1$, which means that the boundary layer thickness is approximately one Larmor radius. If the trigonometric substitution

$$s = \sin 2\Phi \qquad 0 \leq \Phi \leq \frac{\pi}{4} \tag{8.97}$$

is made, (8.96) becomes

$$\frac{dx}{r_c} = \frac{2 \cos 2\Phi d\Phi}{[1 - (1 - \sin^2 2\Phi)^{1/2}]^{1/2}} \tag{8.98}$$

The use of the trigonometric identities $\sin 2\Phi = 2 \sin \Phi \cos \Phi$ and $\cos 2\Phi = 1 - 2\sin^2 \Phi$ reduces (8.98) to

$$\frac{dx}{\sqrt{2} r_c} = \frac{d\Phi}{\sin \Phi} - 2 \sin \Phi d\Phi \tag{8.99}$$

Integration of (8.99) yields

$$\frac{x}{\sqrt{2} r_c} = \ln \left(\tan \frac{\Phi}{2} \right) + 2 \cos \Phi \tag{8.100}$$

Note also that in terms of the variable Φ, (8.92) becomes

$$\frac{B}{B_0} = \sqrt{2} \sin \Phi \tag{8.101}$$

A plot of B/B_0 as a function of $x/\sqrt{2} r_c$ is shown in Figure 8.14.

Now take the more realistic situation and let the electron and ion masses differ, but let them approach the boundary with the same velocity (as shown below, plasma neutrality requires this). Because the electrons are more easily bent by the magnetic field, the ions will penetrate further across the boundary and get ahead of the electrons. This gives rise to an electric field in the $-x$-direction, which pulls the electrons (ions) toward the ions (electrons).

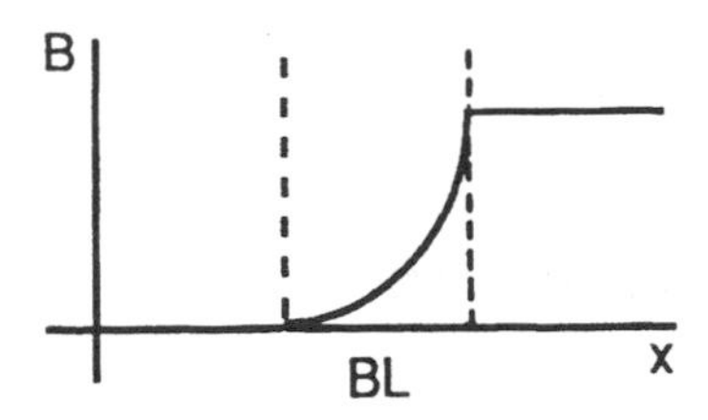

FIGURE 8.14 Behavior of the magnetic field B in the boundary layer.

Since the velocities of positive ions and electrons are equal, we need not make any distinction between the velocities of electrons and those of the heavier positive ions. The Lorentz equation of motion for the electrons and ions are

$$\begin{aligned} m_e v_x \frac{dv_x}{dx} &= -e[E(x) + v_y B] \\ m_i v_x \frac{dv_x}{dx} &= eE(x) \end{aligned} \tag{8.102}$$

where the electric field is a function of x and we have ignored the magnetic force in the ion equation. As will be shown, this is permitted because the thickness of the boundary is much less than the ion Larmor radius. e is the charge of electrons and we assume that the ions are singly charged. The electric field arises from charge separation, but the charge density produced is small and, to first order, plasma approximation holds. Hence

$$n^+(x) \approx n^-(x) \tag{8.103}$$

Particle conservation requires that

$$\begin{aligned} n^+(x)v^+(x) &= n_0 v_0 \\ n^-(x)v^-(x) &= n_0 v_0 \end{aligned} \tag{8.104}$$

Hence (8.103) and (8.104) show that

$$v^+(x) = v^-(x) \tag{8.105}$$

which we have assumed in the model. Eliminate $E(x)$ in (8.102) by combining the two equations and since $m_e/m_i \ll 1$, obtain

$$m_i v_x \frac{dv_x}{dx} = -ev_y B \tag{8.106}$$

Note the appearance of m_i in this equation. This is because the electrons and ions are coupled by means of the electric field and the electric field transfers the ion momentum to the electrons. The other equation of motion of the electrons is

$$m_e \frac{dv_y}{dx} = eB \tag{8.107}$$

The Maxwell equation (8.88) is still valid,

$$\frac{dB^2}{dA} = \mu_0 n e v_y \tag{8.108}$$

We can ignore the ion contribution to the current because this is smaller by m_e/m_i.

Define now (for convenience) two new variables,

$$\begin{aligned} U_x &= v_x \left(\frac{m_i}{m_e}\right)^{1/2} \\ U_0 &= v_0 \left(\frac{m_i}{m_e}\right)^{1/2} \end{aligned} \tag{8.109}$$

Then (8.106) becomes

$$U_x \frac{dU_x}{dx} = \frac{e}{m_e} v_y B \tag{8.110}$$

which has the same form as (8.81). Since the other equations are not changed, the solution of the problem under consideration will have the same form as in the previous case. The energy equation (8.83) becomes

$$m_i v_x^2 + m_e v_y^2 \approx m_i v_0^2 \tag{8.111}$$

since $m_e \ll m_i$. In (8.98), the Larmor radius must now be replaced with

$$r_c^* = \frac{m_e U_0}{eB_0} = \frac{\sqrt{m_e m_i} v_0}{eB_0} \tag{8.112}$$

which is the geometric mean of the electron and ion Larmor radius. The boundary thickness in this case is now estimated to be r_c^*, which is much smaller than the ion Larmor radius. This justifies the neglect of the magnetic force on the ion equation of motion in (8.103).

Note that at the turning point, $v_x = 0$ and (8.111) then shows that

$$\frac{1}{2} m_e v_y^2 = \frac{1}{2} m_i v_0^2 \tag{8.113}$$

This means that the electrons have acquired all of the initial ion energy at this point. The electrons will attain a velocity $v_y = v_0 (m_i/m_e)^{1/2}$ in the y-direction and therefore the electron orbits at the boundary are elongated by $(m_i/m_e)^{1/2}$.

Charge density resulting from charge separation was assumed small. By small we mean that $\rho = (n^+ - n^-)e \ll n_0 e$. This charge density produces an electric field given by $\nabla \cdot \mathbf{E} = \rho/\epsilon_0$ (recall that the ions are turned around by the electric field). At the turning point, the kinetic energy of the ions is approximately equal to the potential energy,

$$eEr_c^* \approx \frac{m_i v_0^2}{2} \tag{8.114}$$

since $v_x = 0$. An estimate of the charge density obtained from $\nabla \cdot \mathbf{E} = \rho/\epsilon_0$ yields

$$\begin{aligned} \rho &\approx \frac{\epsilon_0 E}{r_c^*} \\ &= \frac{\epsilon_0 e B_0^2}{2m_e} \end{aligned} \tag{8.115}$$

where (8.114) was used to eliminate E and (8.112) was used to eliminate r_c^*. Now note that $\mu_0 \epsilon_0 c^2 = 1$ and pressure balance requires that $B_0^2/2\mu_0 = n_0 m_i v_0^2$. Hence

$$\frac{\rho}{n_0 \epsilon_0} \approx \frac{m_i v_0^2}{m_e c^2} \tag{8.116}$$

which shows that $\rho/n_0\epsilon_0 \ll 1$ if $m_i v_0^2/m_e c^2 \ll 1$. $m_e c^2$ is the rest energy of the electron, 510 keV. Thus the kinetic energy of the stream $m_i v_0^2/2$ must be much less than 255 keV if the charge density can be ignored. The kinetic energy of the solar wind is of the order of 1 keV and therefore this requirement is easily satisfied.

The results can be summarized as follows. If a stream of solar wind electrons and ions impinges on a plane magnetic boundary, the ions will penetrate more deeply into the magnetosphere than the electrons because of the greater mass. This produces charge separation and, because a plasma tends to maintain neutrality, a polarization electric field is set up (to oppose the charge separation). Electrons and ions will be accelerated toward each other to minimize charge separation (a small charge separation will always remain). Ions and electrons will move roughly perpendicular to the boundary and both will $\mathbf{E} \times \mathbf{B}$ drift parallel to the boundary. The ions are reflected mainly by this polarization electric field instead of the $\mathbf{v} \times \mathbf{B}$ force and consequently the ion trajectory will be bent sharply. The electrons are accelerated by the electric field and deflected by the magnetic field until they gain a maximum energy that is nearly equal to the original energy of the ions. Thus the electric field transfers energy from the ions to the electrons and the electrons can gain transverse velocity (parallel to the boundary) that can exceed the ion velocity. Hence the current is carried predominantly by the electrons in this model. The electrons that are reflected will be decelerated by the electric field and their energy will be

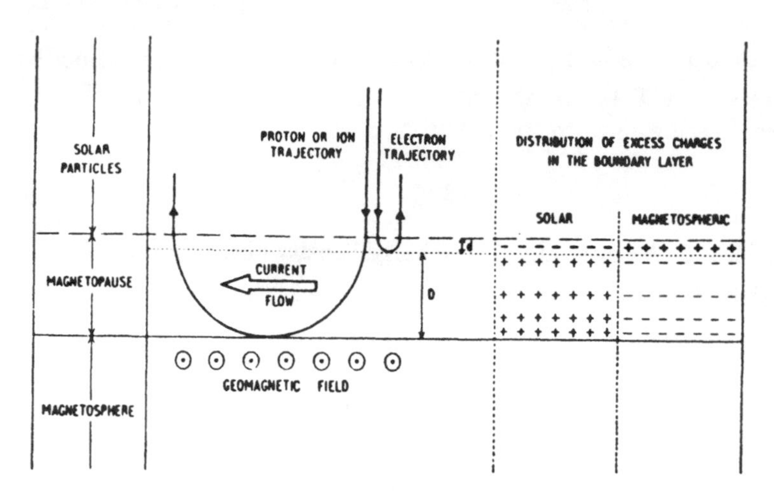

FIGURE 8.15 A schematic diagram of electron and ion trajectories incident normally on a plane boundary in the presence of a polarization electric field produced in response to charge separation. (From Willis, 1971)

transferred back to the reflected ions. The orbits of the particles and the direction of current flow are illustrated schematically in Figure 8.15.

In conclusion, it should be noted that the model of the magnetopause boundary layer considered here ignores the presence of ionospheres. Ionospheres are important because they are a source of particles and any charge separation created by the solar wind can be neutralized by ionospheric particles travelling along the magnetic field. Important consequences of ionospheric particles are that the polarization field will not be set up and therefore the thickness of the boundary layer will increase to about one ion Larmor radius; electron and ion motions will not be coupled and, since inertia is mainly carried by the ions, ions will be the main species that contribute to the magnetopause current.

The question of whether the separation of charges can be totally neutralized depends on a number of factors, such as whether the solar wind is steady or fluctuating, and how quickly neutralization can occur. Our model has also ignored the presence of trapped particles, and these particles will certainly contribute toward magnetopause currents. Finally, we note that the field-aligned current produced by ionospheric particles travelling along the planetary magnetic field brings another complication to the boundary problem because it alters the equilibrium configuration of the magnetopause and the boundary may become unstable.

Recent observations of the magnetopause show that the dimension of the boundary layer is generally 1 to 10 ion Larmor radius, and while this tends to support the idea that neutralization is occurring, the boundary problem is far from being completely understood. Note that the origin of the large boundary thickness (10-ion Larmor radii) is not predicted by our model. One interpretation invokes turbulence to broaden the boundary but this has not been quantitatively demonstrated (turbulence is created by some instability phenomena).

The boundary problem will remain in the forefront of research for a long time, as there are space missions designed specifically to study the structure of boundaries. In 1995, the European Space Agency and the National Aeronautics and Space Administration will launch a cluster of four satellites with identical instrumentation to study the spatial and temporal structures of boundaries. No doubt the boundary problem will receive further experimental and theoretical considerations.

While we have only considered models of "closed" magnetospheres, it should be noted that current research is attempting to identify and establish the existence of "open" magnetospheres. Open magnetospheric models require understanding the trajectories of particles in magnetic geometries that include magnetic field-free regions. This remains one of the most important problems to be solved in space plasmas. Finally, mention must also be made that the boundary problem can be formulated more elegantly, using the Vlasov equation. Although the Vlasov treatment is beyond the scope of our intention and therefore will not be treated here, we note that if one models the boundary as we have done here with the same approximations, the Vlasov results will yield the same results as given above. However, the Vlasov approach is better and permits greater flexibility in the physical models. One last comment concerns using the particle simulation codes to model boundaries and solving the problem from first principles in a self-consistent way. Particle simulation models will no doubt shed new light on this important problem.

Bibliography

Aggson, T.L., P.J. Gambardella and N.C. Maynard, Electric Field Measurements at the Magnetopause, 1. Observations of Large Convective Velocities at Rotational Magnetopause Discontinuities, *J. Geophys. Res.*, **88**, 10,000, 1983. This article examines the behavior of the electric field at the magnetopause, assuming the boundary has a rotational discontinuity.

Alfvén, H., Some Properties of Magnetospheric Neutral Surfaces, *J. Geophys. Res.*, **73**, 4379, 1968. This two-page article uses a capacitor to model how a plasma behaves in a region where the magnetic field changes direction.

Axford, I., Viscous Interaction Between the Solar Wind and the Earth's Magnetosphere, *Planet. Space Sci*, **12**, 45, 1964.

Chapman, S. and V.C.A. Ferraro, A New Theory of Magnetic Storms, *Terr. Magn. Atmos. Elec.*, **36**, 171, 1931. This article and the one by Ferraro represent two of a series of articles written by these authors on magnetic storms.

Cowley, S.W.H., A Self-Consistent Model of a Simple Magnetic Neutral Sheet System Surrounded by a Cold, Collisionless Plasma, *Cosmic Electrodynamics*, **3**, 448, 1973. This author elaborates in more detail some of the consequences of models proposed by H. Alfvén and A. Dessler.

Dessler, A.J., Magnetic Merging in the Magnetospheric Tail, *J. Geophys. Res.*, **73**, 209, 1967. Also, Vacuum Merging: A Possible Source of the Magnetospheric Cross Tail Electric Field, *J. Geophys. Res.*, **76**, 3174, 1971. This article estimates the current required to maintain the neutral sheet in the tail, which is then related to the strength of the convective electric field.

Dungey, J.W., *Cosmic Electrodynamics*, Cambridge University Press, Cambridge, England, 1958. Also, Interplanetary Magnetic Field and the Auroral Zones, *Phys. Rev. Lett.*, **6**, 47, 1961. Dungey was one of the first to recognize that Earth's magnetopause may support an open configuration.

Eastman, T.E., E.W. Hones, Jr., S.J. Bame and J.R. Asbridge, The Magnetospheric Boundary Layer: Site of Plasma, Momentum and Energy Transfer from the Magnetosheath into the Magnetosphere, *Geophys. Res. Lett.*, **3**, 685, 1976. One of the first papers that reported the existence of a boundary layer.

Elphic R. and C. Russell, ISEE-1 and -2 Magnetometer Observations of the Magnetopause, in *Magnetospheric Boundary Layers*, B. Battrick, ed., Rep. ESA SP-148, p. 43, European Space Agency, Paris, France, 1979. This paper shows several examples of observations of normal components of the magnetic field at the Earth's magnetopause. The authors have interpreted these observations in terms of merging the interplanetary and the geomagnetic fields.

Ferraro, V.C.A., On the Theory of the First Phase of a Geomagnetic Storm: A New Illustrative Calculation Based on an Idealized (Plane, not Cylindrical) Model Field Distribution, *J. Geophys. Res.*, **57**, 15, 1952. This paper extends the calculations of the original Chapman-Ferraro magnetic storm model.

Grad, H., Boundary Layer Between a Plasma and a Magnetic Field, *Phys. Fluids*, **4**, 1366, 1961. This is one of the first articles that examined the interaction of charged particles and magnetic fields.

Hess, W.N., *The Radiation Belt and Magnetosphere*, Blaisdell Publishing Co., Waltham, MA, 1968. This book contains a considerable amount of observational data up to the time it was written. References are complete and the chapter on the outer magnetosphere includes a discussion of the Chapman-Ferraro model.

Hudson, P.D., Discontinuities in an Anisotropic Plasma and their Identification in the Solar Wind, *Planet. Space Sci.*, **18**, 1611, 1970. This is the first article that organized the conservation equations of anisotropic MHD fluids and discussed their continuity requirements across the various types of boundaries in a systematic way.

Longmire, C.L., *Elementary Plasma Physics*, Interscience Publishers, New York, NY, 1963. The boundary layer theory presented in this chapter comes from Chapter 5 of this book. Longmire also discusses how the Boltzmann theory can be used in boundary layer studies.

Mozer, F.S., Electric Field Evidence on the Viscous Interaction at the Magnetopause, *Geophys. Res. Lett.*, **11**, 135, 1984. This article discusses examples of Earth's magnetopause during which the electric field data behaved as a tangential discontinuity.

Ogilvie, K.W., R.J. Fitzenreiter and J.D. Scudder, Observations of Electron Beams in the Low-Latitude Boundary Layer, *J. Geophys. Res.*, **89**, 10723, 1984. This research article examines in detail the structure of the magnetopause boundary layer, using the low energy electron plasma data. The interpretation is given that some of the observed features at the magnetopause resemble those observed in the ionosphere.

Parker, E.N., The Solar-Flare Phenomenon and the Theory of Reconnection and Annihilation of Magnetic Fields, *Astrophys. J. Suppl.*, **8**, 177, 1963. The Parker-Sweet model of flow through a region including a magnetic neutral point is developed in this article. This paper's main emphasis is to explain the acceleration of particles in solar flares.

Paschmann, G., N. Sckopke, G. Haerendel, I. Papamastorakis, S.J. Bame, J.R. Asbridge, J.T. Gosling, E.W. Hones, Jr. and E.R. Tech, ISEE Plasma Observations Near the Subsolar Magnetopause, *Space Sci. Rev.*, **22**, 717, 1978. This article reports early results from a plasma experiment that sampled the Earth's magnetopause.

Petschek, H.W., Magnetic Field Annihilation, in the *Physics of Solar Flares,* W.N. Hess, ed., Proceedings of the AAS-NASA symposium held at Goddard Space Flight Center, NASA SP-50, Greenbelt, MD, October, 1964. This article first reviews the Parker-Sweet model, then modification of the model is presented. See also the review article by V.M. Vasyliunas.

Rosenbluth, M.N., *Plasma Physics and Thermonuclear Research,* C.L. Longmire, J.L. Tuck and W.B. Thompson, eds., Pergamon Press, London, England, **2**, 271, 1963. One of the first articles that examined the interaction of charged particles with magnetic fields. Longmire's treatment comes from this paper.

Russell, C.T. and R.C. Elphic, ISEE Observations of Flux Transfer Events at the Dayside Magnetopause, *Geophys. Res. Lett.*, **6**, 33, 1979. The first report of an article on the existence of a normal component of the magnetic field at the magnetopause.

Sonnerup, B.U.Ö., Adiabatic Particle Orbits in a Magnetic Null Sheet, *J. Geophys. Res.*, **76**, 8211, 1971. This article provides comprehensive calculations on the conditions under which adiabatic invariants can be considered still valid in particle orbits through magnetic null fields.

Sonnerup, B.U.Ö. and L.J. Cahill, Jr., Explorer 12 Observations of the Magnetopause Current Layer, *J. Geophys. Res.*, **73**, 1757, 1968. This is one of the first articles that attempted to show that Earth's magnetopause boundary may be supporting a rotational discontinuity.

Sonnerup, B.U.Ö., G. Paschmann, I. Papamastorakis, N. Sckopke, G. Haerendel, S.J. Bame, J.R. Asbridge, J.T. Gosling and C.T. Russell, Evidence for Magnetic Field Reconnection at the Earth's Magnetopause, *J. Geophys. Res.*, **86**, 10049, 1981. Readers interested in magnetic field merging processes should read this research publication. Eleven cases of Earth's magnetopause boundary on the sunlit hemisphere have been analyzed. Observations of high-speed flows have been interpreted as evidence of reconnection as predicted by magnetic field merging models.

Speiser, T.W., Particle Trajectories in Model Current Sheets, 1. Analytical Solutions, *J. Geophys. Res.*, **70**, 1717, 1965. This author has studied the dynamics of particles in current sheets, using the Lorentz equation of motion. This single-particle approach avoids some of the difficulties encountered in the fluid approach. However, information on collective effects from many particles is not obtained because at this juncture the calculations are based on given field geometries (the fields are not self-consistently calculated).

Spreiter, J.R. and A.Y. Alskne, Plasma Flow Around the Magnetosphere, *Rev. Geophys. Space Phys.*, **7**, 11, 1969. The authors have studied MHD equations and this article reviews both observations and theory of closed magnetopauses.

Vasyliunas, V.M., Theoretical Models of Magnetic Field Line Merging, 1, *Rev. Geophys. Space Phys.*, **13**, 303, 1975. This article reviews in detail the physics underlying the concept of magnetic field merging and compares the models used by various researchers.

Willis, D.M., Structure of the Magnetopause, *Rev. Geophys. Space Phys.*, **9**, 953, 1971. This article reviews the status of our understanding of the various types of boundaries and includes a short discussion of observational data available at the time of writing. Although new data have been obtained since this article was written, the problems discussed are still with us today. This article, although written twenty years ago, is by no means obsolete. The bibliography is complete and useful for readers interested in work performed in the early years.

Magnetic Reconnection in Space and Laboratory Plasmas, E.W. Hones, Jr., ed., American Geophysical Union, Washington, D.C., 1984. This monograph contains papers presented in a conference devoted to the subject of magnetic field reconnection. The references on space observations that have been interpreted in terms of open magnetospheric models are fairly complete.

Magnetospheric Boundary Layers, Proceedings of an International Conference, held in Alpach, Austria. Published by the European Space Agency, ESA SP-148, 8 rue Mario-Nikis, 75738 Paris 15, France, August, 1979. A large part of this publication is devoted to theory and observations of boundary and boundary layer phenomena in Earth's magnetosphere. It also includes several articles of other astrophysical systems such as neutron stars and pulsars. Some of the first observations of data obtained from the new spacecraft pair (International Sun-Earth Explorer, ISEE) are published in this monograph.

Questions

1. What percent of the total solar wind energy density is due to the IMF at the Earth's magnetopause position? Use $B_{IMF} = 5nT$, $n = 10^6$ m^{-3}, U_{SW} = 400 km/sec.
2. Estimate the total dawn-dusk electrical potential across Earth's magnetospheric tail due to the solar wind flow. Use U_{SW} = 400 km/sec, $B_{IMF} = 5nT$, and the tail dimension is $20R_E$.
3. Estimate the percent of the total solar wind density due to the IMF at the Earth's magnetopause. Use $B_{IMF} = 5nT$, $n = 10^6$ m^{-3}, U_{SW}= 400 km/sec.
4. Estimate the total dawn-dusk electrical potential across Earth's magnetospheric tail due to the solar wind flow. Use U_{SW} = 400 km/sec, $B_{IMF} = 5nT$, and the tail dimension is $20R_E$.
5. Consider a spherical magnet of radius R that is rotating about its axis with an angular frequency ω. Let an electron be situated at $r (r < R)$ in the sphere.
   - **a.** What is the force on the electron in the non-rotating frame of reference?
   - **b.** The electron in (a) will respond to the force and move until a balancing force results due to charge separation. What is this balancing force for $r < R$?
   - **c.** What is the electric field in the rotating frame for $r < R$?
   - **d.** What is the net space charge density in the sphere (in the rest frame)?
6. **a.** Show that the magnetopause position for a dipole field is given by

$$r^6 = \frac{\mu_0 m^2}{32\pi 2 m_i U_{SW}^2 \cos^2 \chi}$$

b. Estimate the position of Earth's magnetopause on the equatorial plane for a solar wind velocity of 400 km/sec.

7. Show that

a. The induced electric field due to the solar wind flow $\mathbf{U}_{SW} = -U_0\mathbf{x}$ is given by

$$\begin{aligned}\mathbf{E} &= U_0 \sin\theta[B_T(\theta)\cos\phi - B_T(r)\sin\phi]\hat{\boldsymbol{\phi}} \\ &+ U_0\cos\theta B_T(r)\hat{\boldsymbol{\theta}} - U_0\cos\theta B_T(\theta)\hat{\mathbf{r}}\end{aligned}$$

where $B_T(r)$ and $B_T(\theta)$ are the magnitudes of $(r\theta)$ components of $(\mathbf{B})_{Total}$. In deriving the above equation, use $\mathbf{x} = \hat{\mathbf{r}}\sin\theta\cos\phi + \hat{\boldsymbol{\theta}}\sin\theta\sin\phi + \hat{\boldsymbol{\phi}}\cos\theta$.

b. Consider the equatorial plane, $\theta = \pi/2$. Show that the potential of this electric field is

$$V\left(r, \frac{\pi}{2}, \phi\right) = U_0 B_0 r \sin\phi\left[1 + \left(\frac{r_0}{2r}\right)^3\right]$$

Hint:

$$\left(\mathbf{E} = -\nabla V = -\hat{\boldsymbol{\phi}}\left(\frac{1}{r\sin\theta}\right)\frac{\partial V}{\partial\phi}, \quad \text{hence} \quad V = \int Er\sin\theta d\phi\right)$$

Note that far from the planet $r \gg r_0$, $V \approx U_0 B_0 y$, which is just the potential arising from the solar wind flowing across a northward IMF.

8. Derive Equation (8.101) from Equation (8.92).

9. Show for the Chapman-Ferraro boundary layer model that for small Φ (large negative x),

$$B \approx (8)^{1/2} B_0 \exp\left(\frac{x}{\sqrt{2r_c - 2}}\right)$$

10. Estimate the distance by which the ions overshoot the turning point of electrons in the Chapman-Ferraro boundary layer model.

11. The electrostatic potential Φ is in general a function of (x, y, z) or any arbitrary set of independent variables. A convenient choice is then to use (α, β, s) as variables since α and β are constants along $\mathbf{B}$. Here, s is a distance measured along $\mathbf{B}$. Show that

a. Φ is independent of s (use the fact that $\mathbf{E}\cdot\mathbf{B} = 0$, and that $\nabla\alpha\cdot\mathbf{B} = \nabla\beta\cdot\mathbf{B} = 0$).

b. The potential discussed above can also be written as

$$\Phi(\alpha, \beta) = U_0(2B_0R_p)^{1/2}\sin\left(\frac{\beta}{R_p}\right)\alpha^{1/2}$$

where $\beta = R_p\phi$ and

$$\begin{aligned}\alpha &= \left(\frac{R_pB_0}{2}\right)\left(\frac{r\sin\theta}{R_p}\right)^2\left[1 - \left(\frac{r_0}{r}\right)^3\right] \\ &= \left(\frac{R_pB_0}{2}\right)\left(\frac{y}{R_p\sin\phi}\right)^2\left[1 - \left(\frac{r_0}{r}\right)^3\right]\end{aligned}$$

and use $\Phi = U_0B_0y$ to eliminate y.

c. Evaluate $\Phi(\alpha, \beta)$ for the equatorial plane and show that this expression is the same as the one derived in problem 8.2.

12. Derive the equation of

a. Induced electric field by the solar wind flow for a magnetic configuration defined by the superposition of a southward magnetic field and the dipole.

b. Then derive the equation for the potential in the xz-plane.

c. Estimate the potential difference between dawn and dusk induced by a $B_{IMF} = 10nT$, $U_0 = 400$ km/sec, and $B_p = 30,000nT$ for the latitude where the field lines are open (this gives the polar cap potential in the ionosphere).

9

Waves in Space

9.1 Introduction

The purpose of this chapter is to discuss waves observed in space plasmas and the laws that govern their propagation. Waves are time-dependent phenomena that involve both electron and ion dynamics. Consider, for example, waves generated by the perturbation of fluid elements at the magnetopause by a sudden change in the solar wind pressure. This change will move the boundary from its equilibrium position, and the disturbance created in the fluid element will propagate in the form of Alfvén waves. A broad range of wave frequencies is detected in space, from less than a few cycles per hour to several million cycles per second. Disturbances on stars, our Sun, solar wind, shocks, magnetospheres, and ionospheres all contribute to the generation of waves.

Experiments conducted on single spacecraft have provided detailed information on the amplitude and the frequency of waves, but not information on the propagation vector **k**, which is required to identify the wave modes. Clear, unambiguous identification of wave modes is therefore not always possible. However, the nature of waves in plasmas strongly depends on the plasma parameters, and wave modes can be deduced from information provided by the ambient plasma and magnetic field data, together with the theory that predicts which waves can exist in a medium defined by these plasma parameters.

A plasma in a magnetic field supports a variety of waves, which can be classified conveniently in terms of the orientation of the propagation vector **k** relative to the wave electric field **E**. A wave is classified as electrostatic if **k** is parallel to **E**, and as electromagnetic if **k** is perpendicular to **E**. Both classes have been detected in space. For instance, plasma and "whistler" waves detected in Earth's magnetosphere are, respectively, electrostatic and electromagnetic waves. These waves carry a wealth of information on the active plasma processes in the medium in which the waves propagate.

9.2 Electrostatic and Electromagnetic Waves in Space

A survey plot of electrostatic and electromagnetic waves obtained by a spacecraft experiment that travelled through interplanetary space and the magnetosphere is shown in Figure 9.1. The wave data are displayed in discrete frequency intervals. The ordinate for the electrostatic waves (upper panel) is $E^2/\Delta f$, and for the electromagnetic waves (bottom panel) is $B^2/\Delta f$, where E and B are the amplitudes of the electric and the magnetic fields and Δf is the frequency interval in which the waves are detected. The solid lines represent the maximum field strength and the solid black area represents the average strength.

The spacecraft was initially in interplanetary space on the sunward hemisphere and was bound toward Earth. The interval of times when the spacecraft crossed the bow shock and the magnetopause are marked by the dashed lines. Electrostatic wave activity is intense, complex, and observed everywhere. The electrostatic waves in the solar wind (upstream of Earth's bow shock) were bursty and were observed from a few Hz to the highest frequency channel of the wave antenna, 311 kHz. The emissions below 10 kHz are believed to be ion acoustic waves. Intense electromagnetic waves were observed in the vicinity of the bow shock and the magnetopause (note that the highest frequency channel here is 3.11 kHz). The waves observed ahead of the magnetopause are electromagnetic waves propagating in the "whistler" mode. These waves have frequencies below the Larmor frequency of the electrons and have phase velocities larger than the solar wind flow speed, which permits them to propagate outward into the flowing solar wind. The low-frequency electromagnetic wave activity (frequencies below

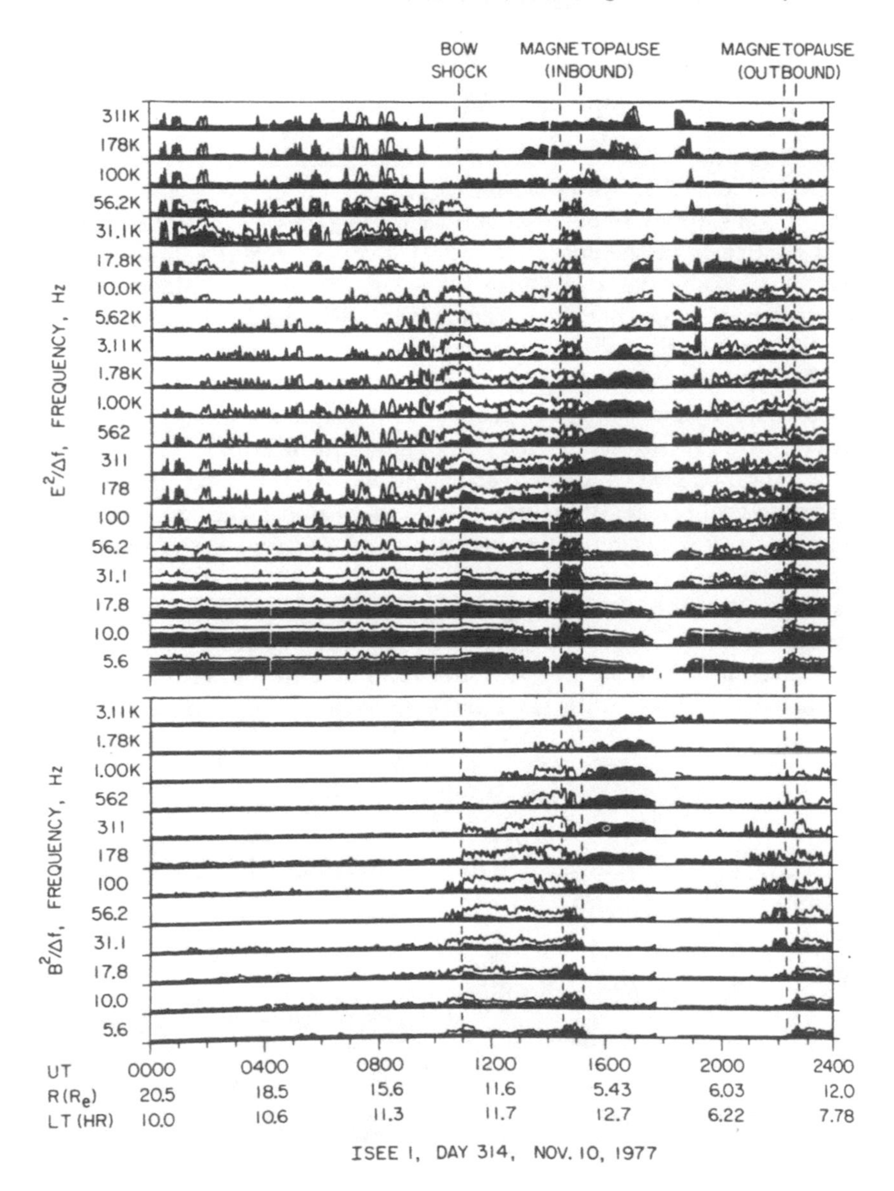

FIGURE 9.1 The plasma wave electric and magnetic field data from ISEE-1 spacecraft for a representative pass through the magnetosphere. The enhanced electric and magnetic field intensities at the inbound and outbound magnetopause crossings are clearly evident. (From Gurnett *et al.*, 1981)

100 Hz) inside the magnetosphere was very much reduced. Note that as the spacecraft crossed the magnetosphere boundary in the outbound portion of its orbit, an increase in wave activity was again observed.

9.2.1 Magnetohydrodynamic (MHD) Waves

Waves generated in a fluid medium in which the MHD description is valid are referred to as MHD waves. MHD waves are low-frequency waves (lower than the ion Larmor frequency) and their existence in the solar wind was established early in the space era. MHD wave activity is enhanced during a disturbed solar wind and some of these waves have been identified with disturbances propagating away from the Sun. Figure 9.2 shows an example of MHD waves in the solar wind observed by the Mariner 5 spacecraft. The three components of the magnetic field, the total intensity, the solar wind velocity components, and the solar wind density, are shown. Wave data obtained in a flowing solar wind are difficult to interpret because spatial and temporal variations are not easily distinguished. For example, consider a magnetic field spatial discontinuity that is created far away. A spacecraft recording this discontinuity as the solar wind flows by it will observe the discontinuity as a time variation.

Proper deconvolution is therefore required to arrive at a correct interpretation of data. This normally involves examining other measurements, such as velocity and density, and understanding how discontinuities propagate in the solar wind. In the example shown, note that while the magnitude

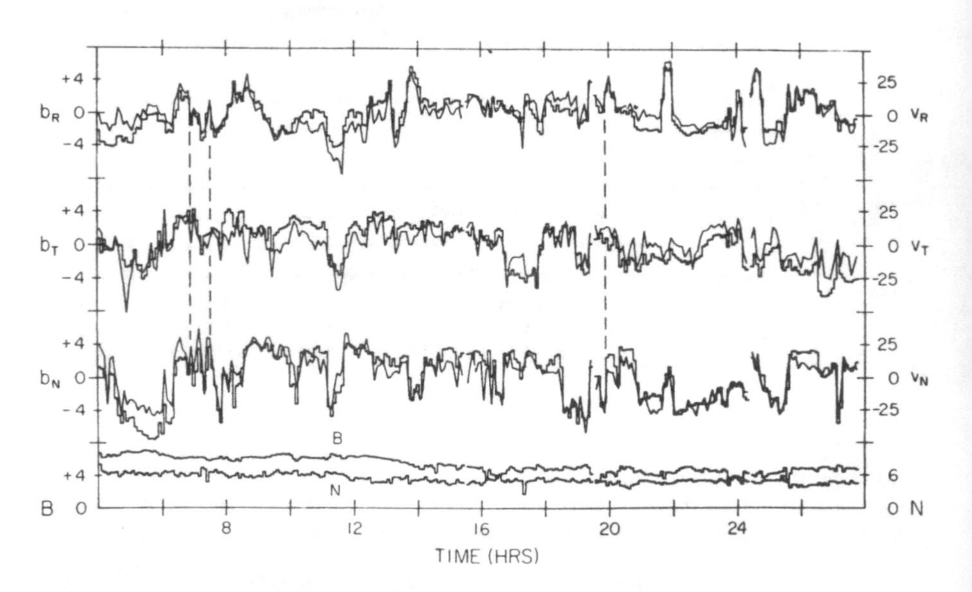

FIGURE 9.2 An example of MHD waves observed in the solar wind. Note that while the magnetic and velocity components fluctuate, the density of protons N and the total magnetic field intensity B remain nearly constant. These wave features are expected of Alfvén waves. (From Davis, 1972)

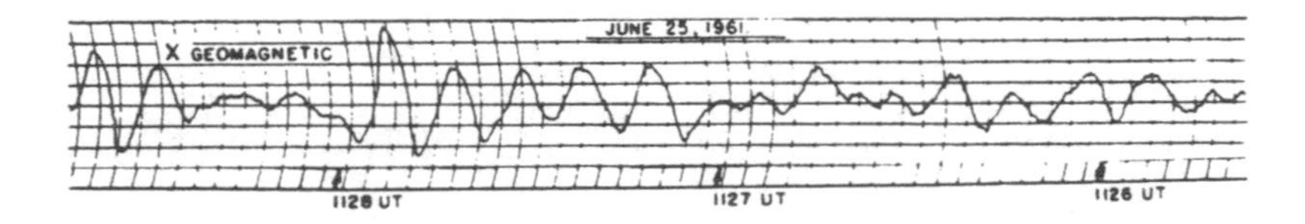

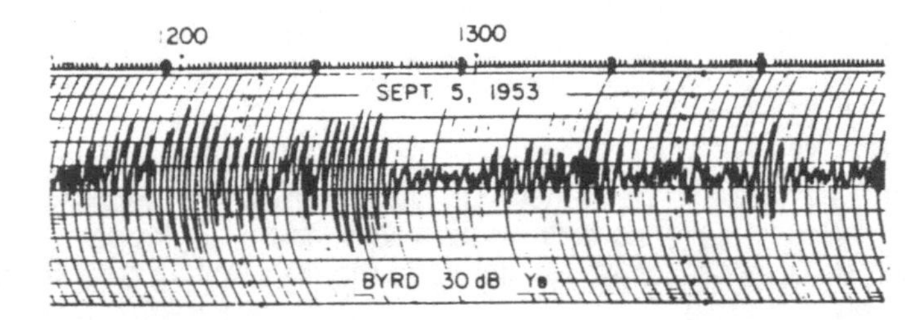

FIGURE 9.3 Examples of p_i (top) and p_c (bottom) events recorded on the ground. (From Campbell, 1967)

of the wave components varied, the total magnetic field intensity and the plasma density remained constant. Also, the variations of the magnetic field components are correlated with the solar wind velocity components. These features are properties of a class of MHD waves called Alfvén waves, and the waves shown have been identified as Alfvén waves.

9.2.2 Waves Detected On the Ground

MHD and electromagnetic waves generated in the magnetosphere can propagate along the ambient magnetic field direction and can be detected by ground-based instruments. For example, micropulsations (waves in the frequency range $\approx$ 0.001 Hz to $\approx$ 1 Hz) were first detected on the ground (Figure 9.3). Micropulsations are observed during both quiet and disturbed geomagnetic conditions and they have amplitudes of a few 10^{-12} T to several 10^{-9} T. Some are very periodic (called continuous pulsations, p_c), while others are irregular (called irregular pulsations, p_i).

Waves detected on the ground cover a wide frequency range. Higher-frequency waves that are detected on the ground include the "whistlers," generated by lightning discharges. These are electromagnetic waves in the audio frequency range of a few kHz to several tens of kHz. The nomenclature of various waves (organized by frequency) that can be detected on the ground is shown in Figure 9.4.

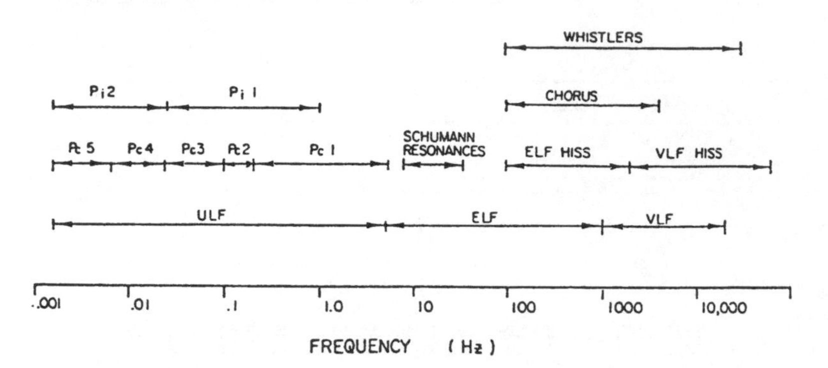

FIGURE 9.4 Nomenclature of AC magnetic fields as a function of the frequency. ULF stands for ultra-low frequency, ELF for extremely-low frequency, VLF for very low frequency, p_c for continuous pulsations, and p_i for irregular pulsations. Schumann resonances are waves set up in the Earth-ionosphere wave guide. (Courtesy of C. Russell)

9.3 Basic Equations and Concepts

The electromagnetic fields and plasma parameters obey the Maxwell equations (which are stated again for convenience),

$$\begin{aligned}
\nabla \times \mathbf{E} &= -\frac{\partial \mathbf{B}}{\partial t} \\
\nabla \times \mathbf{H} &= \mathbf{J} + \frac{\partial \mathbf{D}}{\partial t} \\
\nabla \cdot \mathbf{B} &= 0 \\
\nabla \cdot \mathbf{D} &= \rho_c
\end{aligned} \tag{9.1}$$

These field equations are coupled to the fluid equations

$$\frac{\partial n}{\partial t} + \nabla \cdot n\mathbf{u} = 0$$

$$\frac{d}{dt}(p\rho_m^{-\gamma}) = 0$$

$$\rho_m \frac{d\mathbf{u}}{dt} = -\nabla p + qn(\mathbf{E} + \mathbf{u} \times \mathbf{B}) \tag{9.2}$$

where an adiabatic equation of state for the fluid has been assumed. The formulation of the wave equations begins with the assumption that the plasma pressure is isotropic and therefore p is a scalar.

9.3.1 Plane Waves

Consider small-amplitude oscillations that permit the wave fields to be represented as sinusoidal plane waves. A plane wave is defined as a wave whose direction of propagation and amplitude is the same everywhere. For a monochromatic plane wave disturbance with frequency ω, a sinusoidally varying quantity in space and time is represented by

$$\mathbf{s}(\mathbf{r}, t) = \mathbf{s}_0 e^{i(\mathbf{k}\cdot\mathbf{r}-\omega t)} \tag{9.3}$$

where $\mathbf{s}_0$ is a constant vector defining the amplitude of the wave, i is the imaginary index $\sqrt{-1}$, $\mathbf{k}$ is the wave vector, and t is time. The measurable quantity is understood to be the real part of this complex expression.

The exponent in (9.3) is the phase of the disturbance. The time derivative of the phase is defined as the frequency of the wave

$$\frac{\partial}{\partial t}(\mathbf{k}\cdot\mathbf{r} - \omega t) = -\omega \tag{9.4}$$

The spatial derivative is defined as the wave vector,

$$\frac{\partial}{\partial \mathbf{r}}(\mathbf{k}\cdot\mathbf{r} - \omega t) = \mathbf{k} \tag{9.5}$$

which specifies the direction of wave propagation.

9.3.2 Phase and Group Velocity

The surface of a constant phase (also called the wave surface) is displaced with a phase velocity, which is obtained by taking the total time derivative of the phase and setting it equal to zero. Thus $d(\mathbf{k}\cdot\mathbf{r} - \omega t)/dt = 0$ yields

$$V_{ph} = \frac{\omega}{k} \tag{9.6}$$

or, in vector form,

$$\mathbf{V}_{ph} = \frac{\omega}{k^2}\mathbf{k} \tag{9.7}$$

A complex wave of arbitrary shape (such as a wave packet) can be synthesized by the superposition of the plane waves. Each of the partial waves in the packet may be traveling with a different phase velocity. It is then necessary to define a group velocity for the wave packet, which is

$$\mathbf{V}_{gp} = \frac{\partial \omega}{\partial \mathbf{k}} \tag{9.8}$$

While $\mathbf{V}_{ph}$ of waves can exceed the speed of light c, $\mathbf{V}_{gp}$ is always less than c. $\mathbf{V}_{gp}$ determines the rate at which the energy of the wave is transported.

9.3.3 Index of Refraction

The concept of index of refraction is useful in describing the behavior of electromagnetic wave propagation. The definition of index of refraction is

$$\begin{aligned} n &= \frac{c}{V_{ph}} \\ &= \frac{ck}{\omega} \end{aligned} \tag{9.9}$$

where c is the speed of light in vacuum. The index n is generally frequency- and wave-number dependent, $n(k, \omega)$. As in ordinary optics, propagation of electromagnetic waves occurs in a medium only if the medium is transparent, that is, if the imaginary part of the index of refraction vanishes.

When wave packets are considered, the propagation is described by the group velocity. The group refractive index is defined by

$$\begin{aligned} n_{gp} &= \frac{c}{V_{gp}} \\ &= \frac{c}{(\partial\omega/\partial k)} \end{aligned} \tag{9.10}$$

The relation between the group refractive index and the ordinary refractive index is obtained from $n = ck/\omega$. Differentiating this yields

$$\frac{\partial n}{\partial \omega} = \frac{c}{\omega(\partial\omega/\partial k)} - \frac{ck}{\omega^2} \tag{9.11}$$

which can be solved for $\partial\omega/\partial k$. Equation (9.10) then becomes

$$n_{gp} = n + \frac{\omega \partial n}{\partial \omega} \tag{9.12}$$

The group velocity is

$$V_{gp} = \frac{c}{[n + \omega(\partial n/\partial\omega)]} \tag{9.13}$$

In normal behavior, $\partial n/\partial\omega > 0$ and $n > 1$ and $V_{gp} = c/n_{gp} < c$. However, $\partial n/\partial\omega$ may be large and negative and (9.13) could yield $V_{gp} > c$. This kind of result cannot be studied under the approximations made in defining (9.10). It requires a theory that treats more completely the propagation of complex wave packets.

9.3.4 Dispersion Relation

Phase and group velocities can be calculated if a relation exists between ω and k. This relation, $\omega = \omega(k)$, is known as the dispersion relation. The dispersion relation contains physical parameters of a given medium in which a wave exists and propagates. Hence the dispersion relation contains all relevant information on how the medium responds to a given wave. One of the main tasks in the study of waves involves the derivation of the dispersion relation.

9.3.5 Polarization

Let $\mathbf{s}$ in (9.3) be the electric field $\mathbf{E}$ or the magnetic field $\mathbf{B}$. Then we can write as a measurable quantity

$$\begin{aligned} \mathbf{E} &= Re\mathbf{E}_0 e^{i(\mathbf{k}\cdot\mathbf{r}-\omega t)} \\ \mathbf{B} &= Re\mathbf{B}_0 e^{i(\mathbf{k}\cdot\mathbf{r}-\omega t)} \end{aligned} \tag{9.14}$$

where $\mathbf{E}_0$ and $\mathbf{B}_0$ are, in general, complex vectors. We can write $\mathbf{E}_0$ (or $\mathbf{B}_0$), for example, in the form

$$\mathbf{E}_0 = \mathbf{A}_0 e^{i\delta} = \mathbf{E}_1 + i\mathbf{E}_2 \tag{9.15}$$

where $\mathbf{A}_0$ is a real vector. Consider now a wave propagating in the z-direction. Then we can let

$$\begin{aligned} \mathbf{E}_1 &= \hat{\mathbf{x}} E_1 \sin\delta + \hat{\mathbf{y}} E_2 \cos\delta \\ \mathbf{E}_2 &= -\hat{\mathbf{x}} E_1 \cos\delta + \hat{\mathbf{y}} E_2 \sin\delta \end{aligned} \tag{9.16}$$

For simplicity, let $\delta = \pi/2$. Then, $\mathbf{E}_1 = E_1\hat{\mathbf{x}}$ and $\mathbf{E}_2 = E_2\hat{\mathbf{y}}$. With this choice, $E_0^2 = E_1^2 + E_2^2$ and we obtain

$$\begin{aligned} E_x &= E_1 \sin(\mathbf{k}\cdot\mathbf{r} - \omega t) \\ E_y &= E_2 \cos(\mathbf{k}\cdot\mathbf{r} - \omega t) \end{aligned} \tag{9.17}$$

Equation (9.17) shows that

$$\frac{E_y^2}{E_2^2} + \frac{E_x^2}{E_1^2} = 1 \tag{9.18}$$

which is an equation of an ellipse. Thus, for plane waves, the electric field vector $\mathbf{E}$ at each point in space rotates with an angular frequency ω in a plane perpendicular to the direction of propagation. If $E_1 \neq E_2$, the tip of the vector describes an ellipse and the wave is elliptically polarized. An observer looking in the direction of propagation sees a clockwise sense of rotation. If $E_1 = E_2$, then the electric field $\mathbf{E}$ rotates with a constant magnitude and the wave is circularly polarized. If E_1 or $E_2 = 0$, then the wave is linearly polarized.

9.3.6 Energy Flux

The energy flux in plane waves is given by $\mathbf{S} = \mathbf{E} \times \mathbf{H}$ which is also called the Poynting flux. The energy flux is directed along the direction of propagation of the wave. The total energy density of the wave, using the relations $\mathbf{D} = \epsilon_0 \mathbf{E}$ and $\mathbf{B} = \mu_0 \mathbf{H}$ is

$$\begin{aligned} W &= \frac{\mathbf{E} \cdot \mathbf{D}}{2} + \frac{\mathbf{B} \cdot \mathbf{H}}{2} \\ &= \frac{\epsilon_0 E^2}{2} + \frac{B^2}{2\mu_0} \end{aligned} \tag{9.19}$$

9.3.7 Small-Amplitude Perturbation Equations

If plane harmonic wave solutions are assumed, the time and spatial derivations can be represented by

$$\begin{aligned} \frac{\partial}{\partial t} &\rightarrow -i\omega \\ \nabla &\rightarrow i\mathbf{k} \\ \nabla \cdot &\rightarrow i\mathbf{k} \cdot \\ \nabla \times &\rightarrow i\mathbf{k} \times \end{aligned} \tag{9.20}$$

When only small-amplitude perturbations are considered, the dynamic variables can be expressed in terms of their equilibrium and perturbed parts,

$$\begin{aligned} \mathbf{u} &= \mathbf{u}_0 + \mathbf{u}_1 = \mathbf{u}_1 \\ n &= n_0 + n_1 \\ p &= p_0 + p_1 \\ \mathbf{E} &= \mathbf{E}_0 + \mathbf{E}_1 = \mathbf{E}_1 \\ \mathbf{B} &= \mathbf{B}_0 + \mathbf{B}_1 \end{aligned} \tag{9.21}$$

where the subscripts 0 and 1 refer to the equilibrium and perturbed quantities. In equilibrium, $\mathbf{u}_0 = \mathbf{E}_0 = 0$. $\mathbf{u}$ and $\mathbf{E}$ arise from perturbing an equilibrium fluid and hence they are first-order quantities. Our interest is to retain quantities only to first-order, ignoring higher order terms. If we use (9.20) and (9.21), Maxwell equations to first-order become

$$\begin{aligned} i\mathbf{k} \times \mathbf{E}_1 &= i\omega \mathbf{B}_1 \\ i\mathbf{k} \times \mathbf{H}_1 &= \mathbf{J}_1 - i\omega \mathbf{D}_1 \\ i\mathbf{k} \cdot \mathbf{B}_1 &= 0 \\ i\mathbf{k} \cdot \mathbf{D}_1 &= \rho_c \end{aligned} \tag{9.22}$$

and momentum and continuity equations are

$$-imn_0\omega\mathbf{u}_1 = qn_0(\mathbf{E}_1 + \mathbf{u}_1 \times \mathbf{B}_0) - i\mathbf{k}p_1$$
$$-i\omega n_1 + in_0\mathbf{k}\cdot\mathbf{u}_1 = 0 \quad (9.23)$$

The plasma is initially assumed to be uniform, $\nabla n_0 = 0$ and stationary and cold ($\partial n_0/\partial t = 0$). Since $\mathbf{u}_1$ is a small quantity, $(\mathbf{u}_0 \cdot \nabla)\mathbf{u}_1 = 0$. $\mathbf{J}_1$ is the total perturbed current density. These equations are fundamental for wave studies and will be referred to often in this chapter.

9.3.8 Dielectric Properties of a Plasma

In the study of waves, it is useful to treat plasma as a dielectric material. In a sense, a plasma resembles a system of coupled oscillators and supports a spectrum of normal mode oscillations. These oscillations always have time- and spatial-varying electric fields and the propagation of such fields in plasmas is precisely determined by the dielectric properties of a plasma.

The electrical properties of plasma are defined by the polarization and magnetization vectors,

$$\mathbf{P} = \mathbf{D} - \epsilon_0\mathbf{E}$$
$$\mathbf{M} = \frac{\mathbf{B}}{\mu_0} - \mathbf{H} \quad (9.24)$$

These vectors are meaningful only in a material substance, since in free space they vanish. Substitution of these field quantities into the Maxwell equations yields

$$\nabla\times\mathbf{B} = \mu_0\mathbf{J}_T + \mu_0\epsilon_0\frac{\partial\mathbf{E}}{\partial t}$$
$$\nabla\cdot\mathbf{E} = \frac{\rho_T}{\epsilon_0} \quad (9.25)$$

where

$$\mathbf{J}_T = \mathbf{J} + \frac{\partial\mathbf{P}}{\partial t} + \nabla\times\mathbf{M}$$
$$\rho_T = \rho_c - \nabla\cdot\mathbf{P} \quad (9.26)$$

are total current and charge densities. The terms $\partial\mathbf{P}/\partial t$ and $\nabla\times\mathbf{M}$ are polarization and magnetization currents and $\nabla\cdot\mathbf{P}$ is the polarization charge. Maxwell equations written in this form explicitly show how the material substance contributes to the current and charge densities. $\mathbf{J}$ and ρ_c are "external" sources, while the other terms are "internal" sources dependent on the material substance.

A systematic way to obtain the dispersion relation is through the dielectric properties of the plasma medium. Take the curl of the first equation of (9.1) and use (9.25) to obtain

$$\nabla \times (\nabla \times \mathbf{E}) = -\mu_0 \frac{\partial \mathbf{J}}{\partial t} - \frac{1}{c^2}\frac{\partial^2 \mathbf{E}}{\partial t^2} \tag{9.27}$$

where $\mathbf{J}$ is the total current (the subscript T is now omitted). Application of (9.20) and (9.21) reduces this equation to

$$-\mathbf{k} \times (\mathbf{k} \times \mathbf{E}_1) = i\omega\mu_0 \mathbf{J}_1 + \frac{\omega^2}{c^2}\mathbf{E}_1 \tag{9.28}$$

which is more convenient to work with than (9.22), which contains $\mathbf{D}$ and $\mathbf{H}$ field vectors. $\mathbf{J}_1$ is obtained from

$$\mathbf{J}_1 = \sum q_\alpha n_{\alpha 0} \mathbf{u}_{\alpha 1} \tag{9.29}$$

where $\mathbf{u}_1$ is obtained from (9.23) and we sum over the species α.

Consider now waves propagating in an anisotropic medium. In such a medium

$$D_i = \epsilon_0 \kappa_{ij} E_j \tag{9.30}$$

where D_i is the ith component of the displacement vector, κ_{ij} is the ij^{th} component of the dielectric tensor, and E_j is the jth component of the electric field. Ohm's law in an anisotropic medium is $J_i = \sigma_{ij} E_j$ where J_i is the ith component of the current density, σ_{ij} is the ij^{th} component of the conductivity tensor, and E_j is the jth component of the electric field. Using (9.29) and (9.30), the ith component of (9.28) can be written as

$$\begin{aligned} [\mathbf{k} \times (\mathbf{k} \times \mathbf{E})]_i &= -\frac{\omega^2}{c^2}\left(E_i + \frac{i\sigma_{ij}E_j}{\epsilon_0 \omega}\right) \\ &= -\frac{\omega^2}{c^2}\left(\delta_{ij} + \frac{i\sigma_{ij}}{\epsilon_0 \omega}\right) E_j \end{aligned} \tag{9.31}$$

where δ_{ij} is the unit tensor. The dielectric tensor is given by

$$\kappa_{ij} = \frac{\epsilon_{ij}}{\epsilon_0} = \delta_{ij} + \frac{i\sigma_{ij}}{\epsilon_0 \omega} \tag{9.32}$$

where ϵ_{ij} is the permittivity tensor. The dielectric tensor contains essential information on the medium and how it responds to a specific wave.

If we use (9.32), (9.31) can be written generally as

$$-n^2[\hat{\mathbf{k}} \times (\hat{\mathbf{k}} \times \mathbf{E})] = \boldsymbol{\kappa} \cdot \mathbf{E} \tag{9.33}$$

where $\hat{\mathbf{k}}$ is a unit vector, $n^2 = c^2k^2/\omega^2$, and $\boldsymbol{\kappa}$ is the dielectric tensor. This equation written in tensor notation is

$$[\kappa_{ij} - n^2(\delta_{ij} - \hat{k}_i \hat{k}_j)]E_j = 0 \tag{9.34}$$

A non-vanishing solution of (9.34) exists only if the determinant of the tensor multiplying **E** vanishes. The solution yields the dispersion relation of the waves and the ratios of the components of **E** determine the polarization of these waves.

9.4 Electrostatic Waves

An electrostatic wave is generated, for example, when a perturbation creates a charge imbalance in a neutral fluid element. This charge imbalance will accelerate electrons (ions) in the neighborhood of the charged fluid element, resulting in charges oscillating back and forth. These oscillations involve only the electric field and they are defined as electrostatic waves (the oscillating magnetic field is zero). The first Maxwell equation (9.22) shows the electric field **E** is then parallel to the direction of the wave vector **k**.

9.4.1 Electron Plasma Waves in Cold Plasma

Consider a simple case of a neutral plasma with uniform distribution occupying a large region of space. Assume that the particles have a zero temperature and therefore do not have any thermal motion ($p = 0$). Initially these particles are not flowing and the system can be considered to be in equilibrium. Let the ambient magnetic field **B** be zero or consider only motions along **B** so that $\mathbf{u} \times \mathbf{B}$ effects can be ignored.

Now suppose a small region of this equilibrium plasma is perturbed so that some electrons are displaced from their equilibrium positions (for this problem, ignore the motion of the ions since the mass of the ion is much greater than that of the electrons). An imbalance of charges will be created, giving rise to electrostatic fields, which will pull the electrons back. If the electrons were massless, the electrostatic force would restore the charge neutrality. However, electron inertia keeps the electrons moving (from the higher electron charge region across the electron-depleted region), and the electrons will overshoot. This results in an oscillatory motion of the electrons around the equilibrium position (Figure 9.5).

Let us now quantify this oscillatory motion by assuming that the motion is fluid-like. The first-order momentum equation (9.23) for this example is

$$imn_0\omega\mathbf{u}_1 = en_0\mathbf{E}_1 \tag{9.35}$$

where e, m, and n_0 refer to electrons. This equation is coupled to the electron continuity equation (9.23), $-i\omega n_1 + in_0\mathbf{k} \cdot \mathbf{u}_1 = 0$. We can write

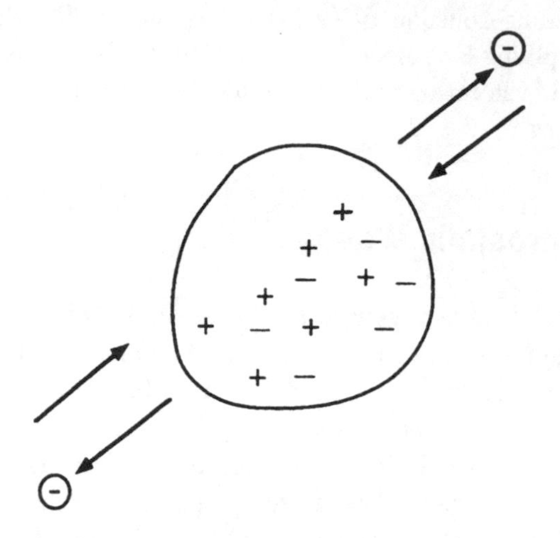

FIGURE 9.5 Excess positive charge causes neighboring electrons to move toward the positive charges and electron oscillations will be set up.

the wave equation (9.28), using (9.29), as

$$\begin{aligned} -\mathbf{k} \times (\mathbf{k} \times \mathbf{E}_1) &= \frac{\omega^2}{c^2}\mathbf{E}_1 + i\omega\mu_0 \sum q_\alpha n_{\alpha 0}\mathbf{u}_{\alpha 1} \\ &= \frac{\omega^2}{c^2}\mathbf{E}_1 - \frac{i\omega n_0 e \mathbf{u}_1}{c^2 \epsilon_0} \\ &= \left(1 - \frac{\omega_p^2}{\omega^2}\right)\frac{\omega^2}{c^2}\mathbf{E}_1 \end{aligned} \tag{9.36}$$

where (9.35) was used to eliminate $\mathbf{u}_1$ and

$$\omega_p^2 = \frac{n_0 e^2}{m\epsilon_0} \tag{9.37}$$

is defined as the electron plasma frequency.

If (9.36) is written in a form suitable for comparison to (9.33),

$$n^2[\mathbf{E}_1 - (\hat{\mathbf{k}} \cdot \mathbf{E}_1)\hat{\mathbf{k}}] = \kappa \mathbf{E}_1 \tag{9.38}$$

it shows the dielectric constant is

$$\kappa = 1 - \frac{\omega_p^2}{\omega^2} \tag{9.39}$$

For electrostatic waves, $\mathbf{k}$ is parallel to $\mathbf{E}$, the left-hand side of (9.38) vanishes, and one obtains

$$\omega^2 = \omega_p^2 \tag{9.40}$$

The frequency of oscillation of n_1 about the equilibrium point is just the plasma frequency. ω_p depends only on the density of the medium (since e, m, and ϵ_0 are given) and therefore, given the plasma frequency, information on the density of the medium can be derived (and vice versa). These plasma waves (also called Langmuir waves) do not propagate, since they do not depend on k and the group velocity $\partial\omega/\partial k = 0$. Plasma waves do not radiate because their Poynting flux vanishes. Plasma waves are local oscillations confined to the region where the perturbations occur.

Another way to demonstrate the plasma oscillations is to note that the induced motion of the electrons gives rise to a polarization charge and therefore the induced dipole moment per unit volume is

$$\mathbf{P} = n_0 e \mathbf{r}_1 \tag{9.41}$$

where $\mathbf{r}_1$, deduced from the equation of motion of the electrons (9.35), is

$$\mathbf{r}_1 = -\frac{e}{m\omega^2}\mathbf{E}_1 \tag{9.42}$$

$\mathbf{r}_1$ represents the perturbed distance measured from the equilibrium position (where the ion is) to the new position of the electron. $\mathbf{E}_1$ is the electric field that arises due to the charge separation. Hence

$$\mathbf{P} = -\frac{n_0 e^2}{m\omega^2}\mathbf{E}_1 \tag{9.43}$$

We can write

$$\begin{aligned} \mathbf{D} &= \epsilon_0 \mathbf{E} + \mathbf{P} \\ &= \left(1 - \frac{n_0 e^2}{m\epsilon_0\omega^2}\right)\epsilon_0 \mathbf{E} \\ &= \epsilon\, \mathbf{E} \end{aligned} \tag{9.44}$$

where ϵ is the permittivity of the material. Since the dielectric constant is

$$\kappa = \frac{\epsilon}{\epsilon_0} \tag{9.45}$$

Equation (9.44) shows

$$\kappa = 1 - \frac{n_0 e^2}{m\epsilon_0\omega^2} \tag{9.46}$$

which is the same as (9.39). Plasma oscillations are obtained by setting the dielectric constant equal to zero.

9.4.2 Electron Plasma Waves in Warm Plasma

If the electron fluid possesses thermal energy, then the pressure term in the momentum equation (9.23) cannot be ignored. Suppose the fluid has a temperature $T_i = T_e = T$ and let the fluid undergo an isothermal perturbation. Then

$$\begin{aligned} \nabla p_e &= \nabla(n_e k_B T) \\ &= k_B T i\mathbf{k} n_1 \end{aligned} \tag{9.47}$$

(k_B here is the Boltzmann constant). Substitute (9.47) into the momentum equation (9.23) and the first-order momentum equation becomes

$$imn_0\omega\mathbf{u}_1 = -en_0\mathbf{E}_1 - k_B T i\mathbf{k} n_1 \tag{9.48}$$

Now, dot this equation with $\mathbf{k}$ and eliminate $\mathbf{k}\cdot\mathbf{u}_1$, using the continuity equation of the second equation of (9.23). To eliminate $\mathbf{k}\cdot\mathbf{E}_1$, note that the last equation of (9.22) can be written as $i\mathbf{k}\cdot\mathbf{E}_1 = -en_1/\epsilon_0$. Equation (9.48) then becomes

$$\begin{aligned} \omega^2 &= \frac{n_0 e^2}{m_e \epsilon_0} + \frac{k_B T}{m_e} k^2 \\ &= \omega_p^2 + k^2 v_{th}^2 \end{aligned} \tag{9.49}$$

where $v_{th}^2 = k_B T_e/m_e$ is the square of the electron thermal speed. This dispersion relation is for warm electron plasma and the second term arises from thermal effects. Plasma waves in warm plasma can propagate since the dispersion relation now has a k dependence. The thermal term arises from the additional restoring force provided by the thermal pressure.

Equation (9.49) was first derived by assuming that the plasma behaved isothermally at all times. However, this is incorrect since temperature variations cannot be ignored when computing the density variations. If an adiabatic equation of state is used instead, the dispersion equation becomes

$$\omega^2 = \omega_p^2 + \frac{5}{3} v_{th}^2 k^2 \tag{9.50}$$

which is different from (9.49) in the coefficient of the thermal term. This expression is applicable if a local Maxwellian equilibrium is always established. Hence the plasma must support collisions that occur frequently enough to maintain a Maxwellian distribution. If collisions are not frequent, as is often the case in space where plasma is tenuous, neither of the above dispersion relations is correct. Instead, it can be shown from the kinetic theory that the correct dispersion relation of electron plasma waves is

$$\omega^2 = \omega_p^2 + 3v_{th}^2 k^2 \tag{9.51}$$

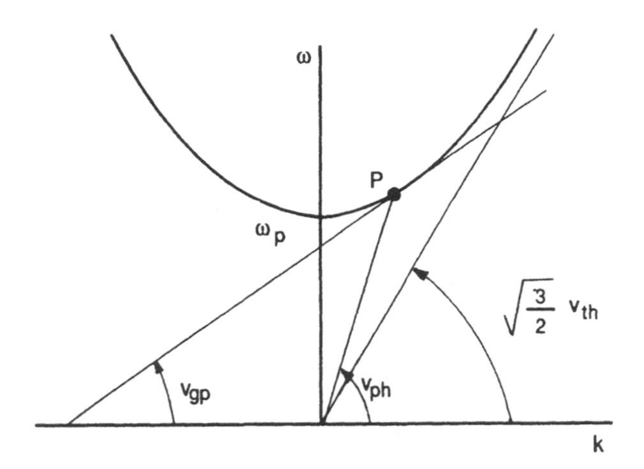

FIGURE 9.6 Dispersion relation of electrostatic electron plasma waves.

This equation was first derived by D. Bohm and E. Gross in 1949 and is known as the Bohm-Gross equation. The group velocity of these waves is

$$\begin{aligned} v_{gp} &= \frac{d\omega}{dk} = \frac{k}{\omega} v_{th}^2 \\ &= \frac{3v_{th}^2}{v_{ph}} \end{aligned} \tag{9.52}$$

which vanishes if the plasma thermal energy is zero.

The dispersion relation given by (9.51) has been plotted in Figure 9.6. The phase velocity ω/k at any point P on the curve is obtained from the slope of the line joining P and the origin O. The slope of the curve at point P gives the group velocity, $\partial\omega/\partial k$. This is always less than $\sqrt{3v_{th}}$, which must always be less than the speed of light, c. Note that at large k (short wavelength) $\omega/k \approx \sqrt{3v_{th}}$, hence the group velocity is essentially the thermal speed. At small k (large wavelength), $\omega/k \approx \sqrt{3v_{th}}/\omega_p$, assuming that $\omega_p \gg kv_{th}$. The group velocity here tends to be even less than v_{th} because with large wavelengths the density gradient is small, and very little net motion results to adjacent regions from the thermal motions.

We conclude this section by discussing once again the electrostatic wave data that were observed in space as presented in a spectrogram format (see Figure 1.6). The ordinate of this diagram is frequency and the abscissa is universal time when the data were recorded. Also shown on the horizontal scale is the position of the spacecraft measured from the center of Earth in units of earth-radii. The fairly sharp emissions detected inside the magnetosphere, which increase in frequency as the satellite approaches Earth, are identified as the plasma oscillations. These oscillations were recorded as the satellite penetrated the plasmasphere, where the electron density is higher than in the other regions of the magnetosphere. Plasma frequency increased (therefore the density) as the satellite penetrated deeper into the plasmasphere.

9.4.3 Ion Acoustic Waves

Thus far, the ions have been assumed at rest and did not take part in the dynamics. However, if ion motions are considered, we will find that sound waves exist in plasmas. Sound waves in ordinary fluids require collisions to transmit energy. In the case of tenuous plasmas where collisions are rare, it would seem that sound waves cannot exist. However, this conclusion is incorrect because plasma particles can still interact (transmit vibrations) with each other by means of their charge, which gives a long-range electrostatic force. Since sound waves involve massive ions, the frequencies of these waves will be lower than those of the electron waves discussed above, and the term involving the displacement current can be ignored.

The ion fluid equation in the absence of an ambient magnetic field is

$$\begin{aligned} m_i n_i \frac{\partial \mathbf{u}_i}{\partial t} + m_i n_i (\mathbf{u}_i \cdot \nabla)\mathbf{u}_i &= e n_i \mathbf{E} - \nabla p_i \\ &= -e n_i \nabla \psi - \gamma_i k_B T_i \nabla n_i \end{aligned} \tag{9.53}$$

where $\mathbf{E} = -\nabla\psi$ and $p_1 = \gamma_i n_i k_B T_i$. As before, expand this equation around the equilibrium and retain the quantities only to first order. Then, (9.53) becomes

$$\omega m_i n_0 \mathbf{u}_{i1} = e n_0 \mathbf{k} \psi_1 + \gamma_i k_B T_i \mathbf{k} n_{i1} \tag{9.54}$$

where $\psi_0 = 0$ since $\mathbf{E}_0 = 0$. The ion equation of continuity yields

$$\omega n_{i1} = n_0 \mathbf{k} \cdot \mathbf{u}_{i1} \tag{9.55}$$

The electrons are very light and they can move around very rapidly. For electrons with a thermal energy $k_B T_e$, the electrons in a scalar potential field ψ_1 requires the electrons to be distributed according to a Maxwellian distribution,

$$n_e = n_0 \exp\left(\frac{e\psi_1}{k_B T_e}\right) \tag{9.56}$$

$$= n_0 \left(1 + \frac{e\psi_1}{k_B T_e} + \cdots\right) \tag{9.57}$$

where n_0 is the equilibrium density in the absence of any disturbance. Ignoring the higher-order terms, the perturbed electron density is

$$n_{e1} = \frac{n_0 e \psi_1}{k_B T_e} \tag{9.58}$$

ψ_1 is a solution of the Poisson equation

$$\nabla \cdot \mathbf{E} = k^2 \psi_1 \tag{9.59}$$

$$= \frac{e}{\epsilon_0}(n_{i1} - n_{e1}) \tag{9.60}$$

Using (9.58) to eliminate n_{e1} in (9.60) yields

$$\psi_1 \left(k^2 + \frac{e^2 n_0}{\epsilon_0 k_B T_e} \right) = \frac{e n_{i1}}{\epsilon_0} \tag{9.61}$$

Now recall that the Debye length is defined by (Chapter 2)

$$\lambda_D^2 = \frac{\epsilon_0 k_B T_e}{n_0 e^2} \tag{9.62}$$

and, therefore, (9.61) reduces to

$$\psi_1 (1 + k^2 \lambda_D^2) = \frac{e n_{i1} \lambda_D^2}{\epsilon_0} \tag{9.63}$$

Dot (9.54) with $\mathbf{k}$ and use (9.55) to eliminate $\mathbf{k} \cdot \mathbf{U}_{i1}$, and use (9.63) to eliminate ψ_1. This yields

$$m_i \omega^2 = k^2 \left[\frac{n_0 e^2 \lambda_D^2}{\epsilon_0 (1 + k^2 \lambda_D^2)} + \gamma_i k_B T_i \right] \tag{9.64}$$

and, rearranging, obtain the dispersion relation of ion acoustic waves,

$$\frac{\omega}{k} = \left[\frac{k_B T_e}{m_i (1 + k^2 \lambda_D^2)} + \frac{\gamma_i k_B T_i}{m_i} \right]^{1/2} \tag{9.65}$$

ω/k is the phase velocity of the ion acoustic waves. For ion acoustic waves, the group velocity equals the phase velocity. Unlike ordinary sound waves, ion acoustic waves still exist even if the ion temperature goes to zero. In the solar wind, $T_i \ll T_e$, and in this situation (9.65) becomes

$$\frac{\omega}{k} = \left[\frac{k_B T_e}{m_i (1 + k^2 \lambda_D^2)} \right]^{1/2} \tag{9.66}$$

Here the speed of sound depends on the electron temperature and on the ion mass. Except for waves with a very short wavelength (large k), the correction term $k^2 \lambda_D^2$ may be $\ll 1$, in which case the ion acoustic waves

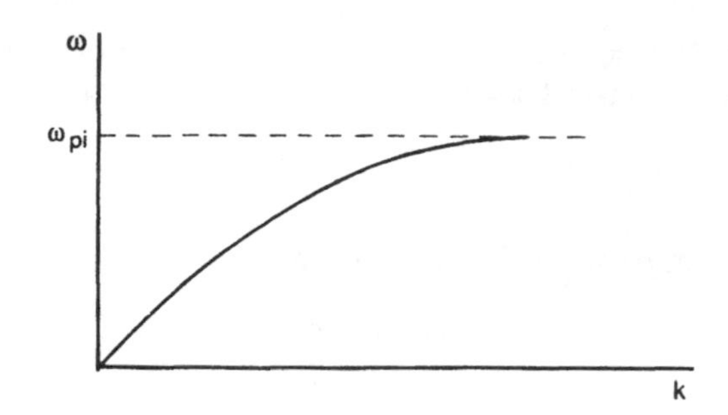

FIGURE 9.7 Dispersion relation of ion acoustic waves.

depend only on the electron temperature and on the ion mass. The broadband bursty emissions below about 10 kHz observed in the solar wind (see Figure 1.6) are believed to be ion acoustic waves.

In the ring current of Earth's magnetosphere, $T_e \ll T_i$, and although our formulation predicts that sound waves can exist, kinetic theory shows that these waves are heavily damped (Landau damping). Consequently, ion acoustic waves are not expected to be detected normally in the ring current regions.

The dispersion curve for the ion acoustic waves is plotted in Figure 9.7. Ion acoustic waves are essentially constant-velocity waves and thermal motion of particles is required for their existence. Note that if $k^2\lambda_D^2 \gg 1$, (9.64) reduces to, in the limit $T_i = 0$,

$$\omega_{pi}^2 = \frac{n_0 e^2}{m_i \epsilon_0} \tag{9.67}$$

which is just the plasma frequency of the ions. Hence at large k (high frequency, short wavelength), ion acoustic waves become constant-frequency waves. In conclusion, it is worthy of note that, except for the thermal correction, the electron plasma waves (9.51) can be regarded also as constant-frequency waves. However, in the limit of large k, these electron waves become constant-velocity waves.

9.4.4 Upper Hybrid Waves

The electron plasma oscillations exist in the absence of the ambient magnetic field $\mathbf{B}_0$. Let us now investigate how this oscillation is modified in the presence of $\mathbf{B}_0$. The ions are massive; therefore, as before, assume that they are at rest. Neglect also the thermal motions of the electrons and let the equilibrium plasma be defined by a uniform and constant density n_0 and a magnetic field $\mathbf{B}_0$. The first-order perturbed equations for the electrons are

$$-i\omega m\mathbf{u}_1 = -e(\mathbf{E}_1 + \mathbf{u}_1 \times \mathbf{B}_0) \tag{9.68}$$

$$i\omega n_1 + n_0 i\mathbf{k} \cdot \mathbf{u}_1 = 0 \tag{9.69}$$

$$i\mathbf{k} \cdot \mathbf{E}_1 = -en_1 \tag{9.70}$$

(The subscript e has been omitted.) Let $\mathbf{B}_0 = B_0\hat{\mathbf{z}}$. The components of (9.68) are

$$\begin{aligned} i\omega m u_{1x} &= eE_x + eu_{1y}B_0 \\ i\omega m u_{1y} &= eE_y - eu_{1x}B_0 \\ i\omega m u_{1z} &= eE_z \end{aligned} \tag{9.71}$$

Assume for simplicity that $\mathbf{E} = E\hat{\mathbf{x}}$. Then $E_y = E_z = 0$, and u_{1x} is easily obtained, yielding (omit the subscript 1)

$$u_x = \frac{eE/im\omega}{1 - \omega_c^2/\omega^2} \tag{9.72}$$

Here $\omega_c = eB/m_e$ is the electron cyclotron frequency. Note that $\omega \neq \omega_c$, since then u_x is not defined (the meaning of the vanishing denominator is due to a cyclotron resonance, which will be discussed in a later section). Consider now the longitudinal waves, $\mathbf{k} \parallel \mathbf{E}$. Since $\mathbf{E} = E\hat{\mathbf{x}}$, $\mathbf{k} = k\hat{\mathbf{x}}$. Then (9.69) yields

$$n_1 = \frac{k}{\omega} n_0 u_x \tag{9.73}$$

Combine (9.73) and (9.72) and insert the result into (9.70) and obtain

$$\left(1 - \frac{\omega_c^2}{\omega^2}\right) E = \frac{\omega_p^2}{\omega^2} E \tag{9.74}$$

This yields the dispersion relation

$$\omega^2 = \omega_p^2 + \omega_c^2 = \omega_{uh}^2 \tag{9.75}$$

ω_{uh} is called the upper hybrid frequency and is associated with electrostatic waves propagating across the ambient magnetic field. The upper hybrid frequency arises because the cyclotron motion couples to the plasma motion of the electrons in an electrostatic field, yielding another mode of oscillation. Note that as $\mathbf{B}_0 \rightarrow 0$, we recover the plasma oscillations and as the density $n_0 \rightarrow 0$, the electrostatic force becomes weaker and we recover the cyclotron frequency.

9.4.5 Electrostatic Ion Cyclotron Waves

Like the electron plasma oscillations, ion acoustic waves will also be modified in the presence of an ambient magnetic field. An educated guess is that a new mode would result from the coupling of the ion acoustic and the ion Larmor motions. This in fact is correct and, as shown below, the dispersion relation of this new mode is a simple modification of the ion acoustic dispersion relation given by (9.65). To simplify the derivation, ignore the thermal motion of the ions. Following the procedure of Section 9.4.3, the ion equation of motion becomes, to first-order,

$$i\omega m_i n_0 \mathbf{u}_i = -ei\mathbf{k}\psi_1 + e\mathbf{u}_i \times \mathbf{B}_0 \tag{9.76}$$

Assume as before that $\mathbf{B}_0 = B_0\hat{\mathbf{z}}$ and the waves are propagating only in the x-direction. Hence, the component equations of (9.76) are

$$\begin{aligned} -i\omega m_i u_x &= -eik\psi_1 + eu_y B_0 \\ -i\omega m_i u_y &= -eu_x B_0 \end{aligned} \tag{9.77}$$

Solve for u_x and obtain

$$u_x = \frac{ek_B\psi_1/m_i\omega}{1-\Omega_c^2/\omega^2} \tag{9.78}$$

where Ω_c is the ion Larmor frequency. The ion continuity equation yields

$$\begin{aligned} n_i &= n_0\frac{k}{\omega}u_x \\ &= \frac{n_0 ek^2\psi_1/m_i\omega^2}{1-\Omega_c^2/\omega^2} \end{aligned} \tag{9.79}$$

Eliminate ψ_1 by using (9.63). In the limit $k^2\lambda_D^2 \ll 1$, we obtain

$$\left(1-\frac{\Omega_c^2}{\omega^2}\right)n_1 = k^2\frac{k_BT_e}{m_i\omega^2}n_1 \tag{9.80}$$

which yields

$$\omega^2 = \Omega_c^2 + k^2\frac{k_BT_e}{m_i} \tag{9.81}$$

The second term is the same as (9.66) in the limit $k^2\lambda_D^2 \ll 1$. Equation (9.81) is the dispersion relation for electrostatic ion cyclotron waves. As mentioned above, these waves arise because the Larmor motion couples to the ion acoustic motion, resulting in this new normal mode. This derivation assumed $k_BT_i = 0$.

The use of (9.63) to eliminate ψ_1 in the above derivation assumes that the electrons are distributed by a Maxwellian distribution (9.56). This can only be achieved if the electrons were permitted to travel along $\mathbf{B}_0$, because it is only along $\mathbf{B}_0$ that electrons can travel freely to maintain an equilibrium distribution. Hence, the derivation of the ion cyclotron waves implicitly assumes that, while the wave vector $\mathbf{k}$ was perpendicular to $\mathbf{B}_0$, there is a small component of $\mathbf{k}$ along $\mathbf{B}_0$ (it can be shown from more rigorous derivation that $k_z/k_x \approx \sqrt{m_e/m_i}$).

9.4.6 Lower Hybrid Waves

If the wave vector $\mathbf{k}$ is forced to maintain its direction entirely perpendicular to $\mathbf{B}_0$, then (9.56) cannot be used. In this case, ψ_1 must be solved directly from the equation of motion of electrons. The ion equations (9.78) and (9.79) still remain valid. However, we need a complimentary set of equations that describe the electron motions. These are

$$u_{ex} = -\frac{ek\psi_1/m_e\omega}{1-\omega_c^2/\omega^2} \tag{9.82}$$

and

$$n_e = n_0\frac{k}{\omega}u_{ex} \tag{9.83}$$

Since the motion is strictly maintained in the perpendicular direction only, we require that charge neutrality be always maintained in this direction by the plasma. That is, we require $n_e = n_i$. In this case, (9.79) and (9.83) indicate that $u_i = u_e$. Then equating (9.78) and (9.82) yields

$$m_i\left(1-\frac{\Omega_c^2}{\omega^2}\right) = -m_e\left(1-\frac{\omega_c^2}{\omega^2}\right) \tag{9.84}$$

which can be solved to obtain

$$\begin{aligned}\omega^2 &= \frac{e^2B_0^2}{m_i m_e}\\ &= \omega_c\Omega_c\end{aligned} \tag{9.85}$$

Hence

$$\omega = (\Omega_c\omega_c)^{1/2} = \omega_{lh} \tag{9.86}$$

ω_{lh} is defined as the lower hybrid frequency. This is the frequency of the electrostatic ion cyclotron waves if $\mathbf{k}$ is maintained strictly perpendicular to $\mathbf{B}_0$. The lower hybrid waves arise from the coupling of the cyclotron motions of ions and electrons and the new mode oscillates with the average frequency of the two. Because of the strict requirement that $\mathbf{k}$ be perpendicular to $\mathbf{B}_0$, the lower hybrid waves are difficult to excite (in principle).

9.5 Electromagnetic Waves

Electromagnetic waves have both oscillating electric and magnetic field components. An example of electromagnetic waves are the ordinary light waves, which are transverse electromagnetic waves in their simplest form. As shown below, there are many more complex electromagnetic wave modes in plasmas.

9.5.1 Ordinary Light Waves ($\mathrm{B}_0 = 0$)

Ordinary light waves are transverse electromagnetic waves in which the wave vector $\mathbf{k}$ is perpendicular to both the oscillating electric field $\mathbf{E}_1$ and the magnetic field $\mathbf{B}_1$ of the wave. Let us first study these waves when the ambient magnetic field $\mathbf{B}_0 = 0$. The electric field of the wave obeys (see (9.28))

$$-\mathbf{k} \times (\mathbf{k} \times \mathbf{E}_1) = i\omega\mu_0\mathbf{J}_1 + \frac{\omega^2}{c^2}\mathbf{E}_1 \tag{9.87}$$

For electromagnetic waves, $\mathbf{k} \times \mathbf{E}$ does not normally vanish. Expand the left-hand side and obtain

$$k^2\mathbf{E}_1 - (\mathbf{k} \cdot \mathbf{E}_1)\mathbf{k} = i\omega\mu_0\mathbf{J}_1 + \frac{\omega^2}{c^2}\mathbf{E}_1 \tag{9.88}$$

The wave electric field drives the electrons into motion (assume ions are at rest) and this gives rise to a current

$$\mathbf{J}_1 = -en_0\mathbf{u}_1 \tag{9.89}$$

where $\mathbf{u}_1$ obtained from the equation of motion is (for $\mathbf{B}_0 = 0$)

$$\mathbf{u}_1 = \frac{e\mathbf{E}_1}{im\omega} \tag{9.90}$$

Combine (9.89) and (9.90) and eliminate $\mathbf{J}_1$ in (9.88). This yields

$$k^2\mathbf{E}_1 - (\mathbf{k} \cdot \mathbf{E}_1)\mathbf{k} = -\frac{\omega_p^2}{c^2}\mathbf{E}_1 + \frac{\omega^2}{c^2}\mathbf{E}_1 \tag{9.91}$$

For transverse waves, $\mathbf{k} \cdot \mathbf{E}_1 = 0$ and we obtain

$$\omega^2 = \omega_p^2 + k^2c^2 \tag{9.92}$$

which is the dispersion relation of transverse electromagnetic waves in a plasma with no static background magnetic field. Note that in a vacuum, $\omega_p = 0$, and (9.92) reduces to a dispersion relation for ordinary light waves

$$\omega^2 = c^2k^2 \tag{9.93}$$

whose phase velocity is c. The introduction of plasma modifies the vacuum dispersion relation because the electron motion in plasma creates a current.

The phase velocity of transverse electromagnetic waves is

$$V_{ph} = \frac{\omega}{k} = c\left(1 + \frac{\omega_p^2}{k^2c^2}\right)^{1/2} \tag{9.94}$$

which is always $> c$, the speed of light in vacuum. The group velocity is

$$V_{gp} = \frac{\partial \omega}{\partial k} = \frac{kc^2}{\omega} = \frac{c^2}{V_{ph}} \tag{9.95}$$

which is less than c since $c/V_{ph} < 1$. The index of refraction is given by

$$n^2 = \frac{c^2k^2}{\omega^2} = 1 - \frac{\omega_p^2}{\omega^2} \tag{9.96}$$

Waves can propagate through a medium only if n^2 is greater than zero. Equation (9.96) shows that the frequency of the propagating transverse electromagnetic waves must be greater than the plasma frequency, $\omega > \omega_p$.

Equation (9.96) shows that $n^2 = 0$ when $\omega = \omega_p$. This defines the cutoff frequency, and for transverse electromagnetic waves the cutoff is at the plasma frequency of the medium. The plasma medium is opaque to waves for frequencies less than ω_p. For $\omega < \omega_p$, the index of refraction is imaginary. If ω is real, k must be imaginary. Since our formulation assumes a spatial k dependence through $e^{i\mathbf{k}\cdot\mathbf{r}}$, imaginary $\mathbf{k}$ means the waves are attenuated. Let

$$\mathbf{k} = \mathbf{k}_R + i\mathbf{k}_I \tag{9.97}$$

where $\mathbf{k}_R$ and $\mathbf{k}_I$ are real. Then the spatial dependence becomes

$$e^{i\mathbf{k}\cdot\mathbf{r}} = e^{i\mathbf{k}_R\cdot\mathbf{r}}e^{-\mathbf{k}_I\cdot\mathbf{r}} \tag{9.98}$$

The skin depth is defined as

$$d = |\mathbf{k}_I|^{-1} \tag{9.99}$$

For the transverse waves in plasma, this is

$$d = \frac{c}{(\omega_p^2 - \omega^2)^{1/2}} \tag{9.100}$$

There are various ways to represent the solutions of the above dispersion equation. In analogy with optical phenomena, we can plot n^2 vs. ω, or as the ionospheric physicist prefers, n^2 vs. $1 - X$ where $X = \omega_p^2/\omega^2$. Yet another representation is to seek the solutions to the equation $c^2k^2/\omega^2 = 1 - \omega_p^2/\omega^2$. These solutions are shown in Figure 9.8.

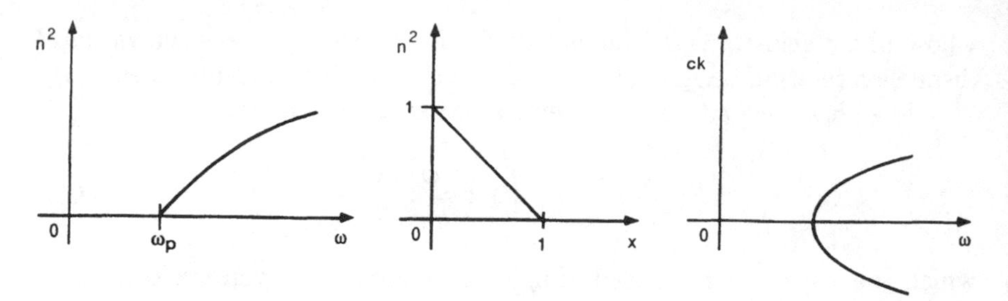

FIGURE 9.8 Dispersion relation of electromagnetic waves in plasma when the ambient static magnetic field is zero.

The fact that an electromagnetic wave will not propagate through a medium unless its frequency is larger than the local plasma frequency enables one to perform an experiment to determine the electron density in Earth's ionosphere. Consider an experimenter with a transmitter at A and a receiver at B (Figure 9.9). Assume that the ionized layer begins at some height h. Let the index of refraction outside the layer be n_i. Consider a plane wave front MN entering the layer. If the electron density is uniform, the wave front MN will simply propagate in a straight line. However, the electron density increases with height and reaches a maximum at the F-layer. This initial increase of electron density causes the phase velocity also to increase with height (because the plasma frequency is increasing). The wave front MN will therefore bend and eventually turn around and return to the ground (the point M catches up with N). Hence the receiver B, properly placed, will receive a signal sometime after the signal is transmitted at A.

The situation just outlined can be described quantitatively by using Snell's law from optics. Let i be the angle of incidence of the signal and ψ the angle of refraction. Then

$$n_i \sin i = n_\psi \sin \psi \tag{9.101}$$

where the right-hand side describes the characteristics of the electromagnetic waves along the ray path in the plasma medium. n_ψ is the index of refraction at any point along the ray path. If the space between the transmitter and the entrance layer of the plasma is assumed to be a vacuum (not a bad approximation), then $n_i = 1$. Total reflection of the wave front MN occurs when ψ is 90°. At this point, $\sin i = n_\psi$, where n_ψ is given by (9.96).

The experiment can be simplified if $i = 0$, that is, if we use a vertical geometry. The transmitter and the receiver are located at the same site. In this case, the total reflection occurs when $n_\psi = 0$, or when the frequency

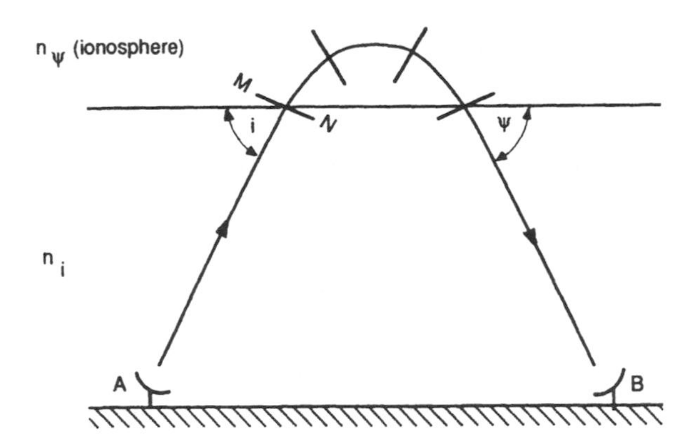

FIGURE 9.9 Propagation experiment of electromagnetic waves through the ionosphere to determine the electron density.

of the signal equals the local plasma frequency. In principle then, if the difference is measured between the times a signal is transmitted and received, the height h from where the signal is totally reflected can be determined. Since this occurs precisely where the signal frequency is equal to the local plasma frequency, the electron density at this turning point is calculated. If this experiment were performed over a broad frequency interval, then a correspondingly broad height would be sampled. Information on the electron density as a function of height is obtained this way.

This simple "sounder" experiment demonstrates that we can probe the ionosphere and learn about the variations of the electron density with height. However, in practice, the situation is more complicated. First, when the geomagnetic field is included, the ray breaks into "ordinary" and "extraordinary" modes whose velocities in the plasma differ. Another complication is that, because the electron density is continuously changing with height, the estimate of the height of reflection made from the time propagation information is not exactly correct. This is because the wave velocity in the medium is continually changing. The experimenter must also take into account that the ionospheric plasma has thermal and collisional effects that could complicate the interpretation of the data.

9.5.2 Ordinary Waves ($\mathbf{k} \perp \mathbf{B}_0$, $\mathbf{E}_1 \parallel \mathbf{B}_0$)

Start again with the electric field equation (9.88)

$$k^2\mathbf{E}_1 - (\mathbf{k} \cdot \mathbf{E}_1)\mathbf{k} = i\omega\mu_0\mathbf{J}_1 + \frac{\omega^2}{c^2}\mathbf{E}_1 \tag{9.102}$$

$\mathbf{J}_1 = -en_0\mathbf{u}_1$ and $\mathbf{u}_1$ is obtained from the equation of motion

$$-imn_0\omega\mathbf{u}_1 = -en_0(\mathbf{E}_1 + \mathbf{u}_1 \times \mathbf{B}_0) \tag{9.103}$$

Without loss of generality, we again let $\mathbf{B}_0 = B_0\hat{\mathbf{z}}$. Since we are considering waves with $\mathbf{E}_1$ parallel to $\mathbf{B}_0$, $\mathbf{E}_1 = E_1\hat{\mathbf{z}}$. $\mathbf{k}$ is perpendicular to $\mathbf{B}_0$, $\mathbf{k} = k_x\hat{\mathbf{x}} + k_y\hat{\mathbf{y}}$, but for simplicity, let $\mathbf{k} = k\hat{\mathbf{x}}$. Hence, $\mathbf{k} \cdot \mathbf{E} = 0$ and using $\mathbf{J}_1 = -en_0\mathbf{u}_1$, (9.102) reduces to

$$(\omega^2 - k^2c^2)\mathbf{E}_1 - \frac{i\omega en_0}{\epsilon_0}\mathbf{u}_1 = 0 \tag{9.104}$$

Since $\mathbf{E}_1 = E_1\hat{\mathbf{z}}$, we are only concerned with the z-component of $\mathbf{u}_1$. Equation (9.103) gives

$$u_1 = -\frac{ieE_1}{m\omega} \tag{9.105}$$

Insert this into (9.104) and obtain

$$(\omega^2 - k^2c^2)E_1 - \frac{n_0e^2}{m\epsilon_0}E_1 = 0 \tag{9.106}$$

or

$$\omega^2 = \omega_p^2 + k^2c^2 \tag{9.107}$$

This is the dispersion relation of an ordinary wave. Recall that the perturbation electric field $\mathbf{E}_1$ is along $\mathbf{B}_0$. Equation (9.107) is identical to the dispersion of light waves (9.92) considered above with $\mathbf{B}_0 = 0$. This is because $\mathbf{J}_1$ included only $\mathbf{u}_1$ along $\mathbf{B}_0$ and $\mathbf{u}_1 \times \mathbf{B}_0$ identically vanishes, giving the same result as though $\mathbf{B}_0 = 0$.

9.5.3 Extraordinary Waves ($\mathbf{k} \perp \mathbf{B}_0$; $\mathbf{E}_1 \perp \mathbf{B}_0$)

If the perturbed electric field $\mathbf{E}_1$ is perpendicular to $\mathbf{B}_0$, electrons will be driven across $\mathbf{B}_0$, and this motion must be taken into account in the calculation of $\mathbf{J}_1$. Since $\mathbf{B}_0 = B_0\hat{\mathbf{z}}$, let $\mathbf{E}_1 = E_x\hat{\mathbf{x}} + E_y\hat{\mathbf{y}}$. The velocity components obtained from (9.103) are

$$\begin{aligned} u_x &= -\frac{ie}{m\omega}(E_x + u_yB_0) \\ u_y &= -\frac{ie}{m\omega}(E_y - u_xB_0) \end{aligned} \tag{9.108}$$

These coupled equations, solved for u_x and u_y, yield

$$
\begin{aligned}
u_x &= \frac{e}{m\omega}\frac{(-iE_x - \omega_c E_y/\omega)}{1-\omega_c^2/\omega^2} \\
u_y &= \frac{e}{m\omega}\frac{(-iE_y + \omega_c E_x/\omega)}{1-\omega_c^2/\omega^2}
\end{aligned}
\tag{9.109}
$$

The current $\mathbf{J}_1 = -en_0\mathbf{u}_1$ is obtained by using these equations. To continue, recall that we are still considering the case $\mathbf{k}$ is perpendicular to $\mathbf{B}_0$. Hence, $\mathbf{k} = k_x\hat{\mathbf{x}} + k_y\hat{\mathbf{y}}$. However, as before, let $\mathbf{k} = k\hat{\mathbf{x}}$. Then the wave electric field equation (9.102) becomes

$$
(\omega^2 - c^2k^2)\mathbf{E}_1 + kE_x\mathbf{k} - i\omega\mu_0 en_0\mathbf{u}_1 = 0 \tag{9.110}
$$

Now use (9.109) for $\mathbf{u}_1$ and, solving for the components of (9.110), obtain

$$
\begin{aligned}
&\left[\omega^2\left(1-\frac{\omega_c^2}{\omega^2}\right) - \omega_p^2\right]E_x + \frac{i\omega_p^2\omega_c}{\omega}E_y = 0 \\
&\left[(\omega^2 - c^2k^2)\left(1-\frac{\omega_c^2}{\omega^2}\right) - \omega_p^2\right]E_y - \frac{i\omega_p^2\omega_c}{\omega}E_x = 0
\end{aligned}
\tag{9.111}
$$

where $\omega_c = eB_0/m$ and $\omega_p^2 = n_0e^2/m\epsilon_0$. The solutions of these simultaneous equations for E_x and E_y are obtained by setting the determinant of the coefficients of E_x and E_y to zero. This yields

$$
\frac{c^2k^2}{\omega^2} = \frac{\omega^2 - \omega_{uh}^2 - [(\omega_p^2\omega_c/\omega)^2/(\omega^2-\omega_{uh}^2)]}{\omega^2-\omega_c^2} \tag{9.112}
$$

Substitute the first ω_{uh}^2 on the right side by its definition $\omega_c^2 + \omega_p^2$ and multiplying through by $\omega^2 - \omega_{uh}^2$ yields

$$
\frac{c^2k^2}{\omega^2} = 1 - \frac{\omega_p^2(\omega^2-\omega_{uh}^2) + (\omega_p^4\omega_c^2/\omega^2)}{(\omega^2 - \omega_c^2)(\omega^2 - \omega_{uh}^2)} \tag{9.113}
$$

which, after a little algebraic manipulation, results in

$$
\frac{c^2k^2}{\omega^2} = 1 - \frac{\omega_p^2(\omega^2-\omega_p^2)}{\omega^2(\omega^2 - \omega_{uh}^2)} \tag{9.114}
$$

This dispersion relation defines the extraordinary mode wave. As mentioned above, this electromagnetic wave propagates perpendicular to $\mathbf{B}_0$ with the perturbation electric field $\mathbf{E}_1$ oscillating perpendicular to $\mathbf{B}_0$. Since this wave has the wave vector $\mathbf{k}$ in the directions parallel and perpendicular to $\mathbf{E}_1$, it is a mixed mode consisting of both longitudinal and transverse components.

The cutoff frequencies occur when the index of refraction is zero, where the phase velocity and the wavelength become infinite. A wave is generally reflected at cutoffs. The cutoff frequencies of the extraordinary mode are obtained by setting k to zero in (9.114) and solving for the ω's. This yields

$$(\omega^2 - \omega_p^2 - \omega_c^2) = \frac{\omega_p^2}{\omega^2}(\omega^2 - \omega_p^2)$$

$$1 - \frac{\omega_c^2}{\omega^2 - \omega_p^2} = \frac{\omega_p^2}{\omega^2}$$

$$1 - \frac{\omega_p^2}{\omega^2} = \frac{\omega_c^2}{\omega^2\left(1 - \omega_p^2/\omega^2\right)}$$

$$1 - \frac{\omega_p^2}{\omega^2} = \pm\frac{\omega_c}{\omega} \tag{9.115}$$

which is a quadratic equation in ω

$$\omega^2 \mp \omega\omega_c - \omega_p^2 = 0 \tag{9.116}$$

Two cutoff frequencies are obtained, corresponding to the two signs. Traditionally, these are labelled ω_R and ω_L where R and L are right-hand and left-hand cutoff frequencies. The two roots are

$$\begin{aligned}\omega_R &= \frac{1}{2}[\omega_c + (\omega_c^2 + 4\omega_p^2)^{1/2}] \\ \omega_L &= \frac{1}{2}[-\omega_c + (\omega_c^2 + 4\omega_p^2)^{1/2}]\end{aligned} \tag{9.117}$$

Since the frequency is chosen to be always positive, we have ignored the minus sign in front of the square root in each case. Here, ω_R and ω_L only depend on the cyclotron and plasma frequencies of the medium.

The resonance of a wave mode occurs when the index of refraction becomes infinite or where the wavelength goes to zero. The resonance frequencies of the extraordinary mode are found by letting $k \to \infty$, which occurs when $\omega^2 \to \omega_{uh}^2$ in the denominator of (9.114). Hence the resonance frequencies are

$$\omega^2 = \omega_{uh}^2 = \omega_p^2 + \omega_c^2 \tag{9.118}$$

Note that (9.118) is exactly the same dispersion for the electrostatic plasma waves propagating across the magnetic field. What is indicated here is that, as the extraordinary wave approaches the resonance, its phase and the group velocities approach zero and the electromagnetic energy is converted into electrostatic oscillations. Mode conversion of electromagnetic into electrostatic waves often characterizes the behavior of waves at resonances. The waves near resonance have small or vanishing velocities, and dissipation of wave energy is expected near resonances. This wave energy is absorbed by the medium.

To plot the dispersion relation of the extraordinary mode, note that in addition to the cutoffs and resonances, $n^2 \to 1$ as $\omega \to \infty$. If $\omega < \omega_{uh}$, $n^2 > 1$ and near ω_{uh}, $n^2 \to \infty$. In the frequency range $\omega_{uh} < \omega < \omega_R$ and $\omega < \omega_L$, there is no wave propagation. This is a stopgap region. Figure 9.10 summarizes these features of the extraordinary mode. The reader is reminded that the formulation assumed that the plasma density and ambient magnetic field are constant. Therefore, ω_c and ω_p in the diagram are fixed.

9.5.4 Right- and Left-Hand Waves ($\mathbf{k} \parallel \mathbf{B}_0$; $\mathbf{E}_1 \perp \mathbf{B}_0$)

This class of electromagnetic waves is important in space plasma environments, since the waves in the magnetosphere can propagate along **B** to the ground. Let $\mathbf{k} = k\hat{\mathbf{z}}$ and let $\mathbf{E}_1$ be transverse, $\mathbf{E}_1 = E_x\hat{\mathbf{x}} + E_y\hat{\mathbf{y}}$. With these conditions, the wave electric field equation (9.102) becomes

$$(\omega^2 - c^2k^2)\mathbf{E}_1 - i\omega\mu_0 c^2 e n_0 \mathbf{u}_1 = 0 \tag{9.119}$$

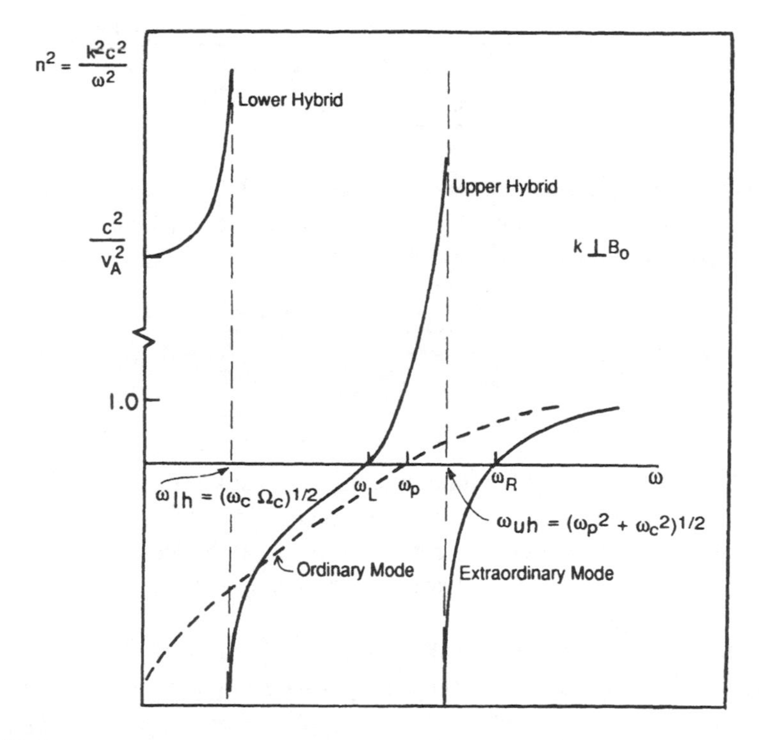

FIGURE 9.10 Dispersion relation of extraordinary waves.

Substitute the equation of motions of the particles (9.109) and obtain

$$\begin{aligned}(\omega^2 - c^2k^2)E_x &= \frac{\omega_p^2}{1-\omega_c^2/\omega^2}\left(E_x - \frac{i\omega_c}{\omega}E_y\right)\\ (\omega^2 - c^2k^2)E_y &= \frac{\omega_p^2}{1-\omega_c^2/\omega^2}\left(E_y + \frac{i\omega_c}{\omega}E_x\right)\end{aligned} \tag{9.120}$$

Equation (9.120) can be rewritten as

$$\begin{aligned}(\omega^2 - c^2k^2 - \alpha)E_x + i\alpha\left(\frac{\omega_c}{\omega}\right)E_y &= 0\\ (\omega^2 - c^2k^2 - \alpha)E_y - i\alpha\left(\frac{\omega_c}{\omega}\right)E_x &= 0\end{aligned} \tag{9.121}$$

where $\alpha = \omega_p^2/(1-\omega_c^2/\omega^2)$. Now set the determinant of the coefficients of E_x and E_y to zero and obtain

$$\omega^2 - c^2k^2 - \alpha = \pm\frac{\alpha\omega_c}{\omega} \tag{9.122}$$

Thus,

$$\begin{aligned}\omega^2 - c^2k^2 &= \alpha\left(1 \pm \frac{\omega_c}{\omega}\right)\\ &= \frac{\omega_p^2(1\pm\omega_c/\omega)}{1-\omega_c^2/\omega^2} = \frac{\omega_p^2(1\pm\omega_c/\omega)}{(1+\omega_c/\omega)(1-\omega_c/\omega)}\\ &= \frac{\omega_p^2}{(1\mp\omega_c/\omega)}\end{aligned} \tag{9.123}$$

As before, the $\mp$ sign indicates there are two different frequencies, corresponding to the two different wave modes that can propagate along B_0. The dispersion relations of these two modes are

$$\begin{aligned}n_R^2 &= \frac{c^2k^2}{\omega^2} = 1 - \frac{\omega_p^2/\omega^2}{1-\omega_c/\omega}\\ n_L^2 &= \frac{c^2k^2}{\omega^2} = 1 - \frac{\omega_p^2/\omega^2}{1+\omega_c/\omega}\end{aligned} \tag{9.124}$$

The R and L denote right- and left-hand waves associated with the circular polarization of these waves. The electric field of the R waves rotates in the same way ordinary screws rotate, with the right-hand thumb pointing along $\mathbf{B}_0$. Viewed along $\mathbf{B}_0$, this rotation is clockwise. The L wave rotates in the opposite direction.

The cutoff frequencies of these waves are obtained by setting the refractive indices in (9.124) at zero. It can be easily shown that the cutoffs occur at frequencies given by

$$\begin{aligned}\omega_R &= \frac{1}{2}[\omega_c + (\omega_c^2 + 4\omega_p^2)^{1/2}] \\ \omega_L &= \frac{1}{2}[-\omega_c + (\omega_c^2 + 4\omega_p^2)^{1/2}]\end{aligned} \tag{9.125}$$

which are identical to (9.117). Note that ω_R is always greater than ω_L.

The R-mode has a resonance when $\omega = \omega_c$. This occurs when the electric field of the R-wave is in phase with the cyclotron motion of the electrons. The electric field rotation couples to the cyclotron motion and the waves accelerate the electrons. This resonance is also called cyclotron resonance. The phase velocity goes to zero at resonance and these waves no longer propagate. The L-mode does not resonate with the electrons. This is because the wave field is rotating in the opposite direction from the electrons. Note that in the above derivation, only the electron motion was considered. Had we included the ion motion, the L-wave would have been found to resonate with the ions at the ion cyclotron frequency, $\omega = \Omega_c$.

The dispersion relation of the R and L waves is shown in Figure 9.11 for $\omega_p \ll \omega_c$ and $\omega_p \gg \omega_c$. The main difference between these two cases is where the cutoffs occur for the left-hand mode. As shown, when $\omega_p \ll \omega_c$, the cutoff is at frequencies less than ω_c, while they are above ω_c for $\omega_p \gg \omega_c$. Given a constant cyclotron frequency, these waves approach $n^2 = 1$ line for very high frequencies. As the frequency decreases, the waves encounter a cutoff at the frequencies given by ω_R and ω_L in (9.125). The R-wave has a stop band between ω_R and ω_c, below which propagation is again allowed. The waves in this low-frequency range are called whistler mode waves and these waves are very important for wave propagation studies in the magnetosphere.

When whistlers propagate in a medium in which ω_c varies, the waves must have frequencies less than the lowest cyclotron frequency along its path. For example, whistlers propagating in Earth's dipole field must have frequencies less than the equatorial cyclotron frequency of the magnetic field. The L-waves do not propagate below ω_L in our formulation. However, if the ion motion were included, another resonance would appear at the ion cyclotron frequency. Propagation of the L-wave is therefore permitted below the ion cyclotron frequency.

9.6 Whistlers in Earth's Magnetosphere

Lightning discharges emit impulsive electromagnetic radiation with frequency components covering a band from a few hundred Hz to more than 10^4 Hz. This radiation enters the ionosphere and then propagates in the whistler mode in the magnetosphere. Figure 9.12 shows some examples of

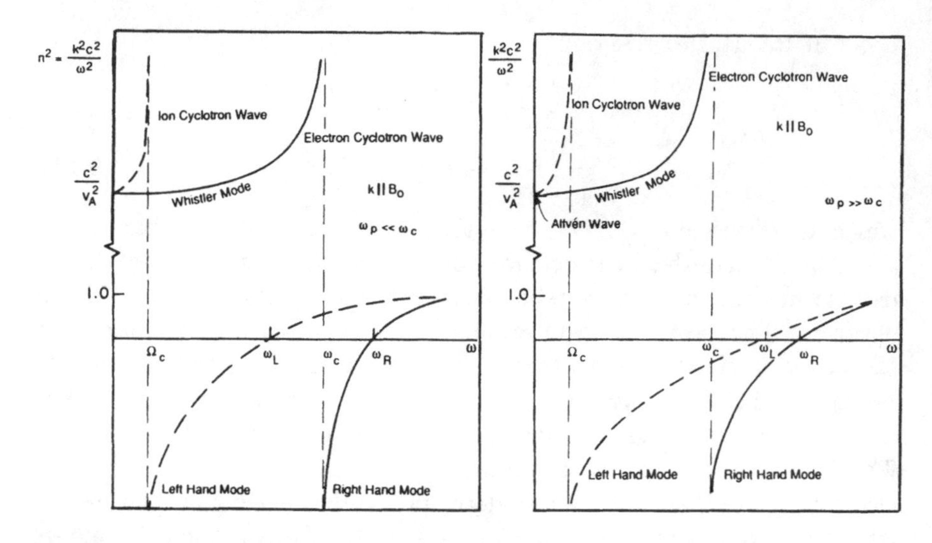

FIGURE 9.11 Dispersion relation of right- and left-hand waves.

data recorded in the southern hemisphere produced by lightning discharges in the northern hemisphere. The data are presented in a dynamic frequency spectral format in which the frequency (ordinate) is plotted as a function of time (abscissa) and the wave intensity forms the third coordinate (out from the page).

A lightning discharge in the northern hemisphere emits a broad frequency spectrum. However, the recorded signal in the southern hemisphere is discrete and shows a frequency dispersion where the high-frequency components arrive first, followed by the lower-frequency waves (gliding tone). This frequency dispersion arises because the group velocity of whistler mode waves is frequency-dependent. V_{gp} deduced from (9.124) is

$$V_{gp} \approx \left[\frac{2c}{(\omega_p/\omega_c)}\right]\left(\frac{\omega}{\omega_c}\right)^{1/2}\left(1-\frac{\omega}{\omega_c}\right)^{3/2} \tag{9.126}$$

where the first term (unity) in the dispersion relation has been ignored. Equation (9.126) shows that V_{gp} is faster for higher-frequency waves. For $\omega \ll \omega_c$, V_{gp} is proportional to $\omega^{1/2}$.

The amount of dispersion increases with increasing propagation path length. Hence whistlers propagating on high-latitude lines of force would show larger dispersion than whistlers propagating at lower latitudes. How-

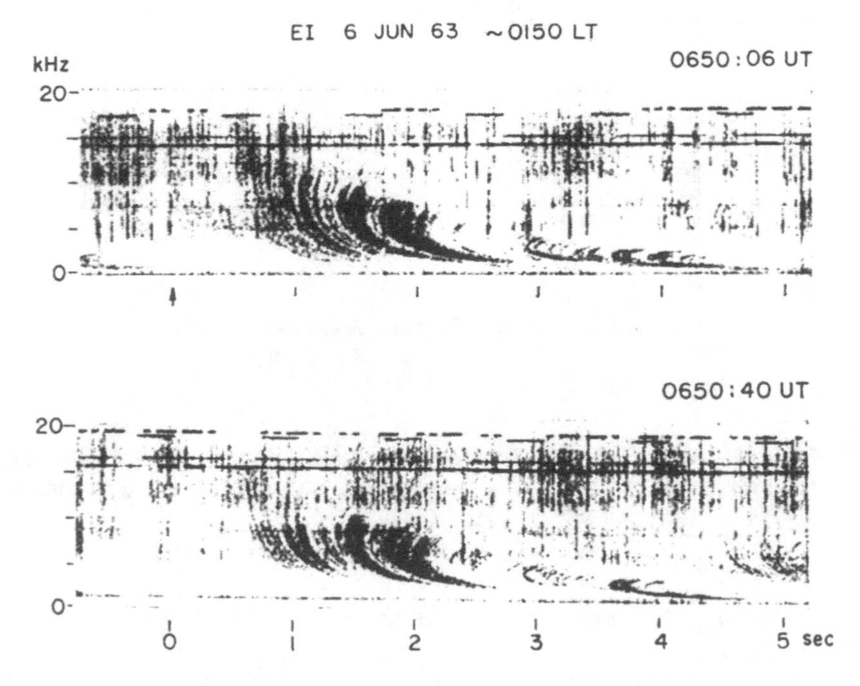

FIGURE 9.12 Frequency spectrograms of two knee whistlers recorded less than one minute apart at Eights, Antarctica. The geomagnetic disturbance level was extremely low at this time. Notice the repeatability of whistler features. (From Angerami and Carpenter, 1966)

ever, the group velocity is also dependent on the plasma frequency (electron density) and proper interpretation of data must take into account the density effects.

9.6.1 Nose Whistlers

The shapes of the detected signal can be quite varied and numerous. Some whistler signals have a "nose" (Figure 9.12) above whose frequency the waves travel slower. This slowing down of the waves above the nose frequency occurs as the waves approach the cyclotron frequency and the cyclotron resonance effects become significant (recall that the phase and group velocities vanish at resonance).

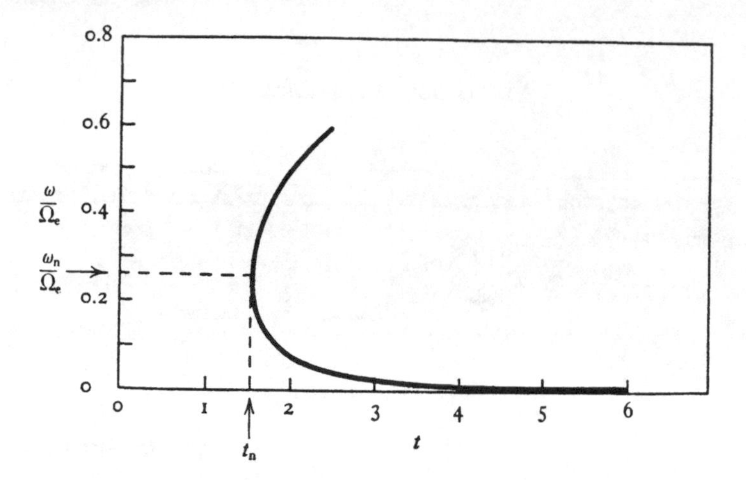

FIGURE 9.13 A plot of (9.129). The frequency scale is normalized to the electron cyclotron frequency. The time scale is in seconds and is represented in units of $(L/c)(\omega_p/\omega_c)$. (From Helliwell, 1965)

To see this, examine the time for the whistlers to travel through a medium given by

$$
\begin{aligned}
t(\omega) &= \int \frac{ds}{V_{gp}} \\
&= \left(\frac{L}{c}\right)\left(\frac{\omega_p}{\omega_c}\right)\left[\frac{1}{2}\left(\frac{\omega}{\omega_c}\right)^{1/2}\left(1-\frac{\omega}{\omega_c}\right)^{3/2}\right]
\end{aligned}
\tag{9.127}
$$

where L represents the total distance travelled. A plot of (9.127) is shown in Figure 9.13, which shows ω/ω_c plotted against t. It can be shown from (9.127) that the frequency of minimum time delay (nose frequency) is

$$\omega_n = \frac{\omega_c}{4} \tag{9.128}$$

and this occurs at

$$t_n = \frac{8}{3}(3)^{1/2}\left(\frac{L}{c}\right)\left(\frac{\omega_p}{\omega_c}\right) \tag{9.129}$$

Hence a measure of the nose frequency gives information on ω_c and, if the path length L is known, ω_p can be deduced from observation of t_n. Recall that these results were derived for a medium in which the magnetic field and the plasma are uniformly distributed.

9.6.2 Discovery of the Plasmapause

Whistlers propagating in realistic ionospheres and magnetospheres are more complicated because of variations arising from ω_c and ω_p. However, detailed studies of whistlers in regions of Earth's magnetosphere where the dipole approximation is valid show that the linear relation between ω_n and ω_c is still valid (instead of 0.25 ω_c, the fit is more like 0.4 ω_c). These studies eventually led to the discovery of Earth's plasmapause (Figure 9.14).

Whistler records on the ground have shown evidence of trains of whistlers, which are due to whistlers that propagate back and forth between the northern and southern hemispheres several times. Also, a given lightning flash can populate whistlers on several magnetic flux tubes. Finally, it should be mentioned that the whistler mode waves can also be generated by unstable magnetospheric plasma distributions.

9.7 Magnetohydrodynamic Waves in Isotropic Fluids

MHD waves, or simply hydromagnetic waves, are produced by the bulk motion of plasma across magnetic fields. These waves are consequences of the motional EMF generated by a conducting fluid interacting with an ambient magnetic field. MHD waves are low-frequency waves (frequency $<$ ion gyrofrequency) and they can possess both longitudinal and transverse modes. The simplest of the hydromagnetic waves, Alfvén waves, are transverse electromagnetic waves with a propagation vector $\mathbf{k}$ along the direction of the ambient magnetic field $\mathbf{B}_0$. Both the fluid velocity and the induced electric field are perpendicular to the magnetic field.

Since motional EMF plays a fundamental role in the generation of these waves, let us first review by a simple example how an EMF is generated. Consider as in Figure 9.15 a stationary U-shaped wire ADCB, whose plane is normal to the magnetic field $\mathbf{B}$, and which is connected to another piece of wire ST that is permitted to move. When ST moves as shown, an EMF equal to $\mathbf{U} \times \mathbf{B}$ is induced and this drives a current flow in the wire loop TDCS. What causes the current to flow in the different parts of the wire? In the moving wire ST, it is the induced EMF, $\mathbf{U} \times \mathbf{B}$. In the stationary wire, it is due to the space charge distribution set up along the wire. If ϕ is the electrostatic potential set up by this space charge, then the electric field driving the current in the stationary wire is $\mathbf{E} = -\nabla\phi$.

Consider now an incompressible conducting fluid that fills a large region of space. Erect a Cartesian coordinate frame of reference and let the magnetic field $\mathbf{B}$ be in the z-direction. The entire plasma is at rest, except that the plasma (in an arbitrarily shaped region) is moving with a velocity $\mathbf{U}$ in the y-direction. According to Faraday's law, this motion induces

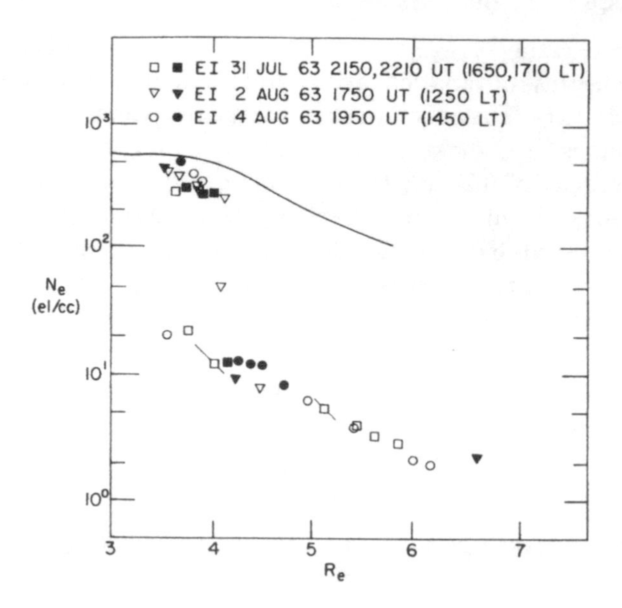

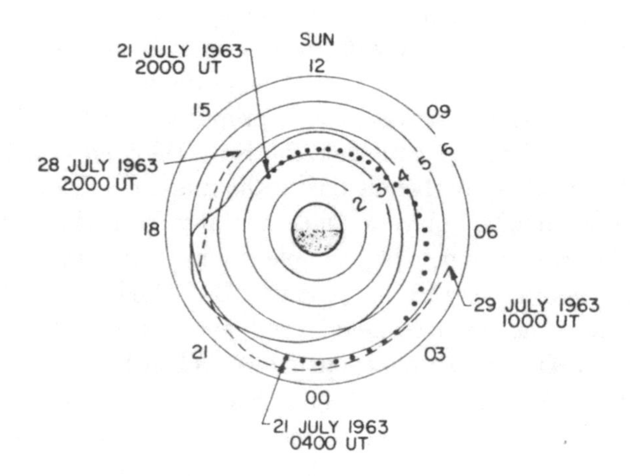

FIGURE 9.14 The top panel shows the density of electrons as a function of the equatorial geocentric distance. A sudden decrease at $\approx 4R_E$ is due to the crossing of the plasmapause boundary. The bottom panel shows the plasmapause boundary at different local times. The plasmapause moves in closer to Earth with increasing geomagnetic activity. (From Carpenter, 1966)

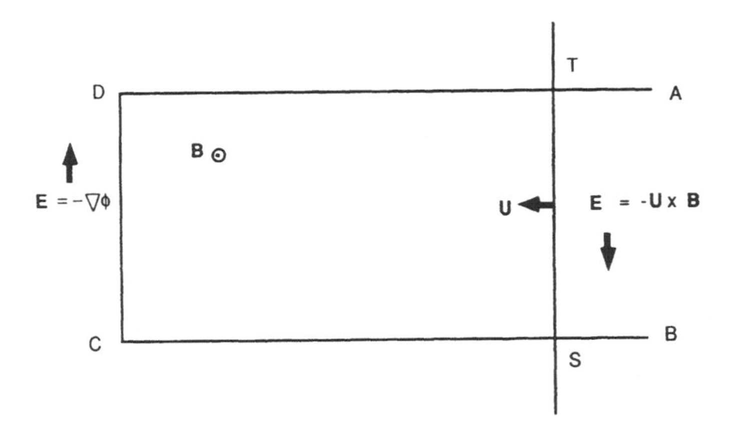

FIGURE 9.15 A schematic diagram to show how an EMF is generated.

an electric field $\mathbf{U} \times \mathbf{B}$ in the x-direction and this field will drive a current flow in the moving plasma. The current flows through the stationary plasma and closes on itself, since in steady-state $\nabla \cdot \mathbf{J} = 0$. The current in the stationary plasma is driven by a distribution of electrostatic charges set up by a charge separation field, $-\nabla\phi$. Hence the total field that drives the current is $\mathbf{E} = -\nabla\phi + \mathbf{U} \times \mathbf{B}$. The charge separation field results from a redistribution of charges induced by the motion. Hence the maximum charge separation field that can be produced is $-\mathbf{U} \times \mathbf{B}$ (for a closed system). This corresponds to the case of infinite conductivity, and the total electric field in the moving frame vanishes.

The induced current will couple with the magnetic field and the motion must now include a $\mathbf{J} \times \mathbf{B}$ force. The $\mathbf{J} \times \mathbf{B}$ force inside the moving plasma is in the $-y$ direction and it is in the $+y$ direction in the surrounding plasma. The $\mathbf{J} \times \mathbf{B}$ force in the $-y$ direction impedes the motion and the plasma that was initially moving will eventually come to rest. On the other hand, the $\mathbf{J} \times \mathbf{B}$ force in the $+y$ direction acting on the surrounding plasma that was initially at rest will make the plasma move. The process repeats itself and the plasma motion originally occurring in a small region propagates in the $\pm z$ direction. This coupling of the motion along the magnetic field direction gives rise to Alfvén waves.

9.7.1 Alfvén Waves

The MHD nature of these waves requires using the fluid equations derived in Chapter 5 to reformulate the problem. Assume the fluid is ideal ($\sigma = \infty$) and incompressible. Together with the momentum equation, this MHD fluid is described by

$$\begin{aligned}
\nabla \cdot \mathbf{U} &= 0 \\
\frac{\partial \mathbf{B}}{\partial t} &= \nabla \times (\mathbf{U} \times \mathbf{B}) \\
\rho_m \frac{\partial \mathbf{U}}{\partial t} &= -\nabla p + \frac{1}{\mu_0}(\nabla \times \mathbf{B}) \times \mathbf{B}
\end{aligned} \tag{9.130}$$

Now the magnetic field $\mathbf{B}$ is the total magnetic field, which includes the perturbation magnetic field $\mathbf{B}_1$ generated by the current $\mathbf{J}$. Hence

$$\mathbf{B}(\mathbf{r}, t) = \mathbf{B}_0 + \mathbf{B}_1(\mathbf{r}, t) \tag{9.131}$$

where as usual the perturbed quantities are assumed small, $\mathbf{B}_1 \ll \mathbf{B}_0$. If only first-order terms are retained, (9.130) reduces to

$$\begin{aligned}
\nabla \cdot \mathbf{U}_1 &= 0 \\
\frac{\partial \mathbf{B}_1}{\partial t} &= \nabla \times (\mathbf{U}_1 \times \mathbf{B}_0) \\
\rho_m \frac{d\mathbf{U}_1}{dt} &= -\nabla \left(p + \frac{B^2}{2\mu_0}\right) + (\mathbf{B}_0 \cdot \nabla)\frac{\mathbf{B}_1}{\mu_0}
\end{aligned} \tag{9.132}$$

where the subscript 1 represents the perturbed time-dependent quantity and the momentum equation has been rewritten by expanding the $(\nabla \times \mathbf{B}) \times \mathbf{B}$ term. The second equation in (9.132) can be rewritten by noting that

$$\begin{aligned}
\nabla \times (\mathbf{U}_1 \times \mathbf{B}_0) &= (\mathbf{B}_0 \cdot \nabla)\mathbf{U}_1 - (\mathbf{U}_1 \cdot \nabla)\mathbf{B}_0 \\
&\quad + \mathbf{U}_1(\nabla \cdot \mathbf{B}_0) - \mathbf{B}_0(\nabla \cdot \mathbf{U}_1) \\
&= (\mathbf{B}_0 \cdot \nabla)\mathbf{U}_1
\end{aligned} \tag{9.133}$$

where the second term on the right side vanishes because $\mathbf{B}_0$ is uniform in space, the third term vanishes because the magnetic field is divergenceless, and the last term vanishes because the fluid is incompressible. Hence the second equation in (9.132) becomes

$$\frac{\partial \mathbf{B}_1}{\partial t} = (\mathbf{B}_0 \cdot \nabla)\mathbf{U}_1 \tag{9.134}$$

Assume now that the ambient magnetic field points in the z-direction, $\mathbf{B}_0 = (0, 0, \mathbf{B}_0)$, and since plane wave solutions are sought, let the variables be independent of x and y, $\partial/\partial x = \partial/\partial y = 0$. Then (9.132) and (9.134) become

$$\frac{\partial \mathbf{B}_1}{\partial t} = B_0 \frac{\partial \mathbf{U}_1}{\partial z}$$

$$\rho_m \frac{\partial \mathbf{U}_1}{\partial t} = \left(\frac{B_0}{\mu_0}\right) \frac{\partial \mathbf{B}_1}{\partial z} - \frac{\partial}{\partial z}\left(p + \frac{B^2}{2\mu_0}\right)\hat{\mathbf{z}} \qquad (9.135)$$

where $\hat{\mathbf{z}}$ is the unit vector along the z-direction. From the conditions that both the fluid velocity and the magnetic field are divergenceless, obtain

$$\frac{\partial U_{1z}}{\partial z} = 0$$

$$\frac{\partial B_{1z}}{\partial z} = 0 \qquad (9.136)$$

This means that both U_{1z} and B_{1z} are uniform in space (in the yz plane) and consistent with the properties of plane waves. Both of these quantities are constants that can be set to zero. Hence the momentum equation in (9.135) does not have a z-component,

$$\frac{\partial}{\partial z}\left(p + \frac{B^2}{2\mu_0}\right) = 0 \qquad (9.137)$$

and the total pressure (fluid plus the magnetic) is constant along the z-direction. Equation (9.135) can thus be written in symmetrical form

$$\frac{\partial \mathbf{B}_1}{\partial t} = B_0 \frac{\partial \mathbf{U}_1}{\partial z}$$

$$\frac{\partial \mathbf{U}_1}{\partial t} = \frac{B_0}{\mu_0 \rho_m} \frac{\partial \mathbf{B}_1}{\partial z} \qquad (9.138)$$

Differentiating the first (second) of these with respect to z and then combining with the second (first) yields

$$\frac{\partial^2 \mathbf{B}_1}{\partial t^2} = \frac{B_0^2}{\mu_0 \rho_m} \frac{\partial^2 \mathbf{B}_1}{\partial z^2}$$

$$\frac{\partial^2 \mathbf{U}_1}{\partial t^2} = \frac{B_0^2}{\mu_0 \rho_m} \frac{\partial^2 \mathbf{U}_1}{\partial z^2} \qquad (9.139)$$

These are equations for plane waves travelling in the z-direction with a phase velocity

$$V_A = \pm \frac{B_0}{(\mu_0 \rho_m)^{1/2}} \qquad (9.140)$$

V_A is defined as the Alfvén velocity of the medium and it depends on B_0 and ρ_m. These are transverse waves, since $\mathbf{B}_1$ and $\mathbf{U}_1$ do not have any z-components. These waves are driven by $(\mathbf{B}_0 \cdot \nabla)\mathbf{B}_1/\mu_0$, the magnetic tension force (see 9.132). The wave equations (9.139) can be solved by assuming plane wave solutions for $\mathbf{U}_1$ and $\mathbf{B}_1$. The derivation of (9.139) assumes that the fluid is incompressible, which can be verified from (9.138).

9.7.2 Waves in Compressible Media

We now treat the more general problem of hydromagnetic waves in compressible fluids. In this case, the first equation in (9.130) does not apply. Instead, assuming that the fluid behaves adiabatically, (5.46) shows that the perturbed pressure p_1 and the perturbed mass density ρ_m are related through

$$\begin{aligned} p_1 &= \frac{\gamma p_0 \rho_1}{\rho_0} \\ &= C_s^2 \rho_1 \end{aligned} \tag{9.141}$$

where the subscript 0 stands for unperturbed and the subscript m has been omitted from the mass density (this should not cause any confusion with the charge density, which is normally designated as ρ_c). C_s is the speed of sound and γ is the ratio of specific heats.

Now use (9.141) in the momentum equation (9.130) and, retaining only first-order terms, obtain

$$\rho_0 \frac{\partial \mathbf{U}_1}{\partial t} + C_s^2 \nabla p_1 - \frac{(\nabla \times \mathbf{B}_1)}{\mu_0} \times \mathbf{B}_0 = 0 \tag{9.142}$$

The continuity equation is

$$\frac{\partial \rho_1}{\partial t} + \rho \nabla \cdot \mathbf{U}_1 = 0 \tag{9.143}$$

and the second equation in (9.132) is a magnetic flux conservation equation that still holds,

$$\frac{\partial \mathbf{B}_1}{\partial t} = \nabla \times (\mathbf{U}_1 \times \mathbf{B}_0) \tag{9.144}$$

Consider a plane wave propagating in the z-direction and let B_0 make an angle θ with respect to the z-axis and confined in the yz plane. Then $\mathbf{B}_0 = (0, B_0 \sin\theta, B_0 \cos\theta)$. As before, assume that all variables depend only on z and t. In this case, the second of (9.136) is still valid, and we can set $B_{z1} = 0$. The component equations of (9.142), (9.143) and (9.144), are

$$\rho_0 \frac{\partial U_x}{\partial t} = \frac{B_0}{\mu_0} \cos\theta \frac{\partial B_x}{\partial z} \tag{9.145}$$

$$\rho_0 \frac{\partial U_y}{\partial t} = \frac{B_0}{\mu_0} \cos\theta \frac{\partial B_y}{\partial z} \tag{9.146}$$

$$\rho_0 \frac{\partial U_z}{\partial t} = -\frac{B_0}{\mu_0} \sin\theta \frac{\partial B_y}{\partial z} - C_s^2 \frac{\partial \rho_1}{\partial z} \tag{9.147}$$

$$\frac{\partial \rho_1}{\partial t} + \rho_0 \frac{\partial U_z}{\partial z} = 0 \tag{9.148}$$

$$\frac{\partial B_x}{\partial t} = B_0 \cos\theta \frac{\partial U_x}{\partial z} \tag{9.149}$$

$$\frac{\partial B_y}{\partial t} = B_0 \cos\theta \frac{\partial U_y}{\partial z} - B_0 \sin\theta \frac{\partial U_z}{\partial z} \tag{9.150}$$

where the subscript 1 has been omitted in U_1 and B_1.

Oblique Alfvén Waves Combining (9.145) and (9.149) yields

$$\begin{aligned} \frac{\partial^2 B_x}{\partial t^2} &= V_A^2 \cos^2\theta \frac{\partial^2 B_x}{\partial z^2} \\ \frac{\partial^2 U_x}{\partial t^2} &= V_A^2 \cos^2\theta \frac{\partial^2 U_x}{\partial t^2} \end{aligned} \tag{9.151}$$

which are wave equations for B_x and U_x. For plane waves, we can replace $\partial/\partial t$ and $\partial/\partial z$ by $-i\omega$ and ik. Equation (9.151) then reduces to

$$\frac{\omega^2}{k^2} = V_{ph}^2 = V_A^2 \cos^2\theta \tag{9.152}$$

$V_A \cos\theta$ is the phase velocity in the z-direction. Equation (9.152) describes an oblique Alfvén wave. The angle θ defines the angle that the wave vector $\mathbf{k}$ makes with the applied field $\mathbf{B}_0$. As before, this wave does not involve the compression of the plasma. At $\theta = \pi/2$ there is no propagation because the magnetic tension force vanishes.

Magnetosonic Waves (Fast and Slow Modes) The variables U_y, U_z, B_y, and ρ_1 are determined from the other equations. Equations (9.150), (9.146), (9.147), and (9.148) become

$$-kB_0U_y \cos\theta + kB_0U_z \sin\theta - \omega B_y = 0 \tag{9.153}$$

$$-\rho_0 \omega U_y - \frac{kB_0B_y \cos\theta}{\mu_0} = 0 \tag{9.154}$$

$$-\rho_0 \omega U_z - \frac{kB_0B_y \sin\theta}{\mu_0} - kC_s^2 \rho_1 = 0 \tag{9.155}$$

$$-\rho_0 k U_z + \omega \rho_1 = 0 \tag{9.156}$$

Use (9.156) to eliminate ρ_1 in (9.155). Then solve for U_z and substitute it into (9.153). The resulting equation has U_y and B_y, which can be solved with the help of (9.154), and one obtains

$$\left(\frac{\omega}{k}\right)^4 - (V_A^2 + C_s^2)\left(\frac{\omega}{k}\right)^2 + V_A^2 C_s^2 \cos^2\theta = 0 \qquad (9.157)$$

Since $\omega/k = V_{ph}$, this equation can be rewritten as

$$V_{ph}^4 - (V_A^2 + C_s^2)V_{ph}^2 + V_A^2 C_s^2 \cos^2\theta = 0 \qquad (9.158)$$

This dispersion equation describes MHD waves in a compressible medium. The solutions of (9.158) are

$$2V_{ph}^2 = (V_A^2 + C_s^2) \pm [(V_A^2 + C_s^2)^2 - 4C_s^2 V_A^2 \cos^2\theta]^{1/2} \qquad (9.159)$$

Since the discriminant of (9.159) is positive definite, there are always two real positive roots for V_{ph}^2. These two solutions correspond to fast (+ sign) and slow (− sign) modes (see below).

When $\theta = \pi/2$, (9.159) reduces to

$$\begin{aligned} V_{ph}^2 &= V_A^2 + C_s^2 \quad (+\,\text{sign}) \\ &= 0 \qquad\qquad\; (-\text{sign}) \end{aligned} \qquad (9.160)$$

When $\theta = 0$, note that the discriminant becomes $|V_A^2 - C_s^2|^2$. Hence, depending on whether $C_s > V_A$ or $V_A > C_s$, we obtain

$$\begin{aligned} V_{ph}^2 &= C_s^2 \quad (+\text{sign}) \quad \text{if } C_s > V_A \\ &= V_A^2 \quad (+\text{sign}) \quad \text{if } C_s < V_A \\ &= C_s^2 \quad (-\text{sign}) \quad \text{if } C_s < V_A \\ &= V_A^2 \quad (-\text{sign}) \quad \text{if } C_s > V_A \end{aligned} \qquad (9.161)$$

Figure 9.16 shows the phase velocities of the two modes in a polar diagram (plotted as a function of θ) for two cases, $C_s < V_A$ and $C_s > V_A$. The modes travelling along the B direction ($\theta = 0$) in (9.161) are referred to as modified sound and Alfvén waves. The mode propagating perpendicular to $\mathbf{B}$ ($\theta = \pi/2$) is called a magnetosonic wave. This mode is faster than the Alfvén speed and it is therefore referred to as a fast mode. The fast mode has a maximum speed when $\theta = \pi/2$. The mode given in (9.152) is referred to as the intermediate mode. There is also a slow mode. This mode is obtained by noting that (9.159) can be rewritten as

$$2V_{ph}^2 = (V_A^2 + C_s^2)\left[1 \pm \left(1 - \frac{4C_s^2 V_A^2 \cos^2\theta}{(C_s^2 + V_A^2)^2}\right)^{1/2}\right] \qquad (9.162)$$

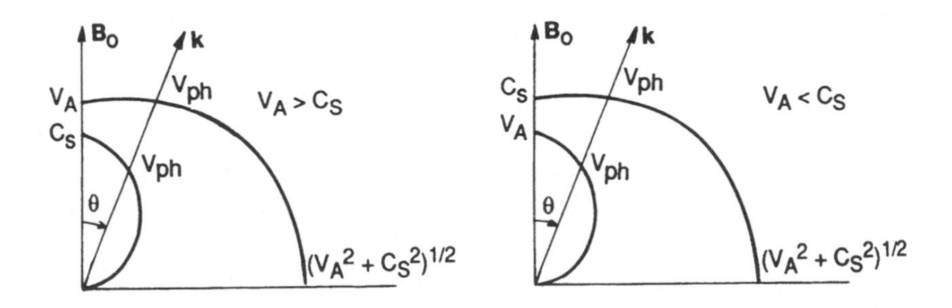

FIGURE 9.16 Phase velocities as a function of the direction of propagation θ.

In the limit that $V_A C_s \cos\theta/(C_s^2 + V_A^2) \ll 1$, either because $\cos\theta \ll 1$ ($\theta \approx \pi/2$), or $C_s \ll V_A$ or $V_A \ll C_s$, (9.162) yields

$$V_{ph}^2 \approx \frac{V_A^2 C_s^2 \cos^2\theta}{V_A^2 + C_s^2} \tag{9.163}$$

or

$$V_{ph}^2 \approx V_A^2 + C_s^2 \tag{9.164}$$

Equation (9.164) is a fast mode and (9.163) a slow mode, since the phase velocity here is less than either V_A or C_s. Like the ordinary Alfvén wave, slow modes cannot propagate across **B**. Fast and slow modes arise because in the former the magnetic and plasma oscillations are in phase and in the latter they are out of phase.

In concluding this section, we note for future reference that the fast and slow modes, which possess both transverse and longitudinal components, are compressional modes and these waves can steepen and form shocks. On the other hand, the intermediate mode is a purely transverse mode and this wave cannot steepen to form shocks.

9.7.3 Magnetohydrodynamic Waves in Anisotropic Fluids

The waves thus far studied are derived from the one-fluid MHD equations, which assume the fluid pressure is a scalar. In more realistic situations, the pressure term is anisotropic because of the magnetic field. If the Alfvén waves are studied by assuming that the pressures along and perpendicular to the direction of the magnetic field are not the same, $p_\parallel \neq p_\perp$, the propagation characteristics of the Alfvén waves will be modified.

The equations of motion when $p_\parallel \neq p_\perp$ are obtained from (7.56)

$$\rho_m \frac{d\mathbf{U}_\parallel}{dt} = -\nabla_\parallel p_\parallel + (p_\perp - p_\parallel)\left(\frac{\nabla B}{B}\right)_\parallel$$

$$\rho_m \frac{d\mathbf{U}_\perp}{dt} = -\nabla_\perp\left(p_\perp + \frac{B^2}{2\mu_0}\right) + \left(\frac{(\mathbf{B}\cdot\nabla)\mathbf{B}}{\mu_0}\right)\left[\frac{p_\perp - p_\parallel}{B^2/\mu_0} + 1\right] \tag{9.165}$$

where the second equation has been rewritten using the identity $\mathbf{J}\times\mathbf{B} = -\nabla B^2/2\mu_0 + (\mathbf{B}\cdot\nabla)\mathbf{B}/\mu_0$. The other equations needed are

$$\frac{\partial \mathbf{B}}{\partial t} = \nabla \times (\mathbf{U}\times\mathbf{B}) \tag{9.166}$$

$$\frac{\partial \rho_m}{\partial t} + \nabla\cdot\rho_m\mathbf{U} = 0 \tag{9.167}$$

These four equations can be used to obtain the dispersion relation of Alfvén waves propagating in an arbitrary direction relative to the magnetic field.

Alfvén Waves Parallel to B Let us begin studying the special case of Alfvén waves propagating along the magnetic field direction. In this case, only (9.165) and (9.166) need to be examined. As before, let $\mathbf{B}(\mathbf{r},t) = \mathbf{B}_0 + \mathbf{B}_1(\mathbf{r},t)$ where $\mathbf{B}_0$ is along the z-axis and $\mathbf{B}_1$ is the perturbed field. The linearized equation of (9.165) is

$$\rho_m \frac{\partial \mathbf{U}_\perp}{\partial t} = -\nabla_\perp\left(p_\perp + \frac{B^2}{2\mu_0}\right) + \frac{B_0}{\mu_0}\alpha\frac{\partial \mathbf{B}_1}{\partial z} \tag{9.168}$$

where $\alpha = [(p_\perp - p_\parallel)/(B^2/\mu_0) + 1]$ is the anisotropy factor. Since solutions in the form of plane waves are sought, let $\partial/\partial x = \partial/\partial y = 0$, and we can ignore the first term on the right of (9.168). Equation (9.166) then becomes

$$\frac{\partial \mathbf{B}_1}{\partial t} = B_0\frac{\partial \mathbf{U}_\perp}{\partial z} \tag{9.169}$$

and using this equation in (9.168) yields

$$\frac{\partial^2 \mathbf{U}_\perp}{\partial t^2} = V_A^2\alpha\frac{\partial^2 \mathbf{U}_\perp}{\partial z^2} \tag{9.170}$$

which is a wave equation nearly identical in form to (9.139) derived earlier, except the phase velocity is modified by the anisotropy factor α. Equation (9.170) yields

$$\begin{aligned} V_{ph}^2 &= \left(\frac{\omega}{k_\parallel}\right)^2 \\ &= \alpha V_A^2 \\ &= \frac{1}{\rho_m}\left(\frac{B^2}{\mu_0} + p_\perp - p_\parallel\right) \end{aligned} \tag{9.171}$$

where the subscript 0 has been omitted from ρ_m and B. If $\alpha = 1$, (9.139) is recovered. In the case of an MHD fluid with anisotropic pressure, the waves can travel faster or slower than fluids with isotropic pressure, depending on whether $p_\perp$ is greater than or less than $p_\parallel$. Note, however, that if α becomes negative, the phase velocity is imaginary and the waves do not propagate. Negative α corresponds to imaginary frequencies, which imply that the solutions have growing modes, $\exp \pm i\omega_i t$. These solutions correspond to an instability, in this case the "firehose" instability, which is driven by the pressure anisotropy (discussed in Chapter 11).

General Case For propagation in directions other than the direction of the magnetic field, let $k = k_\parallel + k_\perp$ and let the plane wave solutions vary as $\exp[i(\omega t - k_\perp x - k_\parallel z)]$. The waves propagate in the xz-plane. In this case, (9.165)–(9.167) yields (after a lot of algebra)

$$\left[\omega^2 \rho_m - k_\perp^2\left(2p_\perp + \frac{B^2}{\mu_0}\right) - k_\parallel^2\left(p_\perp - p_\parallel + \frac{B^2}{\mu_0}\right)\right] u_x + k_\parallel k_\perp p_\perp u_z = 0$$

$$(\omega^2 \rho_m - 3k_\parallel^2 p_\parallel) u_z + k_\parallel k_\perp p_\perp u_x = 0 \tag{9.172}$$

whose solutions are found from the determinant

$$\begin{vmatrix} \omega^2\rho_m - k^2\left(\frac{2p_\perp + B^2}{\mu_0}\right) - k_\parallel^2\left(p_\perp - p_\parallel + \frac{B^2}{\mu_0}\right) & k_\parallel k_\perp p_\perp \\ k_\parallel k_\perp p_\perp & \omega^2 \rho_m - 3k_\parallel^2 p_\parallel \end{vmatrix} = 0 \tag{9.173}$$

Let $k_\parallel = k\cos\theta$ and $k_\perp = k\sin\theta$ where θ is the angle between the propagation vector $\mathbf{k}$ and the magnetic field $\mathbf{B}$ (which is in the z-direction), and $V_{ph} = \omega/k$. Solving the determinant then yields a fourth-order equation

$$V_{ph}^4 - bV_{ph}^2 + c = 0 \tag{9.174}$$

where

$$b = \frac{[B^2/\mu_0 + p_\perp(1+\sin^2\theta) + 2p_\parallel \cos^2\theta]}{\rho_m}$$

$$c = \frac{1}{\rho_m^2}\left\{3p_\parallel \cos^2\theta \left[\frac{B^2}{\mu_0} + p_\perp(1+\sin^2\theta) - p_\parallel\cos^2\theta\right] - p_\perp^2 \sin^2\theta\cos^2\theta\right\} \tag{9.175}$$

The phase velocities of the waves are given by

$$V_{ph}^2 = \frac{1}{2}\rho_m\left\{\frac{B^2}{\mu_0} + p_\perp + 2p_\parallel\cos^2\theta + p_\perp\sin^2\theta \pm \left[\left(\frac{B^2}{\mu_0} + p_\perp(1+\sin^2\theta) - 4p_\parallel\cos^2\theta\right)^2 + 4p_\perp^2\sin^2\theta\cos^2\right]^{1/2}\right\} \tag{9.176}$$

For $\theta = 0$, (9.176) yields

$$\begin{aligned} V_{ph}^2 &= \frac{1}{\rho_m}\left(\frac{B^2}{\mu_0} + p_\perp - p_\parallel\right) \quad (+\,\text{sign}) \\ &= \frac{3p_\parallel}{\rho_m} \quad (-\text{sign}) \end{aligned} \tag{9.177}$$

The first solution is the same as (9.171). The second corresponds to the ion acoustic waves (see Section 9.4.3) where $p_\parallel$ is determined from the distribution function that describes the parallel motion. If the distribution function is Maxwellian, then (9.177) reduces to (9.66).

If $\theta = \pi/2$, (9.176) yields

$$V_{ph}^2 = \frac{1}{\rho_m}\left(\frac{B^2}{\mu_0} + 2p_\perp\right) \tag{9.178}$$

which is the magnetosonic wave discussed above (9.160) with $\gamma = 2$. Recall that all of the quantities ρ_m, $p_\parallel$, $p_\perp$, and B refer to the equilibrium values (for the sake of clarity, the subscript 0 was omitted). Note that at intermediate angles, the waves given by (9.176) are a mixture of magnetosonic and Alfvén waves.

9.8 Concluding Remarks

The various wave modes that we have studied in this chapter are summarized in Table 9.1. The subject of waves in plasmas can be developed much more rigorously (and elegantly) by analyzing the general expression of the dielectric tensor derived from the coupled Maxwell-Vlasov equations. How-

TABLE 9.1
Summary of Waves

Waves	*Dispersion Relation*	*Remarks*
Langmuir (plasma)	$\omega^2 = \omega_p^2 + 3k^2V_{th}^2$	$\mathbf{B}_0 = 0$ or $\mathbf{k} \parallel \mathbf{B}_0$
Ion Acoustic	$\omega^2/k^2 \approx k_B T_e/(m_i(1 + k^2\lambda_D^2))$	$\mathbf{B}_0 = 0$ or $\mathbf{k} \parallel \mathbf{B}_0$
Upper Hybrid	$\omega^2 = \omega_p^2 + \omega_c^2$	$\mathbf{k} \perp \mathbf{B}_0, \mathbf{k} \parallel \mathbf{E}$
Ion Cyclotron	$\omega^2 = \Omega_c^2 + k^2C_s^2$	$\mathbf{k} \perp \mathbf{B}_0, \mathbf{k} \parallel \mathbf{E}$
Lower Hybrid	$\omega^2 = \omega_c\Omega_c$	$\mathbf{k} \perp \mathbf{B}_0, \mathbf{k} \parallel \mathbf{E}$
Bernstein	$\omega_n^2 = n\omega_c^2(1 + \alpha_n)$	$\mathbf{k} \perp \mathbf{B}_0, \mathbf{k} \parallel \mathbf{E}$
Light	$\omega^2 = \omega_p^2 + k^2c^2$	$\mathbf{B}_0 = 0, \mathbf{k} \perp \mathbf{E}$
Ordinary	$\omega^2 = \omega_p^2 + k^2c^2$	$\mathbf{k} \perp \mathbf{B}_0, \mathbf{k} \perp \mathbf{E}$
Extraordinary	$n^2 = 1 - \frac{\omega_p^2}{\omega^2}\frac{\omega^2-\omega_p^2}{\omega^2-\omega_{uh}^2}$	$\mathbf{k}_\perp\mathbf{B}_0, \mathbf{k} \perp \mathbf{E}$
Right-hand (Whistlers)	$n^2 = 1 - \frac{\omega_p^2}{\omega(\omega-\omega_c)}$	$\mathbf{k} \parallel \mathbf{B}_0, \mathbf{k} \perp \mathbf{E}$
Left-hand	$n^2 = 1 - \frac{\omega_p^2}{\omega(\omega+\omega_c)}$	$\mathbf{k} \parallel \mathbf{B}_0, \mathbf{k} \perp \mathbf{E}$
Alfvén	$V_{ph}^2 = V_A^2$	$\mathbf{k} \parallel \mathbf{B}_0, \mathbf{k} \perp \mathbf{E}$
Magnetosonic	$V_{ph}^2 \approx V_A^2 + C_s^2$	$\mathbf{k} \perp \mathbf{B}_0, \mathbf{k} \perp \mathbf{E}$

ever, this requires advanced mathematical techniques for solving complex integral equations and we want to avoid them. Instead, we have chosen to develop this topic from the simpler and less rigorous fluid theory. This approach has the advantage that the algebra is manageable and it gives the reader a better physical feeling for the behavior of the complex waves. Note that we have analyzed the waves assuming a linear perturbation. This approach is valid only for small-amplitude waves. Large-amplitude waves are, however, observed in space. Analysis of these waves would require a non-linear method which is not treated here.

The fluid approach provides an adequate description of the waves when the plasma is sufficiently cold, for then the velocity spread of the particles is small and nearly all of the particles move together as a fluid element. When the temperature is zero, the pressure effects can be ignored and there is no need for the equation of state (recall that closure of the fluid equations is achieved through the equation of state, which is not needed in the Vlasov approach). Hence, in the zero temperature case, the two approaches yield identical results. However, when the temperature is finite, the Vlasov approach yields modes not predicted by the fluid theory. Below we discuss an example of the Bernstein mode whose results from the Vlasov theory are presented without proof (see Krall and Trivelpiece, 1973, cited at the end of the chapter).

The Bernstein mode is of considerable interest in space. This mode propagates perpendicular to $\mathbf{B}_0$ and is almost purely electrostatic, that is, $\mathbf{k}$ is nearly parallel to $\mathbf{E}$. In the limit that $\mathbf{B}_0 \to 0$, this mode degenerates to plasma oscillations at high frequencies and to ion acoustic oscillations at low frequencies. For the general case when $\mathbf{B} \neq 0$, the Bernstein mode at high frequencies becomes upper hybrid waves in the limit of low plasma temperature and long wavelengths ($k\lambda_D \ll 1$).

When the plasma density is low ($\omega_p \ll \omega_c$) as is often the case in space, the Bernstein mode occurs at close to the harmonics of the cyclotron frequency,

$$\omega_n^2 = n\omega_c^2(1 + \alpha_n) \quad n = 1, 2, \ldots \tag{9.179}$$

where n is the harmonic number and

$$\alpha_n = \frac{2\omega_p^2 m}{k^2 k_B T} I_n \left(\frac{k^2 k_B T}{m\omega_c^2} \right) e^{-k^2 k_B T / m\omega_c^2} \tag{9.180}$$

$I_n(k^2 k_B T / m\omega_c^2)$ is the modified Bessel function of the first kind of order n.

Figure 9.17 shows the frequency of the Bernstein modes plotted as a function of k. The modes appear as narrow bands of frequency. This result is very different from the results of the fluid theory, which predicts a cutoff between the lower and the upper hybrid frequencies for waves propagating perpendicular to $\mathbf{B}_0$ (see Figure 9.10). However, the Vlasov theory shows the Bernstein waves propagating in this gap at harmonics of the cyclotron frequency.

The Bernstein mode is different from the extraordinary mode. Whereas the Bernstein mode has $\mathbf{k}$ parallel to $\mathbf{E}$ and $\mathbf{E}$ perpendicular to $\mathbf{B}_0$, the extraordinary mode has $\mathbf{k}$ perpendicular to $\mathbf{E}$ and $\mathbf{E}$ parallel to $\mathbf{B}_0$. The extraordinary wave is nearly a pure electromagnetic mode. Also, note in general that electrostatic waves in plasma are normally damped (Landau damping). However, the Bernstein waves propagating across $\mathbf{B}_0$ are not damped because the Landau damping mechanism is not effective for $\mathbf{k} = \mathbf{k}_\perp$.

Our final comment concerns wave measurements in space. Waves in space have been thoroughly surveyed by the numerous spacecraft that have traversed our solar system. Wave instruments in space can identify whether the waves are electrostatic or electromagnetic. However, as mentioned earlier, wave modes are not usually identified because wave instrumentations flown on single spacecraft only provide information on the amplitude and the frequency ω. Data on both ω and the propagation vector $\mathbf{k}$ are needed to identify a wave mode. Since $\mathbf{k}$ cannot be measured from a single spacecraft, wave modes are often deduced by analyzing the behavior of observed

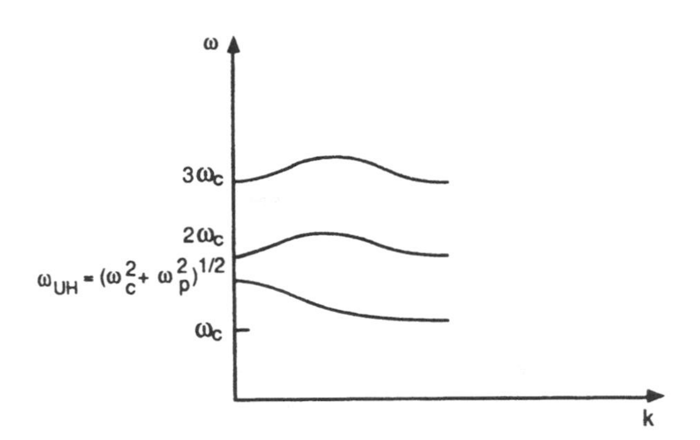

FIGURE 9.17 Dispersion characteristics of Bernstein waves.

wave features and comparing them to the predicted features of a dispersion relation. Whistler mode waves were identified in this way.

Simultaneous measurements of the magnetic field **B**, the total density, and the particle distribution function all help in wave studies. Knowledge of **B** gives information on the cyclotron frequency and the total particle density n gives information on the plasma frequency. Plasma and cyclotron waves were identified by using such information. Although the electrostatic waves in the upstream solar wind region are believed to be ion acoustic waves, this identification is an interpretation (a fairly good one, since no other waves exist in this frequency range). Assuming these waves are indeed ion acoustic waves, observations from two antennas of different lengths estimate their wavelengths to be 75 to 215 meters (see Fuselier and Gurnett, 1984, for how the wavelengths for the upstream waves are determined and Hoppe and Russell, 1983, and LaBelle and Kintner, 1989, for a review of the different techniques for measuring wavelengths).

Normally, waves in fluid media decay away. On the other hand, waves can grow if the fields can extract energy from the fluid. A fundamental problem in the theory of wave origin involves understanding the conditions under which waves grow. This is the domain of instability analyses, which will be briefly discussed in Chapter 11.

Bibliography

Alfvén, H. and C.G. Fälthammer, *Cosmical Electrodynamics, Fundamental Principles*, 2nd ed., Oxford University Press, Oxford, England, 1963. Chapter 3 treats the hydromagnetic waves. As usual, the material is presented in a clear way.

Angerami, J.J. and D.L. Carpenter, Whistler Studies of the Plasmapause in the Magnetosphere, 2. Electron Density and Total Tube Electron Content near the Knee in the

Magnetospheric Ionization, *J. Geophys. Res.*, **71**, 711, 1966. This paper discusses how one can use whistlers to study the electron density of the plasmasphere.

Budden, K.G., *The Propagation of Radio Waves*, Cambridge University Press, Cambridge, England, 1985. This book represents one of the most complete treatments of radio wave propagation phenomena in Earth's ionospheric plasmas.

Campbell, W.H., Geomagnetic Pulsations, in *Physics of Geomagnetic Phenomena*, S. Matsushita and W. Campbell, eds., Academic Press, New York, NY, 1967. A review article on micropulsations observed by ground-based measurements.

Carpenter, D.L., Whistler Studies of the Plasmapause in the Magnetosphere 1. Temporal Variations in the Position of the Knee and some Evidence on Plasma Motions near the Knee, *J. Geophys. Res.*, **71**, 693, 1966. This paper and the paper by Angerami and Carpenter present details about the plasmapause dynamics and electron density properties using whistler data. Carpenter is credited with the discovery of the plasmapause. He and his colleagues published many papers on this subject. Also see another early paper dealing with this subject: Carpenter, D.L., *Radio Science*, **3**, 719, 1968.

Chen, F.F., *Introduction to Plasma Physics*, Plenum Press, New York, NY, 1974. Chapter 4 of this book provides one of the clearer and more coherent discussions of waves in the fluid approximation. This book, one of the many good plasma physics books that treat the subject of plasma waves, is suitable for beginning students.

Clemmow, P.C. and J.P. Dougherty, *Electrodynamics of Particles and Plasmas*, Addison-Wesley Publishing Co., Reading, MA, 1969. This book uses MKS units. Chapters 5 and 11 discuss MHD waves including the case when the pressure is anisotropic.

Davis, L. Jr., The Configuration of the Interplanetary Magnetic Field, in *Solar-Terrestrial Physics/1970*, E.R. Dyer, J.G. Roederer and A.J. Hundhausen, eds., D. Reidel Publishing Co., Dordrecht, Holland, 1972. A review article on the IMF that was presented at an international symposium on solar-terrestrial physics, held in Leningrad, USSR, May 12-19, 1970.

Fuselier, S.A. and D.A. Gurnett, Short Wavelength Ion Waves Upstream of the Earth's Bow Shock, *J. Geophys. Res.*, **89**, 91, 1984. This article uses information obtained by two antennas of different lengths to deduce the wavelength of upstream low-frequency electrostatic waves. The authors also conclude that the observed features are consistent with the properties of ion acoustic waves.

Gurnett, D.A., R.R. Anderson, B.T. Tsurutani, E.J. Smith, G. Paschmann, G. Haerendel, S.J. Bame, and C.T. Russell, Plasma Wave Turbulence at the Magnetopause: Observations from ISEE 1 and 2, *J. Geophys. Res.*, **84**, 7043, 1979. This article investigates plasma wave electric and magnetic field behavior in the neighborhood of Earth's magnetopause.

Helliwell, R.A., *Whistlers and Related Ionospheric Phenomena*, Stanford University Press, Stanford, CA, 1965. This book contains many examples of whistlers recorded on the ground and is one of the main sources of references on this subject.

Hoppe, M.M. and C.T. Russell, Plasma Rest Frame Frequencies and Polarizations of the Low Frequency Upstream Waves: ISEE 1 and 2 Observations, *J. Geophys. Res.*, **88**, 2021, 1983. This article shows how vector magnetic field measurements of low-frequency waves yield information on polarization and frequency, and therefore the wave mode, of upstream electromagnetic waves.

Krall, N.A, and A.W. Trivelpiece, *Principles of Plasma Physics*, McGraw-Hill Book Co., New York, NY, 1973. Chapter 4 gives a clear development of waves in the fluid approximation. This discussion is expanded in Chapter 8, which provides a kinetic description of plasma waves, including the Bernstein mode. This book is for intermediate and advanced students.

LaBelle, J. and P. Kintner, The Measurement of Wavelength in Space Plasmas, *Rev. of Geophys.*, **27**, 495, 1989. This article reviews the current status of plasma wave measurements on sounding rockets and satellites.

Nicholson, D.R., *Introduction to Plasma Theory*, John Wiley and Sons, New York, NY, 1983. This book is a more recent publication on plasma physics. Chapter 7 of this book treats the topic of waves in fluids. This book is for intermediate and advanced students.

Stix, T.H., *The Theory of Plasma Waves*, McGraw Hill, New York, NY, 1962. One of the first comprehensive books on plasma waves.

Storey, L.R.O., An Investigation of Whistling Atmospherics, *Phil. Trans. Roy. Soc.* (London), A, **246**, 113, 1953. Storey conducted a thorough investigation of whistlers and explained correctly that whistlers originated in lightning. This classic paper brings together many observed features and interprets them, using the whistler propagation theory. A popular article, Whistlers, also appeared in *Sci. American*, **194**, 1, 34, 1956.

Questions

1. **a.** Make a plot of the plasma frequency as a function of the density n from $n = 10^4$ particles/m^3 to 10^{14} particles/m^3.
   b. Indicate by arrow the typical plasma frequencies of the solar corona, solar wind, outer magnetosphere, plasmasphere, ionosphere, and the plasma sheet.
   c. Compare these frequencies to the electron cyclotron frequencies of these regions.
2. The dispersion relation of plasma waves for cold plasma is valid for $k\lambda_D \ll 1$. Obtain an estimate of the range of wavelengths in the solar wind, outer radiation belt, plasmasphere, ionosphere, and plasma sheet for which the cold plasma dispersion relation is valid.
3. Show that the expression of plasma frequency for a two component plasma (electrons and ions) is

$$\omega_p^2 = \omega_{p-}^2 + \omega_{p+}^2$$

   where $\omega_{p-}^2 = ne^2/m_-\epsilon_0$ and $\omega_{p+}^2 = ne^2/m_+\epsilon_0$ and m_- and m_+ are electron and positive ion masses.
4. Consider a cold uniform plasma that is drifting with a constant velocity u_0 in the z-direction. To first-order accuracy,
   a. Show that the dispersion relation for this one-dimensional drifting plasma is

$$(\omega - ku_0)^2 = \omega_p^2$$

   Assume as usual that $n_1 \ll n_0$, $u_1 \ll u_0$, and that $T = B_0 = E_0 = 0$.
   b. From the dispersion relation, demonstrate that there are two phase velocities. One is faster and the other slower than the drift velocity u_0. Discuss the meaning of these results.
   c. Apply the results to the solar wind and obtain an estimate of the amount of the "Doppler shift" in the observed frequency.

5. Show that the moving plasma oscillations (problem 4) have a group velocity that is equal to the drift velocity u_0.
6. Consider the ion acoustic waves in the solar wind. Estimate the frequency shift that is due to the finite temperature correction.
7. Show that the wave energy of ion acoustic waves is

$$W_{wave} = \left(1 + \frac{1}{k^2 \lambda_D^2}\right) \epsilon_0 < E^2 >$$

where $<E>$ has been time-averaged.

8. Consider transverse electromagnetic waves propagating through a medium in which collision is occurring. Let ν be the collision frequency and assume that the momentum transport for electrons is modified by the additional "friction" term, $\nu \mathbf{u}_e$. In the absence of an ambient magnetic field ($B_0 = 0$), show that the dielectric constant of this medium for the transverse electromagnetic waves is

$$\epsilon = 1 - \frac{\omega_p^2}{\omega(\omega - i\nu)}$$

9. Derive Equation (9.112).
10. Derive Equation (9.114).
11. Derive Equation (9.126).
12. Derive Equation (9.128).
13. Derive Equation (9.129).
14. Plot the Alfvén velocity as a function of the density for the values given in problem 1 and indicate what the typical values are for the various regions.
15. Estimate the wavelengths of Alfvén waves for the various regions, using the results from problem 14.
16. Consider a flow discontinuity at a plane boundary (for example, the magnetopause) situated along the x-direction. Let $\mathbf{U}$ be parallel to the magnetic field $\mathbf{B}$ on one side of a plane boundary that is also along the x-direction (no flow on the other side).
    a. Show that the dispersion relation of the Alfvén waves becomes

    $$\frac{\omega}{k} = (V_A \pm U)$$

    where V_A is the Alfvén speed.

    b. Is this significant for the solar wind? Why?
17. a. Obtain a plane wave solution to the wave equation (9.139), assuming that $\mathbf{B}_0 = B_0\hat{\mathbf{z}}$ and $\mathbf{k} = k\hat{\mathbf{z}}$ where $\hat{\mathbf{z}}$ is a unit vector in the z-direction.
    b. Derive the expressions of U_y, J_x, E_x and p.
    c. Show that there is equipartition of energy between the kinetic energy of flow and the magnetic energy.
18. In the absence of Alfvén waves, the equation of the straight ambient magnetic line of force is $x = x_0$ and $y = y_0$. Derive the equation of a magnetic line of force perturbed by the Alfvén wave, assuming that the Alfvén waves are given by the solution derived in problem 17a.
19. When Alfvén waves propagating in one region encounter a boundary across which the fluid parameters are different, a part of the wave is reflected and a part is transmitted. Consider the fluid medium is ideal ($\sigma = \infty$) and let the boundary be perpendicular to the ambient magnetic field $\mathbf{B}_0$. Let $z = 0$ be the location of the

boundary and let V_1 and V_2 be the Alfvén wave speeds in the region $z < 0$ and $z > 0$, respectively. Let the magnetic field of the incoming wave be

$$b_y = A \sin \omega \left(t - \frac{z}{V_1} \right) \quad (z < 0)$$

Show that

$$A' = \left(\frac{2V_1}{V_1} + V_2 \right) A$$

$$A'' = \left(\frac{V_1 - V_2}{V_1 + V_2} \right) A$$

where A' and A'' are the amplitudes of the transmitted and reflected waves.

20. In deriving Equation (9.139), an assumption was made that the fluid was ideal ($\sigma = \infty$). Suppose σ is finite. The wave equation (9.139) must then be modified to take into account this finite conductivity.

a. Derive this equation.

b. The wave amplitude of Alfvén waves will be damped due to the finite conductivity. Show that the dispersion relation of Alfvén waves when σ is finite is

$$\omega^2 - \left(V_A^2 + \frac{i\omega}{\mu_0 \sigma} \right) k^2 = 0$$

where i is $(-1)^{1/2}$.

c. Derive the expression of the "skin" depth for the Alfvén waves.

10

Shocks in Space

10.1 Introduction

Boundaries are formed when a flowing fluid encounters an obstacle in its path. In Chapter 8 we learned how to characterize an MHD boundary when the flow speed is subsonic. This discussion will now be extended to flows that are supersonic. It is well known that in this case, shock waves form. Shock waves are consequences of compressibility effects that become important when the flow speed approaches the speed of sound.

Shock waves or simply shocks involve nonlinear processes and shocks in space are produced by collisionless plasmas. For many years, it was debated whether shocks could exist in collisionless plasmas. In 1962, I.A. Axford and P.J. Kellogg independently predicted that a shock wave would form in front of Earth's magnetosphere. They argued that Earth is an obstacle in the path of a supersonic solar wind and, as in ordinary supersonic fluid flows, a shock wave would form. This prediction was confirmed by spacecraft observations in 1963. The discovery of Earth's bow shock and the subsequent observations of other bow shocks (and also of interplanetary shocks) has now firmly established that shocks can be produced by collisionless plasmas. Space plasmas support both steady-state shocks, for example bow shocks produced by the solar wind interacting with planets and transient shocks, produced, for example by solar flares and supernova explosions.

Collisionless and collisional shocks differ in the nature of energy dissipation mechanisms. In ordinary shocks, energy is dissipated by binary collisions, while such collisions are absent in collisionless shocks. A major question concerns how the flow energy is thermalized behind the shock and what collisionless processes dissipate the energy. Although observations indicate that collective effects involving wave-particle interactions play an important role here, the exact mechanisms are not yet identified. Collisionless shocks are an example of macroscopic flow phenomena regulated by microscopic kinetic processes.

10.2 Basic Concepts and Definitions

Earth's collisionless bow shock has received much attention since its discovery in 1963. Figure 10.1 shows several examples of shocks recorded by a magnetometer flown through the sunward hemisphere of the magnetosphere. These shocks exhibit numerous complex structures. Before these features can be discussed, we must understand how a disturbance propagates in moving fluids and how MHD waves behave in supersonic flows. In addition, the study of collisionless shocks is facilitated by the introduction of certain key plasma variables to define shock characteristics and nomenclatures to organize the complex shock features.

10.2.1 Propagation of a Disturbance

If a small perturbation occurs in an ordinary fluid flowing with a velocity **U**, this perturbation will propagate in the fluid as a compressional sound wave. If C_s is the local velocity of the sound wave and **n** a unit vector that defines the direction of propagation, the velocity of the wave relative to a fixed point O (fixed coordinate system) is

$$\mathbf{V} = \mathbf{U} + C_s\mathbf{n} \tag{10.1}$$

since the wave is carried along by the flow **U**. The magnitude of **V** depends on the direction of propagation and is maximum if $C_s\mathbf{n}$ and **U** are in the same direction and minimum if they are in the opposite direction.

A way to illustrate the various magnitudes that **V** can have for other propagation directions is to represent (10.1) in a diagram. Draw the vector **U** originating at O as an arrow whose length is equal to the magnitude, as shown in the left diagram of Figure 10.2. Connect the tip of **U** with $C_s\mathbf{n}$,

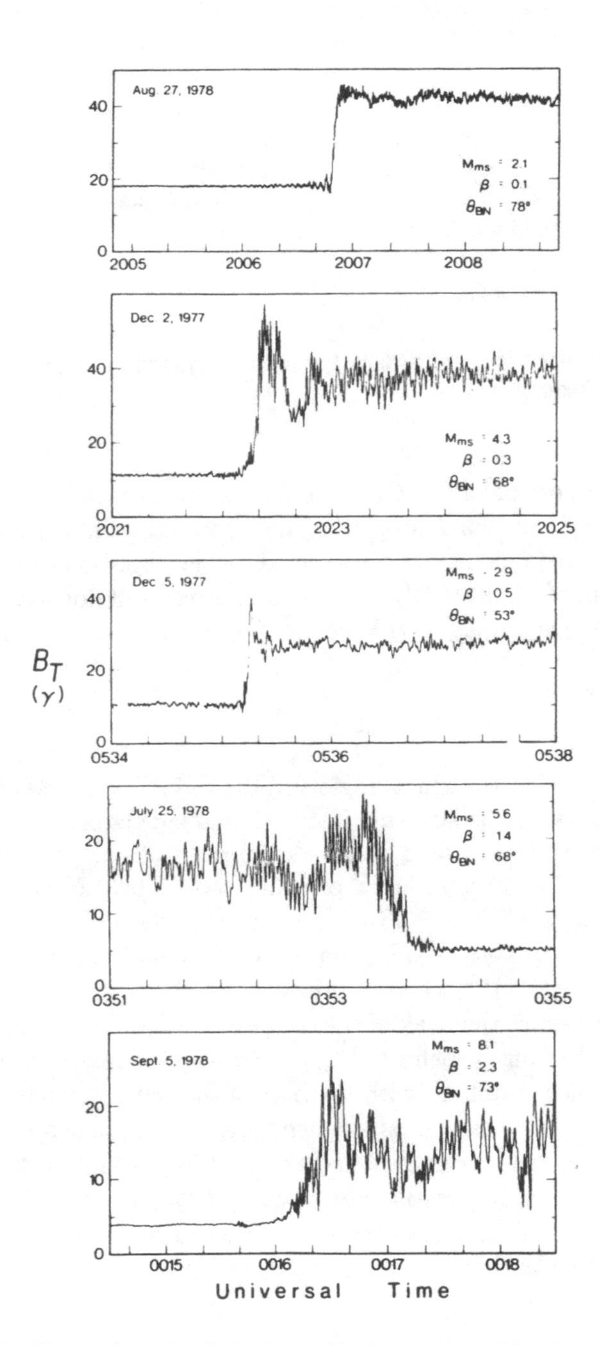

FIGURE 10.1 Examples of collisionless shocks produced by the interaction of the solar wind and Earth's magnetosphere. (From Russell *et al.*, 1982)

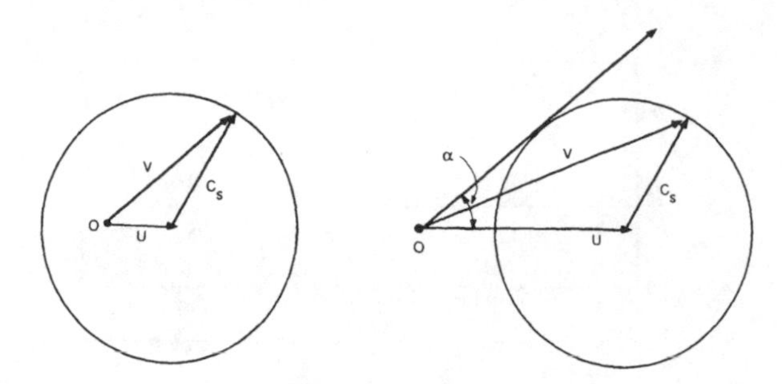

FIGURE 10.2 Illustration of a disturbance propagating in subsonic and supersonic flowing fluids.

and draw a sphere of radius C_s. This diagram illustrates that if $|\mathbf{U}| < C_s$ (subsonic case), waves will propagate and reach all points of the fluid.

A similar diagram can be constructed for the supersonic case, $|\mathbf{U}| > C_s$ (right diagram of Figure 10.2). In this case, waves will not reach all points of the fluid. Wave propagation is confined to within a cone defined by the angle α

$$\sin\alpha = \frac{C_s}{U} \tag{10.2}$$

The angle α is called the Mach angle and the ratio $U/C_s = M$ is called the Mach number. A flow is subsonic if $M < 1$ and supersonic if $M \geq 1$.

Shock waves are created by the steepening of compressional waves. To demonstrate this, consider a pressure profile propagating in a one-dimensional space (Figure 10.3). If the state of the fluid is described by the adiabatic law $p = p_0(\rho/\rho_0)^\gamma$, the speed of sound is $C_s = (\partial p/\partial\rho)^{1/2} = (\gamma p/\rho)^{1/2} = (A\gamma\rho^{\gamma-1})^{1/2}$ where A is a constant. Thus, if $\gamma > 1$, the velocity of sound is greater at the peak of the compressional wave where the density is higher than in front or behind the peak (where the density is lower). The peak will therefore catch up with the part of the wave ahead of it, and the wave steepens. The wave keeps steepening and changes shape as it propagates, until the flow becomes nonadiabatic. Viscous and heat conduction effects then become important and a wave of "permanent" form (shock) develops in the fluid where a balance is achieved between dissipation and steepening. The density, pressure, and velocity of the fluid are different across a shock wave.

10.2.2 Review of MHD Waves

A perturbation in MHD fluids can propagate more wave modes than in ordinary fluids. Our interest is to study the effects of low-frequency waves,

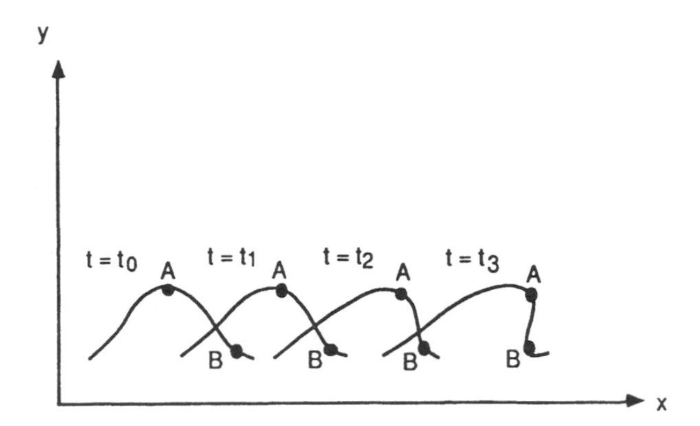

FIGURE 10.3 Illustration of how a compressional wave steepens to form a shock wave.

$\omega \ll \Omega_i$, where Ω_i is the ion Larmor frequency. It was shown in Chapter 9 that linearizing the ideal MHD equations in an infinite fluid medium leads to three plane waves, classified as slow, intermediate, and fast modes. The slow and fast modes are compressive, while the intermediate mode is not. Hence both the slow and the fast modes can steepen to form shock waves in supersonic MHD flows. The assumption of perfect MHD fluid does not include dissipative effects, hence it can only produce an infinitesimally thin shock. Shocks in this case have no structure. However, this formulation can describe steepening of the waves and the changes that occur in the fluid properties across the discontinuities.

The phase speed of the intermediate mode is given by

$$V_i^2 = V_A^2 \cos^2\theta \tag{10.3}$$

where $V_A = B/(\mu_0\rho)^{1/2}$ is the Alfvén speed and θ is the angle that the propagation vector $\mathbf{k}$ makes with the magnetic field $\mathbf{B}$. The phase speed of the other two modes is given by

$$2V_{ph}^2 = (C_s^2 + V_A^2) \pm [(C_s^2 + V_A^2)^2 - 4C_s^2 V_A^2 \cos^2\theta]^{1/2} \tag{10.4}$$

Here $C_s = (\gamma p/\rho)^{1/2}$ is the speed of sound (p, ρ, and γ are the pressure, mass density, and the ratio of specific heats, respectively). V_{ph} depends on the angle θ of the propagation relative to the magnetic field $\mathbf{B}$, and the local sound and the Alfvén speeds. The phase speed of the fast (V_f) and slow (V_s) modes corresponds to the faster (+ sign) and slower (− sign) phase speeds given above. Important properties of these waves include the following:

a. For arbitrary θ, the propagation speed of these waves is distinct and $V_f > V_i > V_s$.

b. The fluid velocities of the three modes are mutually perpendicular to each other.

c. The intermediate mode is a purely transverse mode. The velocity perturbation is transverse to both **B** and **k** and this mode does not create a perturbation in the density, pressure, or ambient magnetic field.

d. The fast and slow waves are compressive modes and they change fluid properties. For example, when the density increases, the magnetic field increases (decreases) for the fast (slow) mode.

e. For both fast and slow modes, the velocity and the magnetic field remain in the plane defined by the magnetic field and the wave normal as they propagate across a discontinuity (discussed further below).

f. When $\theta = 90°$, that is, when the propagation is perpendicular to **B**, $V_i = V_s = 0$ and $V_f = (V_A^2 + C_s^2)^{1/2}$. The waves degenerate to magnetosonic mode and this is the only mode that propagates with a nonvanishing speed.

g. When $\theta = 0$, that is, when the propagation is parallel to **B**, $V_f = V_i$ if $V_A > C_s$ and $V_i = V_s$ if $C_s > V_A$.

10.2.3 Definitions of Shock Parameters

A parameter that is used to organize collisionless shocks is the angle between the magnetic field direction and the shock normal, θ_{Bn}. Shocks are defined as

$$\begin{aligned} &\text{perpendicular shocks} \qquad && \theta_{Bn} = 90° \\ &\text{parallel shocks} && \theta_{Bn} = 0° \end{aligned} \tag{10.5}$$

For θ_{Bn} between 0 and 90°, the shocks are called oblique shocks. Shocks are also referred to as quasiparallel if $0° < \theta_{Bn} < 45°$ and quasiperpendicular if $45° < \theta_{Bn} < 90°$. The examples in Figure 10.1 are all quasiperpendicular shocks.

Another parameter used is the magnetosonic Mach number

$$M_{MS} = \frac{U_{SW}}{U_{MS}} \tag{10.6}$$

where U_{SW} and U_{MS} are, respectively, the solar wind speed and the speed of magnetosonic waves. For perpendicular shocks, (10.6) is

$$M_{MS} = \frac{U_{SW}}{(V_A^2 + C_s^2)^{1/2}} \tag{10.7}$$

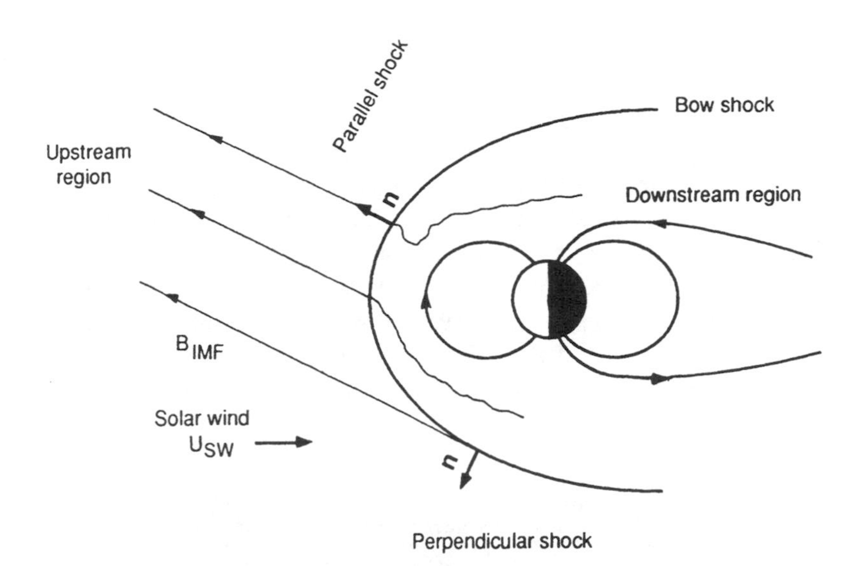

FIGURE 10.4 Schematic of a bow shock to illustrate the regions where parallel and perpendicular shocks are formed.

Use is also made of the parameter β of plasma,

$$\beta = \frac{p}{B^2/2\mu_0} \tag{10.8}$$

where p is the particle pressure and B is the magnetic field. Data shown in Figure 10.1 include examples of both high and low solar wind β and M_{MS}.

In closing this section, we note that the "parabolic" geometry of a planetary bow shock permits detection of both parallel and perpendicular shocks in the environment of a planetary magnetosphere (Figure 10.4). The parallel shock region covers the dawn side and the perpendicular shock the dusk side.

10.2.4 Discussion of Shock Observations

We now return to Figure 10.1. The magnetic field across shocks appears laminar (or quasilaminar) in some shocks, but the majority of shocks in space are turbulent. MHD theory further delineates shocks as subcritical if their Mach number is ≤ 2 to 3 and supercritical if the Mach number is ≥ 2 to 3. The shock shown in the top panel is subcritical and the others are supercritical. Supercritical shock features include a "pedestal" or "foot" immediately ahead of the shock, which is due to particles being reflected from the shock. Note also an overshoot, whose peak value can be nearly a factor of ten of the upstream intensity (second panel). The increase of the magnetic field behind the overshoot is typically a factor of four or less, on the average. The oscillations observed with the supercritical shocks are due to wave-particle interactions. These features also characterize the shocks observed from other planets (Mercury, Venus, Mars, and Jupiter; see Russell, 1985, cited at the end of the chapter).

The behavior of electrons and ions across a supercritical shock is shown in Figure 10.5. The top two panels show electron and ion densities, the third electron (dots) and ion (solid line) temperatures, the fourth bulk speed of the solar wind, the fifth pressure due to electrons, and the bottom two panels magnetic field intensity and direction. The vertical solid line marks the time of the shock and the two dashed lines mark the beginning

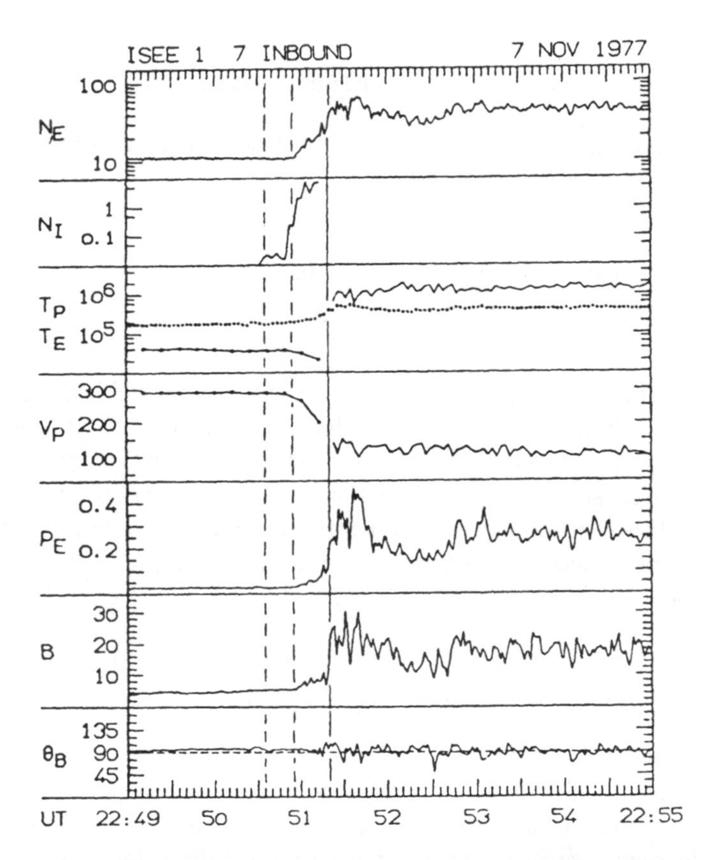

FIGURE 10.5 Earth's quasiperpendicular shock recorded by a plasma instrument on an International Sun Earth Explorer (ISEE) spacecraft. The shock parameters for this event were $M_{\dot{A}} = 8.6$, $\theta_{Bn} = 85°$, $\beta_e = 3.3$, and $\beta_i = 0.6$. Note that while the electron and the ion temperatures are comparable, the measured ion density is about a factor of 40 lower than the electron density. (From Bame *et al.*, 1979)

of increases of electron and ion densities. These particle features are very similar to features observed in the magnetic field.

Although a considerable amount of progress has been made during the last ten years in studying collisionless shocks, it is still not possible at this time to understand all of the observed particle and field features. Our objective is to study limited aspects of collisionless shocks with single-particle concepts and MHD theory developed in the previous chapters. As an introduction to this complicated subject, we will first formulate the shock physics, using the one-fluid MHD equations. This formulation does not identify the specific roles the electrons or ions play in shock dynamics. Information on effects that give rise to distinct electron and ion features (for example, heating of electrons and ions) will require studying two-fluid and kinetic theories. The theory of shock structures is briefly discussed. We conclude this chapter with a short presentation of particle acceleration by the solar wind electric field occurring in the shock frame, and introduce the concept of the foreshock, which is a region in the upstream solar wind that is connected to the bow shock by the IMF.

10.3 Shock Waves in Ordinary Fluids

It is instructive to begin our study with a discussion of shocks in ordinary fluids. In ordinary fluids, we can treat pressure as a scalar and neglect the magnetic field. Then the continuity relations of mass, momentum, and energy across the shocks are given by (see Chapter 8)

$$[\rho U_n] = 0 \tag{10.9}$$

$$[\rho U_n^2 + p] = 0 \tag{10.10}$$

$$[\rho U_n U_t] = 0 \tag{10.11}$$

$$\left[\left(\frac{\rho U^2}{2} + \frac{\gamma p}{(\gamma - 1)}\right) U_n\right] = 0 \tag{10.12}$$

where the fluid is assumed to have an adiabatic equation of state. These relations are known as Rankine-Hugoniot relations and they enable one to obtain the downstream variables in terms of the upstream variables (one normally assumes that the upstream variables are given). Equations (10.9) and (10.11) combined yield $[U_t] = 0$. The tangential component of the flow is continuous in ordinary fluid shocks. This relation permits us to transform to a coordinate system moving along the shock plane. In this frame of reference, $U_t = 0$, hence $U = U_n$.

To study these equations, we introduce new variables

$$\eta = \frac{\rho_2}{\rho_1}$$
$$\xi = \frac{p_2}{p_1} \tag{10.13}$$

to designate the ratios of densities and pressures on the two sides of the shock. The subscripts 1 and 2 denote upstream and downstream regions, respectively. η is called the compression ratio and ξ measures the "strength" of the shock. Use of (10.13) in the continuity relation shows that

$$\eta = \frac{U_1}{U_2} \tag{10.14}$$

If we define the local speed of sound in the upstream region by $c_1^2 = \gamma p_1/\rho_1$ and the upstream Mach number by $M = U_1/c_1$, the conservation of momentum relation across the shock can be rewritten as

$$M_1^2\left(1 - \frac{1}{\eta}\right) = \frac{(\xi - 1)}{\gamma} \tag{10.15}$$

Use of (10.13) and (10.14) reduces the conservation of energy relation to

$$M_1^2\left(1 - \frac{1}{\eta^2}\right) = 2\frac{(\eta/\xi - 1)}{(\gamma - 1)} \tag{10.16}$$

Solve for η, using (10.15) and (10.16) by eliminating ξ. This yields (after some messy algebra) two possible solutions: $\eta = 1$, which implies $U_1 = U_2$ and therefore this is not a shock solution, or

$$\eta = \frac{(\gamma + 1)M_1^2}{(\gamma - 1)M_1^2 + 2} \tag{10.17}$$

Substitution of (10.17) into (10.15) yields

$$\xi = \frac{2\gamma M_1^2 - (\gamma - 1)}{(\gamma + 1)} \tag{10.18}$$

We define shocks as strong when the sonic Mach number ($M = U_1/c_1$) is large. In the limit $M_1^2 \to \infty$, (10.17) and (10.18) yield

$$\frac{\rho_1}{\rho_2} \to \frac{(\gamma + 1)}{(\gamma - 1)} \tag{10.19}$$

$$\frac{U_2}{U_1} \to \frac{(\gamma - 1)}{(\gamma + 1)} \tag{10.20}$$

$$\frac{p_2}{p_1} \to \frac{2\gamma M_1^2}{(\gamma + 1)} \tag{10.21}$$

If we use a Maxwellian distribution function, $\gamma = 5/3$ (see Chapter 5), and in the limit $M_1^2 \to \infty$, $\rho_2/\rho_1 = 4$, and $U_2/U_1 = 1/4$. The downstream fluid is compressed and slowed down by a maximum factor of four of the upstream values.

The shock heats the fluid as it is compressed and slowed down. The amount of heating in the downstream region can be estimated by use of the ideal gas law $p_2 = n_2 k T_2 = (\rho_2/m) k T_2$. In the limit $M_1^2 \to \infty$,

$$\frac{p_2}{\rho_1} \to \frac{2\gamma(\gamma-1)M_1^2 U_1^2}{(\gamma+1)^2} \tag{10.22}$$

The theory predicts no upper limit to heating and that heating increases as $M_1^2 U_1^2$.

10.4 MHD Shocks

MHD shocks can be characterized as in ordinary fluids by using the conservation equations discussed in Chapter 8. All shocks require $U_n \neq 0$ and, in the case of MHD shocks, other factors must be taken into account. This section begins with a discussion of parallel shocks, followed by a discussion of perpendicular shocks. We then discuss oblique shocks and conclude with a discussion of weak MHD shock structures.

10.4.1 Parallel Shocks ($U_n \neq 0$, $B_t = 0$, $B = B_n$)

Parallel shocks refer to a geometry of the flow where the magnetic field $\mathbf{B}$ is parallel to the direction of the shock normal $\mathbf{n}$. Hence $B_t = 0$ and $B = B_n$. These requirements on MHD discontinuity show that all of the continuity equations derived for ordinary shocks apply. In ordinary shocks, collisions dominate and one can assume there is equipartition of energy in all directions and $w_\perp = 2w_\parallel$. This permits us to use the Maxwellian distribution function, which then leads to an adiabatic equation of state. The internal energy is then related to ratios of specific heats and one can write

$$w_\perp = \left(\frac{\gamma-1}{\gamma}\right)\frac{p_\perp}{\rho} \tag{10.23}$$

Here $p_\perp$ only includes the particle pressure, since the magnetic field plays no role. In the limit of strong shocks, (10.19), (10.20), and (10.21) apply and the compression ratio approaches four.

In parallel shocks, one can transform to a shock frame of reference where $U_t = 0$. The velocity $\mathbf{U}$ is then parallel to $\mathbf{B}$. We can write

$$\mathbf{B} = \alpha\rho\mathbf{U} \tag{10.24}$$

Since $\mathbf{U} \parallel \mathbf{B}$, α is a streamline constant. Since $[B] = 0$ and $[\rho U] = 0$, we find that $[\alpha] = 0$. This implies that the passage of parallel shock will not affect the structure of the magnetic field.

10.4.2 Perpendicular Shocks ($G \neq 0$, $B_n = 0$)

Define a reference system in which $\mathbf{V}$ is the velocity of the shock and let $\mathbf{U}$ be the velocity of the fluid. A coordinate system in which the fluid is at rest, $\mathbf{U} = 0$, is called a laboratory system. A coordinate system in which the shock is at rest, $\mathbf{V} = 0$, is called a shock system. The shock system is most convenient for discussing the conservation laws.

Perpendicular shocks refer to a geometry of the flow in which the magnetic field is perpendicular to the direction of the shock normal. Hence $B_n = 0$ and $B = B_t$. Choose a magnetic field $\mathbf{B} = (0, 0, B)$ and let the electric field $\mathbf{E} = (0, E, 0)$ so the shock is confined in the yz plane (Figure 10.6). Let $U = (U, 0, 0)$ and N be the number density. In shocks, $U_n \neq 0$ and, since particles are moving across magnetic fields, U_n is related to an electric field tangent to the shock. For the shock geometry being considered, the electric field is in the y-direction. Maxwell's equation $\nabla \times \mathbf{E} = -\partial \mathbf{B}/\partial t$ then reduces to

$$\frac{\partial E}{\partial x} = -\frac{\partial B}{\partial t} \tag{10.25}$$

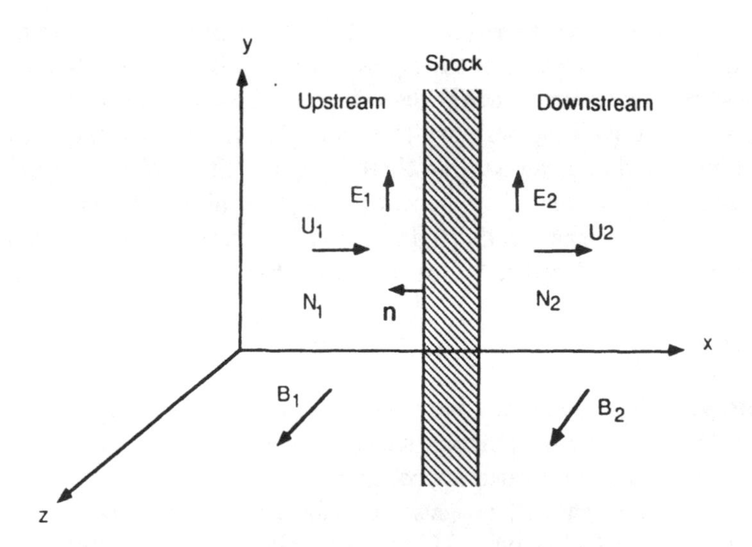

FIGURE 10.6 Geometry of perpendicular shocks.

and integration of this equation across the shock yields

$$E_2 - E_1 = -\frac{\partial \Phi}{\partial t} \tag{10.26}$$

where $\Phi = \int B\,dx$ is the magnetic flux contained in the shock structure and the integration limits are from x_1 to x_2 for a shock with a finite dimension. MHD assumes the shock structure is infinitely thin and therefore $\Phi = 0$ in the shock. (An alternative argument is that even for finite shock dimension, one can assume that $<\Phi> \approx 0$. This ignores the presence of hydromagnetic waves, which is an important consideration in studying shock structures.) We require the electric field to be the same on both sides, $E_2 \approx E_1 = E$.

The fluid velocities on the upstream and downstream sides are then given by

$$\begin{aligned} U_1 &= \frac{E}{B_1} \\ U_2 &= \frac{E}{B_2} \end{aligned} \tag{10.27}$$

Hence

$$\frac{U_1}{U_2} = \frac{B_2}{B_1} \tag{10.28}$$

Conservation of particles requires that

$$N_1 U_1 = N_2 U_2 \tag{10.29}$$

and this equation combined with (10.27) above yields

$$\frac{U_1}{U_2} = \frac{N_2}{N_1} \tag{10.30}$$

Since $U_1 > U_2$, (10.28) and (10.29) show that the density of particles and magnetic fields are compressed by the same amount across the shock.

There must exist a current in the shock if there is a difference of the magnetic field across the shock. Since $B_2 > B_1$, the current must run in the $-y$ direction. Then Maxwell's equation $\nabla \times \mathbf{B} = \mu_0 \mathbf{J}$ for the simple geometry being considered yields

$$\frac{\partial B}{\partial x} = -\mu_0 J_y \tag{10.31}$$

and integration of this equation across the shock yields

$$B_2 - B_1 = \mu_0 \int J_y\,dx \tag{10.32}$$

J_y comes from the magnetization and drift currents. The drift current comes from the drift of the guiding center due to the gradient of the magnetic field across the shock. Curvature current must be added to the gradient current if the shock geometry includes a curvature.

The conservation of momentum equation requires

$$\frac{\partial}{\partial x_j}(\pi_{ij} + T_{ij}) = 0 \tag{10.33}$$

where π_{ij} and T_{ij} are the momentum transfer and the electromagnetic stress tensors, respectively. The indices here represent the components of the Cartesian coordinates (x, y, z). For the plane geometry considered, only the x derivative is nonvanishing, $\partial/\partial x_j = \partial/\partial x$. Hence integration of (10.33) across the shock yields

$$(\pi_{ix} + T_{ix})_1 = (\pi_{ix} + T_{ix})_2 \tag{10.34}$$

The electromagnetic stress tensor is a diagonal for the plane shock geometry; that is, unless $i = x$, $T_{ix} = 0$. We can ignore the E term in the electromagnetic stress tensor because $E_1 = E_2$ since E_{tang} is continuous.

The particle momentum stress tensor $\pi_{ij} = p_{ij} + NmU_iU_j = p_{ix} + NmU_iU_x$ is also diagonal. Thus,

$$\begin{aligned} (\pi_{xx})_1 &= N_1 m U_1^2 + (p_{xx})_1 \\ (\pi_{xx})_2 &= N_2 m U_2^2 + (p_{xx})_2 \end{aligned} \tag{10.35}$$

Define now

$$p^* = p + \frac{B^2}{2\mu_0} \tag{10.36}$$

as the total pressure, and substitution of (10.35), and (10.36) into (10.34) yields

$$\rho_1 U_1^2 + p_1^* = \rho_2 U_2^2 + p_2^* \tag{10.37}$$

where $\rho = Nm$ is the mass density. This equation looks identical in form to the equation of momentum conservation in ordinary shocks except that p^* includes the magnetic pressure. Eliminate ρ_2 in (10.37) by use of the mass conservation relation $\rho_1 U_1 = \rho_2 U_2$. Then (10.37) becomes

$$\rho_1 U_1 (U_1 - U_2) = p_2^* - p_1^* \tag{10.38}$$

To proceed further, we must examine the conservation of energy equation, which, in steady-state, is $\nabla \cdot (\mathbf{q} + \mathbf{S}) = 0$, where $\mathbf{q}$ and $\mathbf{S}$ are the energy flow of particles and the Poynting flux of electromagnetic fields, respectively (see (8.7)). Integration of (10.38) across a plane shock yields

$$(q_x + S_x)_1 = (q_x + S_x)_2 \tag{10.39}$$

The x-component of $\mathbf{S}$ is

$$S_x = \frac{(\mathbf{E} \times \mathbf{B})_x}{\mu_0} = \frac{UB^2}{\mu_0} \quad (10.40)$$

where (10.27) has been used. The definition of q_i is

$$q_i = \int \left(\frac{mv^2}{2}\right) v_i f(r, v) d^3v \quad (10.41)$$

where $v_i = <v_i> + \delta v_i$, $<v_i> = u_i$ which is the velocity averaged over the distribution function, and δv_i is the thermal velocity. (We use δv instead of w to denote the thermal velocity. The letter w is used to designate the internal energy.) Equation (10.41) then becomes

$$q_i = \frac{mn}{2}[u_i u^2 + u_i <\delta v^2> + 2u_i u_j <\delta v_j> + u^2 <\delta v_i> + <\delta v_i \delta v^2> + 2u_j <\delta v_i \delta v_j>] \quad (10.42)$$

The first term represents the flow of macroscopic energy, the second and the sixth the internal energy carried by convection, the third and fourth terms are macroscopic energy carried by the average thermal velocity, and the fifth term is energy carried by heat conduction. Our model is one-dimensional, $u_i = u_j = 0$ except for $i = x$. Hence $u_y = u_z = 0$. Since the integral of any odd function vanishes (the limits of integration are symmetric), $< \delta v_i > = < \delta v_j > = 0$. Thus the third and the fourth terms vanish. Then, ignoring the heat conduction term, (10.42) reduces to

$$q_i = \frac{\rho_m}{2}(u^2 u_i + u_i <\delta v^2> + 2u_j <\delta v_i \delta v_j>) \quad (10.43)$$

where $\rho_m = nm$. Note also that $<\delta v_i \delta v_j> = 0$ for $i \neq j$. For a one-fluid equation, $U \approx u^+ + (m^-/m^+)u^- \approx u^+$ if $u^- \approx u^+$. Then let $u_x = U$ and (10.43) simplifies to

$$q_x = \frac{\rho_m}{2}(U^2 U + U <\delta v^2> + 2U <\delta v_x \delta v_x>) \quad (10.44)$$

The internal energy can be represented by

$$\frac{m <\delta v^2>}{2} = w_\perp + w_\parallel$$
$$\frac{m <\delta v_x \delta v_x>}{2} = w_\perp \quad (10.45)$$

where $w_\perp$ and $w_\parallel$ are, respectively, the internal energy associated with perpendicular and parallel velocities. Substitution of (10.45) into (10.44) yields

$$q_x = \frac{\rho U^2}{2} U + NU(2w_\perp + w_\parallel) \tag{10.46}$$

If we use (10.40) and (10.46), the energy equation across the shock becomes

$$\rho_1 U_1^3 + N_1 U_1 (2w_\perp - w_\parallel)_1 + \frac{B_1^2}{\mu_0} U_1 =$$

$$\rho_2 U_2^3 + N_2 U_2 (2w_\perp + w_\parallel)_2 + \frac{B_2^2}{\mu_0} U_2 \tag{10.47}$$

If we assume that the upstream quantities n_1, B_1, $w_{\perp 1}$, $w_{\parallel 1}$, U_1, and p_1^* are given, the quantities to be determined are the six downstream quantities, n_2, B_2, $w_{\perp 2}$, $w_{\parallel 2}$, U_2, and p_2^*. However, there are only five equations available, (10.27), (10.29), (10.36), (10.37), and (10.47). The sixth equation is obtained from assuming either that $2w_\parallel = w_\perp$, which is valid if collisions dominate, or that in collisionless plasmas the parallel internal energy $w_\parallel$ does not change across a shock,

$$w_{\parallel 1} = w_{\parallel 2} \tag{10.48}$$

We consider for space the case of no collisions. Then (10.48) applies and $w_\parallel$ drops out from the energy equation (10.47). If next we use the mass continuity equation (10.29) and the total pressure (10.36), recognizing that $p_\perp = Nw_\perp$ and $p_\parallel = Nw_\parallel$ (see Equations (7.47) and (7.48)), (10.47) can be rewritten as

$$\rho_1 U_1 (U_1^2 - U_2^2) + 4(U_1 p_1^* - U_2 p_2^*) = 0 \tag{10.49}$$

Use (10.38) to rewrite the first term, and reduce (10.49) to

$$(p_2^* - p_1^*)(U_2 + U_1) + 4(U_1 p_1^* - U_2 p_2^*) = 0 \tag{10.50}$$

Now divide through the equation by $p_1^* U_1$ and obtain

$$\left(\frac{p_2^*}{p_1^*} - 1\right)\left(1 + \frac{U_1}{U_2}\right) + 4\left(\frac{U_1}{U_2} - \frac{p_2^*}{p_1^*}\right) \tag{10.51}$$

Let $\xi = p_2^*/p_1^*$ as the pressure ratio and $\eta = \rho_2/\rho_1 = U_1/U_2 = B_2/B_1$ as the compression ratio. Then (10.51) can be written as

$$(\xi - 1)(\eta + 1) + 4(\eta - \xi) = 0 \tag{10.52}$$

Solve for η and obtain

$$\eta = \frac{(3\xi + 1)}{(\xi + 3)} \tag{10.53}$$

This equation determines the compression ratio of the shock as a function of the pressure ratio in the upstream to downstream regions. A shock is considered strong if ξ is large. In the limit of an infinitely strong shock, $\xi \to \infty$, (10.53) yields $\eta \to 3$. A shock is considered weak if ξ is small. In the limit $\xi \to 1$, there is no shock.

A shock involving the fast mode waves implies that the velocity of the flow is greater than the fast mode wave speed in the upstream region. It is instructive at this juncture to show that perpendicular shocks ($\theta_{Bn} = 90°$) involve only the fast mode MHD waves. We will now demonstrate this for fluids with isotropic pressure. The conservation equations, stated here again for convenience, are

$$[B_n] = 0 \tag{10.54}$$

$$[\rho U_n] = 0 \tag{10.55}$$

$$[(\mathbf{n} \cdot \mathbf{B})\mathbf{U} - (\mathbf{n} \cdot \mathbf{U})\mathbf{B}] = 0 \tag{10.56}$$

$$\left[\rho(\mathbf{U} \cdot \mathbf{n})\mathbf{U} + \left(p + \frac{B^2}{2\mu_0}\right)\mathbf{n} - \frac{B_n \mathbf{B}}{\mu_0}\right] = 0 \tag{10.57}$$

As before, $\mathbf{U}$ is the velocity of the fluid and $\mathbf{n}$ is the unit vector normal to the shock front. Define $G = \rho U_n$ and, if we use the operational definition

$$[AB] = <A> [B] + [A] <B> \tag{10.58}$$

where A and B are field variables on the two sides of the boundary and $<A> = (A_1 + A_2)/2$ and $<B> = (B_1 + B_2)/2$ are the average values, Equations (10.55), (10.56), and (10.58) can be cast in the form (omitting $<>$)

$$G\left[\frac{1}{\rho}\right] - [U_n] = 0 \tag{10.59}$$

$$\frac{G}{\rho}[\mathbf{B}] + \mathbf{B}[U_n] - B_n[\mathbf{U}] = 0 \tag{10.60}$$

$$G[\mathbf{U}] + [p]\mathbf{n} + \frac{\mathbf{B} \cdot [\mathbf{B}]\,\mathbf{n}}{\mu_0} - \frac{B_n[\mathbf{B}]}{\mu_0} = 0 \tag{10.61}$$

Quantities appearing outside the brackets [] are understood to be averaged over upstream and downstream values. Erect a Cartesian coordinate system OXYZ and let $\mathbf{n}$ be along OZ. In this system, the first equation, the continuity equation, is unchanged. The last two equations written out explicitly in the component form are

$$\frac{G}{\rho}[B_x] - B_n[U_x] + B_x[U_n] = 0 \tag{10.62}$$

$$\frac{G}{\rho}[B_y] - B_n[U_y] + B_y[U_n] = 0 \tag{10.63}$$

$$\frac{G}{\rho}[B_n] = 0 \tag{10.64}$$

$$-\frac{B_n}{\mu_0}[B_x] + G[U_x] = 0 \tag{10.65}$$

$$-\frac{B_n}{\mu_0}[B_y] + G[U_y] = 0 \tag{10.66}$$

$$\frac{B_x}{\mu_0}[B_x] + \frac{B_y}{\mu_0}[B_y] + G[U_n] + [p] = 0 \tag{10.67}$$

If (10.62) and (10.65) are combined and $[U_x]$ is eliminated, we obtain

$$\left(\frac{G^2}{\rho} - \frac{B_n^2}{\mu_0}\right)[B_x] + GB_x[U_n] = 0 \tag{10.68}$$

Similarly, elimination of $[U_y]$ by combining (10.63) and (10.66) yields

$$\left(\frac{G^2}{\rho} - \frac{B_n^2}{\mu_0}\right)[B_y] + GB_y[U_n] = 0 \tag{10.69}$$

Solve for $[B_x]$ and $[B_y]$ assuming that the coefficient $(G^2/\rho - B_n^2/\mu_0) \neq 0$. Then use them in (10.67) and obtain

$$\frac{G(B_x^2 + B_y^2)}{\mu_0}[U_n] + G\left(\frac{G^2}{\rho} - \frac{B_n^2}{\mu_0}\right)[U_n] = -\left(\frac{G^2}{\rho} - \frac{B_n^2}{\mu_0}\right)[p] \tag{10.70}$$

If we now eliminate $[U_n]$ by using the continuity relation $G[1/\rho] = [U_n]$, (10.70) yields

$$G^2\left(\frac{G^2}{\rho} - \frac{B^2}{\mu_0}\right)\left[\frac{1}{\rho}\right] = \left(\frac{G^2}{\rho} - \frac{B_n^2}{\mu_0}\right)[p] \tag{10.71}$$

where $B^2 = B_x^2 + B_y^2 + B_n^2$. Rewrite this equation and obtain

$$\rho G^4 + G^2\left(\frac{[p]1/\rho}{[1/\rho]} - \frac{B^2}{\mu_0}\right) - \frac{B_n^2}{\mu_0}\frac{[p]}{[1/\rho]} = 0 \tag{10.72}$$

Note that the $(G^2/\rho - B_n^2/\mu_0) = 0$ case is disregarded, since this corresponds to the rotational discontinuity.

Consider now the propagation of compressional waves. For perturbation of small-amplitude,

$$[p] = \delta p \tag{10.73}$$

$$\left[\frac{1}{\rho}\right] = \delta\left(\frac{1}{\rho}\right) = -\frac{\delta\rho}{\rho^2} \tag{10.74}$$

$$\begin{aligned}\frac{[p]}{[1/\rho]} &= -\frac{\rho^2\delta p}{\delta\rho} \\ &= -\rho^2 C_s^2\end{aligned} \tag{10.75}$$

where C_s^2 is the local speed of sound. Using these relations, (10.72) can be rewritten as

$$\rho^2 U^2\left(\rho U^2 - \frac{B^2}{\mu_0}\right) + \rho^2 C_s^2\left(\rho U^2 - \frac{B_n^2}{\mu_0}\right) = 0 \tag{10.76}$$

Now let $V_A^2 = B^2/\mu_o\rho$ and $V_{An}^2 = B_n^2/\mu_0\rho$ and transform this equation into

$$U^4 - (C_s^2 + U_A^2)U^2 + C_s^2 V_{An}^2 = 0 \tag{10.77}$$

This equation is equivalent to the MHD wave equation derived earlier (9.160). For perpendicular shocks, $G \neq 0$ and $B_n = 0$, and for $\theta = 90°$,

$$U^2 = (C_s^2 + U_A^2) \tag{10.78}$$

which is the speed of the fast mode magnetosonic wave. Note also that for $\theta = 90°$, $U_{slow} = U_{intermediate} = 0$. Hence perpendicular shocks involve only the fast mode waves. This derivation has assumed small-amplitude waves.

10.4.3 Oblique Shocks ($G \neq 0$, $B_n \neq 0$)

Oblique shocks refer to a geometry in which there is mass flow across the shock ($G \neq 0$) and the magnetic field has a component in the direction parallel to the shock normal, $B_n \neq 0$. Choose a straight geometry and let $\mathbf{U}_1 = (U_x, 0, 0)$ in the front of the shock (Figure 10.7). Let $\mathbf{B}$ be the magnetic field in the front of the shock and B_1 its x component. Since $[B_n] = 0$, B_x is the same on both sides of the shock. The magnetic field behind the shock is rotated and has a component in the y-direction $\mathbf{B}_2 = (B_x, B_y, 0)$, generated in the shock by a current $\mathbf{J}$ in the z-direction.

Since $\mathbf{U}_1$ is strictly in the x-direction, an electric field cannot exist in the y- or z-directions in front of the shock (otherwise it is not possible for

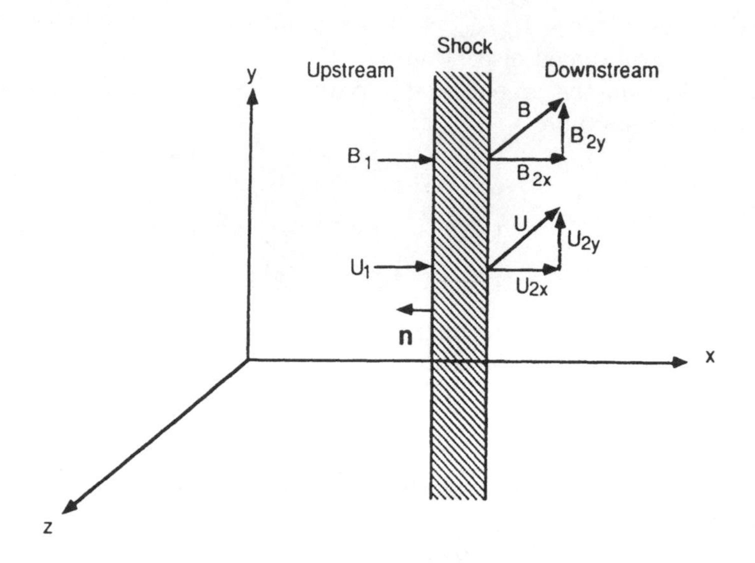

FIGURE 10.7 Geometry of plane oblique shocks.

$\mathbf{U}_1$ to be purely in the x-direction). We cannot have an electric field in the y- or z-direction behind the shock either since $[\mathbf{n} \times \mathbf{E}] = 0$. Moreover, there cannot be an electric field in the x-direction because any charge separation along $\mathbf{B}$ will be neutralized by a flow of current. Hence this model does not include any electric field. $\mathbf{U}_2$ must then be parallel to $\mathbf{B}$ behind the shock, and it also has a y-component, $\mathbf{U}_2 = (U_x, U_y, 0)$. The deflection of the flow is produced by the $\mathbf{J} \times \mathbf{B}$ force where $\mathbf{J}$ is in the z-direction.

Mass conservation requires

$$\rho_1 U_1 = \rho_2 U_{2x} \tag{10.79}$$

and since $\mathbf{U}_2$ is parallel to $\mathbf{B}_2$, we can also write

$$\begin{aligned} \frac{U_{2x}}{U_{2y}} &= \frac{B_{2x}}{B_{2y}} \\ &= \frac{B_1}{B_{2y}} \end{aligned} \tag{10.80}$$

Momentum conservation is expressed by two equations

$$\begin{aligned} (\pi_{xx} + T_{xx})_1 &= (\pi_{xx} + T_{xx})_2 \\ (\pi_{yx} + T_{yx})_1 &= (\pi_{yx} + T_{yx})_2 \end{aligned} \tag{10.81}$$

Note that the components π_{zy} and T_{zx} vanish in the model being considered. The stress tensor for the magnetic field is $T_{ij} = (B^2/2\mu_0)\delta_{ij} - B_iB_j/\mu_0$. This yields

$$(T_{xx})_1 = -\frac{(B_1)^2}{2\mu_0} \tag{10.82}$$

$$(T_{xx})_2 = \frac{(B_{2x}^2 - B_1^2)}{2\mu_0} \tag{10.83}$$

$$(T_{yx})_1 = 0 \tag{10.84}$$

$$(T_{yx})_2 = -\frac{(B_1B_{2y})}{\mu_0} \tag{10.85}$$

To derive an expression of the momentum transfer of particles, we need information on the relation between $w_\perp$ and $w_\parallel$. The argument used earlier about $w_\parallel$ not changing across the shock is not valid here because **B** changes direction in the shock. Information on the relation between $w_\perp$ and $w_\parallel$ can only be obtained by the theory of shock structure (see below). Assume without justification that $w_\perp = 2w_\parallel$ for both sides of the shock. This permits us to write

$$(\pi_{xx})_1 = \rho_1U_1^2 + p_1 \tag{10.86}$$

$$(\pi_{yx})_1 = 0 \tag{10.87}$$

$$(\pi_{xx})_2 = \rho_2U_{2x}^2 + p_2 \tag{10.88}$$

$$(\pi_{yx})_2 = \rho_2U_{2y}U_{2x} \tag{10.89}$$

where p_1 and p_2 include only particle pressures in front and behind the shock. These lead to two equations for the conservation of momentum:

$$\rho_1U_1^2 + p_1 - \frac{B_1^2}{2\mu_0} = \rho_2U_{2x}^2 + \frac{B_{2y}^2}{2\mu_0} - \frac{B_1^2}{2\mu_0} \tag{10.90}$$

$$\rho_2U_{2x}U_{2y} - \frac{B_1B_{2y}}{\mu_0} = 0 \tag{10.91}$$

Finally, note that because the electric field $E = 0$ in this model, the Poynting flux vanishes and the conservation of energy equation is

$$(q_x)_1 = (q_x)_2 \tag{10.92}$$

Using the same procedure as above to derive the expression for q, we can show that

$$\rho_1 U_1^3 + p_1 U_1 = \rho_2 U_{2x}(U_{2x}^2 + U_{2z}^2) + p_2 U_{2x} \tag{10.93}$$

Equations (10.79), (10.80), (10.90), (10.91), and (10.93) define the properties of the oblique shocks. If (10.80) and (10.91) are combined, one obtains

$$\rho_2 U_{2x}^2 = \frac{B_1^2}{\mu_0} \tag{10.94}$$

Use this equation and $p_2^* = p_2 + B_2^2/2\mu_0$ in (10.90) yields the shock velocity

$$U_1 = \left(p_2^* - p_1 + \frac{B_1^2}{\mu_0 \rho_1}\right)^{1/2} \tag{10.95}$$

Now (10.94) can be written as

$$\begin{aligned} \frac{B_1^2}{\mu_0} &= \frac{(\rho_2 U_{2x})^2}{\rho_2} \\ &= \frac{(\rho_1 U_1)^2}{\rho_2} \\ &= \left(\frac{\rho_1}{\rho_2}\right) \rho_1 U_1^2 \end{aligned} \tag{10.96}$$

Now use (10.95) to eliminate $\rho_1 U_1^2$ and, solving for the compression ratio, we obtain

$$\frac{\rho_2}{\rho_1} = 1 + \frac{\mu_0}{B_1^2}(p_2^* - p_1) \tag{10.97}$$

We must now use the energy equation to complete the formulation. In the energy equation (10.93), use (10.80) to eliminate U_{2y}^2. Next use (10.94) to eliminate U_{2x}^2. Then use (10.79) to eliminate U_{2x}. The final result is

$$\rho_1 U_1^2 + 5p_1 = \frac{\rho_1}{\rho_2}\left[\frac{(B_1^2 + B_{2y}^2)}{\mu_0} + 5p_2\right] \tag{10.98}$$

Now use (10.95) to eliminate $\rho_1 U_1^2$, (10.93) to eliminate p_2, and (10.97) to eliminate p_2^*. Then (10.98) yields a solution for the downstream magnetic field energy density as a function of the compression ratio η and other upstream parameters:

$$\frac{3}{2\mu_0} B_{2y}^2 = (\eta - 1)\left[(4 - \eta)\frac{B_1^2}{\mu_0} - 5p_1\right] \tag{10.99}$$

The expression on the right-hand side becomes negative, which is not physical, if η is very large. The largest η that is physically meaningful is obtained by setting the term in the bracket to zero. Solving then for η yields

$$\eta = 4 - \frac{5p_1\mu_0}{B_1^2} \tag{10.100}$$

The shock solution requires that $\eta > 1$. Hence shocks exist only if

$$3 > \frac{5p_1\mu_0}{B_1^2} \tag{10.101}$$

or if

$$p_1 < \frac{3B_1^2}{5\mu_0} \tag{10.102}$$

Note that, since the speed of sound $C_s^2 = \partial p_1/\partial \rho_1$ in the upstream region and the Alfvén speed $V_A^2 = B_1^2/\mu_0\rho_1$ (10.102) shows that for $\gamma = 5/3$, shock formation requires that $V_A > C_s$. If (10.102) is used in (10.97), we obtain

$$p_2 + \frac{B_2^2}{2\mu_0} \approx 4p_1 \tag{10.103}$$

which is the maximum pressure attainable in the downstream region.

B changes direction across the shock. If in general the magnetic field in the upstream region has both tangential and normal components, the magnetic field in the downstream region will also likely have both components. However, the magnitude of the tangential component will be different on the two sides because the tangential magnetic field is not conserved across the shock boundary and a current exists in the shock. If the magnetic field initially consists of only a normal component and the passage across the shock produces a tangential component, this is sometimes referred to as "switch on" shock. Analogously, "switch off" shocks refer to the converse situation, when the shock turns off a magnetic field component across the shock.

Oblique shocks can involve both slow and fast mode waves. To see this, examine the conservation relations (10.59)–(10.61), which are general. Elimination of $[U_n]$ and $[U]$ from the last two equations yields

$$[\mathbf{B}] = kG^2(\mathbf{B} - B_n\mathbf{n}) \tag{10.104}$$

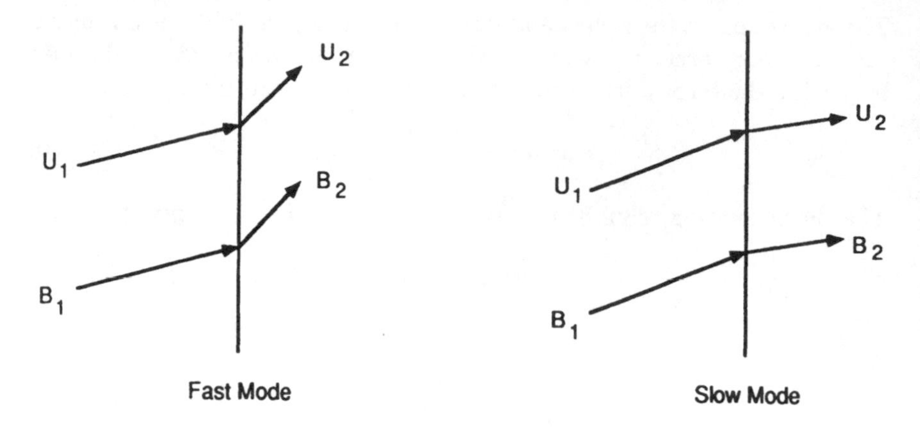

FIGURE 10.8 Rotation of flow and magnetic field across a shock for fast and slow mode waves.

as the "jump" condition for **B**. Similarly, we can show that

$$[\mathbf{U}] = kG\left(\frac{B_n}{\mu_0}\mathbf{B} - \frac{G^2}{\rho}\mathbf{n}\right) \tag{10.105}$$

$$\left[\frac{1}{\rho}\right] = -k\left(\rho U^2 - \frac{B_n^2}{\mu_0}\right) \tag{10.106}$$

Here k is a constant that is determined by the field variables on the downstream side. All shocks involve compression and therefore $[1/\rho] < 0$. Therefore, (10.106) implies that $k > 0$ for fast mode shock and $k < 0$ for slow mode shocks, since $U_{fast} > V_A$, and $U_{slow} < V_A$. It then follows from (10.104) that the magnitude $|\mathbf{B}|$ increases through a fast mode shock and decreases through a slow mode shock. Since $[B_n] = 0$, we can further say that the tangential component of **B** increases (decreases) across fast (slow) mode shocks. A similar conclusion can be made about the behavior of **U** if (10.105) is used. These results indicate that **B** is rotated away from the shock normal **n** across a fast mode shock (which gives a larger tangential component) and toward the normal across a slow mode shock (Figure 10.8). Note that in slow mode shocks, the tangential component of the magnetic field behind the shock can rotate until it changes sign.

10.5 Acceleration of Particles at Shocks

The upstream region of Earth's bow shock is populated with electrons and ions whose energies are much larger than the energies of the solar wind particles. The energies of the upstream particles range from a few hundred eV to several hundred keV. Theory shows (see below) that solar wind particles can be accelerated at the shock and a portion of the upstream particles come from these particles that propagate "backward" along $\mathbf{B}_{IMF}$, into the oncoming solar wind.

To see how a particle encountering a shock is accelerated, consider the solar wind electric field, $\mathbf{E}' = \mathbf{E} + \mathbf{v} \times \mathbf{B}$. We have shown earlier that this equation can be Lorentz transformed so that in the frame of reference moving with the solar wind, $\mathbf{E}' = 0$. However, an electric field $\mathbf{E} = -\mathbf{v} \times \mathbf{B}$ still exists in the rest frame, which is the shock frame of reference. A particle encountering a shock can thus move in the direction of this electric field and be accelerated. If this particle then escapes the shock and propagates into the upstream solar wind region, it will have more energy than before it encountered the shock.

The amount of energy gain that occurs in the interaction process can be conveniently calculated in a reference frame in which the solar wind electric field $\mathbf{E}$ vanishes. $\mathbf{E}$ vanishes if $\mathbf{v}$ is parallel to $\mathbf{B}$ and, in such a frame, the kinetic energy of the particles remains constant. We therefore seek a reference frame in which the solar wind flows parallel to $\mathbf{B}_{IMF}$. This reference frame, also known as a de Hoffmann-Teller (HT) frame, is illustrated in the top diagram of Figure 10.9, which shows the geometry of the shock region. Let $\mathbf{v}_i$ designate the guiding center of the incident solar wind particles and we decompose $\mathbf{v}_i$ into

$$\mathbf{v}_i = \mathbf{v}_{\parallel i} + \mathbf{v}_{HT} \tag{10.107}$$

where $\mathbf{v}_{\parallel i}$ is parallel to $\mathbf{B}$ and $\mathbf{v}_{HT}$ is parallel to the shock surface. $\mathbf{v}_{HT}$ slides along the shock surface and it provides the transformation of $\mathbf{v}_i$ into the HT frame. The bottom diagram shows the geometry of reflected particles. Here $\mathbf{v}_r$ represents the velocity of the reflected particles and, as in the above case of incident particles, $\mathbf{v}_r$ can be decomposed into

$$\mathbf{v}_r = \mathbf{v}_{\parallel r} + \mathbf{v}_{HT} \tag{10.108}$$

where $\mathbf{v}_{\parallel r}$ is the component of the reflected velocity along $\mathbf{B}_{IMF}$.

Independent of any reflection mechanisms (of which there are several, but we need not specify them here), the first invariant may or may not be conserved. If the reflection process conserves the magnetic moment,

$$\mathbf{v}_{\parallel r} = -\mathbf{v}_{\parallel i} \tag{10.109}$$

If the magnetic moment is not conserved, then (10.109) must be replaced with

$$\mathbf{v}_{\parallel r} = -\delta \mathbf{v}_{\parallel i} \tag{10.110}$$

where δ is a positive constant.

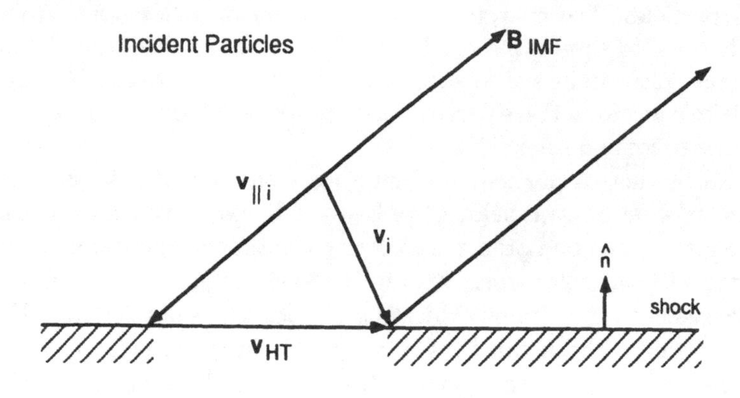

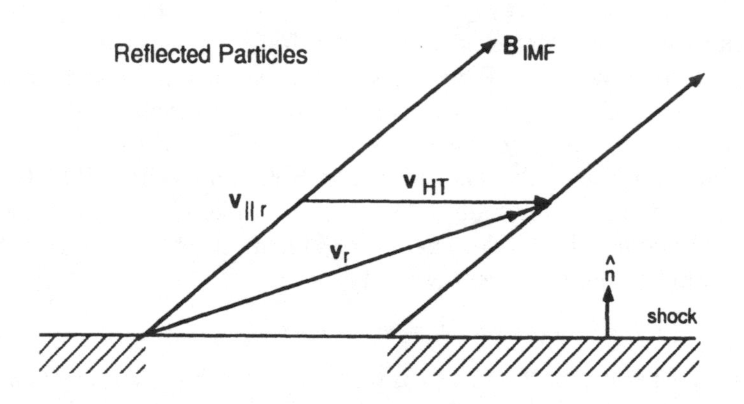

FIGURE 10.9 The top diagram shows how the guiding center of the incident solar wind particle can be decomposed into $\mathbf{v}_{\parallel i}$ parallel to $\mathbf{B}_{IMF}$, and $\mathbf{v}_{HT}$ parallel to the shock front. The bottom diagram shows a reflected particle propagating into the solar wind with $\mathbf{v}_r$ which can be decomposed into $\mathbf{v}_{\parallel r}$ along $\mathbf{B}_{IMF}$ and $\mathbf{v}_{HT}$ along the shock front.

Consider now a three-dimensional geometry (shown in Figure 10.10). The unit vector **n** is normal to the shock surface, **b** is along the IMF, and **v** is along the direction of the incident solar wind. The angle between **b** and **n** is θ_{Bn}, between **v** and **n** is θ_{vn}, and between **b** and **v** is θ_{Bv}. Note that since the vectors **n**, **b**, and **v** are not in general coplanar, θ_{Bv} is not generally equal to $\theta_{Bn} + \theta_{vn}$.

In the *HT* frame, energy is conserved, hence

$$v_{\parallel i}^2 + v_{\perp i}^2 = v_{\parallel r}^2 + v_{\perp r}^2 \tag{10.111}$$

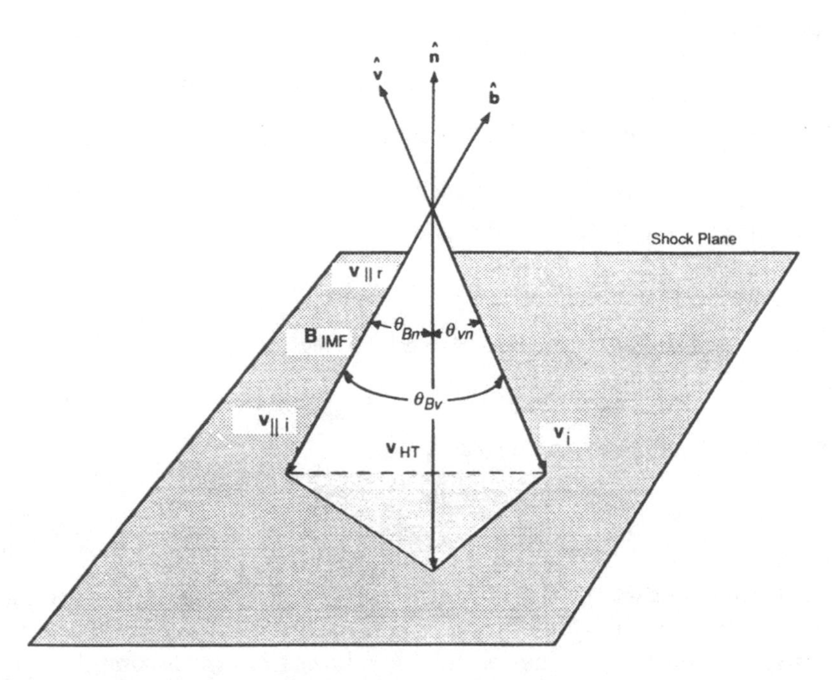

FIGURE 10.10 Same as Figure 10.9, but the velocities are now in a three-dimensional space to illustrate the more general case of solar wind incident on a shock surface. In the HT frame which is obtained by sliding along the shock surface with a velocity $\mathbf{v}_{HT}$, the solar wind electric field $\mathbf{E} = -\mathbf{v}_i \times \mathbf{B}$ vanishes.

where $v_\perp$ are velocity components perpendicular to the magnetic field. If we assume the incident solar wind is "cold," we can ignore the thermal energy and let $v_{\perp i} \approx 0$. Then, using (10.110), we find that

$$\begin{aligned} v_{\perp r}^2 &\approx v_{\parallel i}^2 + v_{\parallel r}^2 \\ &= (1-\delta) v_{\parallel i}^2 \end{aligned} \tag{10.112}$$

This equation shows that $\delta \leq 1$ for it to be physically significant. To evaluate (10.111), obtain from (10.107) and (10.108)

$$\begin{aligned} v_i^2 &= v_{\parallel i}^2 + v_{HT}^2 + 2\mathbf{v}_{\parallel i} \cdot \mathbf{v}_{HT} \\ v_r^2 &= v_{\parallel r}^2 + v_{HT}^2 + 2\mathbf{v}_{\parallel r} \cdot \mathbf{v}_{HT} \end{aligned} \tag{10.113}$$

and, subtracting the bottom equation from the top equation, using (10.110) yields

$$\begin{aligned} v_i^2 - v_r^2 &= v_{\parallel i}^2 - \delta v_{\parallel i}^2 + 2\mathbf{v}_{\parallel i} \cdot \mathbf{v}_{HT} + 2\delta \mathbf{v}_{\parallel i} \cdot \mathbf{v}_{HT} \\ &= (1-\delta^2)\mathbf{v}_{\parallel i}^2 + 2(1+\delta)\mathbf{v}_{\parallel i} \cdot \mathbf{v}_{HT} \end{aligned} \tag{10.114}$$

Now

$$\mathbf{v}_{\parallel i} \cdot \mathbf{v}_{HT} = \mathbf{v}_{\parallel i} \cdot (\mathbf{v}_i - \mathbf{v}_{\parallel i}) \tag{10.115}$$

and the scalar product is evaluated noting from Figure 10.10 that

$$v_{\parallel i} \cos\theta_{Bn} = v_i \cos\theta_{vn} \tag{10.116}$$

Hence (10.115) becomes

$$\mathbf{v}_{\parallel i} \cdot \mathbf{v}_{HT} = v_{\parallel i} v_i \cos\theta_{Bv} - v_{\parallel i}^2 \tag{10.117}$$

and inserting (10.117) into (10.114) and solving for v_r^2 yields

$$v_r^2 = v_i^2 + (1-\delta)^2 v_{\parallel i}^2 - 2(1+\delta) v_{\parallel i} v_i \cos\theta_{Bv} \tag{10.118}$$

Now using (10.110) in (10.111) yields

$$\begin{aligned} v_{\perp r}^2 &= v_{\perp i}^2 + (1-\delta^2) v_{\parallel i} \\ &\approx (1-\delta^2) v_{\parallel i} \end{aligned} \tag{10.119}$$

If $\delta = 1$, we note that the thermal velocities do not change upon reflection. If $\delta \neq 1$, a fraction $(1-\delta^2)$ of the incident energy will be converted to the reflected thermal energy. The second line follows from the assumption that the solar wind is "cold."

The incident and reflected energies of the solar wind in the shock frame (rest frame) are, respectively,

$$\epsilon_i = \frac{m}{2}(v_{\parallel i}^2 + v_{\perp i}^2) \approx \frac{m}{2} v_{\parallel i}^2 \tag{10.120}$$

$$\epsilon_r = \frac{m}{2}(v_{\parallel r}^2 + v_{\perp r}^2) \tag{10.121}$$

We use (10.118) and (10.119) in (10.120) and (10.121) and obtain

$$\frac{\epsilon_r}{\epsilon_i} = 1 + 2(1+\delta)\left(\frac{\cos^2\theta_{vn} - \cos\theta_{vn}\cos\theta_{Bn}\cos\theta_{Bv}}{\cos^2\theta_{Bn}}\right) \tag{10.122}$$

This equation gives the energy of the reflected particles as a function of the incident energy and the shock geometry. As is evident, a large range of energies for the reflected particles is obtained.

The above formulation, which requires finding a frame of reference in which the electric field vanishes, is possible only because the fields are assumed homogeneous and particles behave adiabatically. If the fields are inhomogeneous, the *HT* frame cannot be found and one must resort to full analysis of the Lorentz equation of motion. In this case, particles are still accelerated because they drift in the direction of the electric field and, in

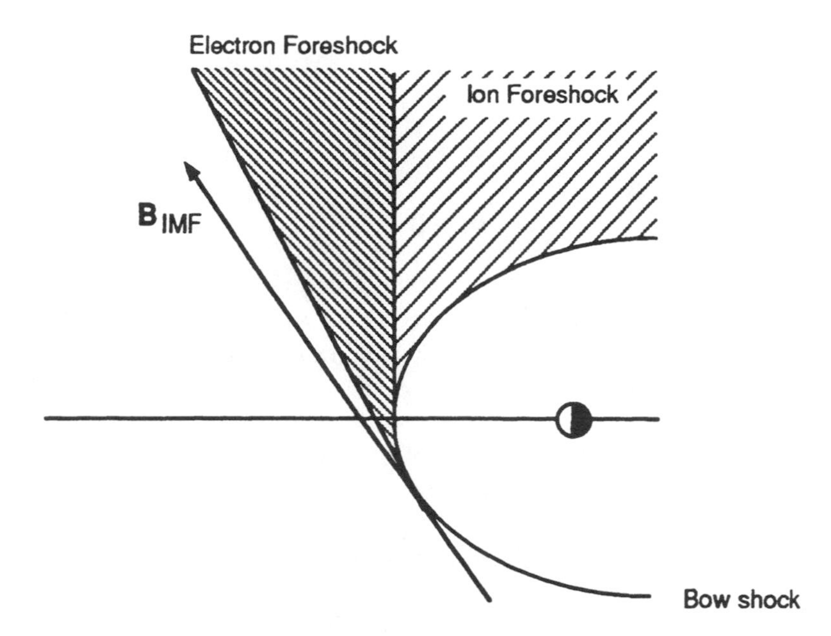

FIGURE 10.11 A schematic diagram of the foreshock region, a region in the upstream solar wind that is connected to the bow shock by the $\mathbf{B}_{IMF}$. Energetic electrons and ions can escape the shock along $\mathbf{B}_{IMF}$ and populate the foreshock region.

the limit of the guiding center approximation, this analysis yields the same result as derived above. However, this analysis can yield new results, one of which is that particle acceleration occurs even for transmitted particles.

There are other particle acceleration mechanisms. For instance, particles can be accelerated by stochastic processes because shocks in nature are turbulent and intense large-amplitude waves permeate shock regions. One stochastic process that has been studied and is applicable for high-energy particles (i.e., cosmic ray particles), invokes the well-known Fermi process, which accelerates particles by repeated encounters with the shock.

10.6 Foreshock Particle Structure and Waves

A region upstream in the solar wind that is connected to the bow shock by the $\mathbf{B}_{IMF}$ is called the foreshock region. Figure 10.11 shows a schematic diagram of the foreshock region. The accelerated particles that escape into the upstream foreshock region are propagating on $\mathbf{B}_{IMF}$ that are being convected with the solar wind. The velocity of the guiding centers of these particles is

$$\begin{aligned} \mathbf{v}_{gc} &= \mathbf{v}_{\|} + \mathbf{v}_{\perp} \\ &= \mathbf{v}_{\|} + \frac{\mathbf{E} \times \mathbf{B}}{B^2} \\ &= \mathbf{v}_{\|} + \mathbf{B} \times \frac{\mathbf{U}_{SW} \times \mathbf{B}}{B^2} \end{aligned} \tag{10.123}$$

where $\mathbf{v}_{\|}$ is the velocity of the particle along the IMF $\mathbf{B}$ (the subscript IMF has been omitted) and $\mathbf{U}_{SW}$ is the solar wind velocity. Note also that $|\mathbf{v}_{\|}| = v_T \cos\alpha$ where v_T is the total speed of the reflected particle and α is its pitch-angle, and $|\mathbf{v}_{\perp}| = U_{SW} \sin\phi$ where ϕ is the angle between $\mathbf{U}_{SW}$ and $\mathbf{B}_{IMF}$.

The guiding center velocity is not directed along $\mathbf{B}_{IMF}$ but makes an angle between $\mathbf{v}_{\|}$ and $\mathbf{B}_{IMF}$ given by

$$\begin{aligned} \tan\gamma &= \frac{|\mathbf{v}_{\perp}|}{|\mathbf{v}_{\|}|} \\ &= \frac{U_{SW} \sin\phi}{v_T \cos\alpha} \end{aligned} \tag{10.124}$$

Since $\tan\gamma$ depends inversely on $v_{\|}$, γ decreases with increasing parallel energy.

If one now assumes that particles of all energies are escaping from a point on the shock, an energy-dependent "particle pattern" will form in space (Figure 10.12). A spacecraft at a given location in space will only have access to particles whose propagation vector (10.123) intercepts the spacecraft. Given that a spacecraft is detecting particles with a certain γ, the spacecraft will also be able to detect particles with smaller γ's (higher parallel energy particles) if these particles are also escaping the shock. Particles with larger γ will not be detected because these particles are swept farther downstream of the spacecraft by the solar wind. Figure 10.13 shows an example of data obtained in the vicinity of Earth's bow shock, which can be interpreted to indicate that the foreshock region is structured with "sheets" of electrons of various energy. The data here can be interpreted in the framework of Equations (10.123) and (10.124) discussed above.

The upstream region is also rich in wave activities. Both electrostatic and electromagnetic waves permeate the shock and the upstream regions. These waves are produced by unstable non-Maxwellian plasma distributions, which are abundant in the shock and the upstream regions. We present below examples of plasma distributions and waves that have been observed in the upstream region. Our discussion will be short and qualitative and limited to observational results. We do not discuss the cause-effect relationships between unstable distributions and the associated

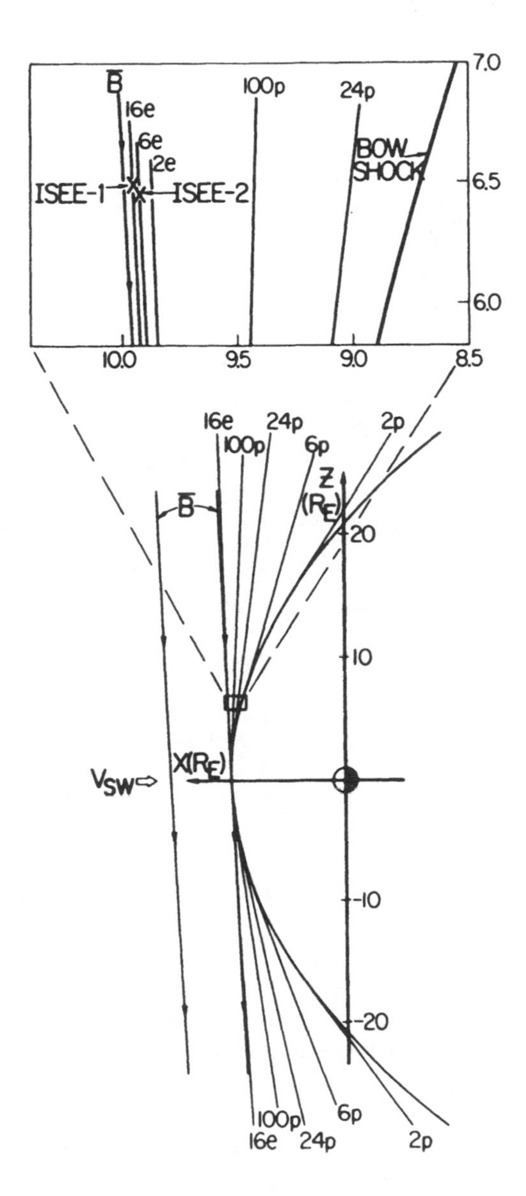

FIGURE 10.12 A plot of Equations 10.125 and 10.126 to show that "sheets" of electrons form in the upstream region. The calculation was made for the electrons that were measured on the ISEE spacecraft. (From Anderson *et al.*, 1979)

waves, which require understanding wave particle interaction and plasma instability phenomena.

Observations have shown that electron and ion distributions in the Earth's bow shock and the foreshock regions include a non-Maxwellian

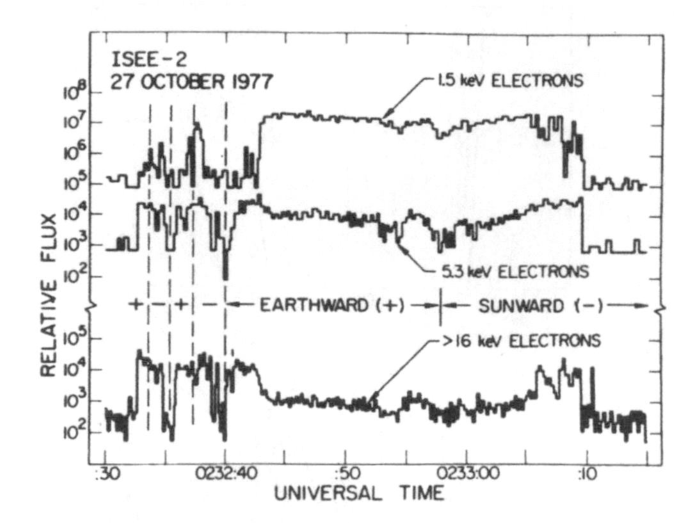

FIGURE 10.13 High-time resolution energetic electron data obtained by an experiment on a spacecraft that observed sheets of electrons. The crossings of the sheet for $>$ 16 keV electrons occurs at 0232:40, and at a later time by 5.3 keV and 1.5 keV electrons. The (+) sign (0232:40 to 0232:58) indicates the motion of the spacecraft with respect to the IMF toward Earth. The reverse motion is indicated by the (−) sign. In the time interval 0232:32 to 0232:40 UT, the motion changed direction several times indicating Earth's bow shock is very dynamic. (From Anderson *et al.*, 1979)

component. For example, two classes of ion distributions populate the foreshock region. One appears in a narrow range of pitch-angles along the $\mathbf{B}_{IMF}$ travelling away from the shock. This ion beam is identified with the reflected solar wind ion particles that have been accelerated (top diagram of Figure 10.14). The beam distribution is contrasted with another distribution that consists of particles over broad pitch-angle ranges. This population has been termed "diffuse" and an example is shown in Figure 10.14 (bottom diagram). Spacecraft have also observed "intermediate" ion distribution, which is a distribution that is not "fully" diffuse (not shown). The diffuse population is always accompanied by waves in the hydromagnetic frequencies (Figure 10.15). These MHD waves are locally produced because their group velocity is less than the solar wind flow speed. Hence, they cannot be produced at the shock and propagate into the upstream region. The correlated wave-diffuse plasma distribution has been interpreted in terms of a beam instability. It is suggested that the diffuse distribution originates from a beam distribution that becomes unstable. Waves produced by the instability pitch-angle scatter the beam particles and broaden the beam distribution resulting in the diffuse population.

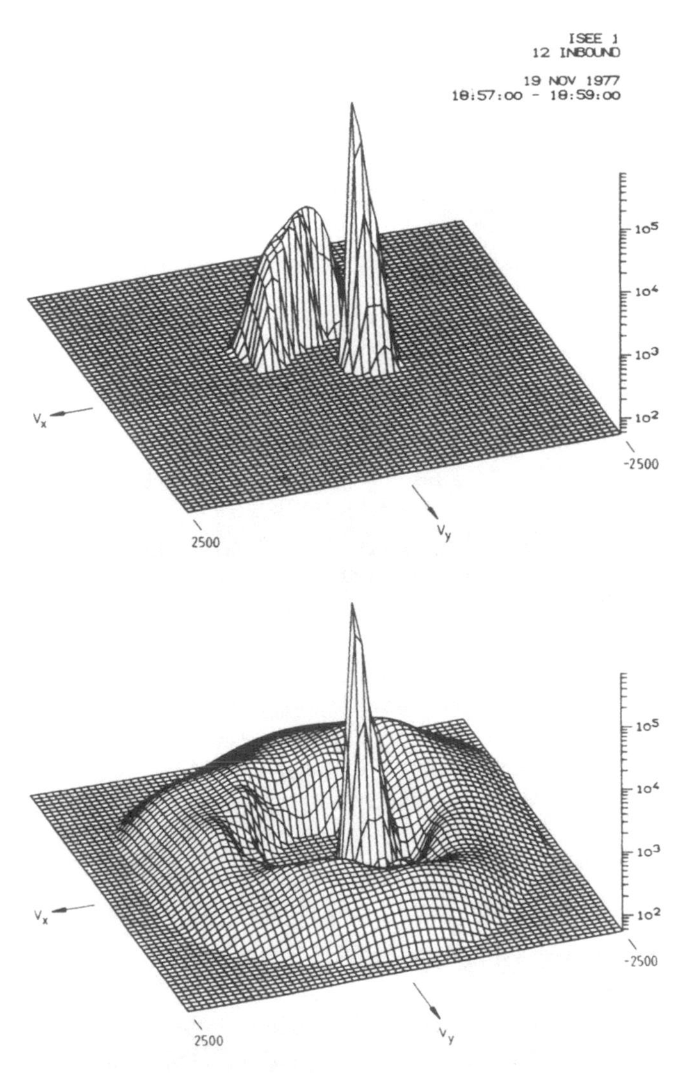

FIGURE 10.14 Ion "beam" distribution observed in the upstream solar wind region (top diagram). Note that the beam is propagating from the shock toward the solar wind. The bottom diagram shows diffuse ion distribution observed in the upstream foreshock region. The diffuse population covers all pitch-angles and is similar to a "ring" distribution. (From Paschmann *et al.*, 1981)

Higher-frequency electrostatic and electromagnetic waves are also observed. As shown in Figure 10.16, wave amplitudes are enhanced across the shock. High-time-resolution data show that detailed correlation exists between electrons and the high-frequency waves. Figure 10.17 shows one such example. Figure 10.18 shows an example of electron distribution obtained

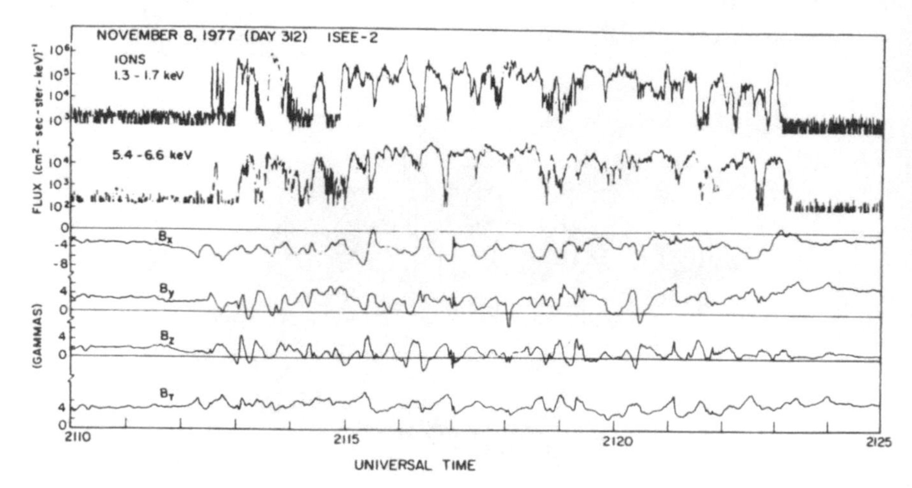

FIGURE 10.15 An example of ion flux variations that are correlated to low-frequency waves observed in magnetic field variations.

near the bow shock. As is evident, there is a "beam component" present in the region between the upstream and the downstream regions. The electrostatic waves are locally produced by unstable plasma distributions, since electrostatic waves to first order do not propagate. The presence of electromagnetic waves in the upstream region indicates that their phase velocities must be larger than the solar wind speed. Hence they are identified as whistler mode waves.

10.7 Structure of Weak MHD Shocks

We conclude this chapter with a discussion of the structure of weak MHD shocks. As shown in Figures 10.1 and 10.5, MHD shocks include a considerable amount of structure. The study of shock structure requires a two-fluid treatment. Assume for simplicity a cold plasma, so the temperature $T = 0$. Let $\partial/\partial t = 0$ and consider a plane geometry ($\partial/\partial y = \partial/\partial z = 0$) with $\mathbf{B} = (0, 0, B)$. The flow approaches the shock with $\mathbf{U} = (U_x, 0, 0)$ and so the theory to be developed applies to perpendicular shocks. Letting n_i and n_e represent the number densities of ions and electrons and $\mathbf{E} = (E_x, E_y, 0)$, $\nabla \cdot \mathbf{E} = \rho/\epsilon_0$ becomes

$$\begin{aligned} \frac{dE_x}{dx} &= \frac{e}{\epsilon_0}(n_i - n_e) \\ \frac{dE_y}{dy} &= \frac{dE_z}{dz} = 0 \end{aligned} \tag{10.125}$$

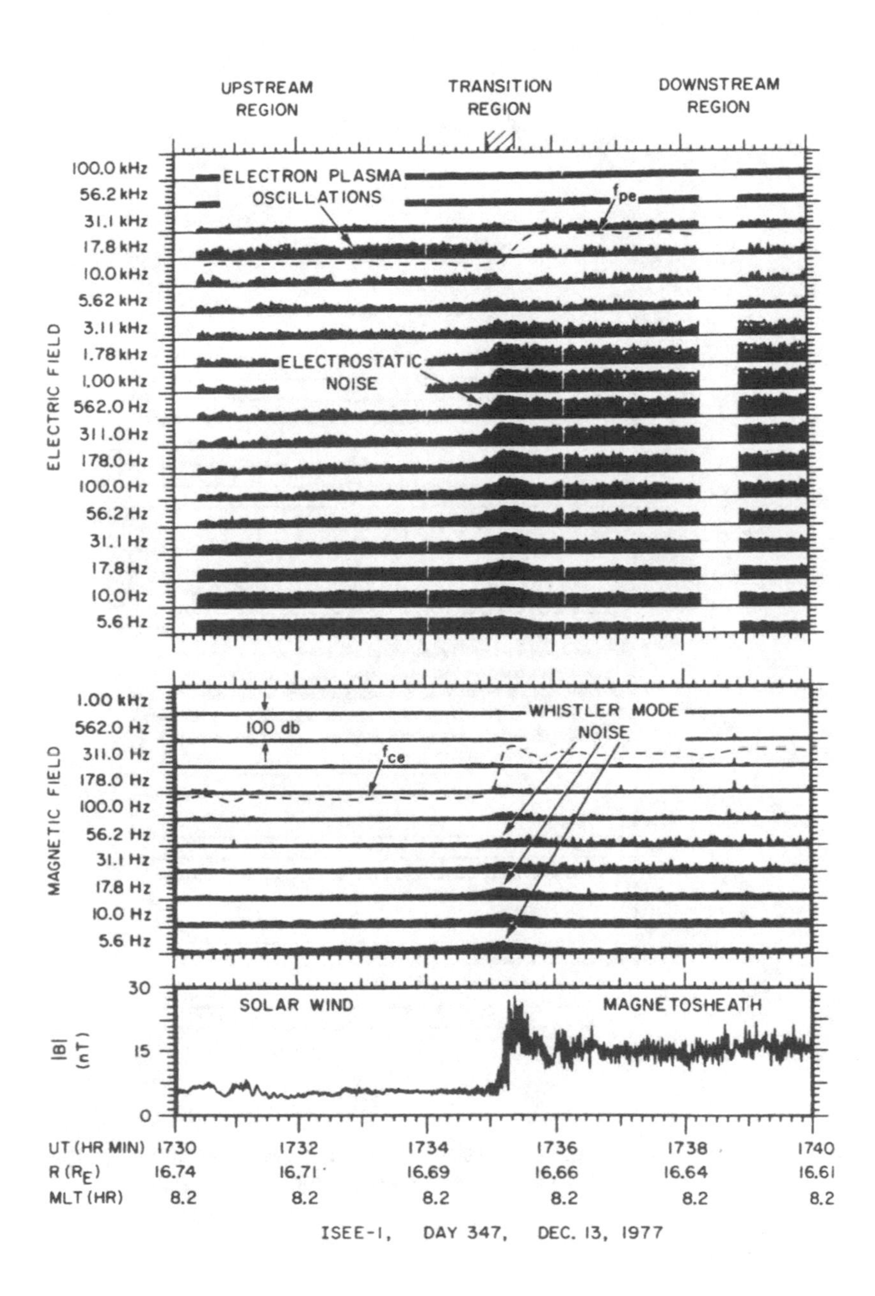

FIGURE 10.16 High-frequency electrostatic and electromagnetic waves observed in the vicinity of the Earth's bow shock. (From Gurnett, 1985)

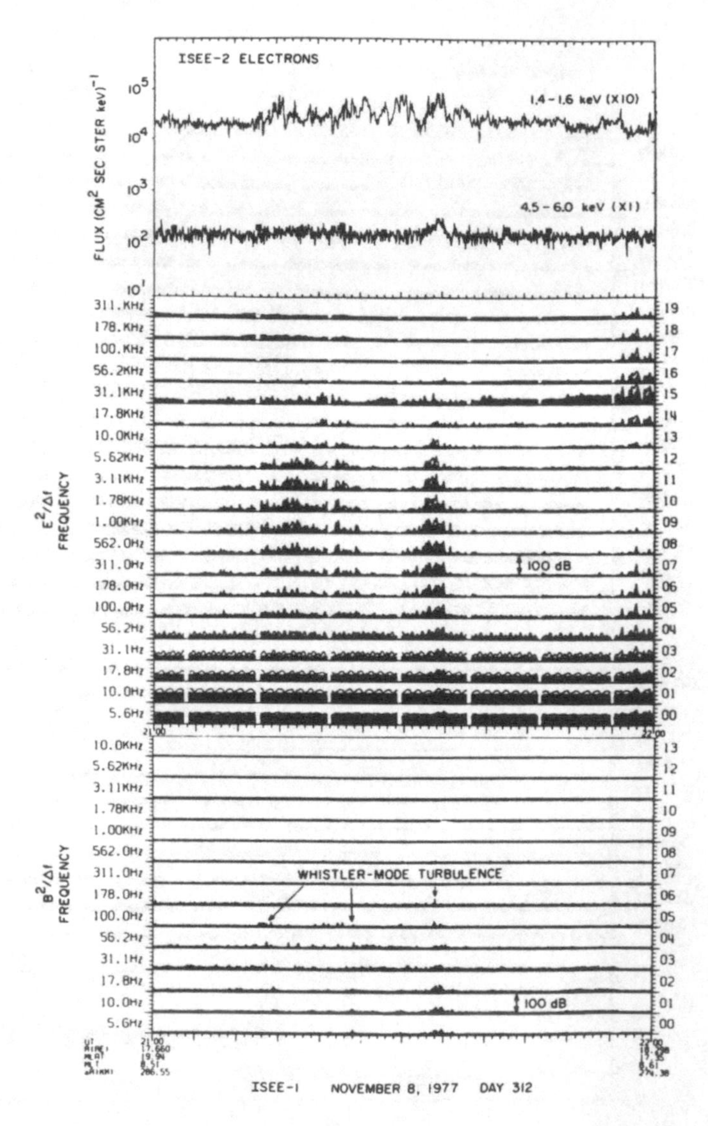

FIGURE 10.17 An example of detailed correlation observed between the keV electrons and electrostatic waves in the upstream foreshock region.

where e is positive. Integration of the y-equation across the shock toward the upstream region yields

$$E_y = U_1 B_1 \tag{10.126}$$

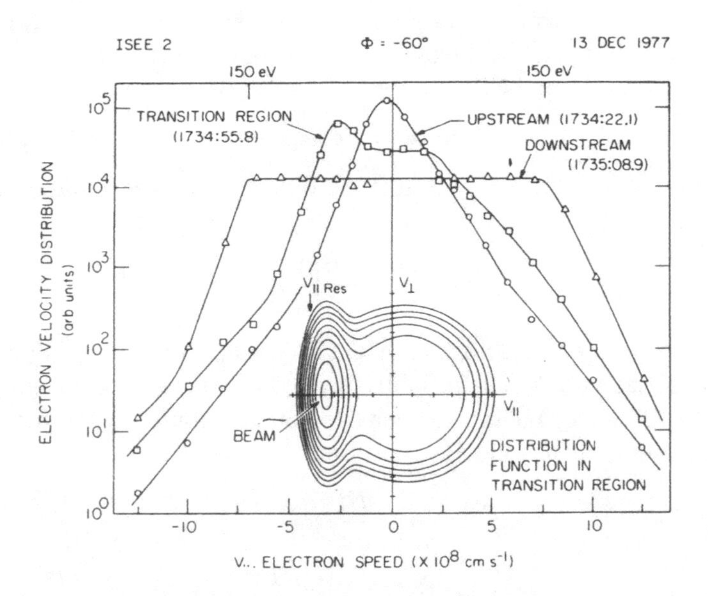

FIGURE 10.18 An example of a two-dimensional electron phase space distribution that shows the existence of a "beam" component in the region just ahead of the shock. These beams could be the source of the observed electrostatic waves. (From Feldman, 1985)

where E_y is a constant. (Here, as before, the subscript 1 denotes upstream.) Now $\nabla \times \mathbf{B} = \mu_0 \mathbf{J} = \mu_0 \sum qn\mathbf{u}$ yields

$$-\frac{dB}{dx} = \mu_0 en(u_{iy} - u_{ey}) \tag{10.127}$$

The mass continuity equation $d(nu_x)/dx = 0$ yields

$$nu_x = n_1 U_1 \tag{10.128}$$

The momentum equation is $mu_x d\mathbf{u}/dx = q(\mathbf{E} + \mathbf{u} \times \mathbf{B})$ where $q = e$ for + ions and $q = -e$ for electrons. The x and y components are

$$\begin{aligned} mu_x \frac{du_x}{dx} &= q(E_x + u_y B) \\ mu_x \frac{du_y}{dx} &= q(E_y - u_x B) \end{aligned} \tag{10.129}$$

Assume charge neutrality condition, $n_i \approx n_e = n$, which also requires that $u_{ix} \approx u_{ex} = u_x$. Now add the x-components of (10.129) for ions and electrons and obtain

$$(m_e + m_i)\frac{u_x du_x}{dx} = eB(u_{iy} - u_{ey}) \tag{10.130}$$

Subtracting the electron equation from the ion equation in (10.129) yields $E_x \approx -Bu_{ey}$ where factors involving m_e/m_i have been ignored. Addition of (10.129) for ions and electrons for the y-component yields

$$m_i u_{iy} = m_e u_{ey} \tag{10.131}$$

Now use (10.126) to (10.128) in (10.130) and integrate the equation across the shock to obtain

$$u_x = -\frac{(B^2 - B_1^2)}{2\mu_0 m_i n_1 U_1} + U_1 \tag{10.132}$$

We next proceed to obtain an equation for u_y. According to (10.131), u_{iy} is smaller than u_{ey} by the ratio m_e/m_i and therefore we will neglect this term in (10.127). This means that ions do not contribute toward producing currents. Solving for u_{ey} yields

$$\begin{aligned} u_{ey} &\approx \frac{1}{\mu_0 e n}\frac{dB}{dx} \\ &= \frac{1}{\mu_0 e n_1 U_1}\left(\frac{dB}{dx}\right) u_x \end{aligned} \tag{10.133}$$

where (10.128) was used to eliminate n. Substitute u_x from (10.132) and obtain

$$u_{ey} = \frac{1}{\mu_0 e n_1 U_1}\frac{dB}{dx}\left[1 - \frac{(B^2 - B_1^2)}{2\mu_0 m_i n_1 U_1^2}\right] \tag{10.134}$$

Now multiply (10.127) by $u_x du_x/dx$ and obtain

$$\begin{aligned} u_x \frac{d}{dx}\left(u_x \frac{dB}{dx}\right) &= \mu_0 e n u_x \frac{d}{dx}(u_x u_{ey}) \\ &= \mu_0 e n_1 U_1 \frac{du_{ey}}{dx} \end{aligned} \tag{10.135}$$

Eliminate du_{ey}/dx by using the second equation in (10.129) and u_x by using (10.132). Then (10.135) can be rewritten as

$$\begin{aligned} &\left(1 - \frac{B^2 - B_1^2}{2\mu_0 n_1 m_i U_1^2}\right)\frac{d}{dx}\left(1 - \frac{B^2 - B_1^2}{2\mu_0 n_1 m_i U_1^2}\right)\frac{dB}{dx} = \\ &\qquad \omega_p^2 (B - B_1)\left[1 - \frac{B(B + B_1)}{2\mu_0 n_1 m_i U_1^2}\right] \end{aligned} \tag{10.136}$$

where $\omega_p^2 = ne^2/m_e\epsilon_0$ is the square of the electron plasma frequency. This equation can be simplified by multiplying through by dB/dx and integrating once to obtain

$$\frac{1}{2}\left(\frac{dB}{dx}\right)^2 + \Phi(B) = 0 \tag{10.137}$$

Here

$$\Phi(B) = \frac{-(dB/dx)_1^2 - \omega_p^2(B - B_1)^2[1 - (B + B_1)^2/4\mu_0 n_1 m_i U_1^2]}{2(1 - (B^2 - B_1^2)/2\mu_0 n_1 m_i U_1^2)^2} \tag{10.138}$$

where $(dB/dx)_1$ is the value of (dB/dx) at $x = x_1$, in the upstream region where $u_x = U_1$, $B = B_1$, $n = n_1$, and $(du_{ey}/dx)_1 = 0$. Insight can be gained by a change in the variables $B \to v$, $x \to t$, and $\Phi \to V$. Then (10.137) becomes

$$\frac{1}{2}\left(\frac{dv}{dt}\right)^2 + V = 0 \tag{10.139}$$

which describes the motion of a particle in a potential V. We can thus obtain information on Φ in the same way one studies particle motion in a potential field.

For a shock in a uniform flow, $(dB/dx)_1$ must vanish. Hence (10.138) shows that in the upstream region where $B = B_1$, $\Phi = 0$ also. The shape of the potential function is shown in Figure 10.19. Here, like a particle entering the potential at x_1, the 'motion' begins at B_1 and goes over the turning point B_2 on the right (where the magnetic field is a maximum) and then returns back to B_1. Since the final state is identical to the initial state, this solution does not represent a shock. This solution corresponds to a pressure pulse travelling in the fluid and is called a "soliton."

At the turning point where the field is a maximum, $(dB/dx)_2 = 0$ and therefore Φ again vanishes. Then (10.138) shows that

$$U_1 = \frac{V_{A1}}{2}\left(1 + \frac{B_2}{B_1}\right) \tag{10.140}$$

where $V_{A1} = B_1/(\mu_0 m_i n_1)$ is the Alfvén speed in the upstream region. In order for a solution to exist at B_2, u_x must be positive and greater than zero. From (10.132) we require

$$U_1 - \frac{(B^2 - B_1^2)}{2\mu_0 n_1 m_i U_1} > 0 \tag{10.141}$$

or by using (10.140), we obtain

$$(B_2^2 + B_1)^2 > 2(B^2 - B_1^2) \tag{10.142}$$

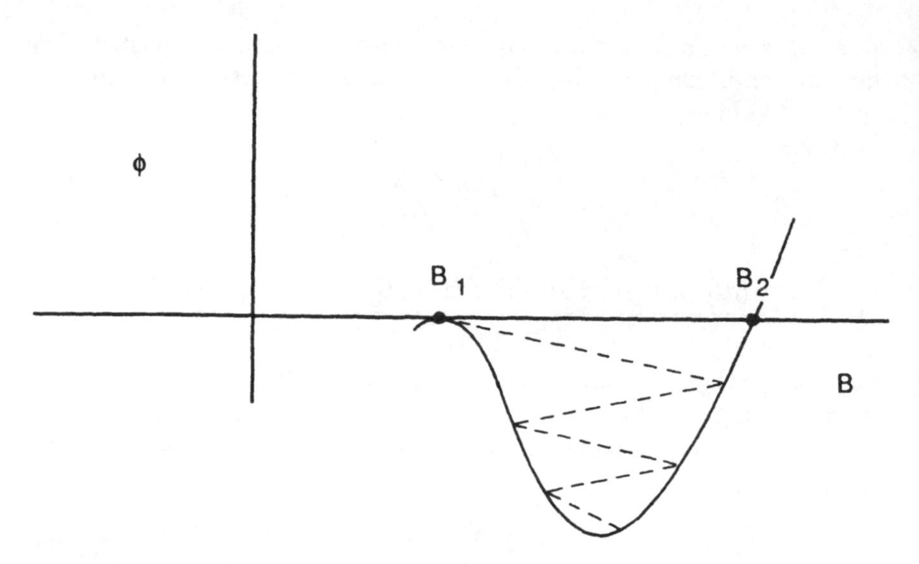

FIGURE 10.19 Shape of electrostatic potential across a weak perpendicular shock.

Hence we find

$$B_2 < 3B_1 \tag{10.143}$$

and from (10.140) we see

$$\begin{aligned} M_A &= \frac{U_1}{V_{A1}} \\ &< 2 \end{aligned} \tag{10.144}$$

where M_A is the Alfvén Mach number. This discussion shows the maximum speed a soliton can achieve is less than $2V_{A1}$. Higher values than this cannot be achieved, since $u_x > 0$ cannot be maintained. Particles are then reflected and $M_A = 2$ is called the critical Mach number (the actual critical Mach number depends on the shock model and can vary between ≈ 2 and 3). Note that these conclusions are not rigorous because once energy dissipation is introduced, the shape of the potential, (10.138), will evolve in time (see Biskamp, 1973, cited at the end of the chapter).

A shock solution requires dissipation of energy. Since collisions are absent in collisionless plasmas, dissipation here involves waves produced by the unstable plasmas in the shock. These waves interact with plasmas and they mimic collisions. One consequence of wave-particle interaction is that energy loss occurs in the potential well and the particles now execute back and forth motion, losing energy, until they settle at the minimum of the potential well. Hence the final state is different from the initial state and this corresponds to a legitimate shock solution (Figure 10.19).

A quantitative formulation of damping requires that we modify the equation of motion by including a collisional term. Assuming that ions are at rest, the term needed is νu_{ey} in (10.129) where ν is the collision frequency. This new term leads to a magnetic field equation

$$\frac{u_x}{U_1}\frac{d^2B}{dx^2} + \frac{\partial \Phi}{\partial B} + \frac{\nu}{U_1^2}u_x\frac{dB}{dx} = 0 \tag{10.145}$$

whose form is exactly the same as the equation of motion of a particle in a potential Φ damped by collisions. The shock magnetic profile will resemble a train of soliton solutions with decreasing amplitudes due to dissipation. The shock now has two scale lengths: one is the soliton width, $\approx c/\omega_p$, and the other is the damping length, U_1/ν.

The theory developed above is for cold plasma and at best is only valid for low β, low Mach number shocks. Hence the theory cannot explain energy dissipation processes that include heating of plasmas. Instabilities were mentioned without discussing any specific mechanism. Qualitatively, it can be easily seen that plasmas in the shock are unstable to a variety of wave modes because there are several sources of free energy (density and temperature gradients, for example).

In conclusion, the theory of high β shocks (not discussed here) indicates that the critical Mach number depends on θ_{Bn}, β, and T_e/T_i, in addition to M_A. Subcritical shocks are also referred to as resistive shocks because heating involving resistive mechanisms is adequate to achieve dissipation. Hot plasma theory further shows that as the critical Mach number is reached, the resistive heating mechanism alone cannot adequately dissipate the available energy and additional mechanisms are required. Supercritical shocks are said to be viscous because viscosity is needed to dissipate the energy. Here, ion reflection plays an important role because in the absence of reflected ions, only solitons (not shocks) can exist. These reflected ions appear as a "foot" in front of shocks with Mach number ≥ 2 (see Figure 10.5).

10.8 Concluding Remarks

Collisionless shocks show a variety of complex features that are not all understood. To introduce and to begin discussion of shocks, we presented simple MHD models to obtain information on macroscopic features such as how mass, momentum, and energy fluxes are transported. MHD fluid theory is not capable of explaining microscopic features such as the "foot," which requires the presence of reflected particles and a kinetic treatment.

However, useful information on the dynamics of the reflected particles was obtained by use of the guiding center theory. We also applied a two-fluid theory (assuming the plasma is cold) to learn about weak shock structures.

There has been a growth of knowledge on collisionless shocks, both observational and theoretical. In spite of these advances, many fundamental aspects of collisionless shocks still remain unknown. For example, it is not yet known how the structure of shocks depends on the upstream plasma and field parameters and how the reflected particles affect the shock structure. What reflection mechanisms operate at the shock? Are they the same for electrons and ions? Why are the structures of quasiparallel shocks so much broader than those of quasiperpendicular shocks? What instabilities are associated with the shock and upstream waves? These are some of the topics for future studies.

Acceleration of particles by shocks is an important element of space plasma phenomena. We briefly discussed how particle acceleration occurs in planetary shocks as is evidenced by the presence of tens of keV particles in the upstream region of Earth's bow shock. Energetic particles are also a part of solar flares, supernovas, and cosmic rays. In these cases, much larger and more dynamic (and possibly stronger) shocks may be involved, since particles are accelerated to tens and hundreds of meV. Understanding the dynamics of these shocks requires comprehensive shock theories that must incorporate high beta and relativistic effects.

Bibliography

Anderson, K.A., R.P. Lin, F. Martel, C.S. Lin, G.K. Parks and H. Rème, Thin Sheets of Energetic Electrons Upstream from the Earth's Bow Shock, *Geophys. Res. Lett.*, **6**, 401, 1979. This article uses high-time-resolution to show that the detection of higher-energy electrons first by a spacecraft can be interpreted in terms of a simple guiding center model in which the electrons from the bow shock propagate along $\mathbf{B}_{IMF}$ that is being convected past the spacecraft by the solar wind. See also the article by P. Filbert and P. Kellogg, *J. Geophys. Res.*, **84**, 1369, 1979.

Biskamp, D., Collisionless Shock Waves in Plasmas, *Nucl. Fusion*, **13**, 719, 1973. One of the first articles that reviewed collisionless shock research conducted during the early years.

Collisionless Shocks in the Heliosphere: A Tutorial Review, R.G. Stone and B.T. Tsurutani, eds., Geophysical Monograph, American Geophysical Union, Washington, DC, 1985. *Collisionless Shocks in the Heliosphere: Reviews of Current Research*, B.T. Tsurutani and R.G. Stone, eds., Geophysical Monograph, American Geophysical Union, Washington, DC, 1985. These two volumes present review articles of research being conducted in collisionless shocks. The articles include discussion of theory and topics of interest in current research. While some of the articles require the reader to have some knowledge about shocks, the articles are very well written and very readable. The references in the articles are fairly complete and of value to interested readers who wish to further pursue shock studies.

Feldman, W.C., Electron Velocity Distributions Near Collisionless Shocks, in *Collisionless Shocks in the Heliosphere: Reviews of Current Research*, B.T. Tsurutani and R.G. Stone, eds., Geophysical Monograph, American Geophysical Union, Washington, D.C., 1985.

Ferraro, V.C.A. and C. Plumpton, *An Introduction to Magneto-Fluid Mechanics*, 2nd ed., Clarendon Press, Oxford, England, 1966. MHD shocks are discussed in Chapter 7 of this book. Their results are incorporated into our discussion of perpendicular shocks.

Gurnett, D.A., Plasma Waves and Instabilities, in *Collisionless Shocks in the Heliosphere: Review of Current Research*, B.T. Tsurutani and R.G. Stone, eds., American Geophysical Union, Geophysical Monograph, Washington, D.C., 1985.

Hasegawa, A., *Plasma Instabilities and Nonlinear Effects*, Springer-Verlag, New York, NY, 1975. Some of the instabilities discussed in this book may contribute toward wave-particle interactions in energy dissipation processes.

Hudson, P.D., Discontinuities in an Anisotropic Plasma and their Identification in the Solar Wind, *Planet. Space Sci.*, **18**, 1611, 1970. This article develops in a systematic way the conservation equations across boundaries for anisotropic plasmas.

Landau, L.D. and E.M. Lifshitz, *Electrodynamics of Continuous Media*, Pergamon Press, Ltd., New York, NY, 1960. Chapter 8 discusses magneto-fluid dynamics including shocks. The classification scheme used by us was originally due to these authors.

Lin, R.P., C.-I. Meng and K.A. Anderson, 30-to 100-keV Protons Upstream from the Earth's Bow Shock, *J. Geophys. Res.*, **79**, 489, 1974. This article extends the observations of the upstream protons, first reported by J. Asbridge, S. Bame and I. Strong, *J. Geophys. Res.*, **73**, 5777, 1968, to high energies.

Livesey, W.A., C.F. Kennel and C.T. Russell, ISEE-1 and -2 Observations of Magnetic Field Strength Overshoots in Quasi-Perpendicular Shocks, *Geophys. Res. Lett.*, **9**, 1037, 1982. One of the first articles that reported on the observed properties of subcritical and supercritical shocks.

Longmire, C.L., *Elementary Plasma Physics*, Interscience Publishers, New York, NY, 1963. The discussion of perpendicular shocks in this chapter is based on Chapter 7 of Longmire's book.

Ness, N.F., C.S. Searce and J.B. Seek, Initial Results of the IMP 1 Magnetic Field Experiment, *J. Geophys. Res.*, **69**, 3531, 1964. One of the first results that showed clearly the collisionless Earth's bow shock.

Paschmann, G., N. Sckopke, J.R. Asbridge, S.J. Bame and J.T. Gosling, Energization of Solar Wind Ions by Reflection from the Earth's Bow Shock, *J. Geophys. Res.*, **85**, 4689, 1980. This article interprets the upstream energetic ions in terms of a reflection mechanism. The model is three-dimensional and extends the calculation originally performed by B.U.Ö. Sonnerup.

Paschmann, G., N. Sckopke, I. Papamastorakis, J.R. Asbridge, S.J. Bame and J.T. Gosling, Characteristics of Reflected and Diffuse Ions Upstream from the Earth's Bow Shock, *J. Geophys. Res.*, **86**, 4355, 1981.

Russell, C.T., Planetary Bow Shocks, in *Collisionless Shocks in the Heliosphere: Reviews of Current Research*, B.T. Tsurutani and R.G. Stone, eds., Geophysical Monograph, American Geophysical Union, Washington, DC, 1985. This article presents and reviews shocks from Mercury, Venus, Earth, Mars, and Jupiter as observed by magnetometers.

Russell, C.T., M.M. Hoppe and W.A. Livesey, Overshoots in Planetary Bow Shocks, *Nature*, **296**, 45, 1982. This article reports observations of overshoots in Earth's supercritical shocks. Figure 10.5 comes from this article.

Schwartz, S.J., M.F. Thomsen and J.T. Gosling, Ions Upstream of the Earth's Bow Shock: A Theoretical Comparison of Alternative Source Populations, *J. Geophys. Res.*, **88**, 2039, 1983. This article extends further the original calculation of Sonnerup's solar wind reflection mechanism. Consequences of various reflection mechanisms that conserve the magnetic moment and those that do not, including specularly reflected particles, are discussed in a systematic way.

Siscoe, G.L., Solar System Magnetohydrodynamics, in *Proceedings of the 1982 Boston College Theory Institute, Solar Terrestrial Physics*, R.L. Carovillano and J.M. Forbes, eds., D. Reidel Publishing Co., Dordrecht, Holland, 1982. The last part of this article discusses shocks.

Sonett, C.P. and I.J. Abrams, The Distant Geomagnetic Field, 3. Disorder and Shocks in the Magnetopause, *J. Geophys. Res.*, **68**, 1233, 1963. Preliminary discussion of magnetic field data that were interpreted in terms of shocks.

Sonnerup, B.U.Ö., Acceleration of Particles Reflected at a Shock Front, *J. Geophys. Res.*, **74**, 1301, 1969. This article shows how particles reflected from a shock can be accelerated by the solar wind electric field and offers this mechanism as an explanation for the observed energetic particles in the upstream region.

Tidman, D.A. and N.A. Krall, *Shock Waves in Collisionless Plasmas*, Interscience Publishers, New York, NY, 1971. Physics of shock structures is developed in this work. An authoritative and thorough treatment of collisionless shocks. For advanced students.

Tsurutani, B. and P. Rodriguez, Upstream Waves and Particles: An Overview of ISEE Results, *J. Geophys. Res.*, **86**, 4314, 1981. This issue includes 22 articles on Earth's bow shock and upstream foreshock observations and theories.

Wu, C.S., Physical Mechanisms for Turbulent Dissipation in Collisionless Shock Waves, *Space Sci. Rev.*, **32**, 83, 1982. This review article discusses the underlying physical processes occurring in parallel and perpendicular shocks.

Questions

1. Consider the electric and magnetic fields of the form (see Chapter 3)

$$\mathbf{E} = E_o\hat{\mathbf{y}}$$

$$\mathbf{B} = B_0\left(\frac{z}{L}\hat{\mathbf{x}} + \delta\hat{\mathbf{z}}\right)$$

Find the velocity of a moving frame in which the electric field vanishes.

2. The solar wind speed $\mathbf{U}_{SW}$ varies from 200 km/sec to 800 km/sec. Compute and plot
   a. The Alfvén Mach number as a function of U_{SW} whose $\mathbf{B}_{IMF}$ varies from 1–50 nT.
   b. The beta of the plasma for this solar wind. Assume a thermal equilibrium and let the number density vary between $n = 10^6/\text{m}^3$–$10^7/\text{m}^3$.
   c. The magnetosonic Mach number as a function of the solar wind temperature, 1 to 100 eV.
3. Consider a perpendicular shock and use the values given in problem 10.1 as the upstream parameters. Determine the downstream
   a. Flow speed as a function of the upstream flow speed for the range of $\mathbf{B}_{IMF}$ if the compression ratio is 2.
   b. Pressure as a function of the upstream pressure for the range of magnetic field if the downstream flow speed is 50 km/sec.
4. Use Equations (10.9), (10.10), and (10.12) and
   a. Derive

$$\frac{p_2}{p_1} = \frac{(\gamma+1)U_1 - (\gamma-1)U_2}{(\gamma+1)U_2 - (\gamma-1)U_1}$$

   b. In the above equation, p_2, p_1, U_1, and U_2 are all positive. Hence this equation can be mathematically satisfied if the numerator and the denominator are both positive or negative. Show, however, that if $\gamma > 1$, both cannot be negative.
   c. Show that if $\gamma > 1$, U_1/U_2 must satisfy the relation

$$\frac{\gamma-1}{\gamma+1} < \frac{U_1}{U_2} < \frac{\gamma+1}{\gamma-1}$$

5. The equation of the ideal gas $p = \rho RT$ implies $T_2/T_1 = p_2\rho_1/p_1\rho_2$. Use this to show that Equation (10.22) is valid.
6. Verify that Equations (10.59)–(10.61) are valid.
7. Derive Equation (10.99), starting from Equation (10.93).
8. Verify that Equation (10.104) through Equation (10.106) are valid.
9. Fill in the missing steps in deriving Equation (10.138) from Equation (10.135).
10. Discuss Equation (10.122)
   a. The conditions under which $\epsilon_r/\epsilon_i = 1$.
   b. What is the significance of $\delta = 0$ and 1 in particle reflection from the shock?
   c. If $\theta_{Bn} \to 90°$, $v_{HT} \to \infty$, which is not physical. Discuss why this occurs and how one can avoid this non-physical solution.
11. Consider the special case where the solar wind is along the normal to the shock surface ($\theta_{vn} = 0$). Let the average "garden hose" angle between the magnetic field and the solar wind velocity direction (θ_{Bv}) be 45°. Plot the ratio ϵ_r/ϵ_i as a function θ_{Bn} for $-90° < \theta_{Bn} < 90°$. (Note that for a parabolic geometry, θ_{Bn} is positive on the dawn side and negative on the dusk side for $\delta = 0$, 1/2, and 1). Do the same for $\theta_{vn} = 45°$.

12. Consider the special shock geometry in which the $\mathbf{B}_{IMF}$ is tangent at the "nose" of a parabolically shaped shock front. Let the angle between the solar wind velocity $\mathbf{U}_{SW}$ and the $\mathbf{B}_{IMF}$ be $45°$.

a. Compute the angle γ for electrons of energies 0.5 keV, 5 keV, 10 keV, and 20 keV. Assume a solar wind velocity of 400 km/sec $0°$ for the pitch-angle of electrons.

b. What is the energy of the proton that will give the same angle γ of 5 keV electrons?

11

Instabilities in Space

11.1 Introduction

The MHD configurations and flow phenomena we have studied so far assume that space plasma systems are in stable equilibrium. Since all physical parameters describing stable systems are not time-dependent, the models were formulated for steady-state conditions. These models are only valid in the average sense and under very quiet solar wind conditions. However, evidence shows that plasma systems in space are often unstable. For example, solar flares and auroras represent unstable space plasma phenomena.

It is not our intention to cover the topic of instabilities in detail but to introduce the basic concepts and methods used so that we can begin to discuss how steady-state equilibrium structures become unstable. Since space is concerned primarily with large-scale phenomena, we will limit discussion to MHD instabilities only. We begin with a formal discussion of how instabilities are classified and analyzed. To quantitatively establish whether a particular system is stable or not requires that we examine how that system reacts to an arbitrary disturbance. Here the steady-state models developed in the previous chapters are a standard of reference.

We study several well-known unstable MHD configurations, the MHD analog of the Rayleigh-Taylor instability, instabilities associated with drift and Kelvin-Helmholtz waves, two-stream instability, and the "firehose" instability. Then we discuss data obtained simultaneously from interplanetary space, the radiation belt, and the geomagnetic tail to demonstrate how a magnetospheric instability works. This chapter concludes with a short discussion of future space-borne experiments (scheduled for launch during the next ten years) that have been designed to study the solar system space plasma phenomena and instabilities.

11.2 Classification of Instabilities

A variety of instabilities can occur in plasma systems. A convenient classification scheme groups the instabilities into two classes: configurational and velocity space instabilities. A configurational space instability involves systems with finite dimensions and the instability lowers the energy state by distorting the shape. These instabilities do not involve the velocity distribution of the particles and they are usually studied with fluid equations.

Velocity space instabilities, on the other hand, involve the distribution function of the particles. These are microscopic instabilities and they arise when the distribution function describing the particles departs from a Maxwellian distribution function. These instabilities are kinetic in nature and they are studied with the Boltzmann-Vlasov equations.

These plasma instabilities can be further subdivided into electrostatic and electromagnetic. An electrostatic instability involves growth of electrostatic waves, induced by the growing accumulation of charges. An electromagnetic instability involves electromagnetic waves, arising from the growing current densities in the plasma.

11.3 Methods of Instability Analysis

A method used to study equilibrium problems imagines the system to undergo a small displacement as the result of the application of an arbitrary force. If the force increases the displacement and thereby deforms the system, the system is said to be unstable. If, however, the effects of the force are damped and the system returns to the initial configuration, the system is considered stable. We can examine whether the perturbation grows or damps by studying the motion of the particles in the immediate neighborhood of an equilibrium point or by calculating the energy of the initial and final states. These two approaches are referred to as normal mode analysis and energy principle methods.

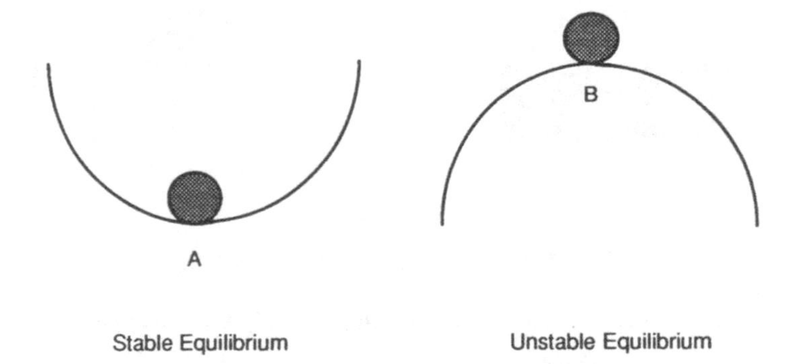

FIGURE 11.1 Two point masses in stable and unstable equilibrium.

Consider the two point masses in a one-dimensional potential field $V(x)$, as shown in Figure 11.1. The masses are initially at rest and therefore, both systems are in equilibrium states. The two states behave very differently under perturbation. A small perturbation applied to point A will cause the mass to oscillate about the equilibrium point while perturbation applied at point B will accelerate the mass away from the equilibrium point. System A is stable and system B is unstable.

Let the coordinate of the equilibrium position be given by x_0 and the force $F(x)$. The equation of motion of the mass m at position x, obtained by Taylor expansion about the point x_0, is

$$\begin{aligned} m\frac{d^2x}{dt^2} &= F(x) \\ &= F(x_0) + F'(x_0)(x - x_0) + \cdots \\ &= F'(x_0)(x - x_0) + \cdots \end{aligned} \tag{11.1}$$

where the prime $'$ denotes the time derivative and, since x_0 is the equilibrium position, $F(x_0) = 0$. We have also ignored higher-order terms, since only small perturbations are considered. Let the displacement be defined by $\xi = x - x_0$. Then the above equation becomes

$$m\frac{d^2\xi}{dt^2} = F'(x_0)\xi \tag{11.2}$$

The solution of this equation is

$$\begin{aligned} \xi &= \xi_0 \exp\left[\frac{F'(x_0)}{m}\right]^{1/2} t \\ &= \xi_0 \exp(i\omega t) \end{aligned} \tag{11.3}$$

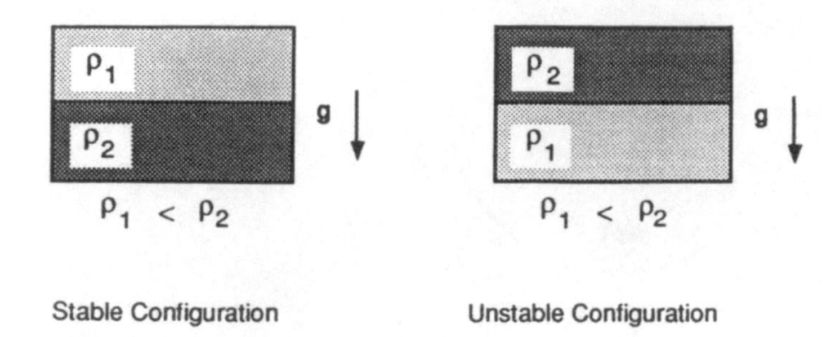

FIGURE 11.2 Vessels containing two fluids with different densities.

where $\omega^2 = -F'(x_0)/m$. For the example given above, $F'(x_0) < 0$ at A, hence ω is real and the solution is oscillatory. $F'(x_0) > 0$ at B and we see that the disturbance grows exponentially in time. Stability is characterized by $\omega^2 > 0$ and instability by $\omega^2 < 0$.

As another example, consider two vessels in which there are two kinds of fluids in a gravitational field (Figure 11.2). Let the fluid on the top in one case (left) be lighter than the one on the bottom, and let the reverse be true in the other case (right). Both systems are initially in equilibrium. Introduce now a small perturbation in the form of waves, to the interfaces of the two fluids. As before, the two systems will react differently to the perturbation. In the system on the left, the waves will oscillate about the equilibrium and will eventually damp out, while the waves in the system on the right will grow, which will lead to the interchange of the positions of the upper and lower fluids. The system on the left will return to its initial configuration, and therefore is considered stable. The system on the right will end up in a different state, and this system is considered unstable. If we ignore dissipative effects, the total energy is conserved,

$$\frac{\rho U^2}{2} + \frac{p}{\gamma - 1} + \frac{B^2}{2\mu_0} = \text{constant} \tag{11.4}$$

In the example above, the potential energy is only a function of the coordinates and the kinetic energy is velocity-dependent. The lower-energy state is reached by lowering the potential energy and increasing the kinetic energy.

11.4 MHD Instabilities

A perturbed MHD system can be studied in terms of waves that are generated by a disturbance. Small disturbances create plasma variations that can be approximated by a travelling plane wave solution of the form

$\exp i(\mathbf{k}\cdot\mathbf{r}-\omega t)$. In Chapter 9, the dispersion relations were examined only for real frequencies. However, we showed that in the case of anisotropic MHD fluids, the frequency of the wave can become complex (see 9.173). As already mentioned, complex frequencies are associated with unstable wave modes. The frequency of a wave can be generally written as $\omega = \omega_R + i\omega_I$, where ω_R and ω_I are real. Then

$$\exp i(\mathbf{k}\cdot\mathbf{r}-\omega t) = \exp i(\mathbf{k}\cdot\mathbf{r}-\omega_R t)\exp(\omega_I t) \tag{11.5}$$

This expression shows that ω_I determines whether a wave grows ($\omega_I > 0$) or decays ($\omega_I > 0$) in time.

To see how an MHD structure becomes unstable, consider an arbitrary plasma boundary across which a density gradient exists. Plasmas on both sides could be flowing or at rest. Let an external force $\mathbf{F}$, a non-magnetic constant force, act perpendicular to the boundary ($\mathbf{F}$ can be due to pressure gradient, curvature force, or gravitational force, for example). The problem is then to examine the conditions under which such configurations become unstable.

11.4.1 Rayleigh-Taylor Instability

The famous Rayleigh-Taylor configuration corresponds to the case in which $\mathbf{F}$ is the gravitational force. A magnetized fluid exists on one side and a magnetic field on the other side (upper diagram of Figure 11.3). Let the plane boundary be in the yz-plane and let there be a density gradient in the $-x$-direction. Let $\mathbf{B}_0$ be in the z-direction. We assume the plasma β is low ($\beta \ll 1$) so that we can let $kT_e \approx kT_i \approx 0$. This implies that there is no diamagnetic current and, since the plasma remains cold, we conclude that the magnetic field will remain uniform.

The system is in equilibrium. The initial static configuration is defined by the equation

$$m_i n_0(\mathbf{u}_0\cdot\nabla)\mathbf{u}_0 = q n_0 \mathbf{u}_0\times\mathbf{B}_0 + m_i\mathbf{g} \tag{11.6}$$

The $(\mathbf{u}_0\cdot\nabla)\mathbf{u}_0$ term can be ignored if $\mathbf{u}_0$ is uniform in space, which is justified if $\mathbf{F}$ is uniform and constant. Taking the cross product of (11.6) with $\mathbf{B}_0$ and solving for $\mathbf{u}_0$ yields

$$\begin{aligned}\mathbf{u}_0 &= m_i\mathbf{g}\times\frac{\mathbf{B}}{qB^2}\\ &= -\frac{g}{\Omega_c}\hat{\mathbf{y}}\end{aligned} \tag{11.7}$$

which is just the guiding center drift of ions acted on by the gravitational force. Here m_i is the ion mass and $\Omega_c = qB/m_i$ is the ion Larmor frequency. We can obtain a similar equation for the electrons that drift in the opposite

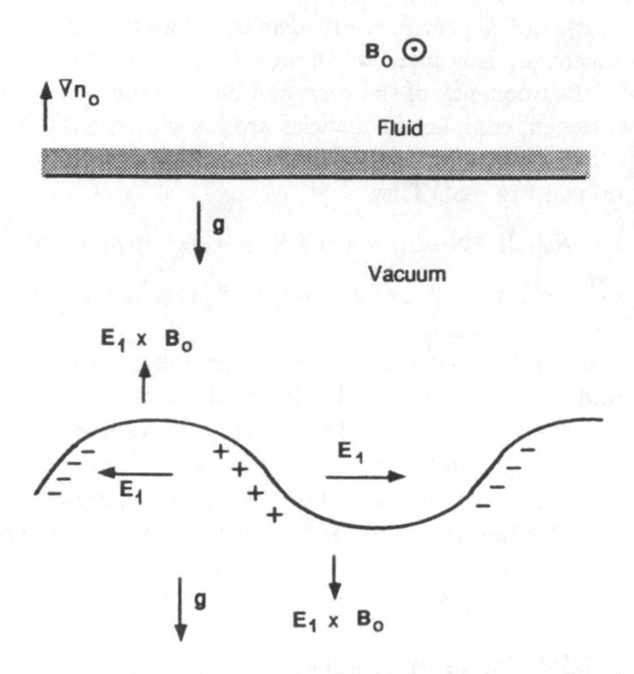

FIGURE 11.3 Fluid supported by a magnetic field in a gravitational field.

direction. However, in the limit $m_e/m_i \to 0$, the electron contribution can be ignored.

Introduce now a small disturbance so that the boundary becomes rippled (lower diagram of Figure 11.3). Because the $\mathbf{g} \times \mathbf{B}$ drift is mass-dependent, the ions will drift faster than the electrons. Hence it can be easily deduced that the drift $\mathbf{u}_0$ of the ions over the rippled surface will cause the charges to build up as shown. This charge separation produces an electric field $\mathbf{E}_1$ and since the charges change sign between the minimum and the maximum of the ripples, $\mathbf{E}_1 \times \mathbf{B}$ is in the x-direction at the minimum and in the $-x$-direction at the peaks. The amplitude of the ripple will thus grow larger and the boundary becomes unstable.

To proceed with the perturbation analysis, let $n = n_0 + n_1$ and $\mathbf{u} = \mathbf{u}_0 + \mathbf{u}_1$ where the subindices 0 and 1 denote equilibrium and perturbed quantities. The equation of motion of ions with perturbation is

$$m_i(n_0 + n_1)\frac{d}{dt}(\mathbf{u}_0 + \mathbf{u}_1) = e(n_0 + n_1)[\mathbf{E}_1 + (\mathbf{u}_0 + \mathbf{u}_1) \times \mathbf{B}_0] + m_1(n_0 + n_1)\mathbf{g} \tag{11.8}$$

Multiply (11.6) with $(1 + n_1/n_0)$ and obtain

$$m_i(n_0 + n_1)(\mathbf{u}_0 \cdot \nabla)\mathbf{u}_0 = e(n_0 + n_1)\mathbf{u}_0 \times \mathbf{B}_0 + m_i(n_0 + n_1)\mathbf{g} \quad (11.9)$$

Subtracting (11.9) from (11.8) and keeping only first-order terms yields

$$m_i n_0 \left[\frac{\partial \mathbf{u}_1}{\partial t} + (\mathbf{u}_0 \cdot \nabla)\mathbf{u}_1\right] = e n_0(\mathbf{E}_1 + \mathbf{u}_1 \times \mathbf{B}_0) \quad (11.10)$$

Note that $\mathbf{g}$ is not explicitly shown here except through $\mathbf{u}_0$.

We now use the same method as in wave analysis and let all perturbations be proportional to $\exp i(\mathbf{k} \cdot r - \omega t)$. In the example being considered, the disturbance is propagating in the y-direction. Hence let the wave vector $\mathbf{k} = k\hat{\mathbf{y}}$. Then the waves have the form $\exp i(ky - \omega t)$. Use the relations $\partial/\partial t \rightarrow -i\omega$ and $\nabla \rightarrow i\mathbf{k}$ in (11.10) and obtain

$$m_i(\omega - k u_0)\mathbf{u}_1 = ie(\mathbf{E}_1 + \mathbf{u}_1 \times \mathbf{B}_0) \quad (11.11)$$

Here $\omega - k u_0$ is the Doppler-shifted frequency of the wave. Assuming that $E_x = 0$ and that $(\omega - k u_0) \ll \Omega_c$, the solutions of (11.11) are

$$\begin{aligned} u_{ix} &= \frac{E_y}{B_0} \\ u_{iy} &= -i\frac{\omega - k u_0}{\Omega_c}\frac{E_y}{B_0} \end{aligned} \quad (11.12)$$

The first equation is the usual electric field drift. The second equation is the polarization drift in the ion frame of reference.

A similar analysis for the perturbed electron motion in the approximation $m_e/m_i \rightarrow 0$ yields to first-order

$$\begin{aligned} u_{ex} &= \frac{E_y}{B_0} \\ u_{ey} &= 0 \end{aligned} \quad (11.13)$$

To eliminate E_y from these equations, we need to examine the perturbed continuity equation of ions. Assuming charge neutrality condition $n_{i1} \approx n_{e1} = n_1$ and expansion of the continuity equation in perturbed quantities yields

$$\begin{aligned} &\frac{\partial n_1}{\partial t} + \nabla \cdot (n_0\mathbf{u}_0) + (\mathbf{u}_0 \cdot \nabla)n_1 + n_1\nabla \cdot \mathbf{u}_0 + (\mathbf{u}_1 \cdot \nabla)n_0 \\ &\qquad + n_0\nabla \cdot \mathbf{u}_1 + \nabla \cdot (n_1\mathbf{u}_1) = 0 \end{aligned} \quad (11.14)$$

The term $\nabla \cdot (n_0 \mathbf{u}_0)$ vanishes because $\mathbf{u}_0$ is perpendicular to ∇n_0 and $\nabla \cdot \mathbf{u}_0$ vanishes if $\mathbf{u}_0$ is constant (incompressible fluid). The remaining terms in the first-order approximation yield (after the plane wave analysis)

$$-i\omega n_1 + iku_0 n_1 + u_{ix} \frac{\partial n_0}{\partial x} + ikn_0 u_{iy} = 0 \tag{11.15}$$

Note that the subindex 0 is kept but 1 is now omitted from the perturbed velocity u. For electrons, the continuity equation yields

$$-i\omega n_1 + u_{ex} \frac{\partial n_0}{\partial x} = 0 \tag{11.16}$$

where we assume $\mathbf{u}_{eo} = 0$ and $u_{ey} = 0$. Combine (11.12) and (11.15) and obtain

$$(\omega - ku_0)n_1 + i\frac{E_y}{B_0}\frac{\partial n_0}{\partial x} + ikn_0 \frac{\omega - ku_0}{\Omega_c}\frac{E_y}{B_0} \tag{11.17}$$

Equations (11.13) and (11.16) yield

$$\omega n_1 + i\frac{E_y}{B_0}\frac{\partial n_0}{\partial x} = 0 \tag{11.18}$$

Hence

$$\frac{E_y}{B_0} = i\omega n_1 \left(\frac{\partial n_0}{\partial x}\right)^{-1} \tag{11.19}$$

If (11.19) is used to eliminate E_y in (11.17), we obtain

$$(\omega - ku_0)n_1 - \left(\frac{\partial n_0}{\partial x} + kn_0 \frac{\omega - ku_0}{\Omega_c}\right)\frac{\omega n_1}{(\partial n_0/\partial x)} = 0 \tag{11.20}$$

Equation (11.20) reduces to

$$\omega(\omega - ku_0) + \frac{u_0 \Omega_c}{n_0}\frac{\partial n_0}{\partial x} = 0 \tag{11.21}$$

u_0 is given by (11.7) and use of this in (11.21) yields

$$\omega^2 - ku_0\omega - \frac{g}{n_0}\frac{\partial n_0}{\partial x} \tag{11.22}$$

The solutions of this quadratic equation are

$$\omega = \frac{ku_0}{2} \pm \left[\frac{k^2 u_0^2}{4} + \frac{g}{n_0}\left(\frac{\partial n_0}{\partial x}\right)\right]^{1/2} \tag{11.23}$$

Complex ω corresponds to a growing solution and therefore the system is unstable if

$$-\frac{g}{n_0}\frac{\partial n_0}{\partial x} > \frac{k^2 u_0^2}{4} \tag{11.24}$$

Unless the density gradient $\partial n_0/\partial x$ is in the opposite direction of g, the solutions yield real ω corresponding to stable configurations.

Although we worked the above problem assuming a gravitational force for $\mathbf{F}$, it could represent as well other inertial forces. The driving mechanism is the gravitational drift. This implies that whatever the force, it must produce a drift parallel to the boundary, which causes a charge separation, which in turn produces an electric field in such a direction to give rise to growing amplitudes once perturbation is introduced. Noting this, consider a plasma configuration that, instead of a plane boundary, supports a curved boundary. This geometry is more appropriate and models space configurations such as the magnetopause. In this case, the guiding centers will be subject to curvature and gradient drifts (Chapter 4)

$$\mathbf{w} = \frac{m}{qB^4}\left(v_\parallel^2 + \frac{v_\perp^2}{2}\right)\left[\mathbf{B} \times \nabla\left(\frac{B^2}{2}\right)\right] \tag{11.25}$$

This drift is parallel to the boundary and the charge dependence of this drift gives rise to charge separation electric fields, which will make the perturbations grow when the boundary is rippled, as in the above example. Now

$$\frac{(\mathbf{B}\cdot\nabla)}{B}\frac{\mathbf{B}}{B} = -\frac{\boldsymbol{\rho}}{\rho^2} \tag{11.26}$$

where ρ is the radius of curvature and if $\nabla \times \mathbf{B} = 0$, then

$$\frac{\nabla_\perp B}{B} = -\frac{\boldsymbol{\rho}}{\rho^2} \tag{11.27}$$

Then, using the results from the gravitational example, we let

$$g \to \left(v_\parallel^2 + \frac{v_\perp^2}{2}\right)\frac{\boldsymbol{\rho}}{\rho^2} \tag{11.28}$$

to obtain the condition for the criterion of the instability when the boundary has a curved magnetic field. Whether the boundary is stable or not depends on the curvature. If $\boldsymbol{\rho}$ is in the same direction as the density gradient, the configuration is stable, while if it is in the opposite direction, the configuration is unstable. This also implies that if the magnetic field curves away from the plasma, the electric field produced by the charge-dependent drift will damp the oscillation and restore the equilibrium. On the other hand, if the magnetic field points into the plasma, the electric field produced by the drift will make the perturbations grow and the boundary is unstable (Figure 11.4). Ripples at the boundary resemble "flutes," and these instabilities are also referred to as flute instabilities.

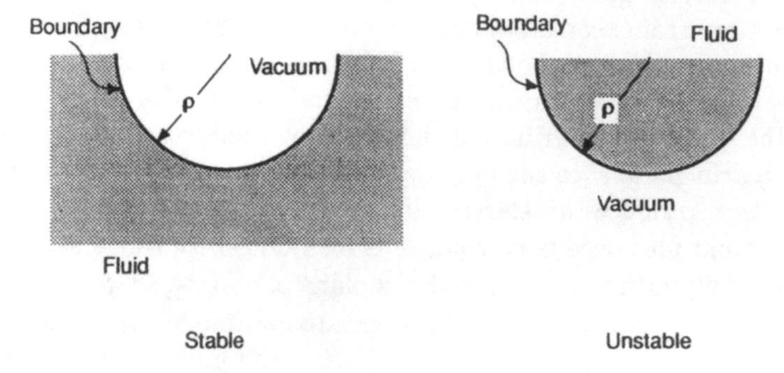

FIGURE 11.4 Stable and unstable plasma-magnetic field configurations.

11.4.2 Drift Wave Instability

Consider now the case of a plasma boundary configuration in which, unlike the previous example, the plasma is warm. The driving force **F** for this problem is due to the pressure gradient, but for simplicity let the temperature be constant. The equilibrium and perturbed configurations of this problem are shown in Figure 11.5. For a density gradient in the $-x$-direction and a magnetic field in the z-direction, the zeroth order diamagnetic drifts of ions and electrons are given by

$$\begin{aligned} \mathbf{v}_{i0} &= \frac{kT_i}{eB_0}\frac{\nabla n_0}{n_0}\hat{\mathbf{y}} \\ \mathbf{v}_{e0} &= -\frac{kT_e}{eB_0}\frac{\nabla n_0}{n_0}\hat{\mathbf{y}} \end{aligned} \tag{11.29}$$

Drift waves are essentially ion acoustic waves modified by the presence of the density gradient. Hence, there is a finite **k** along $\mathbf{B}_0$, k_z, and we let $\mathbf{k} = k_y\hat{\mathbf{y}}+k_z\hat{\mathbf{z}}$. The perturbation in this case is presented by $\exp i(k_y y+k_z z-\omega t)$.

Electrons flowing along $\mathbf{B}_0$ are thermalized and the perturbed density is given by the Boltzmann relation

$$n_{e1} \approx \frac{n_0 e\psi}{kT_e} \tag{11.30}$$

where $\mathbf{E} = -\nabla\psi$. The perturbed continuity equation is

$$\frac{\partial n_1}{\partial t} + u_{1x}\frac{\partial n_0}{\partial x} = 0 \tag{11.31}$$

where we assume that u_{1x} is uniform in the x-direction and that $k_z/k_y \ll 1$. This is the same as saying that the fluid is incompressible and we ignore the term $n_0\nabla\cdot\mathbf{u}_1$. Note that the displacement of plasma in the direction of the density gradient produces a density change. This density change occurs

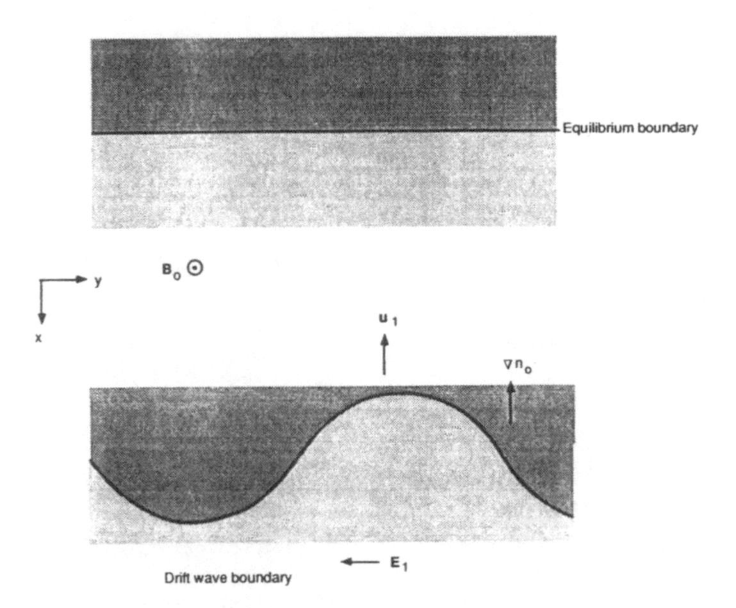

FIGURE 11.5 Plasma configuration to produce drift waves.

because the perturbation $\mathbf{E}_1 \times \mathbf{B}$ drift brings a volume of dense plasma into regions of lower equilibrium density (and vice versa). A drift wave involves the motion of the fluid back and forth in the x-direction (direction of the density gradient) while the wave travels in the y-direction.

The motion of the fluid in the x-direction is

$$\begin{aligned} u_{1x} &= \frac{E_y}{B_0} \\ &= \frac{-ik_y\psi}{B_0} \end{aligned} \tag{11.32}$$

We use (11.30) and (11.32) to rewrite (11.31) as

$$\begin{aligned} -i\omega n_1 &= \frac{i\omega n_0 e\psi}{kT_e} \\ &= ik_y \frac{\psi}{B_0} \frac{\partial n_0}{\partial x} \end{aligned} \tag{11.33}$$

Solving for the phase velocity ω/k_y yields

$$\frac{\omega}{k_y} = -\frac{kT_e}{eB_0} \frac{1}{n_0} \frac{\partial n_0}{\partial x} \tag{11.34}$$

These waves travel with the diamagnetic drift speed of electrons and they are called drift waves. They are therefore coupled to the electron drift

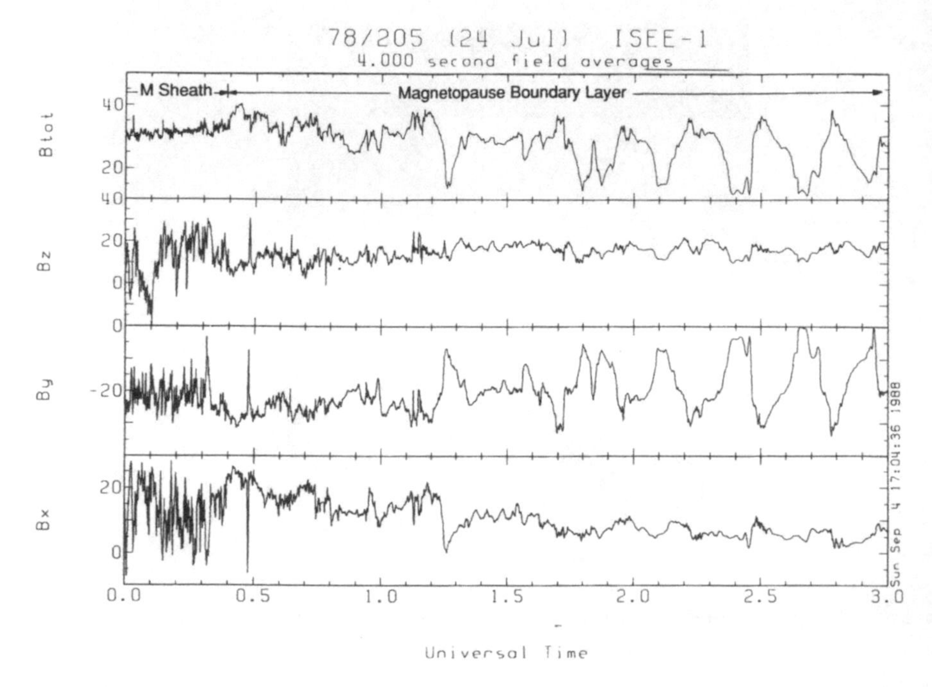

FIGURE 11.6 An example of magnetopause oscillations observed in magnetic field data. The vertical scale is intensity in nT and the horizontal scale is in hours. These waves may be unstable Kelvin-Helmholtz waves. (Courtesy of R. Elphic and C. Russell)

motion and they can grow and become unstable when energy is exchanged between the plasma and the wave. To see exactly how drift waves become unstable requires a kinetic treatment, which is beyond our scope (Ichimaru, 1973, and Krall and Trivelpiece, 1973, cited at the end of the chapter).

11.4.3 Kelvin-Helmholtz Instability

Low-frequency MHD waves that are excited at boundaries by velocity shears are called Kelvin-Helmholtz waves. A tangential discontinuity supports velocity shears and waves observed at magnetopause boundaries (Figure 11.6) may be Kelvin-Helmholtz waves. Kelvin-Helmholtz waves are low-frequency surface waves and they can grow and become unstable under certain conditions (to be studied below). Therefore, this instability is important in solar wind-planetary interactions.

Consider a boundary across which there is a sheared flow and let us assume for simplicity that the two fluids are ideal ($\sigma = \infty$), incompressible, and the pressure is isotropic. Let the subindex 0 and δ represent the equilibrium and perturbed quantities:

$$\begin{aligned}\mathbf{u} &= \mathbf{U}_0 + \delta\mathbf{U} \\ p &= p_0 + \delta p \\ \mathbf{B} &= \mathbf{B}_0 + \delta\mathbf{B}\end{aligned} \tag{11.35}$$

Erect a Cartesian coordinate system and let the z-axis be directed along the normal to the plane of the discontinuity. For a tangential discontinuity, $\mathbf{U}_0$ and $\mathbf{B}_0$ lie in the xy-plane. Let the perturbation produce waves propagating in the xy-plane but allow the waves to decay in strength away from the xy-plane, in the z-direction. Perturbed parameters are then represented by

$$\exp i(k_x x + k_y y - \omega' t) - k_z z \tag{11.36}$$

Here ω' is the Doppler-shifted frequency of the wave measured by a stationary observer in the frame of the boundary. The reciprocal of k_z corresponds to the decay length. Thus, $k_z < 0$ for $z < 0$ and $k_z > 0$ where $z > 0$. Using (11.36), $\partial/\partial t \to -i\omega'$, $\nabla \to i\mathbf{k}_t + k_z\hat{\mathbf{z}}$, $\nabla\cdot \to (i\mathbf{k}_t + k_z\hat{\mathbf{z}})$, and $\nabla\times \to (i\mathbf{k}_t + k_z\hat{\mathbf{z}})\times$, where $\mathbf{k}_t = (k_x\hat{\mathbf{x}} + k_y\hat{\mathbf{y}})$.

The MHD equations to be perturbed are

$$\frac{\partial \mathbf{B}}{\partial t} = -\nabla \times \mathbf{E} \tag{11.37}$$

$$\mathbf{E} = -\mathbf{U} \times \mathbf{B} \tag{11.38}$$

$$\rho_m \frac{d\mathbf{U}}{dt} = -\nabla p + \mathbf{J} \times \mathbf{B} \tag{11.39}$$

where we assume for simplicity a scalar pressure. The magnetic field and the current density are related by $\nabla \times \mathbf{B} = \mu_0 \mathbf{J}$. Plane wave analysis of (11.37) yields

$$\begin{aligned}-i\omega'\delta\mathbf{B} &= (\mathbf{B}\cdot\nabla)\mathbf{U} - (\mathbf{U}\cdot\nabla)\mathbf{B} \\ &= (\mathbf{B}_0\cdot\nabla)\delta\mathbf{U} - (\mathbf{U}_0\cdot\nabla)\delta\mathbf{B} \\ &= i\mathbf{k}_t\cdot(B_0\delta\mathbf{U} - U_0\delta\mathbf{B})\end{aligned} \tag{11.40}$$

Use $\omega' = \omega + \mathbf{k}\cdot\mathbf{U}_0$ and rewrite the above equation as

$$\omega\delta\mathbf{B} = -(\mathbf{B}_0\cdot\mathbf{k}_t)\delta\mathbf{U} \tag{11.41}$$

The incompressible fluid condition $\nabla\cdot\mathbf{U} = 0$ yields

$$\boldsymbol{\kappa}_\pm \cdot \delta\mathbf{U} = 0 \tag{11.42}$$

where $\boldsymbol{\kappa}_\pm = (k_x\hat{\mathbf{x}} + k_y\hat{\mathbf{y}}) \pm k_z\hat{\mathbf{z}}$ is the complex wave number and $\boldsymbol{\kappa}_+$ is for $z > 0$ and $\boldsymbol{\kappa}_-$ for $z < 0$.

The equation of motion involves $\rho_m d\mathbf{U}/dt = -i\omega\rho_m\delta\mathbf{U}$, $\nabla p = \boldsymbol{\kappa}_\pm \delta p$, and $(\nabla \times \mathbf{B}) \times \mathbf{B} = (\mathbf{B}\cdot\nabla)\mathbf{B} - \nabla B^2/2$. To first-order, the last term reduces to $(\mathbf{B}_0 \cdot \nabla)\mathbf{B}_1 + \nabla(B_0^2/2 + \mathbf{B}_0 \cdot \delta\mathbf{B}) = (\mathbf{B}_0 \cdot \boldsymbol{\kappa}_\pm)\delta\mathbf{B} - i\boldsymbol{\kappa}_\pm(\mathbf{B}_0 \cdot \delta\mathbf{B})$. If we combine these, (11.39) becomes

$$\omega\rho_m\delta\mathbf{U} - \boldsymbol{\kappa}_\pm\delta p = -\frac{(\mathbf{B}_0 \cdot \mathbf{k}_t)\delta\mathbf{B}}{\mu_0} + \frac{(\mathbf{B}_0 \cdot \delta\mathbf{B})\boldsymbol{\kappa}_\pm}{\mu_0} \tag{11.43}$$

The total pressure is $p^* = p + B^2/2\mu_0$ and, to first-order,

$$\delta p^* = \delta p + \frac{\mathbf{B}_0 \cdot \delta\mathbf{B}}{\mu_0} \tag{11.44}$$

Use of this equation in (11.43) cancels the $\boldsymbol{\kappa}_\pm(\mathbf{B}_0 \cdot \delta\mathbf{B})/\mu_0$ term. Then using (11.41) reduces (11.43) to

$$\boldsymbol{\kappa}_\pm\delta p^* = \frac{\rho_m}{\omega}\left[\omega^2 - \frac{(\mathbf{B}_0 \cdot \mathbf{k}_t)^2}{\mu_0\rho_m}\right]\delta\mathbf{U} \tag{11.45}$$

To obtain the properties of the surface waves, scalar multiply $\boldsymbol{\kappa}_\pm$ with (11.45) and, noting that $\boldsymbol{\kappa}_\pm \cdot \delta\mathbf{U} = 0$, obtain

$$\boldsymbol{\kappa}_\pm^2\delta p^* = 0 \tag{11.46}$$

This equation is satisfied by either

$$\delta p^* = 0 \tag{11.47}$$

which yields the intermediate mode Alfvén waves, $\omega/k = \pm V_A$ where $V_A = B_0\lambda/\sqrt{\mu_o\rho_m}$ (see (11.45)), or

$$\boldsymbol{\kappa}_\pm^2 = 0 \tag{11.48}$$

The second option (11.48) is specific to surface waves and yields

$$k_t^2 = k_z^2 \tag{11.49}$$

Equation (11.49) is a condition of the decay length for the perturbation away from the surface of discontinuity, $k_z^{-1} = k_t^{-1}$.

Scalar multiplying (11.45) with $\hat{\mathbf{z}}$ yields

$$\delta p_t\boldsymbol{\kappa}_\pm \cdot \hat{\mathbf{z}} = \rho_m\left[\omega^2 - \frac{(\mathbf{B}_0 \cdot \mathbf{k}_t)^2}{\mu_0\rho}\right]\delta\mathbf{U}\cdot\hat{\mathbf{z}} \tag{11.50}$$

Note that

$$\boldsymbol{\kappa}_\pm \cdot \hat{\mathbf{z}} = \pm ik_z \tag{11.51}$$

and

$$\begin{aligned}\delta\mathbf{U}\cdot\hat{\mathbf{z}} &= \frac{d\delta z}{dt}\\ &= \frac{\partial\delta z}{\partial t}+(\mathbf{U}_0\cdot\nabla)\delta z\\ &= i(\omega'-kU_0)\delta z\\ &= i\omega\delta z\end{aligned} \tag{11.52}$$

since $\omega'=\omega+\mathbf{U}_0\cdot\mathbf{k}$. Hence (11.50) becomes

$$\pm k_z\delta p^* = [\omega^2-(\mathbf{V}_A\cdot\mathbf{k})^2]\rho_m\delta z \tag{11.53}$$

where $\mathbf{V}_A=\mathbf{B}_0/(\mu_0\rho_m)^{1/2}$ is the Alfvén velocity along $\mathbf{B}_0$.

The total pressure across a tangential discontinuity is continuous, $[p^*]=0$. Also, the displacement of the two fluids in the z-direction must be continuous in order to avoid separation or interpenetration of the fluids, $[\delta z]=0$. Equation (11.53) is in the form $\delta p^*=A\delta z$. Hence $[\delta p^*]=[A\delta z]=[A][\delta z]=0$ implies $[A]=0$. With this, (11.53) written out explicitly for both sides of the boundary is

$$\rho_1[\omega_1^2-(\mathbf{V}_{A1}\cdot\mathbf{k})^2]+\rho_2[\omega_2^2-(\mathbf{V}_{A2}\cdot\mathbf{k})^2]=0 \tag{11.54}$$

The frequency measured by an observer is $\omega'=\omega_1+\mathbf{U}_1\cdot\mathbf{k}=\omega_2+\mathbf{U}_2\cdot\mathbf{k}$ and it is the same on both sides. Substitution of ω_1 and ω_2 in terms of ω' into (11.54) and solving for ω' yields

$$\begin{aligned}\omega'=\frac{\rho_1\mathbf{U}_1\cdot\mathbf{k}+\rho_2\mathbf{U}_2\cdot\mathbf{k}}{(\rho_1+\rho_2)} &\pm\frac{1}{(\rho_1+\rho_2)}\{(\rho_1+\rho_2)[\rho_1(\mathbf{V}_{A1}\cdot\mathbf{k})^2+\rho_2(\mathbf{V}_{A2}\cdot\mathbf{k})^2]\\ &-\rho_1\rho_2(\Delta\mathbf{U}\cdot\mathbf{k})^2\}^{1/2}\end{aligned} \tag{11.55}$$

where $\Delta\mathbf{U}=\mathbf{U}_1-\mathbf{U}_2$ is the flow shear and the subscripts 0 and m have been omitted from $\mathbf{U}_0$ and ρ_m.

Here we have made use of the relation $\mathbf{V}_{A1}=\mathbf{B}_1\cdot\mathbf{k}/(\mu_0\rho_1)^{1/2}$ and $\mathbf{V}_{A2}=\mathbf{B}_2\cdot\mathbf{k}/(\mu_0\rho_2)^{1/2}$. $\mathbf{B}_1$ and $\mathbf{B}_2$ are the zeroth-order magnetic fields on the two sides (the subscript 0 has been omitted). Equation (11.55) is the dispersion relation for the Kelvin-Helmholtz waves. The Kelvin-Helmholtz waves are unstable when ω' is imaginary. Equation (11.55) shows that when

$$\rho_1\rho_2(\Delta\mathbf{U}\cdot\mathbf{k})^2>\frac{1}{\mu_0}(\rho_1+\rho_2)[(\mathbf{B}_1\cdot\mathbf{k})^2+(\mathbf{B}_2\cdot\mathbf{k})^2] \tag{11.56}$$

the imaginary part of (11.55), $Im(\omega')>0$. Recall that $\mathbf{U}$, $\mathbf{B}$, and $\mathbf{k}$ all lie in the xy-plane.

A shear "threshold" is required to produce an instability. This threshold is required because tension in the magnetic field will resist any force acting on it to stretch it. Therefore, wave growth can only occur when the velocity shear overcomes the magnetic tension. Equation (11.56) can be satisfied by a variety of conditions. If $\mathbf{B}_1$ and $\mathbf{B}_2$ are perpendicular to $\mathbf{k}$, then the right-hand side vanishes and $(\Delta\mathbf{U}\cdot\mathbf{k})^2 > 0$ implies the boundary is unstable to an arbitrarily small shear across the boundary. For other orientations, the growth rate depends on the relative directions of $\mathbf{B}$, $\mathbf{k}$, and $\Delta\mathbf{U}$. Note that $(\Delta\mathbf{U}\cdot\mathbf{k})^2$ is a maximum if $\Delta\mathbf{U}$ and $\mathbf{k}$ are parallel. However, the actual magnitude depends on the relative directions of $\mathbf{B}_1$ and $\mathbf{B}_2$ with respect to $\mathbf{k}$. Let α be the angle between $\Delta\mathbf{U}$ and $\mathbf{k}$, and θ_1 and θ_2 be the angles between $\mathbf{k}$ and $\mathbf{B}_1$ and $\mathbf{B}_2$. Then the instability condition can written as

$$\Delta U^2\cos^2\alpha > \frac{1}{\mu_0}\left(\frac{1}{\rho_1}+\frac{1}{\rho_2}\right)(B_1^2\cos^2\theta_1 + B_2^2\cos^2\theta_2) \qquad (11.57)$$

We conclude the discussion of the Kelvin-Helmholtz instability by noting that since $\mathbf{B}_0$ and $\mathbf{U}_0$ are restricted to be in the xy-plane, this instability is likely to occur at the dawn-dusk flanks of the magnetopause. (For the application of the Kelvin-Helmholtz instability in the Earth's magnetopause, see Belmont and Chanteur, 1989, cited at the end of the chapter.)

11.4.4 Firehose Instability

We showed in Chapter 9 that the phase velocity of waves propagating along the direction of the magnetic field when the pressure is anisotropic is

$$\frac{\omega}{k} = V_A\left(1-\frac{p_\parallel - p_\perp}{B^2/\mu_0}\right)^{1/2} \qquad (11.58)$$

This relation shows that for low β plasma ($p_\parallel$ or $p_\perp \ll B^2/\mu_0$) or for nearly isotropic plasmas ($p_\parallel \approx p_\perp$), the phase velocity reduces to the Alfvén speed, V_A. Otherwise, the speed is larger than V_A if $p_\parallel < p_\perp$. However, note that ω becomes imaginary if $p_\parallel > p_\perp + B^2/\mu_0$. This is the "firehose" instability and the growth rate is

$$\omega_I = 1\frac{k}{\sqrt{\rho_m}}\left(p_\parallel - p_\perp - \frac{B^2}{\mu_0}\right)^{1/2} \qquad (11.59)$$

where $p_\parallel = n\int mv_\parallel^2 f d^3v$, and $p_\perp = n\int (mv_\perp^2/2) f d^3v$.

To see how this instability arises, consider a perturbation that creates slight bends in the magnetic field lines. The particles that are tied to the field lines and those that flow along $\mathbf{B}$ ($p_\parallel$ term) will now exert centrifugal force on the magnetic field. The centrifugal force drives the perturbation and is very similar to the action of water that flows in the firehose. At the same time, this action is resisted by the perpendicular pressure ($p_\perp$) and the tension of the magnetic field (B^2/μ_0), which are stabilizing factors. The intensity of the bent magnetic field is not uniform and since the particles in the perpendicular direction conserve the magnetic moment, the perpendicular energy density becomes greater where the magnetic field is more intense. Hence increased perturbation will be resisted by the local concentration of the perpendicular energy density. An instability arises when the perturbation driven by $p_{\parallel}$ cannot be balanced by the stabilizing forces of $p_{\perp}$ and B^2/μ_0.

11.4.5 Two-Stream Instability

Two-stream instability involves high-frequency plasma oscillations. Consider a plasma system in which the ions are at rest and the electrons are streaming with a velocity $\mathbf{u}_0$ relative to the ions. Let the plasma be cold for both ions and electrons, hence $kT_e = kT_i = 0$. To ignore the effects of the magnetic field, let the electrons stream along the direction of the magnetic field $\mathbf{B}$. The equations of motion for the ions and the electrons are, to first-order,

$$m_i n_0 \frac{\partial \mathbf{u}_{i1}}{\partial t} = e n_0 \mathbf{E}_1 \tag{11.60}$$

$$m_e n_0 \left[\frac{\partial \mathbf{u}_{e1}}{\partial t} + (\mathbf{u}_0 \cdot \nabla)\mathbf{u}_{e1} \right] = -e n_0 \mathbf{E}_1 \tag{11.61}$$

where the subscripts have their usual meaning. The term $(\mathbf{u}_0 \cdot \nabla)\mathbf{u}_i$ does not appear because we have assumed that $\mathbf{u}_{i0} = 0$. The $(\mathbf{u}_{e1} \cdot \nabla)\mathbf{u}_0$ term has been dropped because $\mathbf{u}_0$ is assumed to be uniform in space. $\mathbf{E}_1$ is electrostatic and it has the form $\mathbf{E}_1 = E \exp i(kz - \omega t)\mathbf{z}$ where $\mathbf{z}$ is the direction of $\mathbf{u}_0$. Performing the plane wave analysis on these equations yields

$$-i\omega m_i n_0 \mathbf{u}_{i1} = e n_0 \mathbf{E}_1 \tag{11.62}$$

$$i m_e n_0 (-\omega + k u_0)\mathbf{u}_{e1} = -e n_0 \mathbf{E}_1 \tag{11.63}$$

The perturbed ion and electron velocities obtained from these equations are

$$\begin{aligned} \mathbf{u}_{i1} &= \frac{ieE}{m_i \omega}\hat{\mathbf{z}} \\ \mathbf{u}_{e1} &= -\frac{ie}{m_e}\frac{E\hat{\mathbf{z}}}{\omega - k u_0} \end{aligned} \tag{11.64}$$

The continuity equations for the ions and electrons are

$$\frac{\partial n_{i1}}{\partial t} + n_0 \nabla \cdot \mathbf{u}_{i1} = 0 \tag{11.65}$$

$$\frac{\partial n_{e1}}{\partial t} + n_0 \nabla \cdot \mathbf{u}_{e1} + (\mathbf{u}_0 \cdot \nabla) n_{e1} = 0 \tag{11.66}$$

Our model assumes $\mathbf{u}_0 = \nabla n_0 = 0$ and, therefore, $n_{i1} \nabla \cdot \mathbf{u}_{i0} = (\mathbf{u}_{i0} \cdot \nabla) n_{i1} = (\mathbf{u}_{i1} \cdot \nabla) n_{i0} = 0$. Performing the usual plane wave analysis and solving for the perturbed densities yields

$$\begin{aligned} n_{i1} &= \frac{k n_0 u_{i1}}{\omega} \\ n_{e1} &= \frac{n_0 k}{\omega - k u_0} u_{e1} \end{aligned} \tag{11.67}$$

If we use (11.64) to eliminate these velocities, these equations become

$$\begin{aligned} n_{i1} &= \frac{i e n_0 k E}{m_i \omega^2} \\ n_{e1} &= -\frac{i e n_0 k E}{m_e (\omega - k u_0)^2} \end{aligned} \tag{11.68}$$

There will be a charge imbalance here because high frequencies are involved. The perturbed electric field and the perturbed densities are related by the Poisson equation $\epsilon_0 \nabla \cdot \mathbf{E}_1 = e(n_{i1} - n_{e1})$. With (11.68), this yields

$$-\epsilon_0 i k E = e(i e n_0 k E) \left[\frac{1}{m_i \omega^2} + \frac{1}{m_e (\omega - k u_0)^2} \right] \tag{11.69}$$

which can be rewritten as

$$1 = \frac{\omega_{p+}^2}{\omega^2} + \frac{\omega_{p-}^2}{(\omega - k u_0)^2} \tag{11.70}$$

where ω_{p+}^2 and ω_{p-}^2 are the square of the ion and electron plasma frequencies. Equation (11.70) is a dispersion relation and the equation is fourth order in ω. In principle, this dispersion equation can be solved for arbitrary values of k. The frequency will be generally complex ($\omega = \omega_R + i\omega_I$), $\omega_I > 0$ corresponds to growing solutions, $\omega_I = 0$ to solutions that remain constant in time, and $\omega_I < 0$ to damped solutions.

Equation (11.70) can be simplified by letting $m_e/m_i \to 0$. The first term can then be ignored, since $\omega_{p+}^2 \to 0$. The solution of (11.70) for this case is

$$\omega = k u_0 \pm \omega_{p-} \tag{11.71}$$

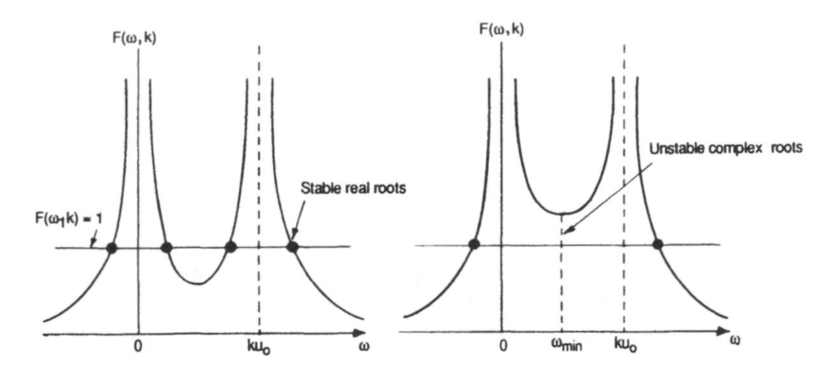

FIGURE 11.7 Plots of Equation (11.72) for stable (left curve) and unstable (right curve) modes.

If $m_e/m_i \neq 0$, then (11.70) must be examined in its entirety. A convenient way to study the dispersion equation is to display (11.70) graphically. Define the right-hand side of (11.70) as $F(\omega, k)$,

$$F(\omega, k) = \frac{\omega_{p+}^2}{\omega^2} + \frac{\omega_{p-}^2}{(\omega - ku_0)^2} \tag{11.72}$$

and plot $F(\omega, k)$ as a function of ω for given values of ku_0. In Figure 11.7, the ordinate is $F(\omega, k)$ and ω is the abscissa. Note that $F(\omega, k)$ has singularities at $\omega = 0$ and $\omega = ku_0$. The line $F(\omega, k) = 1$ gives the values satisfying the dispersion relation. Equation (11.70) yields four roots and the streams are stable when all the frequencies are real (left diagram). However, we are only guaranteed two real roots. The other two can be complex. This occurs, for example, when the minimum value of $F(\omega, k) > 1$ (right diagram). These complex roots correspond to unstable waves.

$F_{min}(\omega, k)$ is determined from (11.72) by computing $\partial F/\partial\omega = 0$. This yields

$$\omega_{min} \approx \left(\frac{m_e}{m_i}\right)^{1/3} ku_0 \tag{11.73}$$

assuming $\omega/ku_0 \gg 1$. Then inserting (11.73) into (11.72) yields

$$F_{min}(\omega, k) \approx \frac{\omega_{p+}^2}{(m_e/m_i)^{2/3}} k^2 u_0^2 + \frac{\omega_{p-}^2}{k^2 u_0^2} \tag{11.74}$$

Instability is predicted whenever $|k_u| < \omega_{p-}$. This predicts a broad range of unstable wave numbers,

$$-\frac{\omega_{p-}}{u_0} < k < \frac{\omega_{p-}}{u_0} \tag{11.75}$$

11.5 Magnetospheric Instability

Solar flares, magnetic storms, and auroras are examples of unstable plasma phenomena in our solar system. These instabilities are produced by a dynamic and complex system of interacting plasmas, magnetic fields, and elcctrical currents. Many years of observations have advanced our understanding of these phenomena, but key details concerning the specific plasma processes still remain unknown. Below, a phenomenological description is given of a sequence of events that involves the terrestrial aurora.

An instability that affects the entire magnetosphere is called a magnetospheric substorm. An aurora is a product of this substorm. Substorms involve both microscopic and macroscopic instabilities. Here we discuss only the macroscopic features. A substorm starts with a chain of interacting events beginning in the solar wind with the turning of the IMF from a northward to a southward direction. In response to this, the magnetospheric topology changes (Chapter 3) and the solar wind becomes more efficiently coupled across the magnetopuase (Chapter 8). Magnetospheric convection (Chapter 6) is then induced and plasma from the geomagnetic tail is transported closer toward the planet. Some of the particles are accelerated in this process, and a small amount of particles near the loss cone has been observed to precipitate into the ionosphere (Chapter 4). Precipitation increases ionospheric conductivity, which permits more currents to flow between the ionosphere and the magnetosphere (Chapter 7). This precipitation process grows exponentially and an intense auroral event is observed in the ionosphere, in conjunction with the "collapse" of the geomagnetic tail and injection and acceleration of the outer trapped radiation. Subsequently, the magnetosphere recovers, and the whole process repeats itself. An average-size substorm on Earth dissipates about 10^{21} ergs of energy in $\approx$ 10 to 30 minutes and it occurs once every 1 to 3 hours, depending on the level of solar wind disturbance.

The actual boundary interactions that allow the efficient entry of the solar wind energy into the magnetosphere and the mechanisms that "trigger" a substorm instability are still unknown. However, it is instructive to familiarize the reader with the general features of a substorm phenomenon by using data obtained from a variety of sources. We discuss below an event that occurred on March 4, 1979, which was observed by five spacecraft, fortuitously situated in the "right" regions of space. Additional data were also available from a high-altitude balloon-borne experiment instrumented to study the aurora, from an all-sky camera that provided pictures of the aurora from stations located in northern Scandinavia, and from auroral zone magnetometers that provided information on auroral ionospheric currents.

11.5.1 Interplanetary Magnetic Field

It has been known since the turn of the century that auroral activity is closely tied to solar activity. Space observations have since shown that an even closer connection exists between the behavior of the solar wind and the numerous magnetospheric activities. A particularly striking correlation has been found between the direction of the $\mathbf{B}_{IMF}$ and the auroral activity. Active auroras are observed predominantly when the solar wind is disturbed, when the $\mathbf{B}_{IMF}$ has a southward-pointing component.

The top curve of Figure 11.8 shows the behavior of the north-south component of the $\mathbf{B}_{IMF}$ obtained by IMP-8 and ISEE-3 spacecraft. At the time these data were taken, IMP-8 was located at (in solar magnetospheric (SM) coordinate system) $X_{SM} = -3.0R_E$, $Y_{SM} = -32E_E$, and $Z_{SM} = -15.3R_E$, and ISEE-3 was at L_1 (X_{SM} for ISEE-3 was about $230R_E$ toward the Sun). The $\mathbf{B}_{IMF}$ component turned southward (negative B_z) approximately at 2117 UT, and except for a brief northward return around 2200 UT, the $\mathbf{B}_{IMF}$ remained southward for about two hours, until around 2310 UT. The maximum intensity of the negative $\mathbf{B}_{IMF}$ reached was around $10nT$.

The negative $\mathbf{B}_{IMF}$ component evidently enhances the coupling between the solar wind and the magnetosphere. Let us now examine the behavior of data in the magnetosphere and the ionosphere to learn what particle and field effects were observed.

11.5.2 Auroral Electrojet

The global behavior of the auroral current system (electrojet) in response to changes of the $\mathbf{B}_{IMF}$ direction is shown on the bottom part of Figure 11.8. The onset of the positive magnetic activity at Fort Churchill was nearly coincident (accurate to about five minutes) with the southward turning of the $\mathbf{B}_{IMF}$. This indicates that the eastward current of the electrojet activity increased as the $\mathbf{B}_{IMF}$ became more southward. The magnetic field of this current recorded on the ground reached a maximum value of about $400nT$. Correlation between the $\mathbf{B}_{IMF}$ and the electrojet activity even extends to the short time interval centered around 2200 UT when the $\mathbf{B}_{IMF}$ momentarily turned northward.

Correlated features can also be observed between the $\mathbf{B}_{IMF}$ and the westward electrojet activity monitored by the magnetometer at Kanin-nos located near local midnight. The large negative "bay," which begins around 2237 UT, 80 minutes after the $\mathbf{B}_{IMF}$ turned southward and during the time the negative B_z of $\mathbf{B}_{IMF}$ was at maximum, signals the onset of a global auroral breakup (magnetospheric substorm). A detailed reconstruction of the electrojet activity, using magnetic records from several stations, indi-

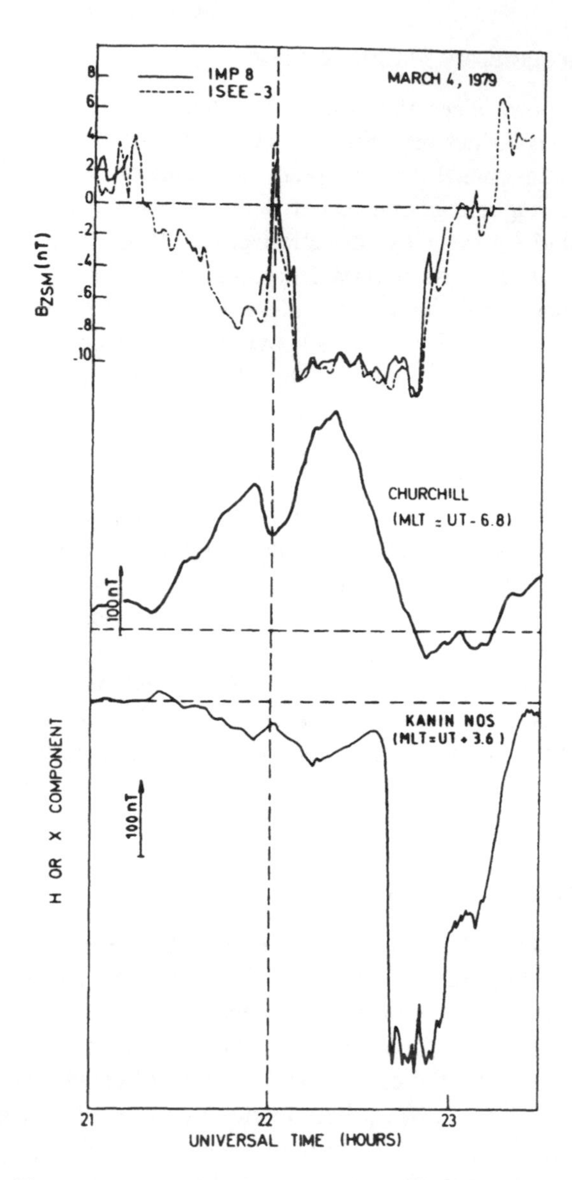

FIGURE 11.8 From top to bottom; north-south component of the $\mathbf{B}_{IMF}$ observed by IMP-8 and ISEE-3 spacecraft, in solar magnetospheric (SM) coordinates; east-west X-component of the ground auroral zone magnetogram at Fort Churchill; and the horizontal H-component of the magnetogram at Kanin-Nos. Magnetic deviations are due to enhanced ionospheric currents. (From Sauvaud *et al.*, 1987)

cates that this substorm was accompanied by a northward expansion of the electrojet (for details, see Sauvaud *et al.*, 1987, cited at the end of the chapter).

11.5.3 Outer Magnetosphere

A geostationary satellite on the equator at $6.6R_E$ around local midnight is very sensitive to topological changes that occur in association with the orientation of the $\mathbf{B}_{IMF}$. The center and bottom panels of Figure 11.9 show electron and magnetic field data obtained by experiments on the European geostationary satellite, GEOS-2, which was located near local midnight (MLT = UT + 2.4) (MLT = Magnetic Local Time) during the interval of interest. The GEOS magnetic field data show that, accompanying the southward turning of the $\mathbf{B}_{IMF}$, the magnetic topology became more tail-like at GEOS-2 (decrease of the H- and increase of the V-components). At the same time, there was evidence of a growing field-aligned current (D-component). This persisted until 2237 UT, when the tail topology collapsed. The magnetospheric topology was then more dipole-like (increase of the H-component) and this dipolarization was accompanied by an injection of energetic electrons. The dipolarization coincided with the onset of a negative bay in the auroral zone (see Figure 11.8).

11.5.4 Geomagnetic Tail

The ISEE spacecraft were in the geomagnetic tail on this day, traversing the plasma sheet at a distance of about $22.4R_E$. Figure 11.10 shows the 1.5 keV and 6 keV plasma sheet electrons from ISEE-2 and their behavior can be compared to the $\mathbf{B}_{IMF}$ and GEOS magnetic field data, shown again in the upper panel for easy reference. Also shown are the projections of ISEE and GEOS positions at 2200 UT in the solar magnetospheric equatorial plane.

Prior to about 2000 UT, the ISEE spacecraft were in the central plasma sheet. Within a few minutes of the southward turning of the $\mathbf{B}_{IMF}$ at 2117 UT, the plasma sheet became more dynamic. The electron fluxes began to decrease around 2145 UT. This has been given the interpretation that the ISEE spacecraft were moving toward the high-latitude boundary of the plasma sheet. The spacecraft between 2145 and 2237 UT were in the plasma sheet boundary layer (indicated by large fluctuations of the fluxes) and crossed into the high-latitude lobe region at 2237 UT. Recall that this time is also the time of the onset of the substorm (negative bay at Kanin-Nos).

Around 2325 UT, after the $\mathbf{B}_{IMF}$ returned to the northward direction, the plasma sheet was observed to recover. The recovery was extremely rapid: a change of two orders of magnitude in fluxes occurred in less than one minute. Here the interpretation is that a magnetic flux tube filled with particles convected rapidly over the spacecraft at speeds > 100 km/sec. Note also that the fluxes of 6 keV following the recovery were much more

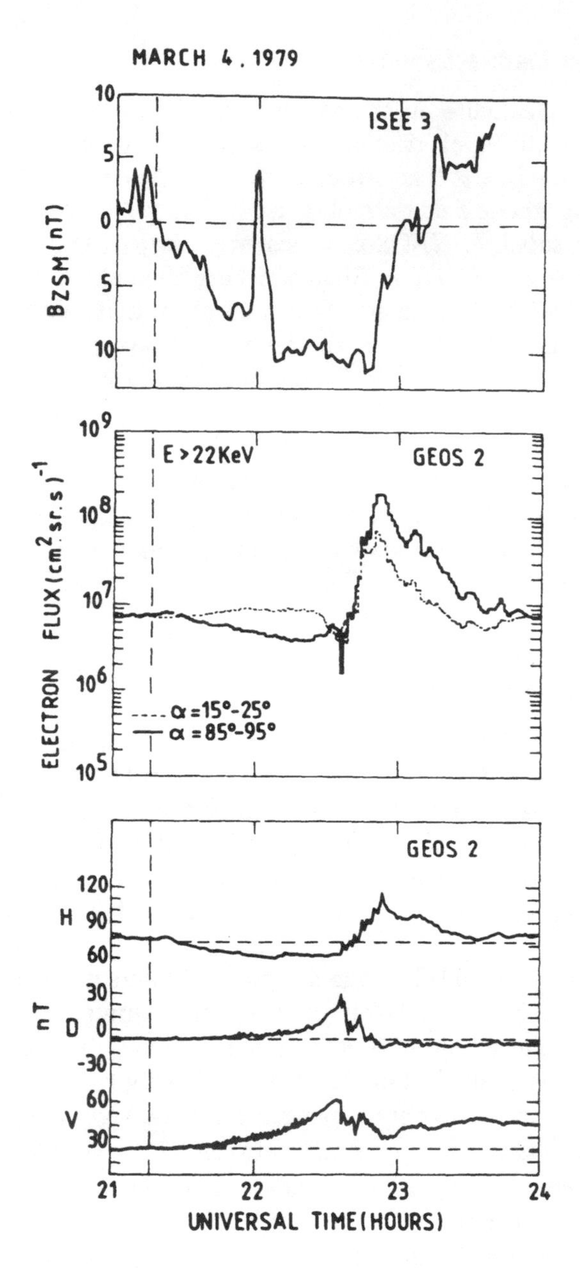

FIGURE 11.9 Top panel: $\mathbf{B}_{IMF}$ B_z component. Middle panel: Integral fluxes of electrons with energies greater than 22 keV for two pitch-angles outside the loss cone measured by GEOS-2 in the night sector. Bottom panel: Three components of the magnetic field measured on GEOS-2. H is the horizontal component and is defined positive along the nominal dipole field direction, V is the vertical component and is defined as positive toward the Earth, and D completes the third component and is defined as positive in the eastward direction. To aid the eye, dashed lines have been added to indicate the times of the southward turning of the $\mathbf{B}_{IMF}$. (From Sauvaud *et al.*, 1987)

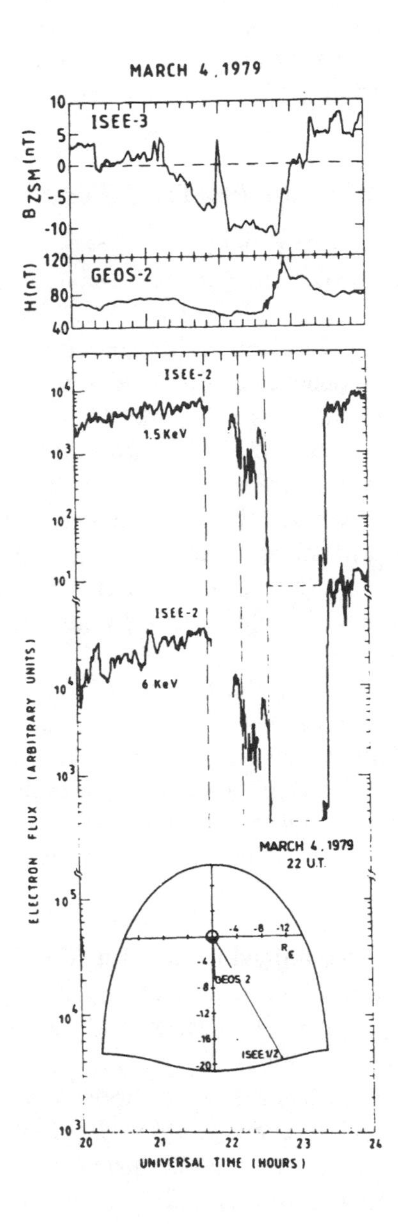

FIGURE 11.10 Comparison of data from the interplanetary medium, geostationary orbit near local midnight, and in the plasma sheet. From top to bottom: North-south component of the $\mathbf{B}_{IMF}$, H-component of the magnetic field measured at the geostationary orbit, and 1.5 and 6 keV electron fluxes measured on ISEE spacecraft. ISEE spacecraft at 2200 UT were located at a distance of about $22.4 R_E$ from the center of Earth in the tailward direction. The inset shows the projection of GEOS-2 and ISEE in the solar magnetospheric equatorial plane. (From Sauvaud *et al.*, 1987)

intense than before the spacecraft left the plasma sheet, suggesting that electrons were accelerated in the plasma sheet during the substorm.

11.5.5 Ionospheric Electric Field and Electron Precipitation

The behavior of the electric field and X-rays that were obtained by a balloon-borne experiment from northern Scandinavia, near the magnetic conjugate of GEOS-2 is shown in Figure 11.11. This figure shows the time history of the horizontal electric field (the data from the ISEE spacecraft are again shown in the top panel to facilitate comparison). Two reversals of the electric field directions were observed, one at about 2210 UT, from south to north, and the other at about 2230 UT, from north to south. Electric field amplitudes reached values as large as 60 millivolts/meter. The 2210 UT reversal coincided with the time of maximum southward $\mathbf{B}_{IMF}$ intensity and with the development of the eastward electrojet. The 2230 UT is probably associated with the pre-breakup activity. All sky camera records further indicated that these electric field reversals were correlated with motions of the aurora (not shown).

Figure 11.12 shows that the turning of the $\mathbf{B}_{IMF}$ to the southward direction induced electron precipitation, detected by means of their atmospheric bremsstrahlung X-rays (third panel). Although the observed fluxes are low, the measured fluxes are statistically significant and well above the background flux level. This low-level precipitation lasted until the onset of the expansion at 2237 UT, when very intense precipitation accompanied the injection.

The energy spectra of the original precipitating electrons responsible for the X-rays have been obtained by inverting the X-ray data (atmospheric correction and bremsstrahlung theory are used). The soft spectrum is for the electrons accompanying the injection and the hard spectrum describes the precipitation observed before the breakup. The parent electron spectra in both cases have been assumed to be exponential in form. The *e*-folding energy for the pre-breakup event is about 100 keV, which is roughly five times that of the electrons precipitated during the substorm (which is typically 20 keV). The pre-onset precipitation fluxes are interpreted as precipitation of trapped ambient electron fluxes induced by the reconfiguration of the magnetosphere.

11.5.6 Unanswered Questions

The example of correlated observations shown above demonstrates that plasma processes are active from the interplanetary space to the ionosphere. The coupling of the $\mathbf{B}_{IMF}$ with the magnetosphere and the subse-

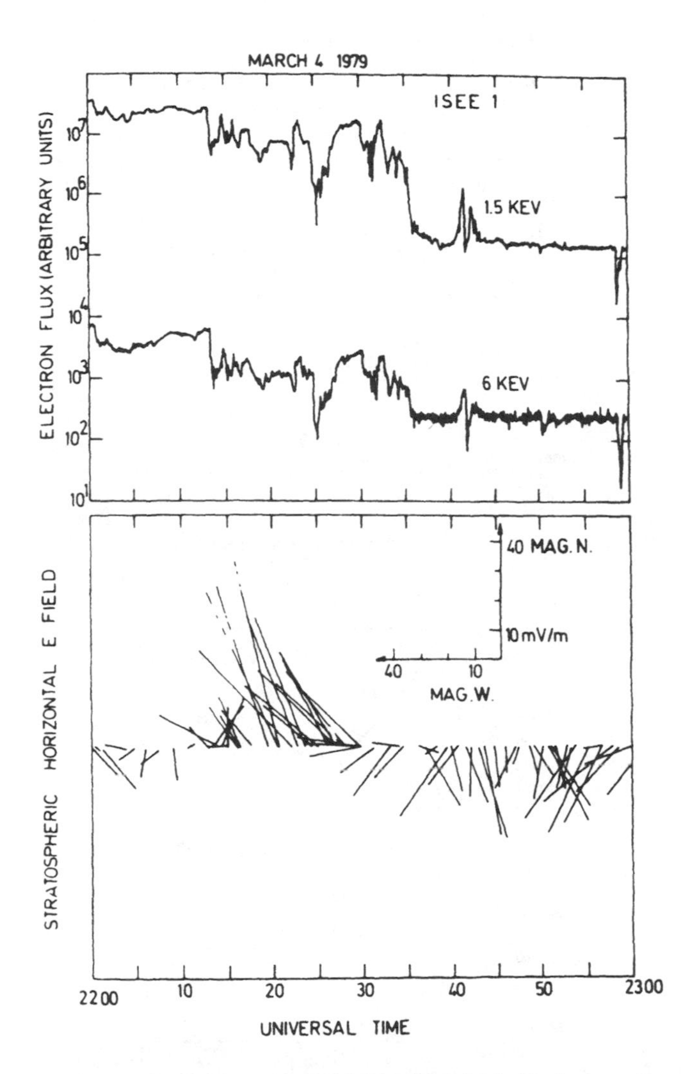

FIGURE 11.11 Comparison between the 1.5 keV and 6 keV electron fluxes measured on ISEE spacecraft against the ionospheric vector electric field measured on a high-altitude balloon. (From Sauvaud *et al.*, 1987)

quent responses observed in the geomagnetic tail, the outer magnetosphere, and the ionosphere are clearly evident. Many questions, however, remain unanswered. The large-scaled interaction between the solar wind and the magnetosphere depends on a series of plasma interactions that occur in a number of different spatial regions. Moreover, the processes appear to be highly time-dependent. How do we begin to try to understand the cause and effect relations in such a complex chain of interactive processes? In this

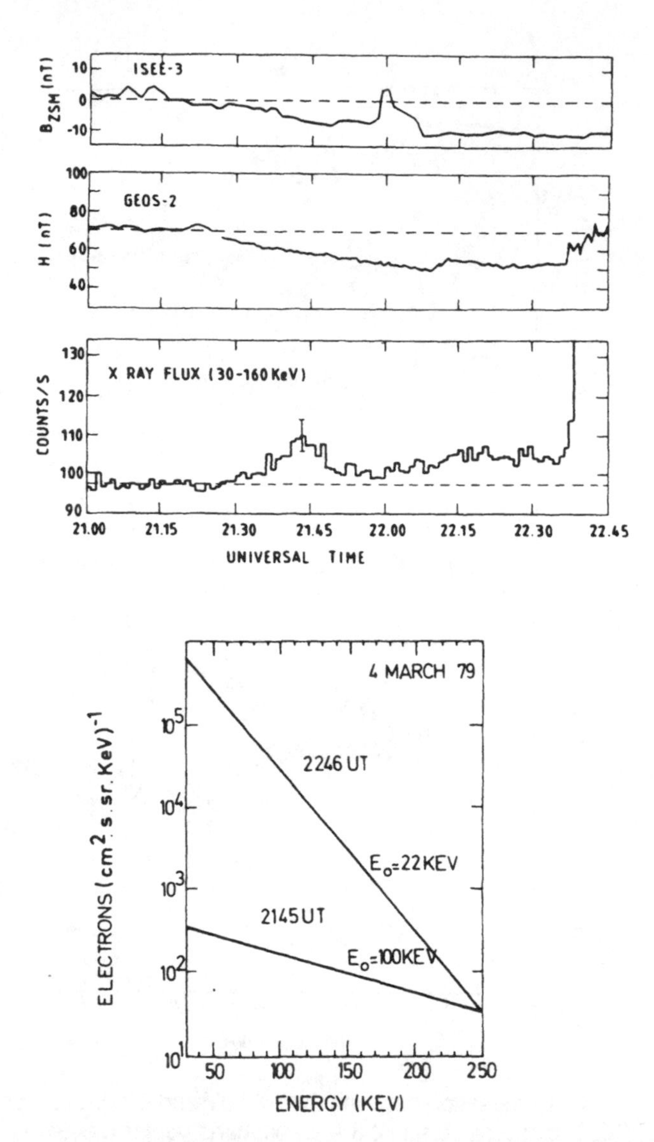

FIGURE 11.12 Top panel: North-south component of the $\mathbf{B}_{IMF}$. Second panel: H-component magnetic field measured on GEOS-2. Third panel: Integral fluxes of 30–160 keV X-rays measured by a balloon-borne X-ray detector. The observed X-rays have been inverted and the parent electron spectra obtained from the X-rays are also shown. (From Sauvaud *et al.*, 1987)

section, a partial list of scientific questions and observations that will help to understand these problems is given.

Since the processes of coupling, transport, and dissipation of energy all involve a variety of plasma interactions, it is appropriate to view the plasma populations in space as both tracers and carriers of the energy flow. Hence the plasma source regions must be identified and relevant mechanisms for storage of plasma and energy must be understood. We must also understand the energization processes and the transport processes that control the entry, coupling, and loss of energy between and within the source and the storage regions.

There are two plasma sources in space, the solar wind and the ionosphere; and two storage regions, the geomagnetic tail and the inner plasma sheet region, where trapped particles encircling the magnetosphere contribute to the ring current. Thus, the general physical problem involves understanding the source regions, the storage regions, the transport processes, the energization processes, and the communication links that dynamically tie the system together (Figure 11.13).

Sources: Measurements in specific regions where hot plasmas are transported into the magnetosphere must investigate *in situ* the physics responsible for the entry. Observations must characterize the plasmas as they enter the magnetosphere so that large-scale transport, storage, acceleration and loss in other regions of the magnetosphere can be assessed by using simultaneous observations of similar information obtained from other regions of space. The relative contributions of ionospheric and solar wind populations can be studied by using charge state information about ions.

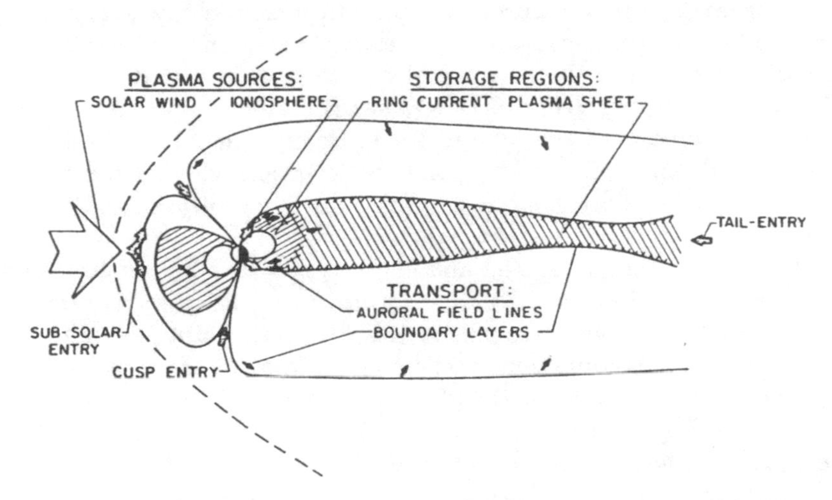

FIGURE 11.13 A schematic diagram to illustrate solar wind and ionospheric sources, ring current and tail storage regions and some of the transport paths which tie the system together. (Courtesy of NASA Global Geospace Science Program)

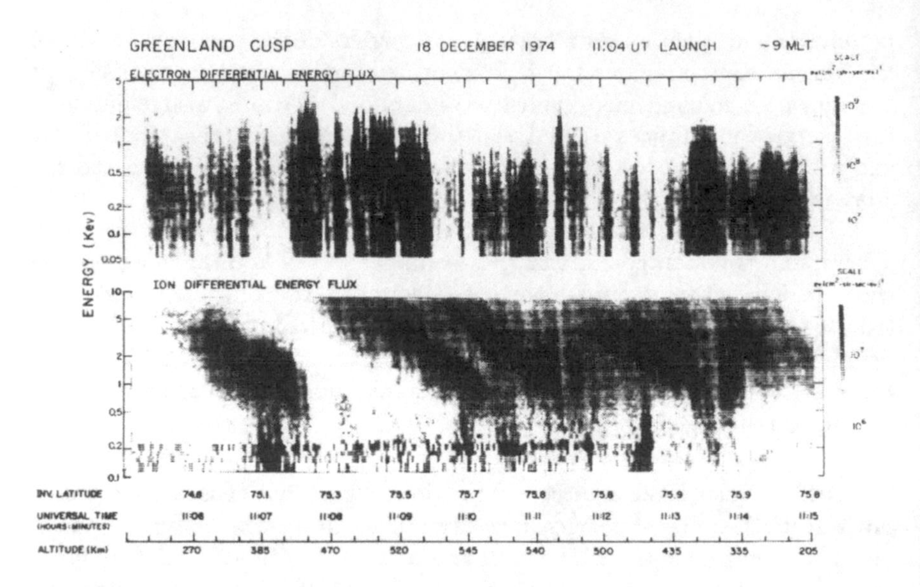

FIGURE 11.14 Spectrogram of particle data obtained by a sounding rocket experiment on December 18, 1974 from Sondre/Stromfjord, Greenland. The ordinate is energy per charge and darker areas represent more intense energy flux. (From Carlson and Torbert, 1980)

Ions possess certain features that make them suitable for studying the source and transport mechanisms. For example, helium is a useful tracer of plasma origin, since the solar wind helium is doubly charged (He^{++}) while helium from the Earth is singly charged (He^{+}). Similarly, oxygen ions (O^{+}) are tracers of ionospheric origin. By studying and following the different distributions, it is also possible to determine whether the acceleration, transport, and loss mechanisms are mass-, momentum-, and energy-dependent.

Figure 11.14 shows electron and ion data obtained by a rocket-borne experiment obtained at an altitude of approximately 500 km from the surface of Sondre/Stromfjord, Greenland (74.5° invariant latitude). The data are shown in a differential energy spectrogram format. Here the ordinate represents energy per charge, the abscissa is time, and the grey scale represents the intensity, with darker areas indicating the more intense energy flux. A feature that stands out is the several ion precipitation bands that sweep in time from high to low energies. Five such bands can be identified in which a decrease in differential energy flux by as much as 100 was observed.

These bands are produced by a burst of ions injected into the loss cone at a localized region above the rocket. The ions then experienced a time of flight dispersion such that faster ions arrived earlier than the slower ones. Non-relativistic ions of mass m and energy E injected at a distance L along the magnetic field direction (0° pitch-angle) would arrive at the rocket after a time delay $T = L(m/2E)^{1/2}$. The spectrometer (electrostatic analyzer) selects ions according to energy per charge E/q, so in terms of the measured quantity, the dispersion equation becomes $(E/q)^{1/2} = (2q/m)^{1/2}T/L$. The arrival time for ions of a given E/q depends both on the injection distance L and the ion charge to mass ratio.

Representative energy spectra illustrating the peaks of different ion species are shown in the upper panels of Figure 11.15. These spectra are associated with the second and third dispersion bands. The primary and the secondary peaks on the right-hand panel arise because the third injection started before the second injection ended. The dispersion curves shown in the bottom panel of Figure 11.15 were obtained by first determining the energies per charge of the multiple peaks in each spectrum sweep and then plotting $(E/q)^{1/2}$ versus time to obtain the slope $(2q/m)^{1/2}L$. For convenience, E/q was plotted as the ordinate. According to the injection hypothesis, this analysis would yield a family of straight lines; the results support the proposed injection hypothesis. Note that all ions project to a common injection time, suggesting one injection source for these ions. The distance L from the rocket to the source deduced from all of the curves gives $12.5 \pm 1 R_E$.

The injection point is high up in the magnetopause. Although injection mechanisms for these ions remain unknown, the dispersion curves suggest that the ions originate from the magnetosheath (since only He^{++} were detected). It appears that solar wind ions were injected into the loss cone in bursts and then travelled adiabatically on the geomagnetic field to the ionosphere where they were detected (for details, see Carlson and Torbert, 1980, cited at the end of the chapter).

Transport: Transport processes transform the magnetosphere from a quiescent system into a dynamic network of highly time-dependent elements. A number of well-known mechanisms, such as the $\mathbf{E} \times \mathbf{B}$ drift, field-aligned flow, cross-field diffusion, and merging, can carry plasma flow from one region to another. In some cases, the transport processes also act as acceleration mechanisms. In other cases, dissipation or loss can occur because the plasma in transport becomes unstable and the waves generated by the unstable plasma can pitch-angle scatter the particles into the loss cone.

At the present time, there is very little information on these coupled processes in the different regions as a function of the dynamic state of the solar wind-magnetosphere system. We need information on the coupling among the source and storage and acceleration regions, and on the loss of plasma from the system to the atmosphere and the solar wind. Investigation

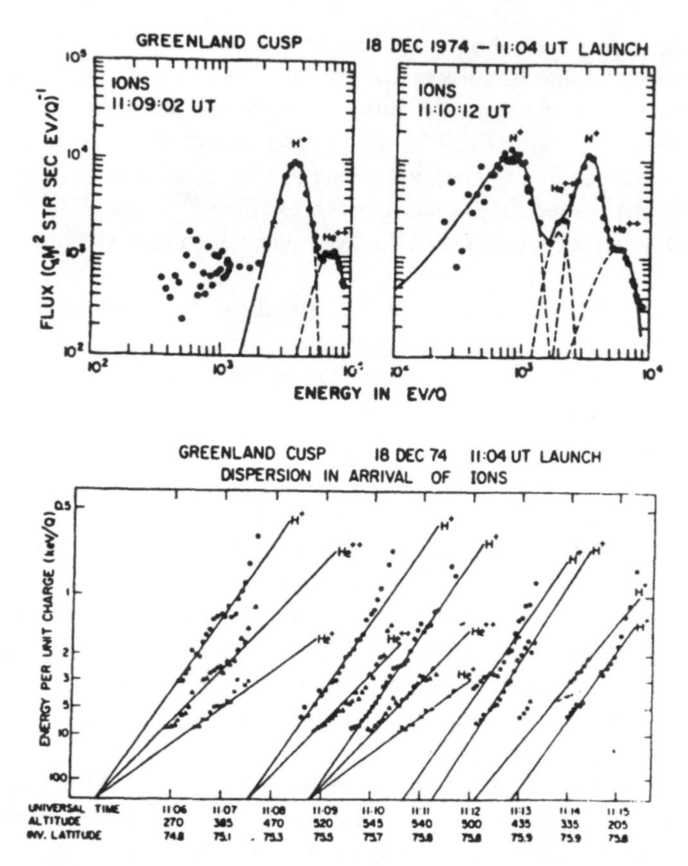

FIGURE 11.15 Two representative energy spectra illustrating the peaks due to different ion species (top diagram). Gaussian fits to individual peaks plus the composite fit are shown as broken and solid lines, respectively. Curves of inverse velocity (labeled by corresponding energy per charge) versus arrival times correspond to spectral peaks. The He^+ lines are the least square fit to the dominant peaks. The He^{++} and He^+ lines are theoretical curves with slopes of 1/œ2 and 1/œ4 of the corresponding H^+ line. Slopes correspond to an injection distance of about $12R_E$ from the rocket. (From Carlson and Torbert, 1980)

in time scales of the dynamics of the various transport processes will permit studies of the origin of observed plasmas, whether the ions are from the solar wind or the ionosphere.

Storage: Energy is stored in the magnetic field of the magnetotail, in the inner and the outer plasma sheet, and the radiation belts and the ring current. During magnetospheric substorms, the plasmas in these regions are redistributed by high-speed flows and plasma turbulence. Plasma heating and acceleration occur throughout the entire magnetosphere. We still do not understand how and by what processes the vast amounts of energy (in excess of 10^{18} ergs/sec in a typical aurora) are released into the atmosphere during substorms. During magnetic storms, particles stored in the ring current are dissipated by precipitation into the auroral zones, by charge exchange processes, and by generation of electromagnetic radiation. The temporal evolution of these storage regions in relation to the dynamical behavior of the space system as a whole is not yet understood (see Figure 7.7).

Energization: Cosmic plasmas are frequently accelerated to high energies. There are many examples of fundamental acceleration processes in the terrestrial magnetosphere where hot plasmas and energetic particles are found. Although acceleration processes such as the conversion of magnetic energy to plasma energy by means of reconnection and acceleration by large-scale induction electric field have been suggested for a long time, *in situ* measurements have not yet been made with instruments capable of providing the details to study the microscopic and macroscopic energy conversion mechanisms. Study is needed of other processes active in phenomena such as particle acceleration along the magnetic field by parallel electric field and transverse to the magnetic field direction by, for example, lower hybrid waves.

An important acceleration mechanism has been observed to occur at altitudes of 1 to $3R_E$ over discrete auroral arcs. This region displays signatures characteristic of field-aligned electrostatic acceleration of electrons and ions and comprises a major contribution to the current system that couples the ionosphere to the outer magnetosphere. It is thought, for example, that the distant magnetotail current closes in the ionosphere and that this same current may be associated with reconnection. Reconnection may also occur sporadically in the near-Earth tail region, and this near-Earth reconnection is one of the principal features of the reconnection substorm model. The conditions that cause such reconnection from start to end must be identified along with the microscopic plasma processes that operate in these regions.

11.6 Concluding Remarks

A magnetosphere is established in the solar wind by collisionless MHD convective flows and large-scale currents. This MHD configuration is unstable and its free energy is dissipated periodically. The substorm event discussed above demonstrates that an instability is initiated by the $\mathbf{B}_{IMF}$ which induces a sequence of dynamic interactions in particles and fields in the outer magnetosphere, geomagnetic tail, and the ionosphere. While many details are still unknown, observations from previous experiments have permitted us to ask important questions about the substorm dynamics.

The physics of solar terrestrial plasmas will receive much attention in the decade of the 1990s. A major, coordinated effort is being made by research communities of the world to study the Sun and magnetospheric substorm phenomena. The Sun will be at its maximum activity in 1991 and the plan is to study the solar system plasma systematically. Planned programs include ground-based observations coordinated with observations made on high-altitude balloons, sounding rockets, and satellites.

A complement of particle and field experiments covering the wide energy and frequency ranges (from eV to MeV and < 1 Hz to hundreds of kHz) will be flown on several coordinated spacecraft into key regions of space to study the Earth's magnetosphere and the substorm phenomena. In addition, a cluster of four spacecraft identically instrumented will be launched in the mid-nineties to study the dynamics of boundary structures. A spacecraft with the most up-to-date instrumentation will also be launched in the mid-nineties to observe the Sun. For the first time in the history of space science, a coordinated data set from Sun to Earth's ionosphere will be obtained. These will be exciting times for space physicists who will study and model the complex space plasma phenomena.

Bibliography

Carlson, C.W. and R.B. Torbert, Solar Wind Ion Injections in the Morning Auroral Oval, *J. Geophys. Res.*, **85**, 2903, 1980.

Chen, F.F., *Introduction to Plasma Physics*, Plenum Press, New York, NY, 1974. Chapter 11 of this book treats the topic of instabilities. This book can be read by beginning students.

Hasegawa, A., *Plasma Instabilities and Nonlinear Effects*, Springer-Verlag, New York, NY, 1975.

Ichimaru, S., *Basic Principles of Plasma Physics, A Statistical Approach*, W.A. Benjamin Publishing Co., Inc., Reading, MA, 1973. Chapters 7 and 8 of this book treat plasma instability phenomena. This book is for advanced students but beginning and intermediate students may find it useful.

Krall, N.A. and A.W. Trivelpiece, *Principles of Plasma Physics*, McGraw Hill Book Co., New York, NY, 1973. The topic of instability is treated in several chapters. For intermediate and advanced students.

Morozov, A.I. and L.S. Solov'ev, Motion of Charged Particles in Electromagnetic Fields, in *Reviews of Plasma Physics*, Volume 2, M.A. Leontovich, ed., Consultants Bureau, a division of Plenum Publishing Corp., New York, NY, 1966. This article reviews the motion of particles in electromagnetic fields with an emphasis on understanding laboratory fusion plasmas. Several examples of plasmoid motions are discussed.

Sauvaud, J.A., J.P. Treilhou, A. Saint-Marc, J. Dandouras, H. Rème, A. Korth, G. Kremser, G.K. Parks, A.N. Zaitzev, V. Petrov, L. Lazutine and R. Pellinen, Large Scale

Response of the Magnetosphere to a Southward Turning of the Interplanetary Field, *J. Geophys. Res.*, **92**, 2365, 1987.

Instabilities responsible for magnetospheric substorms, while not fully known, involve waves over a broad frequency interval and particles covering a broad energy range. A very large number of papers have been published (and are still being published) on this topic and interested readers will find substorm-related articles in the *Journal of Geophysical Research (Space), Planetary and Space Sciences, Journal of Atmospheric and Terrestrial Physics, Space Science Review, Review of Geophysics and Space Physics*, and in books devoted to the publications of the proceedings of meetings. A partial list of review papers and books on various aspects of substorm dynamics is given below.

Achievements of the International Magnetospheric Study (IMS), Proceedings of an International Symposium, Grax, Austria, June, 1984. European Space Agency, ESA SP-217, 1984.

Akasofu, S.-I., *Polar and Magnetospheric Substorms*, Springer Verlag, New York, NY, 1968.

Akasofu, S.-I., *Physics of Magnetospheric Substorms*, D. Reidel Publishing Co., Dordrecht, Holland, 1977.

Akasofu, S.-I., ed., *Dynamics of the Magnetosphere*, D. Reidel Publishing Co., Dordrecht, Holland, 1980.

Akasofu, S.-I., Energy Coupling Between the Solar Wind and the Magnetosphere, *Space Sci. Rev.*, **28**, 121, 1981.

Alfvén, H., Hydromagnetics of the Magnetosphere, *Space Sci. Rev.*, **2**, 862, 1963.

Arnoldy, R., Auroral Particle Precipitation and Birkeland Currents, *Rev. Geophys. Space Phys.*, **12**, 217, 1974.

Atkinson, G., Duality of the Magnetic Flux Tube and Electric Current Descriptions of Magnetospheric Plasma and Energy Flow, *Rev. Geophys. Space Phys.*, **19**, 617, 1981.

Belmont, G. and G. Chanteur, Advances in Magnetopause Kelvin-Helmholtz Instability Studies, *Physica Scripta*, **40**, 124, 1989. This article presents a concise review of the status of research on Kelvin-Helmholtz instabilities. The references are fairly complete.

Brown, R.R., Electron Precipitation in the Auroral Zone, *Space Sci. Rev.*, **6**, 311, 1966.

Burch, J.L., Observations of Interactions Between Interplanetary and Geomagnetic Fields, *Rev. Geophys. Space Phys.*, **12**, 363, 1974.

Chamberlain, J.W., Electric Acceleration of Auroral Particles, *Rev. Geophys. Space Phys.*, **7**, 461, 1969.

Chappell, C.R., Recent Satellite Measurements of the Morphology and Dynamics of the Plasmasphere, *Rev. Geophys. Space Phys.*, **10**, 951, 1972.

Cole, K.D., Magnetic Storms and Associated Phenomena, *Space Sci. Rev.*, **6**, 699, 1966.

Cowley, S.W.H., Plasma Populations in a Simple Open Model Magnetosphere, *Space Sci. Rev.*, **26**, 217, 1980.

Dessler, A.J., Vacuum Merging: A Possible Source of the Magnetospheric Cross Tail Electric Field, *J. Geophys. Res.*, **76**, 3174, 1971.

Donahue, T.M., Polar Ion Flow: Wind or Breeze?, *Rev. Geophys. Space Phys.*, **9**, 1, 1971.

Dungey, J.W., Neutral Sheets, *Space Sci. Rev.*, **17**, 173, 1975.

Feldstein, Y.I. and Y.I. Galperin, The Auroral Luminosity Structure in the High-Latitude Upper Atmosphere: Its Dynamics and Relationship to the Large-Scale Structure of the Earth's Magnetosphere, *Rev. Geophys. Space Phys.*, **23**, 217, 1985.

Galeev, A.A., M.M. Kuznetsova and L.M. Zeleny, Magnetopause Stability Threshold for Patchy Reconnection, *Space Sci. Rev.*, **44**, 1, 1986.

Gendrin, R., Substorm Aspects of Magnetic Pulsations, *Space Sci. Rev.*, **11**, 54, 1970.

Gendrin, R., General Relationships Between Wave Amplification and Particle Diffusion in a Magnetoplasma, *Rev. Geophys. Space Phys.*, **19**, 171, 1984.

Goertz, C.K., Double Layers and Electrostatic Shocks in Space, *Rev. Geophys. Space Phys.*, **17**, 418, 1979. See also Comment on Double Layers and Electrostatic Shocks in Space by J.R. Kan, *Rev. Geophys. Space Phys.*, **18**, 337, 1980, and the reply to Comment on "Double Layers and Electrostatic Shocks in Space" by J.R. Kan, *Rev . Geophys. Space Phys.*, **19**, 230, 1981.

Gold, T., Magnetic Storms, *Space Sci. Rev.*, **1**, 100, 1962.

Gurevich, A.V., A.L. Krylov and E.E. Tsedilina, Electric Fields in the Earth's Magnetosphere and Ionosphere, *Space Sci. Rev.*, **19**, 59, 1976.

Hill, T.W., Origin of the Plasma Sheet, *Rev. Geophys. Space Phys.*, **12**, 379, 1974.

Hones, E.W. Jr., Transient Phenomena in the Magnetotail and their Relation to Substorms, *Space Sci. Rev.*, **23**, 393, 1979.

Hones, E.W. Jr., ed., *Magnetic Reconnection in Space and Laboratory Plasmas*, American Geophysical Union, Washington, D.C., 1984.

Horwitz, J.L., The Ionosphere as a Source for Magnetospheric Ions, *Rev. Geophys. Space Phys.*, **20**, 929, 1982.

Kan, J.R., Towards a Unified Theory of Discrete Auroras, *Space Sci. Rev.*, **31**, 71, 1982.

Kaufmann, R.L., What Auroral Electron and Ion Beams Tell Us About Magnetosphere-Ionosphere Coupling, *Space Sci. Rev.*, **37**, 313, 1984.

Kim, J.S., D.A. Graham and C.S. Wang, Inductive Electric Fields in the Ionosphere-Magnetosphere System, *Rev. Geophys. Space Phys.*, **17**, 2049, 1979.

Lemaire, J. and M. Scherer, Kinetic Models of the Solar and Polar Winds, *Rev. Geophys. Space Phys.*, **11**, 427, 1973.

Lin, C.S. and R.A. Hoffman, Observations of Inverted-V Electron Precipitation, *Space Sci. Rev.*, **33**, 415, 1982.

Low, B.C., Nonlinear Force-Free Magnetic Fields, *Rev. Geophys. Space Phys.*, **20**, 145, 1982.

McPherron, R.L., Magnetospheric Substorms *Rev. Geophys. and Space Phys.*, **17**, 657, 1979.

Meng, C.-I., Electron Precipitation and Polar Auroras, *Space Sci. Rev.*, **22**, 223, 1978.

Mishin, V.M., High-Latitude Geomagnetic Variations and Substorms, *Space Sci. Rev.*, **6**, 621, 1977.

Moore, T.E., Superthermal Ionospheric Outflows, *Rev. Geophys. Space Phys.*, **22**, 264, 1984.

Nishida, A., IMF Control of the Earth's Magnetosphere, *Space Sci. Rev.*, **34**, 185, 1983.

Obayashi, T., The Streaming of Solar Flare Particles and Plasma in Interplanetary Space, *Space Sci. Rev.*, **3**, 79, 1964.

Obayashi, T. and A. Nishida, Large-Scale Electric Field in the Magnetosphere, *Space Sci. Rev.*, **8**, 3, 1968.

Parks, G.K., G. Laval, and R. Pellat, Behavior of Outer Radion Zone and a New Model of Magnetospheric Substorm, *Planet. Space Sci.*, **20**, 1291, 1972.

Pudovkin, M.I., Electric Fields and Currents in the Ionosphere, *Space Sci. Rev.*, **16**, 727, 1974.

Pudovkin, M.I. and V.S. Semenov, Magnetic Field Reconnection Theory and the Solar Wind-Magnetosphere Interaction: A Review, *Space Sci. Rev.*, **41**, 1, 1985.

Rostoker, G., Polar Magnetic Substorms, *Rev. Geophys. Space Phys.*, **10**, 157, 1972.

Rostoker, G., S.-I. Akasofu, W. Baumjohann, Y. Kamide and R.L. McPherron, The Holes of Direct Input of Energy from the Solar Wind and Unloading of Stored Magnetotail Energy in Driving Magnetospheric Substorms, *Space Sci. Rev.*, **46**, 93, 1987.

Russell, C.T., and R.L. McPherron, The Magnetotail and Substorms, *Space Sci. Rev.*, **15**, 205, 1973.

Schardt, A.W. and A.G. Opp, Particles and Fields: Significant Achievements, 1, *Rev. Geophys. Space Phys.*, **5**, 411, 1967.

Schardt, A.W. and A.G. Opp, Particles and Fields: Significant Achievements, 2, *Rev. Geophys. Space Phys.*, **7**, 799, 1969.

Schindler, K., Plasma and Fields in the Magnetospheric Tail, *Space Sci. Rev.*, **17**, 589, 1975.

Schindler, K. and J. Birn, Magnetotail Theory, *Space Sci. Rev.*, **44**, 307, 1986.

Shebanski, V.P., Magnetospheric Processes and Related Geophysical Phenomena, *Space Sci. Rev.*, **8**, 366, 1968.

Stern, D.P., The Origins of Birkeland Currents, *Rev. Geophys. Space Phys.*, **21**, 125, 1983.

Stern, D.P., Energetics of the Magnetosphere, *Space Sci. Rev.*, **39**, 193, 1984.

Swift, D.W., Mechanisms for Auroral Precipitation: A Review, *Rev. Geophys. Space Phys.*, **19**, 185, 1981.

Troshichev, O.A., Polar Magnetic Disturbances and Field-Aligned Currents, *Space Sci. Rev.*, **32**, 275, 1982.

Walt, M., Radial Diffusion of Trapped Particles and Some of its Consequences, *Rev. Geophys. Space Phys.*, **9**, 11, 1971.

Williams, D.J. and G.D. Mead, eds., Magnetospheric Physics, *Rev. of Geophys.*, **7**, 1969.

Wolf, R.A., Ionosphere-Magnetosphere Coupling, *Space Sci. Rev.*, **17**, 537, 1975.

Questions

1. Consider two dipoles $\mathbf{M}_1$ and $\mathbf{M}_2$ in a plane separated by a distance r. $\mathbf{M}_1$ is fixed and $\mathbf{M}_2$ is free to rotate about its center. Show that for equilibrium, $\tan\theta_1 = 2\tan\theta_2$ where θ_1 and θ_2 are the angles between r and $\mathbf{M}_1$ and $\mathbf{M}_2$ respectively.
2. Consider a model of a boundary geometry in which $\mathbf{B} = B\hat{\mathbf{z}}$ and which has a density gradient in the x-direction. Now apply an arbitrary force $\mathbf{F} = F\hat{\mathbf{x}}$ in the x-direction. The unperturbed distribution can be represented by

$$f(x, \mathbf{v}) = \left(\frac{m}{2\pi kT}\right)^{3/2}\left[1 + \alpha\left(x + \frac{v_y}{\Omega}\right)\right]\exp - \left(\frac{mv^2}{2kT} + \frac{Fx}{kT}\right)$$

where α is a parameter that characterizes the spatial variation of the density, and the rest of the symbols have their usual meaning. Note that in the absence of the density and the force, the distribution is simply Maxwellian.

   a. Show that the total energy of a particle is

$$w = \frac{mv^2}{2} - Fx$$

   a constant of motion.

   b. Show that the canonical moments in the y- and z-directions

$$\begin{aligned} p_y &= mv_y + qBx \\ p_z &= mv_z \end{aligned}$$

   are also constants of motion. Give a physical reason why these canonical moments are conserved.

   c. Use the above distribution function and show that at $x = 0$,

$$\alpha = \frac{1}{n}\frac{dn}{dx} - \frac{F}{kT}$$

   d. Show that the drift velocity resulting from this distribution is given by (also at $x = 0$)

$$\begin{aligned} \mathbf{v}_d &= \frac{\alpha T}{m\Omega}\hat{\mathbf{y}} \\ &= \left(\frac{T}{m\Omega}\frac{1}{n}\frac{dn}{dx} - \frac{F}{m\Omega}\right)\hat{\mathbf{y}} \end{aligned}$$

   where $\hat{\mathbf{y}}$ is a unit vector in the y-direction. Explain the origin of the two terms.
3. Use Equation (11.24) to estimate the instability growth rate for the plasma
   a. In the Earth's ionosphere at 100 km. Information on $\partial n_0/\partial x$ can be obtained from scale height of ionospheric particles, which is about 5 km. Assume a number density of $10^{12}/\text{m}^3$ and $u_0 \approx 100$ m/s and recall that $(\omega - ku_0) \ll \Omega_c$.
   b. In the Sun's corona. Estimate a reasonable scale height, using the continuity equation and assuming that the coronal number density is $10^{12}/\text{m}^3$.
4. Using data given in Table 8.1 in Chapter 8, estimate
   a. The strength of the drift current at the magnetopause.
   b. The phase velocity of the drift waves.
5. Start from the one-fluid momentum equation and show the missing steps in the derivation of Equation (11.45).
6. Use the data shown in Table 8.1 of Chapter 8 to predict whether the Earth's magnetopause is stable or unstable to the growth of Kelvin-Helmholtz waves. Estimate

7. Derive Equations (11.73) and (11.74) for the two-stream instability.

8. An isolated "small" volume of highly conducting plasma (plasmoid) in a weakly inhomogeneous magnetic field can be regarded as a dipole with a magnetic moment

$$\mu = -\kappa \mathbf{B}$$

where κ is a form factor that depends on the plasmoid geometry. If the plasmoid is spherical, $\kappa = r^2/2$ where r is the radius (see Morozov and Solov'ev, 1966).

a. If the magnetic field is irrotational, show that the force acting on the plasmoid is given by

$$\mathbf{F} = -\frac{\kappa}{2}\nabla B^2$$

This problem shows that if κ is constant, the motion of a plasmoid is equivalent to a particle moving in a potential $\phi = (\kappa B^2/2)$.

b. What is the total energy of the plasmoid if the plasmoid velocity is U?

c. Discuss the motion of plasmoids for a magnetotail field given by $\mathbf{B} = (B_0 z/L)\hat{\mathbf{x}}$, where B_0 is a constant (see Chapter 3).

(Plasmoids are a part of a phenomenological model of substorms discussed by Hones, 1979, cited in the reference list at the end of this chapter.)

APPENDIX A
Useful Constants and Units

TABLE A.1

		MKS	CGS
c	velocity of light	3×10^8 m/sec	3×10^{10} cm/sec
e	electronic charge	1.6×10^{-19} coulombs	4.8×10^{-10} esu
m_e	electron rest mass	9.1×10^{-31} kg	9.1×10^{-28} g
m_p	proton rest mass	1.67×10^{-27} kg	1.67×10^{-24} g
h	Planck's constant	6.63×10^{-34} Joule sec	6.63×10^{-27} erg sec
k	Boltzmann's constant	1.38×10^{-23} Joule/°K	1.38×10^{-16} erg/°K
G	Universal Gravitational Constant	6.67×10^{11} newton m²/kg²	6.67×10^{-8} dyne cm²/g²
eV	electron Volt	1.6×10^{-19}Joule	1.6×10^{-12} ergs
eV	kT	$11,600$°K	$11,600$°K
B	magnetic induction	Tesla (T)	Gauss = 10^{-4}T
I	ampere of current	Coulomb/sec	3×10^9 esu/sec (statamp)
J	current density	amp/m²	statamp/cm²
E	electric field	Volt/m	statVolt/cm
μ_0	magnetic permeability of free space	$4\pi 10^{-7}$ newton/amp²	1
ϵ_0	dielectric constant in vacuum	8.85×10^{-12} farads/m	1
f_{pe}	electron plasma frequency (Hz)	$8.98 n_e^{1/2}$	$8.98 \times 10^3 n_e^{1/2}$
f_{pi}	proton plasma frequency (Hz)	$2.1 \times 10^{-1} n_p^{1/2}$	$2.1 \times 10^2 n_p^{1/2}$
f_{ce}	electron cyclotron frequency (Hz)	$2.8 \times 10^{10} B$	$2.8 \times 10^6 B$
f_{ci}	proton cyclotron frequency (Hz)	$1.5 \times 10^7 B$	1.53×10^3B
λ_D	Debye length	6.9×10^1 T$^{1/2} n^{-1/2}$	6.9T$^{1/2} n^{-1/2}$ cm

TABLE A.1 (continued)

AU	Astronomical unit	1.5×10^{11}m	1.5×10^{13}cm
$M_\odot$	mass of Sun	2×10^{30} kg	2×10^{33}g
$R_\odot$	radius of Sun	7×10^{8}m	7×10^{10}cm
$T_\odot$	rotation period of Sun	27 days	27 days
M_E	mass of Earth	5.98×10^{24}kg	5.98×10^{27}g
M	magnetic moment of Earth	8×10^{22} Amp m^2	8×10^{25} gauss/cm^3
R_E	equatorial radius of Earth	6.38×10^{6}m	6.38×10^{8}cm
	temperature of solar corona	$\approx 10^{6}$°K	$\approx 10^{6}$ °K
	temperature of Earth's ionosphere	$\approx 10^{4}$°K	$\approx 10^{4}$ °K

APPENDIX B

Average Plasma Properties of Earth's Environment

TABLE B.1

SOLAR WIND	
Origin	solar coronal atmosphere
Flow speed	$\sim$ 300–800 km/sec
Flow direction	nearly parallel to Earth-Sun line
Alfvén Mach number	$\sim$ 5–10
Particle composition	e^- and p^+ and a few percent He^{++}
Particle number density	$\sim$ 3–20/cc
Average thermal energy	$kT_e \lesssim 100$ eV; $kT_p \lesssim 50$ eV
Intensity of B_{IMF}	$\sim$ 1–30 nT
Average direction of B_{IMF}	$\sim 50°$ from Earth-Sun line in ecliptic plane
BOW SHOCK	
Position (from center of Earth)	~ 10–$15R_E$ (noon), ~ 15–$20R_E$ (dawn-dusk)
Standoff distance	~ 3–$5R_E$ (noon)
Magnetic field compression	$\sim$ 2–4 (near noon)
Density compression	$\sim$ 2–4 (near noon)
Particle density	$\sim$ 10–20/cc
Waves	broadband electrostatic, electromagnetic, hydromagnetic
Wave frequency	$\lesssim 1$ Hz to $\gtrsim 300$ kHz
Hydromagnetic wave amplitudes	$\sim$ 5–20 nT
$\Delta B/B$	0.1–1

TABLE B.1 (continued)

UPSTREAM FORESHOCK REGION	
Energetic particles	e^- and p^+ ~ 1 keV to several hundreds of keV
Origin of energetic particles	bowshock reflected particles, Fermi acceleration
Energetic particle distribution	beams and diffuse particle distribution
Waves and particles	hydromagnetic with diffuse particle distribution
	ion acoustic with protons
	plasma with electrons
MAGNETOSHEATH	
Flow speed	$\sim$ 50 km/sec (bowshock side)
	$\sim$ 100 km/sec Earth side
	$\sim$ 300–800 km/sec at dawn and dusk
Particle number density	$\sim 3\text{–}20 \times 10^6$ m^{-3}
Thermal energy	$\sim$ a few keV (shocked solar wind)
Energetic particles	a few keV to several hundreds of keV
Origin of energetic particles	bowshock, magnetosphere and local source
MAGNETOSPHERE	
Magnetopause	$\sim 8\text{–}13R_E$ (local noon), $\gtrsim 15R_E$ (dawn-dusk)
Particle density	$\sim 0.1\text{–}1 \times 10^6$ m^{-3}
Thermal energy	$\sim 0.1\text{–}1$ keV
Convective E-field	$\sim 0.5\text{–}10$ mV/meter
Origin of particles	ionosphere and solar wind
Ring current	
particle energy density	$\sim$ 10–20 keV/cc
composition	e^-, p^+, and O^+
Total energy in the ring current	$\sim 10^{23}$ ergs
Plasmapause	$\sim 5R_E$
Plasmaspheric density	$\sim 10^8$ m^{-3}
Plasmaspheric thermal energy	1 eV
Plasmaspheric waves	whistlers, plasma oscillations
Length of geomagnetic tail	$\gtrsim 200R_E$
Plasmasheet inner boundary	$\gtrsim 6\text{–}15R_E$ (tail direction)
Plasmasheet outer boundary	geomagnetic latitudes $\gtrsim 68^\circ$
Plasmasheet thermal energy	$kT_e \sim 0.5$ keV; $kT_p \approx 1\text{–}5$ keV
Plasmasheet number density	$\sim 1\text{–}5 \times 10^6$ m^{-3}
Lobe density	$\lesssim 10^6$ m^{-3}

TABLE B.1 (continued)

AURORAL IONOSPHERE	
Location	$\gtrsim$ 60 km from surface of Earth
Auroral oval	$\sim$ 65°–75° geomagnetic latitude
Electron number density	$\sim 10^5$–10^7/cc at 300 km
Thermal energy	$\lesssim$ 1 eV
Energy of precipitated particles	$\gtrsim$ 10 eV to hundreds of keV
Precipitated particle distribution	field-aligned and loss cone
Frequency of auroral occurrence	$\sim$ 1–3 hrs.
Auroral duration	$\sim$ 0.5 hr
Total particle energy in an aurora	$\sim 10^{21}$ ergs
Field-aligned current density	$\sim 10^{-7}$ amps/m^2
Field-aligned E-field	$\lesssim$ 1–10 mV/meter

APPENDIX C

Useful Vector and Tensor Formulas

$\alpha, \phi, \psi =$ scalars

$\mathbf{A}, \mathbf{B} =$ vectors

$\mathbf{T} =$ tensor

$$\mathbf{A} + \mathbf{B} = \mathbf{B} + \mathbf{A}$$

$$\mathbf{A} \cdot \mathbf{B} = \mathbf{B} \cdot \mathbf{A}$$

$$\mathbf{A} \times \mathbf{B} = -\mathbf{B} \times \mathbf{A}$$

$$\mathbf{A} \cdot (\mathbf{B} \times \mathbf{C}) = \mathbf{C} \cdot (\mathbf{A} \times \mathbf{B}) = \mathbf{B} \cdot (\mathbf{C} \times \mathbf{A})$$

$$\mathbf{A} \times (\mathbf{B} \times \mathbf{C}) = (\mathbf{A} \cdot \mathbf{C})\mathbf{B} - (\mathbf{A} \cdot \mathbf{B})\mathbf{C}$$

$$(\mathbf{A} \times \mathbf{B}) \cdot (\mathbf{C} \times \mathbf{D}) = (\mathbf{A} \cdot \mathbf{C})(\mathbf{B} \cdot \mathbf{D}) - (\mathbf{A} \cdot \mathbf{D})(\mathbf{B} \cdot \mathbf{C})$$

$$\nabla(\alpha\phi) = \nabla(\phi\alpha) = \alpha\nabla\phi + \phi\nabla\alpha$$

$$\nabla \cdot (\phi\mathbf{A}) = \phi\nabla \cdot \mathbf{A} + (\mathbf{A} \cdot \nabla)\phi$$

$$\nabla \times (\phi\mathbf{A}) = \phi\nabla \times \mathbf{A} + (\nabla\phi) \times \mathbf{A}$$

$$\nabla \cdot (\mathbf{A} \times \mathbf{B}) = \mathbf{B} \cdot \nabla \times \mathbf{A} - \mathbf{A} \cdot \nabla \times \mathbf{B}$$

$$\nabla \times (\mathbf{A} \times \mathbf{B}) = \mathbf{A}(\nabla \cdot \mathbf{B}) - \mathbf{B}(\nabla \cdot \mathbf{A}) + (\mathbf{B} \cdot \nabla)\mathbf{A} - (\mathbf{A} \cdot \nabla)\mathbf{B}$$

$$\nabla \cdot \nabla\phi = \nabla^2\phi$$

$$\nabla^2\mathbf{A} = \nabla(\nabla \cdot \mathbf{A}) - \nabla \times (\nabla \times \mathbf{A})$$

$$\nabla \cdot (\phi\mathbf{T}) = \nabla\phi \cdot \mathbf{T} + \phi\nabla \cdot \mathbf{T}$$

$$\nabla \times (\nabla\phi) = 0$$

$$\nabla \cdot (\nabla \times \mathbf{A}) = 0$$

$$\int_V \nabla\phi dV = \int_s \phi d\mathbf{s}$$

$$\int_V \nabla \cdot \mathbf{A} dV = \int_s \mathbf{A} \cdot d\mathbf{s}$$

$$\int_V \nabla \times \mathbf{A} dV = \int_s d\mathbf{s} \times \mathbf{A}$$

$$\int_s \nabla \times \mathbf{A} \cdot ds = \oint_c \mathbf{A} \cdot d\mathbf{l}$$

$$\int_s ds \times \nabla\phi = \oint_c \phi d\mathbf{l}$$

C.1 Vector Operators

Many derivations in this book use vector algebra involving operations of gradient on scalars and divergence and curl on vectors. The most commonly used coordinate frames are the Cartesian, polar, cylindrical, and spherical. We list below expressions that are useful.

C.1.1 Cartesian Coordinate System

The Cartesian coordinate system is defined by the coordinates (x, y, z). Let Ψ be a scalar which is a function of (x, y, z) and $\mathbf{B}$ a vector with components (B_x, B_y, B_z). Define $(\mathbf{i}, \mathbf{j}, \mathbf{k})$ as unit vectors in the directions (x, y, z).

Gradient

$$\nabla\Psi = \mathbf{i}\frac{\partial\Psi}{\partial x} + \mathbf{j}\frac{\partial\Psi}{\partial y} + \mathbf{k}\frac{\partial\Psi}{\partial z}$$

Divergence

$$\nabla \cdot \mathbf{B} = \frac{\partial B_x}{\partial x} + \frac{\partial B_y}{\partial y} + \frac{\partial B_z}{\partial z}$$

Curl

$$\nabla \times \mathbf{B} = \begin{pmatrix} \mathbf{i} & \mathbf{j} & \mathbf{k} \\ \frac{\partial}{\partial x} & \frac{\partial}{\partial y} & \frac{\partial}{\partial z} \\ B_x & B_y & B_z \end{pmatrix}$$

C.1.2 Spherical Polar Coordinates

Define a spherical coordinate system by (r, θ, ϕ) and the unit vectors in these directions $(\mathbf{r}, \boldsymbol{\theta}, \boldsymbol{\phi})$. Let $\Psi = \Psi(r, \theta, \phi)$ be a scalar and $\mathbf{B} = (B_r, B_\theta, B_\phi)$.

Gradient

$$\nabla\Psi = \frac{\partial\Psi}{\partial r}\hat{\mathbf{r}} + \frac{1}{r}\frac{\partial\Psi}{\partial\theta}\hat{\boldsymbol{\theta}} + \frac{1}{r\sin\theta}\frac{\partial\Psi}{\partial\phi}\hat{\boldsymbol{\phi}}$$

Divergence

$$\nabla \cdot \mathbf{B} = \frac{1}{r^2}\frac{\partial}{\partial r}(r^2 B_r) + \frac{1}{r\sin\theta}\frac{\partial}{\partial\theta}(\sin\theta B_\theta) + \frac{1}{r\sin\theta}\frac{\partial}{\partial\phi}B_\phi$$

Curl

$$\nabla \times \mathbf{B} = \frac{1}{r^2\sin\theta}\begin{pmatrix} \hat{\mathbf{r}} & r\hat{\boldsymbol{\theta}} & r\sin\theta\hat{\boldsymbol{\phi}} \\ \frac{\partial}{\partial r} & \frac{\partial}{\partial\theta} & \frac{\partial}{\partial\phi} \\ B_r & rB_\theta & r\sin\theta B_\phi \end{pmatrix}$$

C.1.3 Cylindrical Coordinate System

Define cylindrical coordinates with (r, ϕ, z) and the unit vectors in these directions by $(\mathbf{r}, \boldsymbol{\phi}, \mathbf{z})$. A scalar Ψ and a vector $\mathbf{B}$ are functions of these variables, $\Psi = \Psi(r, \phi, z)$ and $\mathbf{B} = (B_r, B_\phi, B_z)$.

Gradient

$$\nabla\Psi = \hat{\mathbf{r}}\frac{\partial\Psi}{\partial r} + \left(\frac{\hat{\boldsymbol{\phi}}}{r}\right)\frac{\partial\Psi}{\partial\phi} + \hat{\mathbf{z}}\frac{\partial\Psi}{\partial z}$$

Divergence

$$\nabla \cdot \mathbf{B} = \frac{\partial B_r}{\partial r} + \frac{B_r}{r} + \frac{1}{r}\frac{\partial B_\phi}{\partial_\phi} + \frac{\partial B_z}{\partial z}$$

Curl

$$\nabla \times \mathbf{B} = \frac{1}{r}\begin{pmatrix} \hat{\mathbf{r}} & r\hat{\boldsymbol{\phi}} & \hat{\mathbf{z}} \\ \frac{\partial}{\partial r} & \frac{\partial}{\partial\phi} & \frac{\partial}{\partial z} \\ B_r & rB_\phi & B_z \end{pmatrix}$$

It is sometimes useful in vector operations to use tensor notations. The following shows the equivalence between vectors and tensors.

Vector	Tensor
$\mathbf{A}$	A_i
$\mathbf{A} \cdot \mathbf{B}$	$A_i B_i$
$\mathbf{A} \times \mathbf{B}$	$\epsilon_{ijk} A_i B_k$
$\nabla\phi$	$\frac{\partial\phi}{\partial r_i}$
$\nabla \cdot \mathbf{A}$	$\frac{\partial A_i}{\partial r_i}$

$$\nabla \times \mathbf{A} \qquad \epsilon_{ijk}\frac{\partial A_k}{\partial r_j}$$

$$\nabla^2 \phi \qquad \frac{\partial^2 \phi}{\partial r_i \partial r_j}$$

ϵ_{ijk} is called the permutation tensor
$\epsilon_{ijk} = 1$ if i, j, k are a cyclic permutation of 1,2,3
$= -1$ if i, j, k are non-cyclic
$= 0$ if any two subscripts are the same
$\epsilon_{ijk}\epsilon_{ilm} = -\epsilon_{ijk}\epsilon_{lim} = \epsilon_{ijk}\epsilon_{lmi} = \delta_{ij}\delta_{km} - \delta_{jm}\delta_{kl}$
δ_{ij} = Kronecker delta is a second-order tensor (with value 1 when $i = j$, otherwise 0).

A scalar ϕ is a zero-order tensor with one element. A vector $\mathbf{A}$ is a first-order tensor with $3^1 = 3$ components in a three dimensional space, $\mathbf{A} = (A_1, A_2, A_3)$. Maxwell stress or pressure tensor is a second-order tensor with $3^2 = 9$ components,

$$(T_{ij}) = \begin{pmatrix} T_{11} & T_{12} & T_{13} \\ T_{21} & T_{22} & T_{23} \\ T_{31} & T_{32} & T_{33} \end{pmatrix}$$

Some tensors, like the pressure tensor, are symmetric. Hence, $p_{ij} = p_{ji}$ but this is not general. p_{ij} represents the (ij) component of the tensor.

C.2 Change of Variables

Cartesian coordinates are defined by (x, y, z) and the unit vectors $(\mathbf{i}, \mathbf{j}, \mathbf{k})$ where $\mathbf{i}$ is in the x-direction, $\mathbf{j}$ is in the y-direction, and $\mathbf{k}$ is in the z-direction.

C.2.1 Cylindrical Coordinates (r, ϕ, z)

$$\begin{aligned} x &= r \sin \phi \\ y &= r \cos \phi \\ z &= z \end{aligned}$$

where (r, ϕ) are the usual variables in the cylindrical system.

C.2.2 Unit Vectors

$$
\begin{aligned}
\mathbf{e}_r &= \mathbf{i}\cos\phi + \mathbf{j}\sin\phi \\
\mathbf{e}_\phi &= -\mathbf{i}\sin\phi + \mathbf{j}\cos\phi \\
\mathbf{e}_z &= \mathbf{k}
\end{aligned}
$$

where $\mathbf{e}_r$ and $\mathbf{e}_\phi$ are unit vectors in (r, ϕ) directions.

C.2.3 Differential Operators

$$
\begin{aligned}
\frac{\partial}{\partial x} &= \cos\phi \frac{\partial}{\partial r} - \frac{\sin\phi}{r}\frac{\partial}{\partial \phi} \\
\frac{\partial}{\partial y} &= \sin\phi \frac{\partial}{\partial r} + \frac{\cos\phi}{r}\frac{\partial}{\partial \phi} \\
\frac{\partial}{\partial z} &= \frac{\partial}{\partial z}
\end{aligned}
$$

Similarly,

$$
\begin{aligned}
\frac{\partial}{\partial r} &= \cos\phi \frac{\partial}{\partial x} + \sin\phi \frac{\partial}{\partial y} \\
\frac{\partial}{\partial \phi} &= -r\sin\phi \frac{\partial}{\partial x} + r\cos\phi \frac{\partial}{\partial y} \\
\frac{\partial}{\partial z} &= \frac{\partial}{\partial z}
\end{aligned}
$$

C.2.4 Spherical Coordinates (r, θ, ϕ)

$$
\begin{aligned}
x &= r\sin\theta\cos\phi \\
y &= r\sin\theta\sin\phi \\
z &= r\cos\theta
\end{aligned}
$$

C.2.5 Unit Vectors

$$
\begin{aligned}
\mathbf{e}_r &= \mathbf{i}\sin\theta\cos\phi + \mathbf{j}\sin\theta\sin\phi + \mathbf{k}\cos\theta \\
\mathbf{e}_\theta &= \mathbf{i}\cos\theta\cos\phi + \mathbf{j}\cos\phi\sin\phi - \mathbf{k}\sin\theta \\
\mathbf{e}_\phi &= -\mathbf{i}\sin\theta + \mathbf{j}\cos\phi
\end{aligned}
$$

C.2.6 Differential Operators

The variables (r, θ, ϕ) are related to (x, y, z) through

$$\frac{\partial}{\partial r} = \frac{\partial}{\partial x}\frac{\partial x}{\partial r} + \frac{\partial}{\partial y}\frac{\partial y}{\partial r} + \frac{\partial}{\partial z}\frac{\partial z}{\partial r}$$

$$\frac{\partial}{\partial \theta} = \frac{\partial}{\partial x}\frac{\partial x}{\partial \theta} + \frac{\partial}{\partial y}\frac{\partial y}{\partial \theta} + \frac{\partial}{\partial z}\frac{\partial z}{\partial \theta}$$

$$\frac{\partial}{\partial \phi} = \frac{\partial}{\partial x}\frac{\partial x}{\partial \phi} + \frac{\partial}{\partial y}\frac{\partial y}{\partial \phi} + \frac{\partial}{\partial z}\frac{\partial z}{\partial \phi}$$

Using the relationship $x = x(r, \theta, \phi)$, $y = y(r, \theta, \phi)$, $z = z(r, \theta, \phi)$ for spherical coordinates, one obtains

$$\frac{\partial}{\partial x} = \sin\theta\cos\phi\frac{\partial}{\partial r} + \cos\theta\cos\frac{\phi}{r}\frac{\partial}{\partial \theta} - \frac{\sin\phi}{r\sin\theta}\frac{\partial}{\partial \phi}$$

$$\frac{\partial}{\partial y} = \sin\theta\sin\phi\frac{\partial}{\partial r} + \cos\theta\sin\frac{\phi}{r}\frac{\partial}{\partial \theta} + \frac{\cos\phi}{r\sin\theta}\frac{\partial}{\partial \phi}$$

$$\frac{\partial}{\partial z} = \cos\theta\frac{\partial}{\partial r} - \frac{\sin\theta}{r}\frac{\partial}{\partial \phi}$$

APPENDIX D
Magnetic Indices

Magnetic indices are used to characterize the disturbances of a magnetized planet. The three most commonly used magnetic indices for Earth are A_E, D_{st}, and K_p. A_E and D_{st} measure respectively, the disturbance levels of auroral electrojets and magnetic storms. K_p, a planetary index, is tied to the disturbance level of the solar wind. These indices are derived from ground-based magnetic records obtained from the polar to equatorial regions.

Ground-based magnetic records either use the (H, D, Z) or the (X, Y, Z) system. The H-component represents the horizontal component of the magnetic field tangent to the surface. The Z-component represents the vertical component defined as positive in the downward direction. D is the declination measured from true north (positive around by east). The relationships between the (H, D, Z) and the (X, Y, Z) components are: $X = H \cos D$, $Y = H \sin D$, and $Z = Z$. Note also that $B_\lambda = +X$, $B_\phi = +Y$, and $B_r = -Z$.

D.1 A_E Index

This index provides information on the auroral activity and the index is constructed from magnetic records obtained by magnetometers located throughout the auroral zone. The magnetic records, arranged in Universal Time (Greenwich Time), are superimposed and two traces are then drawn, creating an envelope of positive and negative variations. The A_E represents the value of the magnetic field between the upper and the lower envelopes. In constructing the A_E index, it is important to use as many magnetograms from as many local time regions as possible because auroral currents have strong local time dependence.

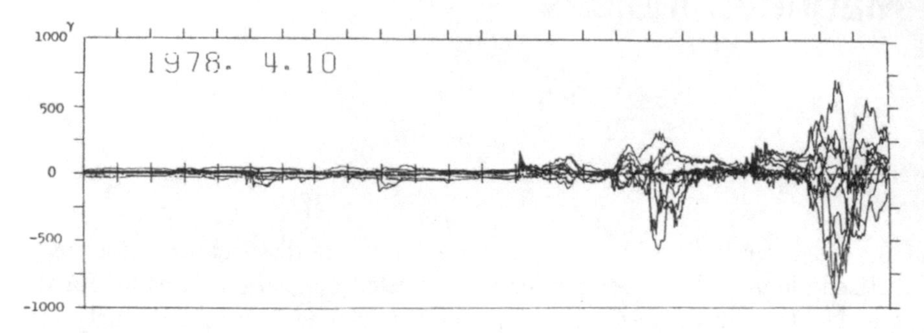

FIGURE D.1 Superposition of 12 auroral zone magnetic records to construct the A_E index.

D.2 K_p index

K_p is intended to be a "quantitative" measure of the planetary disturbance level. K_p ranges from 0 to 9 with plus and minus designations given between two intervals (for example, 1−, 1, 1+). Higher K_p values mean more intense disturbance levels and the scale is logarithmic (the corresponding linear index is designated as A_p). The K_p index is based upon the spread during the stated time interval between the largest upward and downward excursions of the recorded trace relative to a smooth curve obtained on a magnetically quiet day for a given station. For example, Fredericksburg is one of many magnetic observatories that issues K values. The limits used by this station are:

TABLE D.1

K	0	1	2	3	4	5	6	7	8	9
min	0	5	10	20	40	70	120	200	330	500(nT)
max	4	0	19	39	69	119	199	329	499	

The construction of K_p values involves K indices from a number of stations which are collected and regional peculiarities are removed. The average of the "standardized" values represents the geomagnetic planetary K_p index.

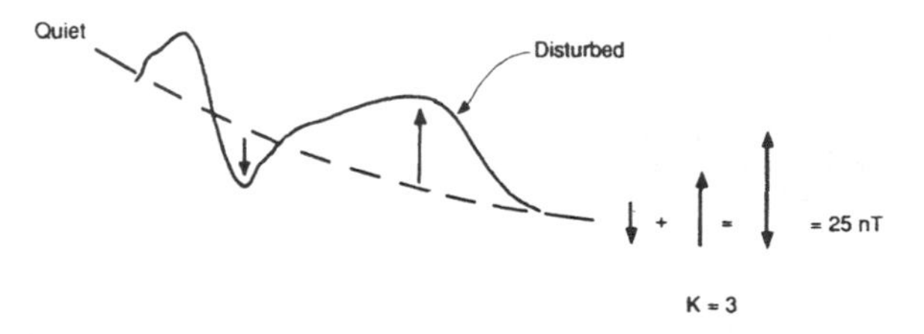

FIGURE D.2 The computation of the K index from Fredericksburg.

D.3 $\mathrm{D_{st}}$ Index

The D_{st} index provides information on magnetic storms and is constructed from world-wide mid-latitude and equatorial magnetograms. The number of stations that recorded the onset of a storm (generally determined by the Sudden Commencement) is noted, and then values of the magnetic field from each station are written out in rows, commencing with the storm. The values are then averaged from all the stations. The average effect of a storm time variation is a reduction of the horizontal component of the geomagnetic field. D_{st} is negative and a larger negative D_{st} means a more intense storm. An example of D_{st} values is shown below. During this period, there was one major and two minor storms. A magnetic storm lasts for about a week.

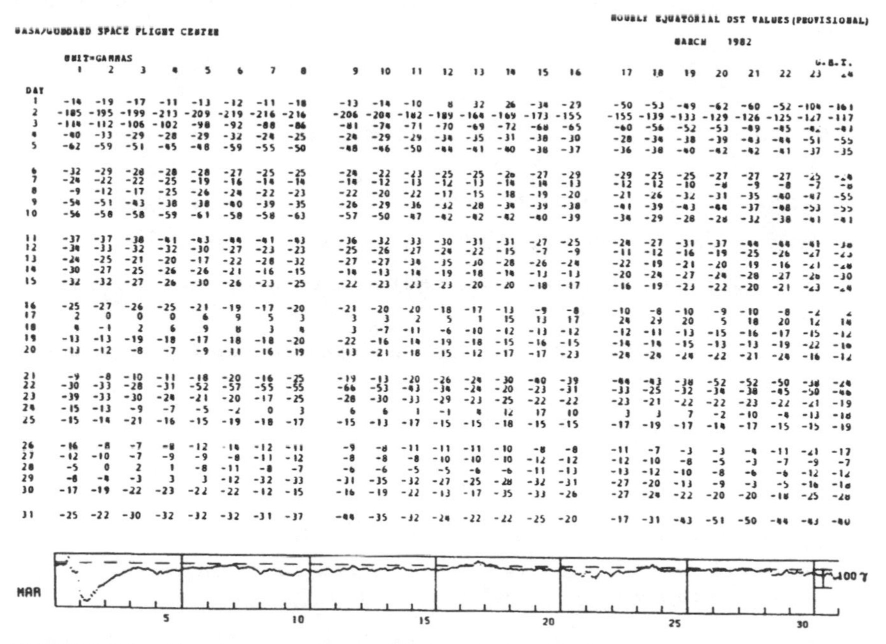

NASA/GODDARD SPACE FLIGHT CENTER

HOURLY EQUATORIAL DST VALUES (PROVISIONAL)

MARCH 1982

UNIT=GAMMAS

U.T.

DAY	1	2	3	4	5	6	7	8
1	-14	-19	-17	-11	-13	-12	-11	-18
2	-185	-195	-199	-213	-209	-219	-216	-216
3	-114	-112	-106	-102	-98	-92	-88	-86
4	-40	-33	-29	-28	-29	-32	-24	-25
5	-62	-59	-51	-45	-48	-59	-55	-50
6	-32	-29	-28	-28	-28	-27	-25	-25
7	-24	-22	-22	-25	-19	-16	-14	-14
8	-9	-12	-17	-25	-26	-24	-22	-23
9	-54	-51	-43	-38	-38	-40	-39	-35
10	-56	-58	-58	-59	-61	-58	-58	-63
11	-37	-37	-38	-41	-43	-44	-41	-43
12	-34	-33	-32	-32	-30	-27	-23	-23
13	-24	-25	-21	-20	-17	-22	-28	-32
14	-30	-27	-25	-26	-26	-21	-16	-15
15	-32	-32	-27	-26	-30	-26	-23	-25
16	-25	-27	-26	-25	-21	-19	-17	-20
17	2	0	0	0	6	9	5	3
18	4	-1	2	6	9	8	3	4
19	-13	-13	-19	-18	-17	-18	-18	-20
20	-13	-12	-8	-7	-9	-11	-16	-19
21	-9	-8	-10	-11	-18	-20	-16	-25
22	-30	-33	-28	-31	-52	-57	-55	-55
23	-39	-33	-30	-24	-21	-20	-17	-25
24	-15	-13	-9	-7	-5	-2	0	3
25	-15	-14	-21	-16	-15	-19	-18	-17
26	-16	-8	-7	-8	-12	-14	-12	-11
27	-12	-10	-7	-9	-9	-8	-11	-12
28	-5	0	2	1	-8	-11	-8	-7
29	-8	-4	-3	3	3	-12	-32	-33
30	-17	-19	-22	-23	-22	-22	-12	-15
31	-25	-22	-30	-32	-32	-32	-31	-37

DAY	9	10	11	12	13	14	15	16
1	-13	-14	-10	8	32	26	-34	-29
2	-206	-204	-182	-189	-164	-169	-173	-155
3	-81	-74	-71	-70	-69	-72	-68	-65
4	-24	-29	-29	-34	-35	-31	-38	-30
5	-48	-46	-50	-44	-41	-40	-38	-37
6	-24	-22	-23	-25	-25	-26	-27	-29
7	-14	-12	-13	-12	-13	-14	-14	-13
8	-22	-20	-22	-17	-15	-18	-19	-20
9	-26	-29	-36	-32	-28	-34	-39	-38
10	-57	-50	-47	-42	-42	-42	-40	-39
11	-36	-32	-33	-30	-31	-31	-27	-25
12	-25	-26	-27	-24	-22	-15	-7	-9
13	-27	-27	-34	-35	-30	-28	-26	-24
14	-14	-13	-14	-19	-18	-14	-13	-13
15	-22	-23	-23	-23	-20	-20	-18	-17
16	-21	-20	-20	-18	-17	-13	-9	-8
17	3	3	2	5	1	15	13	17
18	3	-7	-11	-6	-10	-12	-13	-12
19	-22	-16	-14	-19	-18	-15	-16	-15
20	-13	-21	-18	-15	-12	-17	-17	-23
21	-19	-13	-20	-26	-24	-30	-40	-39
22	-66	-53	-43	-34	-24	-20	-23	-31
23	-28	-30	-33	-29	-23	-25	-22	-22
24	6	6	1	-1	4	12	17	10
25	-15	-13	-17	-15	-15	-18	-15	-15
26	-9	-8	-11	-11	-11	-10	-8	-8
27	-8	-8	-8	-10	-10	-10	-12	-12
28	-6	-6	-5	-5	-6	-6	-11	-13
29	-31	-35	-32	-27	-25	-28	-32	-31
30	-16	-19	-22	-13	-17	-35	-33	-26
31	-44	-35	-32	-24	-22	-22	-25	-20

DAY	17	18	19	20	21	22	23	24
1	-50	-53	-49	-62	-60	-52	-104	-161
2	-155	-139	-133	-129	-126	-125	-127	-117
3	-60	-56	-52	-53	-49	-45	-42	-41
4	-28	-34	-38	-39	-43	-44	-51	-55
5	-36	-38	-40	-42	-42	-41	-37	-35
6	-29	-25	-25	-27	-27	-27	-25	-24
7	-12	-12	-10	-8	-9	-8	-7	-8
8	-21	-26	-32	-31	-35	-40	-47	-55
9	-41	-39	-43	-44	-37	-48	-53	-55
10	-34	-29	-28	-28	-32	-38	-41	-41
11	-24	-27	-31	-37	-44	-44	-41	-38
12	-11	-12	-16	-19	-25	-26	-27	-23
13	-22	-19	-21	-20	-19	-16	-21	-24
14	-20	-24	-27	-24	-28	-27	-26	-30
15	-16	-19	-23	-22	-20	-21	-23	-24
16	-10	-8	-10	-9	-10	-8	-2	2
17	24	29	20	5	18	20	12	14
18	-12	-11	-13	-15	-16	-17	-15	-12
19	-14	-14	-15	-13	-13	-19	-22	-18
20	-24	-24	-24	-22	-21	-24	-16	-12
21	-44	-43	-38	-52	-52	-50	-38	-24
22	-33	-25	-32	-34	-38	-45	-50	-46
23	-23	-21	-22	-22	-23	-22	-21	-19
24	3	3	7	-2	-10	-4	-13	-18
25	-17	-19	-17	-14	-17	-15	-15	-19
26	-11	-7	-3	-3	-4	-11	-21	-17
27	-12	-10	-8	-5	-3	-7	-9	-7
28	-13	-12	-10	-8	-6	-6	-12	-12
29	-27	-20	-13	-9	-3	-5	-16	-16
30	-27	-24	-22	-20	-20	-18	-25	-28
31	-17	-31	-43	-51	-50	-44	-43	-40

FIGURE D.3 Hourly D_{st} values published by NASA/Goddard Space Flight Center for the month of March, 1982. The unit $\gamma = 1nT$.

BIBLIOGRAPHY

Aggson, T.L., Probe Measurements of Electric Fields in Space, in *Atmospheric Emissions*, B.M. McCormac and A. Omholt, eds., Van Nostrand Reinhold, New York, NY, 1969.

Aggson, T.L., P.J. Gambardella and N.C. Maynard, Electric Field Measurements at the Magnetopause, 1. Observations of Large Convective Velocities at Rotational Magnetopause Discontinuities, *J. Geophys. Res.*, **88**, 10,000, 1983.

Akasofu, S.-I., *Polar and Magnetospheric Substorms*, Springer Verlag, New York, NY, 1968.

Akasofu, S.-I., *Physics of Magnetospheric Substorms*, D. Reidel Publishing Co., Dordrecht, Holland, 1977.

Akasofu, S.-I., ed., *Dynamics of the Magnetosphere*, D. Reidel Publishing Co., Dordrecht, Holland, 1980.

Akasofu, S.-I., Energy Coupling Between the Solar Wind and the Magnetosphere, *Space Sci. Rev.*, **28**, 121, 1981.

Akasofu, S.-I. and L.J. Lanzerotti, The Earth's Magnetosphere, *Physics Today*, **28**, 12, 1975.

Alfvén, H., The Theory of Magnetic Storms and Auroras, *Nature*, **167**, 984, 1951.

Alfvén, H., Hydromagnetics of the Magnetosphere, *Space Sci. Rev.*, **2**, 862, 1963.

Alfvén, H., On the Importance of Electric Fields in the Magnetosphere and Interplanetary Space, *Space Sci. Rev.*, **7**, 140, 1967.

Alfvén, H., Some Properties of Magnetospheric Neutral Surfaces, *J. Geophys. Res.*, **73**, 4379, 1968.

Alfvén, H., Plasma Physics, Space Research, and the Origin of the Solar System, *Science*, **172**, 991, 1971.

Alfvén, H., *Cosmic Plasma*, D. Reidel Publishing Co., Dordrecht, Holland, 1981.

Alfvén, H., The Plasma Universe, *Physics Today*, **22**, September, 1986.

Alfvén, H. and C.G. Fälthammar, *Cosmical Electrodynamics, Fundamental Principles*, 2nd ed., Oxford University Press, Oxford, England, 1963.

Anderson, K.A., Energetic Particles in the Earth's Magnetic Field, *Annual Rev. Nuclear Sci.*, **16**, 291, 1966.

Anderson, K.A., H.K. Harris and R.J. Padi, Energetic Electron Fluxes in and beyond the Earth's Outer Magnetosphere, *J. Geophys. Res.*, **70**, 1039, 1965.

Anderson, K.A., R.P. Lin, F. Martel, C.S. Lin, G.K. Parks and H. Rème, Thin Sheets of Energetic Electrons Upstream from the Earth's Bow Shock, *Geophys. Res. Lett.*, **6**, 401, 1979.

Anderson, K.A. and R.P. Lin, Observations of Interplanetary Field Lines in the Magnetotail, *J. Geophys. Res.*, **74**, 3953, 1969.

Angerami, J.J. and D.L. Carpenter, Whistler Studies of the Plasmapause in the Magnetosphere, 2. Electron Density and Total Tube Electron Content Near the Knee in the Magnetospheric Ionization, *J. Geophys. Res.*, **71**, 711, 1966.

Arnoldy, R.L., Auroral Particle Precipitation and Birkeland Currents, *Rev. Geophys. Space Phys.*, **12**, 217, 1974.

Atkinson, G., Duality of the Magnetic Flux Tube and Electric Current Descriptions of Magnetospheric Plasma and Energy Flow, *Rev. Geophys. Space Phys.*, **19**, 617, 1981.

Axford, W.I., Viscous Interaction Between the Solar Wind and the Earth's Magnetosphere, *Planet. Space Sci.*, **12**, 45, 1964.

Axford, W.I. and C.O. Hines, A Unifying Theory of High-Latitude Geophysical Phenomena and Magnetic Storms, *Can. J. Phys.*, **39**, 1433, 1961.

Bame, S.J., A.J. Hundhausen, J.R. Asbridge and I.B. Strong, Solar Wind Ion Composition, *Phys. Rev. Lett.*, **20**, 393, 1968.

Belcher, J.W., H.S. Bridge, F. Bagenal, B. Coppi, O. Divers, A. Eviatar, G.S. Gordon, Jr., A. J. Lazarus, R.L. McNutt, Jr., *et al.*, Plasma Observations Near Neptune: Initial Results from Voyager 2, *Science*, **246**, 1478, 1989.

Belmont, G. and G. Chanteur, Advances in Magnetopause Kelvin-Helmholtz Instability Studies, *Physica Scripta*, **40**, 124, 1989.

Berko, F.W., L.J. Cahill, Jr. and T.A. Fritz, Protons as the Prime Contributors to Storm Time Ring Current, *J. Geophys. Res.*, **80**, 3549, 1975.

Biermann, L., Observed Dynamical Processes in Interplanetary Space, in *Plasma Dynamics*, F.H. Clauser, ed., Addison-Wesley Publishing Co., Reading, MA, 1960.

Birmingham, T.J. and F.C. Jones, Identification of Moving Magnetic Field Lines, 2. Application to a Moving Nonrigid Conductor, *J. Geophys. Res.*, **76**, 1849, 1971.

Biskamp, D., Collisionless Shock Waves in Plasmas, *Nucl. Fusion*, **13**, 719, 1973.

Boström, R., Electrodynamics of the Ionosphere, in *Cosmical Geophysics*, A. Egeland, Ø. Holter and A. Omholt, eds., Scandinavian University Books, UniversitetsforLaget, Oslo, Sweden, 1973.

Bridge, H.S., J.W. Belcher, B. Coppi, A.J. Lazarus, R.L. McNutt, Jr., S. Olbert, J.D. Richardson, M.R. Sands, R.S. Selesnick, J.D. Sullivan, *et al.*, Plasma Observations Near Uranus: Initial Results from Voyager 2, *Science*, **233**, 89, 1986.

Broadfoot, A.L., F. Herbert, J.B. Holberg, D.M. Hunten, S. Kumar, B.R. Sandel, D.E. Shemansky, G.R. Smith, R.V. Yelle, D.F. Strobel, *et al.*, Ultraviolet Spectrometer Observations of Uranus, *Science*, **233**, 74, 1986.

Broadfoot, A.L., S.K. Atreya, J.L. Bertaux, J.E. Blamont, A.J. Dessler, T.M. Donahue, W.T. Forrester, D.T. Hall, *et al.*, Ultraviolet Spectrometer Observations of Neptune and Triton, *Science*, **246**, 1459, 1989.

Brown, R.R., Electron Precipitation in the Auroral Zone, *Space Sci. Rev.*, **6**, 311, 1966.

Büchner, J. and L.M. Zelenyi, Regular and Chaotic Charged Particle Motion in Magnetotaillike Field Reversals. 1. Basic Theory of Trapped Motion, *J. Geophys. Res.*, **94**, 11821, 1989.

Budden, K.G., *The Propagation of Radio Waves*, Cambridge University Press, Cambridge, England, 1985.

Burch, J.L., Observations of Interactions Between Interplanetary and Geomagnetic Fields, *Rev. Geophys. Space Phys.*, **12**, 363, 1974.

Cahill, L.J., Jr., Inflation of the Inner Magnetosphere, in *Physics of the Magnetosphere*, R.L. Carovillano, J.F. McClay and H.R. Radoski, eds., D. Reidel Publishing Co., Dordrecht, Holland, 1968.

Cahill, L.J., Jr. and P.G. Amazeen, The Boundary of the Geomagnetic Field, *J. Geophys. Res.*, **68**, 1835, 1963.

Campbell, W.H., Geomagnetic Pulsations, in *Physics of Geomagnetic Phenomena Vol. 2*, S. Matsushita and W.H. Campbell, eds., Academic Press, New York, NY, 1967.

Carlson, C.W. and R.B. Torbert, Solar Wind Ion Injections in the Morning Auroral Oval, *J. Geophys. Res.*, **85**, 2903, 1980.

Carovillano, R.L. and J.J. Maguire, Magnetic Energy Relationships in the Magnetosphere, in *Physics of the Magnetosphere*, R.L. Carovillano, J.F. McClay and H.R. Radoski, eds., D. Reidel Publishing Co., Dordrecht, Holland, 1968.

Carovillano, R.L. and G.L. Siscoe, Energy and Momentum Theorems in Magnetospheric Processes, *Rev. Geophys. Space Phys.*, **289**, 11, 1973.

Carpenter, D.L., Whistler Studies of the Plasmapause in the Magnetosphere 1. Temporal Variations in the Position of the Knee and Some Evidence on Plasma Motions Near the Knee, *J. Geophys. Res.*, **71**, 693, 1966.

Carpenter, D.L., Recent Research on the Magnetospheric Plasmapause, *Radio Science*, **3**, 719, 1968.

Chamberlain, J.W., Electric Acceleration of Auroral Particles, *Rev. Geophys. Space Phys.*, **7**, 461, 1969.

Chandrasekhar, S., *Plasma Physics*, The University of Chicago Press, Chicago, IL, 1960.

Chapman, S., Historical Introduction to Aurora and Magnetic Storms, *Annales de Geophysique*, **24**, 497, 1968.

Chapman, S. and J. Bartels, *Geomagnetism*, Clarendon Press, Oxford, England, 1940.

Chapman, S. and T.G. Cowling, *The Mathematical Theory of Non-Uniform Gases*, 2nd ed., Cambridge University Press, Cambridge, England, 1952.

Chapman, S. and V.C.A. Ferraro, A New Theory of Magnetic Storms, *Terr. Magn. Atmos. Elec.*, **36**, 171, 1931.

Chappell, C.R., Recent Satellite Measurements of the Morphology and Dynamics of the Plasmasphere, *Rev. Geophys. Space Phys.*, **10**, 951, 1972.

Chappell, R., K.K. Harris and G.W. Sharp, A Study of the Influence of Magnetic Activity on the Location of the Plasmapause as Measured by OGO 5, *J. Geophys. Res.*, **75**, 50, 1970.

Chen, F.F., *Introduction to Plasma Physics*, Plenum Press, New York, NY, 1974.

Chen, J. and P.J. Palmadesso, Chaos and Nonlinear Dynamics of Single-Particle Orbits in a Magnetotaillike Magnetic Field, *J. Geophys. Res.*, **91**, 1499, 1986.

Chernikov A.A., R.Z. Sagdeev and G.M. Zaslavsky, Chaos: How Regular Can it Be?, *Physics Today*, **27**, November, 1988.

Clemmow, P.C. and J.P. Dougherty, *Electrodynamics of Particles and Plasmas*, Addison-Wesley Publishing Co., Reading, MA, 1969.

Cole, K.D., Magnetic Storms and Associated Phenomena, *Space Sci. Rev.*, **6**, 699, 1966.

Conrath, B., F.M. Flasar, R. Hanel, V. Kunde, W. Maguire, J. Pearl, J. Pirraglia, R. Samuelson, P. Gierasch, A. Weir, B. Bezard, D. Gautier, D. Cruikshank, *et al.*, Infrared Observations of the Neptunian System, *Science*, **246**, 1454, 1989.

Cowley, S.W.H., A Self-Consistent Model of a Simple Magnetic Neutral Sheet System Surrounded by a Cold, Collisionless Plasma, *Cosmic Electrodynamics*, **3**, 448, 1973.

Cowley, S.W.H., Plasma Populations in a Simple Open Model Magnetosphere, *Space Sci. Rev.*, **26**, 217, 1980.

Cowling, T.G., On Alfvén's Theory of Magnetic Storms and of the Aurora, *Terrest. Magnetism and Atmos. Elec.*, **47**, 209, 1942.

Davis, L., Jr., The Configuration of the Interplanetary Magnetic Field, in *Solar Terrestrial Physics, 1970*, E.R. Dyer, J.G. Roederer and A.J. Hundhausen, eds., D. Reidel Publishing Co., Dordrecht, Holland, 1972.

DeForest, S.S. and C.E. McIlwain, Plasma Clouds in the Magnetosphere, *J. Geophys. Res.*, **76**, 3587, 1971.

Dessler, A.J., Solar Wind and Interplanetary Magnetic Fields, *Rev. Geophys.*, **5**, 1, 1967.

Dessler, A.J., Magnetic Merging in the Magnetospheric Tail, *J. Geophys. Res.*, **73**, 209, 1967.

Dessler, A.J., Vacuum Merging: A Possible Source of the Magnetospheric Cross Tail Electric Field, *J. Geophys. Res.*, **76**, 3174, 1971.

Donahue, T.M., Polar Ion Flow: Wind or Breeze?, *Rev. Geophys. Space Phys.*, **9**, 1, 1971.

Dungey, J.W., *Cosmic Electrodynamics*, Cambridge University Press, Cambridge, England, 1958.

Dungey, J.W., Interplanetary Magnetic Field and the Auroral Zones, *Phys. Rev. Letters*, **6**, 47, 1961.

Dungey, J.W., Neutral Sheets, *Space Sci. Rev.*, **17**, 173, 1975.

Dunne, J.A., Mariner 10 Mercury Encounter, *Science*, **185**, 141, 1974.

Eastman, T.E., E.W. Hones, Jr., S.J. Bame and J.R. Asbridge, The Magnetospheric Boundary Layer: Site of Plasma, Momentum and Energy Transfer from the Magnetosheath into the Magnetosphere, *Geophys. Res. Lett.*, **3**, 685, 1976.

Elphic, R. and C. Russell, ISEE-1 and -2 Magnetometer Observations of the Magnetopause, in *Magnetospheric Boundary Layers*, B. Battrick, ed., Rep. ESA SP-148, p. 43, European Space Agency, Paris, France, 1979.

Evans, D.S., Precipitating Electron Fluxes Formed by a Magnetic Field Aligned Potential Difference, *J. Geophys. Res.*, **79**, 2853, 1974.

Fairfield, D.H., Solar Wind Control of the Magnetotail, in *Solar Wind-Magnetospheric Coupling*, Y. Kamide and J.A. Slavin, eds., Terra Scientific Publishing Co., Tokyo, Japan, 1986.

Feldman, W.C., Electron Velocity Distributions Near Collisionless Shocks, in *Collisionless Shocks in the Heliosphere: Review of Current Research*, B.T. Tsurutani and R.G. Stone, eds., Geophysical Monograph, American Geophysical Union, Washington, D.C., 1985.

Feldstein, Y.I. and Y.I. Galperin, The Auroral Luminosity Structure in the High-Latitude Upper Atmosphere: Its Dynamics and Relationship to the Large-Scale Structure of the Earth's Magnetosphere, *Rev. Geophys. Space Phys.*, **23**, 217, 1985.

Ferraro, V.C.A., On the Theory of the First Phase of a Geomagnetic Storm: A New Illustrative Calculation Based on an Idealized (Plane, not Cylindrical) Model Field Distribution, *J. Geophys. Res.*, **57**, 15, 1952.

Ferraro, V.C.A. and C. Plumpton, *An Introduction to Magneto-Fluid Mechanics*, 2nd. ed., Clarendon Press, Oxford, England, 1966.

Filbert, P. and P. Kellogg, Electrostatic Noise at the Plasma Frequency Beyond the Earth's Bow Shock, *J. Geophys. Res.*, **84**, 1369, 1979.

Frank, L.A., On the Extraterrestrial Ring Current During Geomagnetic Storms, *J. Geophys. Res.*, **72**, 3753, 1967.

Freden, S.C. and R.S. White, Particle Fluxes in the Inner Radiation Belt, *J. Geophys. Res.*, **65**, 1377, 1960.

Fuselier, S.A. and D.A. Gurnett, Short Wavelength Ion Waves Upstream of the Earth's Bow Shock, *J. Geophys. Res.*, **89**, 91, 1984.

Galeev, A.A., M.M. Kuznetsova and L.M. Zeleny, Magnetopause Stability Threshold for Patchy Reconnection, *Space Sci. Rev.*, **44**, 1, 1986.

Gehrels, T., ed., *Jupiter*, University of Arizona Press, Tucson, AZ, 1976.

Gendrin, R., Substorm Aspects of Magnetic Pulsations, *Space Sci. Rev.*, **11**, 54, 1970.

Gendrin, R., General Relationships Between Wave Amplfiication and Particle Diffusion in a Magnetoplasma, *Rev. Geophys. Space Phys.*, **19**, 171, 1984.

Goertz, C.K., Double Layers and Electrostatic Shocks in Space, *Rev. Geophys. Space Phys.*, **17**, 418, 1979.

Gold, T., Magnetic Storms, *Space Sci. Rev.*, **1**, 100, 1962.

Grad, H., Boundary Layer Between a Plasma and a Magnetic Field, *Phys. Fluids*, **4**, 1366, 1961.

Gringauz, K.I., The Structure of the Plasmasphere on the Basis of Direct Measurements, in *Solar Terrestrial Physics 1970*, E.R. Dyer, ed., D. Reidel Publishing Co., Dordrecht, Holland, 1972.

Gurevich, A.V., A.L. Krylov and E.E. Tsedilina, Electric Fields in the Earth's Magnetosphere and Ionosphere, *Space Sci. Rev.*, **19**, 59, 1976.

Gurnett, D.A., Plasma Waves and Instabilities, in *Collisionless Shocks in the Heliosphere: Reviews of Current Research*, B.T. Tsurutani and R.G. Stone, eds., Geophysical Monograph, American Geophysical Union, Washington, D.C., 1985.

Gurnett, D.A., R.R. Anderson, B.T. Tsurutani, E.J. Smith, G. Paschmann, G. Haerendel, S.J. Bame and C.T. Russell, Plasma Wave Turbulence at the Magnetopause: Observations from ISEE 1 and 2, *J. Geophys. Res.*, **84**, 7043, 1979.

Gurnett, D.A., R.R. Anderson, B.T. Tsurutani, E.J. Smith, G. Paschmann, G. Haerendel, S.J. Bame and C.T. Russell, Plasma Wave Turbulence at the Magnetopause: Observations from ISEE 1 and 2, *J. Geophys. Res.*, **84**, 7043, 1979.

Gurnett, D.A., W.S. Kurth, F. L. Scarf and R.L. Poynter, First Plasma Wave Observations at Uranus, *Science*, **233**, 106, 1986.

Gurnett, D.A., W.S. Kurth, R.L. Poynter, L.J. Granroth, I.H. Cairns, W.M. Macek, S.L. Moses, F.V. Coroniti, C.F. Kennel and D.D. Barbosa, First Plasma Wave Observations at Neptune, *Science*, **246**, 1494, 1989.

Hall, C.F., Pioneer 10, *Science*, **183**, 301, 1974.

Hall, C.F., Pioneer 10 and 11, *Science*, **188**, 445, 1975.

Hamlin, D.A., R. Karplus, R.C. Vik and K.M. Watson, Mirror and Azimuthal Drift Frequencies for Geomagnetically Trapped Particles, *J. Geophys. Res.*, **66**, 1, 1961.

Hanel, R., B. Conrath, F.M. Flasar, V. Kunde, W. Maguire, J. Pearl, J. Pirraglia, R. Samuelson, D. Cruikshank, D. Gautier, P. Gierasch, L. Horn, *et al.*, Infrared Observations of the Uranian System, *Science*, **233**, 70, 1986.

Harris, K.K., G.W. Sharp and C.R. Chappell, Observations of the Plasmapause from OGO 5, *J. Geophys. Res.*, **75**, 219, 1970.

Hasegawa, A., *Plasma Instabilities and Nonlinear Effects*, Springer-Verlag, New York, NY, 1975.

Hasegawa, A. and T. Sato, Generation of Field Aligned Current During Substorm, in *Dynamics of the Magnetosphere*, S.-I. Akasofu, ed., D. Reidel Publishing Co., Dordrecht, Holland, 1980.

Heikkila, W.J., Aurora, *EOS*, **54**, 764, 1973.

Helliwell, R.A., *Whistlers and Related Ionospheric Phenomena*, Stanford University Press, Stanford, CA, 1965.

Herzberg, L., Solar Optical Radiation and its Role in Upper Atmospheric Processes, in *Physics of the Earth's Upper Atmosphere*, C.O. Hines, I. Paghis, T.R. Hartz and J.A. Fejer, eds., Prentice-Hall, Inc., Englewood Cliffs, NJ, 1965.

Hess, W.N., *The Radiation Belt and Magnetosphere*, Blaisdell Publishing Co., Waltham, MA, 1968.

Hill, T.W., Origin of the Plasma Sheet, *Rev. Geophys. Space Phys.*, **12**, 379, 1974.

Holzer, T.E. and W.I. Axford, The Theory of Stellar Winds and Related Flows, *Annual Rev. of Astronomy and Astrophysics*, **8**, 31, 1970.

Hones, E.W., Jr., Transient Phenomena in the Magnetotail and their Relation to Substorms, *Space Sci. Rev.*, **23**, 393, 1979.

Hones, E.W., Jr., ed., *Magnetic Reconnection in Space and Laboratory Plasmas*, American Geophysical Union, Washington, D.C., 1984.

Hoppe, M.M. and C.T. Russell, Plasma Rest Frame Frequencies and Polarizations of the Low-Frequency Upstream Waves: ISEE 1 and 2 Observations, *J. Geophys. Res.*, **88**, 2021, 1983.

Horwitz, J.L., The Ionosphere as a Source for Magnetospheric Ions, *Rev. Geophys. Space Phys.*, **20**, 929, 1982.

Hudson, P.D., Discontinuities in an Anisotropic Plasma and their Identification in the Solar Wind, *Planet. Space Sci.*, **18**, 1611, 1970.

Hundhausen, A.J., Composition and Dynamics of the Solar Wind Plasma, *Rev. of Geophys. and Space Phys.*, **8**, 729, 1970.

Hundhausen, A.J., J.R. Asbridge, S.J. Bame, H.E. Gilbert and I.B. Strong, Vela 3 Satellite Observations of Solar Wind Ions: A Preliminary Report, *J. Geophys. Res.*, **72**, 87, 1967.

Hundhausen, A.J., S. Bame and N. Ness, Solar Wind Thermal Anisotropies: Vela 3 and IMP 3, *J. Geophys. Res.*, **72**, 5265, 1967.

Ichimaru, S., *Basic Principles of Plasma Physics, A Statistical Approach*, W.A. Benjamin Publishing Co., Inc., Reading, MA, 1973.

Jackson, J.D., *Classical Electrodynamics*, 2nd ed., John Wiley and Sons, Inc., New York, NY, 1975.

Jones, F.C. and Birmingham, T.J., Identification of Moving Magnetic Field Lines, *J. Geophys. Res.*, **76**, 1849, 1971.

Kamide, Y., *Electrodynamic Processes in the Earth's Ionosphere and Magnetosphere*, Kyoto Sangyo University Press, Kyoto, Japan, 1988.

Kamide, Y. and G. Rostoker, The Spatial Relationship of Field-Aligned Currents and Auroral Electrojets to the Distribution of Nightside Auroras, *J. Geophys. Res.*, **82**, 5589, 1977.

Kan, J.R., Comment on Double Layers and Electrostatic Shocks in Space, *Rev. Geophys. Space Phys.*, **18**, 337, 1980.

Kan, J.R., Reply to Comment on "Double Layers and Electrostatic Sheets in Space"by C.K. Goertz, *Rev. Geophys. Space Phys.*, **19**, 230, 1981.

Kan, J.R., Towards a Unified Theory of Discrete Auroras, *Space Sci. Rev.*, **31**, 71, 1982.

Katz, L. and D. Smart, Measurements on Trapped Particles Injected by Nuclear Detonations, *Space Res.*, **4**, 646, 1963.

Kaufmann, R.L., What Auroral Electron and Ion Beams Tell Us About Magnetosphere-Ionosphere Coupling, *Space Sci. Rev.*, **37**, 313, 1984.

Kavanagh, L.D., Jr., Discussion of Paper by M.P. Nakada and G.D. Mead, "Diffusion of Protons in the Outer Radiation Belt," *J. Geophys. Res.*, **72**, 6120, 1967.

Kim, J.S., D.A. Graham and C.S. Wang, Inductive Electric Fields in the Ionosphere-Magnetosphere System, *Rev. Geophys. Space Phys.*, **17**, 2049, 1979.

Knecht, D.J. and B.M. Shuman, The Geomagnetic Field, in *Handbook of Geophysics and the Space Environments*, ADA167000, A.S. Jursa, ed., Air Force Geophysics Laboratory, National Technical Information Service, Springfield, VA, 1985.

Krall, N.A. and A.W. Trivelpiece, *Principles of Plasma Physics*, McGraw Hill Book Co., New York, NY, 1973.

Krimigis, S.M., T.P. Armstrong, W.I. Axford, A.F. Cheng, G. Gloeckler, D.C. Hamilton, E.P. Keath, L.J. Lanzerotti, *et al.*, The Magnetosphere of Uranus: Hot Plasma and Radiation Environment, *Science*, **233**, 97, 1986.

Krimigis, S.M., T.P. Armstrong, W.I. Axford, C.O. Bostrom, A.F. Cheng, G. Gloeckler, D.C. Hamilton, E.P. Keath, *et al.*, Hot Plasma and Energetic Particles in Neptune's Magnetosphere, *Science*, **246**, 1483, 1989.

LaBelle, J. and P. Kintner, The Measurement of Wavelength in Space Plasmas, *Rev. Geophys.*, **27**, 495, 1989.

Landau, L.D. and E.M. Liftschitz, *Statistical Physics*, Pergamon Press, Ltd., London, England, 1958.

Landau, L.D. and E.M. Liftshitz, *Fluid Mechanics*, Pergamon Press, Ltd., London, England, 1959.

Landau, L.D. and E.M. Liftshitz, *Electrodynamics of Continuous Media*, Pergamon Press, Ltd., New York, NY, 1960.

Lane, A.L., C.W. Hord, R.A. West, L.W. Esposito, K.E. Simmons, R.M. Nelson, B.D. Wallis, B.J. Buratti, *et al.*, Photometry from Voyager 2: Initial Results from the Uranian Atmosphere, Satellites, and Rings, *Science*, **233**, 65, 1986.

Lane, A.L., R.A. West, C.W. Hord, R.M. Nelson, K.E. Simmons, W.R. Pryor, L.W. Esposito, L.J. Horn, *et al.*, Photometry from Voyager 2: Initial Results from the Neptunian Atmosphere, Satellites, and Rings, *Science*, **246**, 1450, 1989.

Lanzerotti, L.J., Geospace—The Earth's Plasma Environment, *Astronautics and Aeronautics*, April, 1981.

Lemaire, J. and M. Scherer, Kinetic Models of the Solar and Polar Winds, *Rev. Geophys. Space Phys.*, **11**, 427, 1973.

Lichtenberg, A.J. and M.A. Lieberman, *Regular and Stochastic Motion*, Springer-Verlag, New York, NY, 1983.

Lin, C.S. and R.A. Hoffman, Observations of Inverted-V Electron Precipitation, *Space Sci. Rev.*, **33**, 415, 1982.

Lin, R.P., C.-I. Meng and K.A. Anderson, 30-to 100-keV Protons Upstream from the Earth's Bow Shock, *J. Geophys. Res.*, **79**, 489, 1974.

Livesey, W.A., C.F. Kennel and C.T. Russell, ISEE-1 and -2 Observations of Magnetic Field Strength Overshoots in Quasi-Perpendicular Bow Shocks, *Geophys. Res. Lett.*, **9**, 1037, 1982.

Longmire, C.L., *Elementary Plasma Physics*, Interscience Publishers, New York, NY, 1963.

Low, B.C., Nonlinear Force-Free Magnetic Fields, *Rev. Geophys. Space Phys.*, **20**, 145, 1982.

Lyons, L.R. and D.J. Williams, *Quantitative Aspects of Magnetospheric Physics*, D. Reidel Publishing Co., Boston, MA, 1984.

McDonald, K.L., Topology of Steady Current Magnetic Fields, *Am. J. Phys.*, **22**, 586, 1954.

McFadden, J., C.W. Carlson, M.H. Boehm and T.J. Hallinan, Field-Aligned Electron Flux Oscillations that Produce Flickering Aurora, *J. Geophys. Res.*, **92**, 11133, 1987.

McIlwain, C.E., Coordinates for Mapping the Distribution of Magnetically Trapped Particles, *J. Geophys. Res.*, **66**, 3681, 1961.

McPherron, R.L., Magnetospheric Substorms, *Rev. Geophys. and Space Phys.*, **17**, 657, 1979.

Mead, G.D. and D.B. Beard, Shape of the Geomagnetic Field Solar Wind Boundary, *J. Geophys. Res.*, **69**, 1169, 1964.

Meng, C.-I., Electron Precipitation and Polar Auroras, *Space Sci. Rev.*, **22**, 223, 1978.

Merrill, R.T. and M.W. McElhinny, *The Earth's Magnetic Field*, Academic Press, London, England, 1983.

Mishin, V.M., High-Latitude Geomagnetic Variations and Substorms, *Space Sci. Rev.*, **6**, 621, 1977.

Mizera, P.F. and J.F. Fennell, Signatures of Electric Fields from High and Low Altitude Particle Distributions, *Geophys. Res. Lett.*, **4**, 311, 1977.

Moore, T.E., Superthermal Ionospheric Outflows, *Rev. Geophys. Space Phys.*, **22**, 264, 1984.

Morozov, A.I. and L.S. Solov'ev, The Structure of Magnetic Fields, in *Reviews of Plasma Physics, Volume 2*, M.A. Leontovich, ed., Consultants Bureau, a division of Plenum Publishing Corp., New York, NY, 1966.

Morozov, A.I. and L.S. Solov'ev, Motion of Charged Particles in Electromagnetic Fields, in *Reviews of Plasma Physics, Volume 2*, M.A. Leontovich, ed., Consultants Bureau, a division of Plenum Publishing Corp., New York, NY, 1966.

Mozer, F.S., Electric Field Evidence on the Viscous Interaction at the Magnetopause, *Geophys. Res. Lett.*, **11**, 135, 1984.

Mozer, F.S., R.B. Torbert, U.V. Fahleson, C.G. Fälthammar, A. Gonfalone and A. Pedersen, Measurements of Quasi-Static and Low-Frequency Electric Fields with Spherical Double Probes on the ISEE-1 Spacecraft, *IEEE Transactions on Geoscience Electronics*, **GE-16**, No. 3, 1978.

Ness, N.F., Magnetometers for Space Research, *Space Sci. Rev.*, **11**, 459, 1970.

Ness, N.F., Interaction of the Solar Wind with the Moon, in *Solar Terrestrial Physics*, 1970, E.R. Dyer, ed., D. Reidel Publishing Co., Dordrecht, Holland, 1972.

Ness, N.F., K.W. Behannon, R.P. Lepping, Y.C. Whang and K.H. Schatten, Magnetic Field Observations near Mercury: Preliminary Results from Mariner 10, *Science*, **185**, 151, 1974.

Ness, N.F., C.S. Searce and J.B. Seek, Initial Results of the IMP-1 Magnetic Field Experiment, *J. Geophys. Res.*, **69**, 3531, 1964.

Ness, N.F., M.H. Acuña, K.W. Behannon, L.F. Burlaga, J.E.P. Connerney, R.P. Lepping and F.M. Neubauer, Magnetic Fields at Uranus, *Science*, **233**, 85, 1986.

Ness, N.F., M.H. Acuña, L.F. Burlaga, J.E.P. Connerney, R.P. Lepping, F.M. Neubauer, Magnetic Fields at Neptune, *Science*, **246**, 1473, 1989.

Neugebauer, M. and C.W. Snyder, Mariner 2 Observations of the Solar Wind, 1. Average Properties, *J. Geophys. Res.*, **71**, 4469, 1966.

Newcomb, W.A., Motion of Magnetic Lines of Force, *Annals of Physics*, **3**, 347, 1958.

Nicholson, D.R., *Introduction to Plasma Theory*, John Wiley and Sons, New York, NY, 1983.

Nishida, A., Formation of Plasmapause, or Magnetospheric Plasma Knee, by the Combined Action of Magnetospheric Convection and Plasma Escape from the Tail, *J. Geophys. Res.*, **71**, 5669, 1966.

Nishida, A., IMF Control of the Earth's Magnetosphere, *Space Sci. Rev.*, **34**, 185, 1983.

Northrop, T.G., The Guiding Center Approximation to Charged Particle Motion, *Annals of Physics*, **15**, 209, 1961.

Northrop, T.G., *The Adiabatic Motion of Charged Particles*, Interscience Publishers, New York, NY, 1963.

O'Brien, B.J., Interrelations of Energetic Charged Particles in the Magnetosphere, in *Solar Terrestrial Physics*, J.W. King and W.S. Newman, eds., Academic Press, Inc., London, England, 1967.

Obayashi, T., The Streaming of Solar Flare Particles and Plasma in Interplanetary Space, *Space Sci. Rev.*, **3**, 79, 1964.

Obayashi, T. and A. Nishida, Large-Scale Electric Field in the Magnetosphere, *Space Sci. Rev.*, **8**, 3, 1968.

Ögelman, H. and J.R. Wayland, eds., *Introduction to Experimental Techniques of High Energy Astrophysics*, NASA Publication, Washington, D.C., 1970.

Ogilvie, K.W., R.J. Fitzenreiter and J.D. Scudder, Observations of Electron Beams in the Low-Latitude Boundary Layer, *J. Geophys. Res.*, **89**, 10723, 1984.

Olson, W.P., ed., *Quantitative Modeling of Magnetospheric Processes*, American Geophysical Union, Washington, D.C., 1979.

Parker, E.N., *Interplanetary Dynamical Processes*, Interscience Publishers, New York, NY, 1963.

Parker, E.N., The Solar-Flare Phenomenon and the Theory of Reconnection and Annihilation of Magnetic Fields, *Astrophys. J. Suppl.*, **8**, 177, 1963.

Parker, E.N., *Cosmical Magnetic Fields, Their Origin and Their Activity*, Oxford University Press, New York, NY, 1979.

Parks, G.K., G. Laval and R. Pellat, Behavior of Outer Radion Zone and a New Model of Magnetospheric Substorm, *Planet. Space Sci.*, **20**, 1291, 1972.

Paschmann, G., N. Sckopke, G. Haerendel, I. Papamastorakis, S.J. Bame, J.R. Asbridge, J.T. Gosling, E.W. Hones, Jr. and E.R. Tech, ISEE Plasma Observations Near the Subsolar Magnetopause, *Space Sci. Rev.*, **22**, 717, 1978.

Paschmann, G., N. Sckopke, J.R. Asbridge, S.J. Bame and J.T. Gosling, Energization of Solar Wind Ions by Reflection from the Earth's Bow Shock, *J. Geophys. Res.*, **85**, 4689, 1980.

Paschmann, G., N. Sckopke, I. Papamastorakis, J.R. Asbridge, S.J. Bame and J.T. Gosling, Characteristics of Reflected and Diffuse Ions Upstream from the Earth's Bow Shock, *J. Geophys. Res.*, **86**, 4355, 1981.

Peck, E.R., *Electricity and Magnetism*, McGraw-Hill Book Co., Inc., New York, NY, 1953.

Petschek, H.E., Magnetic Field Annihilation, in *The Physics of Solar Flares*, W.N. Hess, ed., Proceedings of the AAS-NASA Symposium held at Goddard Space Flight Center, NASA SP-50, Greenbelt, MD, October 1964.

Potemra, T., ed., *Magnetospheric Currents*, Geophysics Monograph, American Geophysical Union, Washington, D.C., 1984.

Pudovkin, M.I., Electric Fields and Currents in the Ionosphere, *Space Sci. Rev.*, **16**, 727, 1974.

Pudovkin, M.I. and V.S. Semenov, Magnetic Field Reconnection Theory and the Solar Wind-Magnetosphere Interaction: A Review, *Space Sci. Rev.*, **41**, 1, 1985.

Roederer, J.G., *Dynamics of Geomagnetically Trapped Radiation*, Springer-Verlag, New York, NY, 1970.

Roederer, J.G., The Earth's Magnetosphere, *Science*, **183**, 37, 1974.

Roeloff, *Physics of Solar Planetary Environments*, American Geophysical Union, Washington, D.C., 1976.

Rosenbluth, M.N., *Plasma Physics and Thermonuclear Research*, C.L. Longmire, J.L. Tuck and W.B. Thompson, eds., Pergamon Press, London, England, **2**, 271, 1963.

Rossi, B.B. and S. Olbert, *Introduction to the Physics of Space*, McGraw Hill Book Co., New York, NY, 1970.

Rostoker, G., Polar Magnetic Substorms, *Rev. Geophys. Space Phys.*, **10**, 157, 1972.

Rostoker, G., S.-I. Akasofu, W. Baumjohann, Y. Kamide and R.L. McPherron, The Holes of Direct Input of Energy from the Solar Wind and Unloading of Stored Magnetotail Energy in Driving Magnetospheric Substorms, *Space Sci. Rev.*, **46**, 93, 1987.

Rothwell, P.L. and G.K. Yates, Global Single Ion Effects Within the Earth's Plasma Sheet, in *Magnetic Reconnection in Space and Laboratory Plasmas*, E.W. Hones, Jr., ed., American Geophysical Union, Washington, D.C., 1984.

Russell, C.T., Planetary Bow Shocks, in *Collisionless Shocks in the Heliosphere: Reviews of Current Research*, B.T. Tsurutani and R.G. Stone, eds., Geophysical Monograph, American Geophysical Union, Washington, DC, 1985.

Russell, C.T. and R.L. McPherron, The Magnetotail and Substorms, *Space Sci. Rev.*, **15**, 205, 1973.

Russell, C.T. and R.C. Elphic, ISEE Observations of Flux Transfer Events at the Dayside Magnetopause, *Geophys. Res. Lett.*, **6**, 33, 1979.

Russell, C.T., M.M. Hoppe and W.A. Livesey, Overshoots in Planetary Bow Shocks, *Nature*, **296**, 45, 1982.

Salingaros, N., On Solutions of the Equation $\nabla \times \mathbf{a} = k\mathbf{a}$, *J. Physics A*, **19**, L101, 1986.

Sauvaud, J.A., J.P. Treilhou, A. Saint-Marc, J. Dandouras, H. Rème, A. Korth, G. Kremser, G.K. Parks, A.N. Zaitzev, V. Petrov, L. Lazutine and R. Pellinen, Large Scale Response of the Magnetosphere to a Southward Turning of the Interplanetary Field, *J. Geophys. Res.*, **92**, 2365, 1987.

Schardt, A.W. and A.G. Opp, Particles and Fields: Significant Achievements, 1, *Rev. Geophys. Space Phys.*, **5**, 411, 1967.

Schardt, A.W. and A.G. Opp, Particles and Fields: Significant Achievements, 2, *Rev. Geophys. Space Phys.*, **7**, 799, 1969.

Schield, M.A. and L.A. Frank, Electron Observations Between the Inner Edge of the Plasma Sheet and the Plasmasphere, *J. Geophys. Res.*, **75**, 5401, 1970.

Schindler, K., Plasma and Fields in the Magnetospheric Tail, *Space Sci. Rev.*, **17**, 589, 1975.

Schindler, K. and J. Birn, Magnetotail Theory, *Space Science Rev.*, **44**, 307, 1986.

Schmidt, G., *Physics of High Temperature Plasmas, An Introduction*, Academic Press, New York, NY, 1966.

Schulz, M., Interplanetary Sector Structure and the Heliomagnetic Equator, *Astrophys. and Space Sci.*, **24**, 371, 1973.

Schulz, M. and L.J. Lanzerotti, *Particle Diffusion in the Radiation Belts*, Springer-Verlag, Berlin, Germany, 1974.

Schwartz, S.J., M.F. Thomsen and J.T. Gosling, Ions Upstream of the Earth's Bow Shock: A Theoretical Comparison of Alternative Source Populations, *J. Geophys. Res.*, **88**, 2039, 1983.

Shebanski, V.P., Magnetospheric Processes and Related Geophysical Phenomena, *Space Sci. Rev.*, **8**, 366, 1968.

Silsbee, H. and E. Vestini, Geomagnetic Bays, Their Frequency and Current Systems, *Terr. Mag. and Atmos. Elec*, **47**, 195, 1942.

Siscoe, G.L., Towards a Comparative Theory of Magnetospheres, in *Solar System Plasma Physics*, E.N. Parker, C.F. Kennel and L.J. Lanzerotti, eds., North Holland Publishing Co., Amsterdam, the Netherlands, 1979.

Siscoe, G.L., Solar System Magnetohydrodynamics, in *Proceedings of the 1982 Boston College Theory Institute in Solar-Terrestrial Physics*, R.L. Carovillano and J.M. Forbes, eds., D. Reidel Publishing Co., Dordrecht, Holland, 1982.

Smith, B.A., L.A. Soderblom, R. Beebe, D. Bliss, J.M. Boyce, A. Brahic, G.A. Briggs, R.H. Brown, S.A. Collins, A.F. Cook II, *et al.*, Voyager 2 in the Uranian System: Imaging Science Results, *Science*, **233**, 43, 1986.

Smith, B.A., L.A. Soderblom, D. Banfield, C. Barnet, A.T. Basilevsky, R.F. Beebe, K. Bollinger, J.M. Boyce, A. Brahic, G.A. Briggs, *et al.*, Voyager 2 at Neptune: Imaging Science Results, *Science*, **246**, 1422, 1989.

Smith, E.J., B.T. Tsurutani and R.L. Rosenberg, Observations of the Interplanetary Sector Structure up to Heliographic Latitudes of 16°: Pioneer 11, *J. Geophys. Res.*, **83**, 717, 1978.

Sonett, C.P. and I.J. Abrams, The Distant Geomagnetic Field, 3. Disorder and Shocks in the Magnetopause, *J. Geophys. Res.*, **68**, 1233, 1963.

Sonnerup, B.U.Ö., Acceleration of Particles Reflected at a Shock Front, *J. Geophys. Res.*, **74**, 1301, 1969.

Sonnerup, B.U.Ö., Adiabatic Particle Orbits in a Magnetic Null Sheet, *J. Geophys. Res.*, **76**, 8211, 1971.

Sonnerup, B.U.Ö. and L.J. Cahill, Jr., Explorer 12 Observations of the Magnetopause Current Layer, *J. Geophys. Res.*, **73**, 1757, 1968.

Sonnerup, B.U.Ö., G. Paschmann, I. Papamastorakis, N. Sckopke, G. Haerendel, S.J. Bame, J.R. Asbridge, J.T. Gosling, and C.T. Russell, Evidence for Magnetic Field Reconnection at the Earth's Magnetopause, *J. Geophys. Res.*, **86**, 10049, 1981.

Speiser, T.W., Particle Trajectories in Model Current Sheets, 1. Analytical Solutions, *J. Geophys. Res.*, **70**, 4219, 1965.

Spitzer, L., *Physics of Fully Ionized Gases*, 2nd ed., Interscience Publishers, New York, NY, 1962.

Spjeldvik, W.N. and P.L. Rothwell, The Radiation Belts, in *Handbook of Geophysics and the Space Environments*, ADA167000, A.S. Jursa, ed., Air Force Geophysics Laboratory, National Technical Information Service, Springfield, VA, 1965.

Spreiter, J.R. and A.Y. Alskne, Plasma Flow Around the Magnetosphere, *Rev. Geophys. Space Phys.*, **7**, 11, 1969.

Stern, D.P., The Motion of Magnetic Field Lines, *Space Sci. Rev.*, **6**, 147, 1966.

Stern, D.P., Euler Potentials, *Am. J. of Phys.*, **38**, 494, 1970.

Stern, D.P., One Dimensional Models of Quasi-Neutral Parallel Electric Fields, *J. Geophys. Res.*, **86**, 5839, 1981.

Stern, D.P., The Origins of Birkeland Currents, *Rev. Geophys. Space Phys.*, **21**, 125, 1983.

Stern, D.P., Energetics of the Magnetosphere, *Space Sci. Rev.*, **39**, 193, 1984.

Stix, T.H., *The Theory of Plasma Waves*, McGraw Hill, New York, NY, 1962.

Stone, E.C. and A.L. Lane, Voyager 1 Encounter with the Jovian System, *Science*, **204**, 945, 1979.

Stone, E.C. and A.L. Lane, Voyager 2 Encounter with the Jovian System, *Science*, **206**, 925, 1979.

Stone, E.C. and E.D. Miner, Voyager 1 Encounter with the Saturnian System, *Science*, **212**, 159, 1981.

Stone, E.C. and E.D. Miner, Voyager 2 Encounter with the Saturnian System, *Science*, **215**, 499, 1982.

Stone, E.C. and E.D. Miner, The Voyager 2 Encounter with the Uranian System, *Science*, **233**, 39, 1986.

Stone, E.C., J.F. Cooper, A.C. Cummings, F.B. McDonald, J.H. Trainor, N. Lal, R. McGuire, D.L. Chenette, Energetic Charged Particles in the Uranian Magnetosphere, *Science*, **233**, 93, 1986.

Stone, E.C. and E.D. Miner, The Voyager 2 Encounter with the Neptunian System, *Science*, **246**, 1417, 1989.

Stone, E.C., A.C. Cummings, M.D. Looper, R.S. Selesnick, N. Lal, F.B. McDonald, J.H. Trainor and D.L. Chenette, Energetic Charged Particles in the Magnetosphere of Neptune, *Science*, **246**, 1489, 1989.

Stone, R.G. and B.T. Tsurutani, eds., *Collisionless Shocks in the Heliosphere: A Tutorial Review*, Geophysical Monograph, American Geophysical Union, Washington DC, 1985.

Storey, L.R.O., An Investigation of Whistling Atmospherics, *Phil. Trans. Roy. Soc.* (London), A, **246**, 113, 1953.

Storey, L.R.O., Whistlers, *Sci. American*, **194**, 1, 34, 1956.

Swift, D.W., Mechanisms for Auroral Precipitation: A Review, *Rev. Geophys. Space Phys.*, **19**, 185, 1981.

Thompson, W.B., *An Introduction to Plasma Physics*, Addison-Wesley Publishing Co., Inc., Reading, MA, 1962.

Tidman, D.A. and N.A. Krall, *Shock Waves in Collisionless Plasmas*, Interscience Publishers, New York, NY, 1971.

Troshichev, O.A., Polar Magnetic Disturbances and Field-Aligned Currents, *Space Sci. Rev.*, **32**, 275, 1982.

Tsurutani, B.T. and P. Rodriguez, Upstream Waves and Particles: An Overview of ISEE Results, *J. Geophys. Res.*, **86**, 4319, 1981.

Tsurutani, B.T. and R.G. Stone, eds., *Collisionless Shocks in the Heliosphere: Reviews of Current Research*, Geophysical Monograph, American Geophysical Union, Washington DC, 1985.

Tyler, G.L., D.N. Sweetnam, J.D. Anderson, J.K. Campbell, V.R. Eshleman, D.P. Hinson, G.S. Levy, G.F. Lindal, *et al.*, Voyager 2 Radio Science Observations of the Uranian System: Atmosphere, Rings, and Satellites, *Science*, **233**, 79, 1986.

Tyler, G.L., D.N. Sweetnam, J.D. Anderson, S.E. Borutzki, J.K. Campbell, V.R. Eshleman, D.L. Gresh, E.M. Gurrola, D.P. Hinson, *et al.*, Voyager Radio Science Observations of Neptune and Triton, *Science*, **246**, 1466, 1989.

Van Allen, J.A., Particle Description of the Magnetosphere, in *Physics of the Magnetosphere*, R.L. Carovillano, J.F. McClay and H.R. Radoski, eds., D. Reidel Publishing Co., Dordrecht, Holland, 1968.

Vasyliunas, V.M., A Survey of Low Energy Electrons in the Evening Sector of the Magnetosphere with OGO 1 and OGO 3, *J. Geophys. Res.*, **73**, 2839, 1968.

Vasyliunas, V.M., Magnetospheric Plasma, in *Solar Terrestrial Physics 1970*, E.R. Dyer and J.G. Roederer, eds., D. Reidel Publishing Co., Dordrecht, Holland, 1972.

Vasyliunas, V.M., Nonuniqueness of Magnetic Field Line Motion, *J. Geophys. Res.*, **77**, 6271, 1972.

Vasyliunas, V.M., Theoretical Models of Magnetic Field Line Merging, 1, *Rev. Geophys. Space Phys.*, **13**, 303, 1975.

Voigt, G.H., Magnetospheric Equilibrium Configurations and Slow Adiabatic Convection, in *Solar Wind-Magnetospheric Coupling*, Y. Kamide and J.A. Slavin, eds., Terra Scientific Publishing Co., Tokyo, Japan, 1986.

Walt, M., Radial Diffusion of Trapped Particles and some of its Consequences, *Rev. Geophys. Space Phys.*, **9**, 11, 1971.

Warwick, J.W., D.R. Evans, J.H. Romig, C.B. Sawyer, M.D. Desch, M.L. Kaiser, J.K. Alexander, T.D. Carr, D.H. Staelin, S. Gulkis, R.L. Poynter, M. Aubier, *et al.*, Voyager 2 Radio Observations of Uranus, *Science*, **233**, 102, 1986.

Warwick, J.W., D.R. Evans, G.R. Peltzer, R.G. Peltzer, J.H. Romig, C.B. Sawyer, A.C. Riddle, A.E. Schweitzer, M.D. Desch, M.L. Kaiser, W.M. Farrell, *et al.*, Voyager Planetary Radio Astronomy at Neptune, *Science*, **246**, 1498, 1989.

Wilcox, J. and N. Ness, Quasi-Stationary Corotating Structure in the Interplanetary Medium, *J. Geophys. Res.*, **70**, 1233, 1965.

Williams, D.J. and G.D. Mead, Nightside Magnetospheric Configuration as Obtained from Trapped Electrons at 1100 Kilometers, *J. Geophys. Res.*, **70**, 3017, 1965.

Williams, D.J. and G.D. Mead, eds., Magnetospheric Physics, *Rev. of Geophys.*, **7**, 1969.

Willis, D.M., Structure of the Magnetopause, *Rev. of Geophys. Space Phys.*, **9**, 953, 1971.

Wolf, R.A., Ionosphere-Magnetosphere Coupling, *Space Sci. Rev.*, **17**, 537, 1975.

Woltjer, L., A Theorem on Force-Free Magnetic Fields, *Proc. Nat. Acad. Sci.*, **44**, 489, 1958.

Wu, C.S., Physical Mechanisms for Turbulent Dissipation in Collisionless Shock Waves, *Space Sci. Rev.*, **32**, 83, 1982.

Yoshida, S., G.H. Ludwig and J.A. Van Allen, Distribution of Trapped Radiation in the Geomagnetic Field, *J. Geophys. Res.*, **65**, 807, 1960.

Zumuda, A.J., J.H. Martin and F.T. Heuring, Transverse Magnetic Disturbances at 1100 Kilometers in the Auroral Region, *J. Geophys. Res.*, **71**, 5033, 1966.

Achievements of the International Magnetospheric Study (IMS), Proceedings of an International Symposium, Grax, Austria, June, 1984. European Space Agency, ESA SP-217, 1984.

Plasmas and Fluids, National Research Council, National Academy Press, Washington, D.C., 1986. This 318 page book is one of *Physics Through the 1990s* series of the U.S. National Academy of Sciences.

Plasma Science, special issues on space and cosmic plasmas published by IEEE, Inc.

IEEE Transactions, *Plasma Science*, **PS-14**, December, 1986 and **PS-17**, April, 1989.

Magnetospheric Boundary Layers, Proceedings of an International Conference, held in Alpach, Austria. Published by the European Space Agency, ESA SP-148, 8 rue Mario-Nikis, 75738 Paris 15, France, August, 1979.

INDEX